The Biophysical Foundations of Human Movement

Bruce Abernethy
Vaughan Kippers
Laurel Traeger Mackinnon
Robert J. Neal
Stephanie Hanrahan

The University of Queensland, Australia

Library of Congress Cataloging-in-Publication Data

The biophysical foundations of human movement / Bruce Abernethy . . .
[et al.].
p. cm.
Includes bibliographical references and index.
ISBN 0-88011-732-X
1. Human mechanics. 2. Biophysics. I. Abernethy, Bruce, 1958-

QP303.B586 1997 96-50009
612.7'6--dc21 CIP

ISBN: 0-88011-732-X

First published in Australia in 1996 by Macmillan Education Australia of 107 Moray Street, South Melbourne, Victoria 3205. (ISBN 0-7329-3048-0, paper; ISBN 0-7329-3047-2, cloth)

This international version published in 1997 by Human Kinetics Publishers, Inc.

Managing Editor: Lynn M. Hooper; **Text and Cover Designer:** Jan Schmoeger; **Typesetting:** Typeset Gallery, Malaysia; **Indexer:** Kathleen Gray; **Printer:** United Graphics

Printed in the United States of America 10 9 8 7 6 5 4 3 2 1

Human Kinetics
Web site: http://www.humankinetics.com/

United States: Human Kinetics, P.O. Box 5076, Champaign, IL 61825-5076
1-800-747-4457
e-mail: humank@hkusa.com

Canada: Human Kinetics, Box 24040, Windsor, ON N8Y 4Y9
1-800-465-7301 (in Canada only)
e-mail: humank@hkcanada.com

Europe: Human Kinetics, P.O. Box IW14, Leeds LS16 6TR, United Kingdom
(44) 1132 781708
e-mail: humank@hkeurope.com

Australia: Human Kinetics, 57A Price Avenue, Lower Mitcham, South Australia 5062
(08) 277 1555
e-mail: humank@hkaustralia.com

New Zealand: Human Kinetics, P.O. Box 105-231, Auckland 1
(09) 523 3462
e-mail: humank@hknewz.com

Contents

Preface

Knowledge about the biophysical foundations of human movement is important for a number of reasons. A thorough understanding of human movement is fundamental to professional practice in a range of fields such as exercise and sport science, physical education, medicine, physiotherapy, occupational therapy, nursing, and other rehabilitation and health-science professions. Furthermore, the study of human movement is also of value in itself as a means of examining key biological phenomena such as maturation, adaptation and the interactions between hereditary and environmental factors, which underpin growth and development.

Despite the central importance of movement, in an applied sense, to professional preparation in the health sciences and, in a basic sense, to the understanding of human biology, in teaching an introductory subject on the biophysical foundations of human movement we have been surprised at the absence of a suitable introductory textbook to cover this field. This text was written in an attempt to redress that shortfall.

The *Biophysical Foundations of Human Movement* was written with three main purposes in mind, namely to:

- provide an introduction to key concepts concerning the anatomical, mechanical, physiological, neural and psychological bases of human movement
- provide an overview of the multidimensional changes in movement and movement potential that occur throughout the lifespan with the processes of growth, development, maturation and ageing
- provide an overview of the multidimensional changes in movement and movement potential that occur as an adaptation to training, practice and other lifestyle factors.

Fulfilling the first purpose involves consideration of the biophysical dimensions of the field of study known variously as *human movement studies, human movement science, kinesiology*, or *sport and exercise science* and examination of the discipline–profession links in this field. Gaining an overview of the structure of the field is especially important for students taking courses in human movement studies as a means of providing an entrée to more detailed study within one or more of the subdisciplines of human movement studies and as a means of laying the foundations for integrative, multidisciplinary and crossdisciplinary studies.

Fulfilling the second and third purposes is important as a means of exposing readers to fundamental issues in biology itself and in positioning the study of movement as a major topic within human biology. We have deliberately selected *lifespan changes* and *adaptation* as key organising themes for this book because of their centrality to biology itself. Gaining knowledge about the processes of growth, development, maturation and ageing aids understanding of the key changes in movement potential throughout the lifespan that, because of their inevitability, impact directly on all of us. Although the processes of maturation and ageing are inevitable, adaptation (through training, practice and lifestyle decisions) offers humans some degree of control of their own destiny and capabilities. It is our sincere hope that the clear message about the important role physical activity plays in the maintenance of health, which arises from consideration of adaptation within each of the biophysical subdisciplines of human movement, will be one that readers of the text will not only take to heart themselves in making personal lifestyle decisions but will also communicate to others.

The Biophysical Foundations of Human Movement has been intentionally structured into three major parts to broadly reflect its three main purposes. Part 1 provides a general introduction to human movement studies as a field of study. It does this by examining the disciplinary and professional structure of contemporary human movement studies (Chapter 1) and by providing a brief overview of the historical origins of the current academic field (Chapter 2).

Part 2, the largest part of the book, provides an introduction to basic concepts, to lifespan changes, and to adaptations arising in response to training, within each of the five major biophysical subdisciplines of human movement. These are the subdisciplines of functional anatomy, biomechanics, exercise physiology, motor control and sport and exercise psychology. These subdisciplines represent respectively the anatomical, mechanical, physiological, neural and psychological bases of human movement.

If a crude analogy was drawn between the human body and a motor vehicle (crude in the sense that the analogy would grossly underestimate the complexity, dynamics and adaptive potential of the human machine), then the various biophysical subdisciplines recognised within human movement studies have reasonable analogues in terms of the minimal 'clusters' of knowledge needed to explain and predict the functioning and 'behaviour' of the human 'motor vehicle'. The scientist attempting to understand the human body and its potential for movement is faced with synthesising some of the same general classes of knowledge that the automotive engineer needs in order to understand an inanimate motor vehicle and its potential for movement. To understand and optimise the performance of a motor vehicle the automotive engineer needs specific knowledge about the vehicle's material structure (its anatomical basis), its mechanical design characteristics (its mechanical basis), its motor's capacity and fuel consumption (its physiological basis), its electrical wiring and steering mechanisms (its neural basis), and the characteristics and capabilities of its driver (its psychological basis). In the case of the human 'vehicle' this information is provided by the subdisciplines of functional anatomy, biomechanics, exercise physiology, motor control and sport and exercise psychology and the interactions between them.

Within Part 2, each of the five biophysical subdisciplines of human movement studies is discussed in a separate section, with the structure of each of the sections being broadly the same. Each section begins with a brief introduction, which defines the subdiscipline and provides information on its historical development, the typical issues and problems it addresses, the level(s) of analysis it uses and relevant professional training and organisations. This introduction is then followed by a number of chapters that overview the basic concepts within each of the subdisciplines and is then followed by separate chapters devoted to consideration of lifespan changes and adaptation. The exercise physiology section also contains an additional chapter devoted to specific applications of fundamental principles of exercise physiology to health.

Part 3 of the book is devoted to multidisciplinary and crossdisciplinary approaches to human movement and provides some examples of contemporary issues in which the application and integration of knowledge from a number of the biophysical subdisciplines are fundamental to understanding. The material presented in this part of the text is designed to demonstrate the importance of integration of information from different subdisciplines as being the essential strength and prospective direction for the academic discipline of human movement studies and for practice in those professions grounded in the knowledge base of human movement studies. A glossary of terms is supplied at the end of Part 3 to assist in the comprehension and revision of new terms introduced at various points in the text.

Throughout the book we have included a number of boxed sections, peripheral to the main body of the text, to highlight some key individuals and studies that have contributed to our understanding of human movement. Although the body of the text does not, as a general rule, focus on specific research studies or methods (a select list of further reading is given at the end of each section rather than detailed reference citations within each chapter), the boxed sections have been designed with the intent of providing the reader with some feel for the flavour of research methods in this field plus more detailed exposition of some of the pivotal studies that have underpinned current knowledge. In this way these boxed sections provide students with a taste (albeit a limited one) of the kind of research methods and details they can expect to encounter in more detailed studies of each of the subdisciplines of human movement studies. Such studies are undertaken typically in the second and subsequent years of formal courses in human movement studies. The boxed sections contain research examples from around the world. This was done deliberately to both highlight the importance of positioning research and knowledge in a relevant social content and to illustrate the genuinely international nature of research activity in our field.

As the title of this book indicates, our intention in writing this book has been to provide an introduction to the biophysical fundamentals of human movement. We have, therefore, by necessity, deliberately tried to provide a coverage of the topic that is broad and illustrative rather than detailed and exhaustive. An in-depth coverage of each of the biophysical subdisciplines is clearly well beyond the scope of this text and students with an interest in more detailed study in any of the areas should be directed to the many excellent specialised texts now available for each of the subdisciplines. (A number of these are detailed in the Further Reading listings that follow each of the major sections within this book.) We have also deliberately restricted our coverage within the book to the biophysical foundations of human movement and have not attempted coverage of the equally important sociocultural subdisciplines of human movement studies, such as social psychology, history, philosophy, pedagogy and sociology. These are the focus of a companion text, *The Sociocultural Foundations of Human Movement*.

The study and understanding of human movement present an exciting challenge for students, scientists and practitioners alike. Given how central an understanding of human movement and its enhancement is to a wide range of human endeavours, it is our hope that *The Biophysical Foundations of Human Movement* will serve as a readable introduction for both students and professionals involved in sport and exercise science, physical education, kinesiology, ergonomics, music and performing arts, physiotherapy, occupational therapy, nursing, medicine, health education and health promotion and other rehabilitation and health sciences, and will help to convey to our readers some of the magnetism that this subject matter holds for us.

Acknowledgments

Our sincere appreciation is extended to a number of people, without whose contribution this book could never have been completed. In particular we wish to express appreciation to:

- Louise Blood, for the excellent original drawings and artwork throughout the book
- Scott McLean, for the remaining graphical illustrations
- Judy Land, Jan Holmes and Debbie Noon for typing the majority of the manuscript
- Frank and Janet Pyke, for early discussion about the structure of the text and for

arranging contact with our initial publisher, Macmillan Education Australia

- David Aitken, Don Bailey, Robin Burgess-Limerick, Richard Carson, Bob Faulkner, Mark Forwood, Holger Gabriel, Piero Giorgi, David Jenkins, Mariken Leurs, Ian Jobling, Stephan Riek, and Wally Wood for supplying information and constructive commentary on sections of the book
- Macmillan Education Australia and especially Publishing Director, Peter Debus, for their patience and expertise throughout the whole writing and production stage of the original version of this book
- Rainer Martens and his staff at Human Kinetics, who skilfully aided in the transformation of an Australian text into an international one
- Our families, friends, and colleagues for their support, understanding, and tolerance while this book was being written.

The introductory chapters (Chapters 1 and 2) plus Chapters 16–19 were written by Bruce Abernethy; Vaughan Kippers wrote Chapters 3–6; Rob Neal Chapters 7–11; Laurel Mackinnon Chapters 12–15; and Stephanie Hanrahan Chapters 20–23.

Some Notes for North American and European Readers

The Biophysical Foundations of Human Movement was originally published by Macmillan Education Australia and was written with the Australian context and Australian readers in mind. In revising the book for distribution by Human Kinetics throughout Europe and North America, we have been conscious of the need to use terms and examples which are universally understood; at the same time, we remain cognisant that some terms will nevertheless convey somewhat different meanings to readers from different parts of the world.

To assist readers from other cultures, we have prepared (below) a brief explanation of some general Australian terms and expressions that appear within the book but which may have subtly different meanings for either North American or European readers. We trust these explanations are useful.

Disciplinary and Professional Terms

We use the term *human movement studies* to refer to the body of knowledge or discipline concerned with human movement. Many North American readers will be more familiar with the synonymous term *kinesiology*. Likewise North American readers may wish to substitute the term *physical therapy* for *physiotherapy* at various points in the text.

Some Sporting Jargon

Given the nature of our subject matter, we make frequent use throughout the text of examples from sport to illustrate some basic biophysical phenomena of movement. Terms associated with sport are much more culturally-specific than scientific terms. For example, any reference we make throughout the text to *training* should be considered to encompass not only physical conditioning but also skill practice.

Where we refer to *football* and *football codes* readers may wish to consider any of the sports of soccer, American football, Gaelic football, Australian football, rugby union, or rugby league; when we refer to *hockey* we intend reference to field hockey rather than ice hockey; and when we make reference to *basketballers* and *footballers* we mean basketball players and football players. By *social sport* we refer to recreational sport.

Finally, when we use examples from *cricket* and *netball* we are referring to sports which will be familiar to residents of former British Commonwealth countries but relatively few others. Cricket is a sport that requires the player to accurately hit (with a bat) a fast moving, bouncing ball about the same size as a baseball, while netball is a team sport broadly similar to basketball but with much tighter rules constraining player and ball movement.

Part 1

Introduction to human movement studies

CHAPTER 1

Human movement studies as a discipline and a profession

- What is human movement studies and why is it important?
- What are disciplines and professions?
- Is human movement studies a discipline?
- How might a discipline of human movement studies be structured?
- What should the discipline of human movement studies be called?
- Professions based on the discipline of human movement studies
- Relationships between the discipline and professions

The purpose of this chapter is to:

- provide a definition and description of the field of human movement studies
- draw distinction between the discipline and the professions of human movement studies
- examine the justification for, and possible structure of, a discipline of human movement studies
- examine briefly the vexing question of what the discipline of human movement studies should be most appropriately called
- examine what is meant by the terms multidisciplinary, interdisciplinary and crossdisciplinary
- introduce the major professions and professional associations relevant to human movement studies
- explore (desirable) relationships between disciplines and professions.

What is human movement studies and why is it important?

Human movement studies is the comprehensive and systematic study of human movement. It is that field of academic inquiry concerned with understanding how *and* why *people move and the factors that limit and enhance our capacity to move.*

Two key points need to be emphasised within this very simple definition of the field of human movement studies. The first is that the unique focus of the field is on human movement. This is true regardless of whether the movement to be ultimately understood is one performed, for example, in the context of undertaking a fundamental daily skill (such as walking, speaking or reaching and grasping), executing a highly practised sport or musical skill, exercising for health, or regaining the function of an injured limb. The study of human movement is important in, and of, itself, as movement is a central biological and social phenomenon.

The study of movement is central to the understanding of human biology as movement is a fundamental property, indeed indicator, of life (remembering that biology is literally the study of life). Human movement, as we have noted in the preface, offers a valuable medium for the study of biological phenomena fundamental to developmental changes across the lifespan (changes that occur with ageing as a consequence of internal body processes), to adaptation (changes that occur as an accommodation or adjustment to environmental processes) and to the interactions of genetic and environmental factors (nature and nurture) that dictate human phenotypic expression.

Human movement, especially that which occurs in collective settings such as organised sport, exercise settings and health and physical education classes, also clearly has an essential social and cultural component, which also warrants its intensive study. Understanding individual and group motives and opportunities and barriers to involvement in different types of human movement, for instance, provides an important window into the nature of human society just as understanding the mechanisms of individual human movement provides an important window into the understanding of human biology. Movement, in short, plays a fundamental role in human existence and what it means to be human and for these reasons warrants our very best efforts to understand it.

The second key point that needs to be highlighted within our definition of human movement studies is the importance of a systematic, research-based approach to the generation of knowledge. As many aspects of the current practice of human movement involve practices based as much on fads, folklore, tradition and intuition as on sound, logical theory substantiated by systematically collected, reproducible data, it is imperative that the knowledge base for human movement studies be one based on research conducted with a methodological rigour equivalent to that of other established biological, physical and social sciences. Only through such an approach can fact be separated from fiction and a sound basis for best practice in the profession, based on the knowledge base of human movement studies, be established.

In the true tradition of the scientific method the field of human movement studies aims to not only describe key phenomena from within its domain but also to move beyond description to understanding through explanation and prediction. Human movement studies, as a systematic field of study, thus carries the twin goals of all fields of science, namely the generation of knowledge through understanding of basic phenomena, and the application of knowledge for the benefit of society. The basic understanding of human movement, for which the field strives, has applicability to all of the many areas and professions that deal with the enhancement of our capacity to move and to adopt healthy lifestyles. Obvious areas of application of the knowledge base of human movement studies are to sport, exercise, health and physical education, the workplace and rehabilitation; obvious professions that rely on such information include professionals in the health and fitness industry, sports scientists, physical educators, health promoters, doctors, nurses, and therapists.

Given that knowledge about human movement is clearly important in its own right as well as in its application to the practice of many professional groups, an important question that follows is 'how is knowledge about human movement organised and how is this knowledge translated into the practices of relevant professions?'. Answering such a question requires consideration of the discipline and professions of human movement studies.

What are disciplines and professions?

According to the pioneering American physical educator Franklin Henry:

> . . . an academic discipline is an organized body of knowledge collectively embraced in a formal course of learning. The acquisition of such knowledge is assumed to be an adequate and worthy objective as such, without any demonstration or requirement of practical application. The content is theoretical and scholarly as distinguished from technical and professional.
>
> *(Henry 1964)*

The principal function of a discipline is therefore to develop a coherent body of knowledge that describes, explains and predicts key phenomena from the domain of interest (or subject matter).

In contrast, professions, as a general rule, try to improve the conditions of society by providing a regulated service in which practices and educational/training programs are developed that are in accordance with knowledge available from one or more relevant disciplines. Practice in the profession of engineering, for example, is based on application of knowledge from disciplines such as physics, mathematics, chemistry and computer science, and practice in the medical profession is based on knowledge from disciplines such as anatomy, physiology, pharmacology, biochemistry and psychology. Established professions share a number of characteristics including:

- an identified set of jobs or service tasks over which they have jurisdiction
- organisation under the framework of a publicly recognised association
- identified educational competencies and formalised training and education criteria
- political recognition, usually through acts of government legislation
- a code of ethics defining minimal standards of acceptable practice.

Disciplines therefore seek to understand subject matter and professions to implement change based on this understanding. The emphases within disciplines and professions are often characterised as theory/research versus application/practice but such a distinction is overly simplistic and potentially misleading. Applied research (including research on aspects of professional practice) is now an accepted part of the business of the discipline just as the profession may frequently be the site for original research.

Is human movement studies a discipline?

Remembering our earlier definition of human movement studies and Henry's definition of a discipline, an important practical as well as philosophical question is whether there is a unique, organised body of knowledge on how and why people move (to satisfy the criteria for a discipline of human movement studies) or whether human movement studies is simply the application of knowledge from other disciplines, such as anatomy, physiology and so on (thereby making it, by definition, more a profession than a discipline)? This question has been the source of much debate both within and beyond the field for at least the past three decades, with the extent of the uniqueness and collective coherence of the knowledge within the field being the source of most contention. The establishment of university departments of human movement studies (or the like), independent of traditional professions such as physical education teacher training, is predicated on the assumption that human movement studies possesses an organised body of knowledge in much the same way as do 'traditional' disciplines such as physics, chemistry and psychology and that human movement studies is more than simply a loose collection of the applications of knowledge from other fields.

The foreparents of the modern field of human movement studies were, as we will see in the next chapter, primarily physical educators. A number of these, most particularly the physiologist/psychologist Franklin Henry and the motor developmentalist Lawrence Rarick, created strong arguments in the 1960s for both the importance and the existence of a discipline of human movement studies, claiming that such a field (or its precursor physical education) asked questions that would not have arisen from the cognate (parent) disciplines.

Rarick, in a paper in the journal *Quest*, argued that:

> ... most certainly human movement is a legitimate field of study and research. We have only just begun to explore it. There is need for a well-organised body of knowledge about how and why the human body moves, how simple and complex motor skills are acquired and executed, and how the effects (physical, psychological and emotional) of physical activity may be immediate or lasting.
>
> *(Rarick 1967)*

In a similar vein, Henry, in a much-quoted 1964 paper, claimed the pre-existence of a discipline base for the study of human movement. Henry wrote then that:

> ... There is indeed a scholarly field of knowledge basic to physical education. It is constituted of certain portions of such diverse fields as anatomy, physics and physiology, cultural anthropology, history and sociology, as well as psychology. The focus of attention is on the study of man as an individual, engaging in the motor performances required by his daily life and in other motor performances yielding aesthetic values or serving as expressions of his physical and competitive nature, accepting challenges of his capability in putting himself against a hostile environment, and participating in the leisure time activities that have become of increasing importance in our culture.
>
> *(Henry 1964)*

Henry's definition of the discipline base of our field has generally stood well the test of time save for some relatively minor changes, most obviously: (i) the substitution of the term human movement studies for physical education (now typically defined in a narrower professional sense), (ii) the extension of the focus of the field beyond solely the study of the person as an individual to also incorporate the study of the person as an element of a social system, and (iii) the alteration of the language of expression to recognise the equal applicability of the knowledge base to females as to males.

How might a discipline of human movement studies be structured?

If we consider again our earlier definition of human movement studies as that field concerned with understanding how and why people move and the factors that limit and enhance our capacity to move it becomes apparent, in keeping with Henry's propositions, that a human movement studies discipline must include selected aspects of a broad range of other disciplines and discipline groups. This follows because movement potential and performance are known to be influenced by, among other things, biological factors such as maturation, ageing, training and lifestyle; health factors such as disease, disuse and injury; and social factors such as motivation, incentive and opportunity. A discipline of human movement studies must therefore draw heavily, but not exclusively, on the methods, theories and knowledge of a wide range of other disciplines and provide them with an integrative focus on human movement. Information of relevance for a discipline of human movement studies may be gleaned from biological science disciplines such as anatomical science, physiology and biochemistry; physical science disciplines such as physics, chemistry, mathematics and computer science; social science disciplines such as psychology, sociology and education; and disciplines in the humanities such as history and philosophy.

Figure 1.1 presents one possible way of conceptualising the organisation of knowledge within a discipline of human movement studies. In this conceptualisation the discipline of human movement studies consists of the collective knowledge contained within and between each of the subdisciplines of functional anatomy,

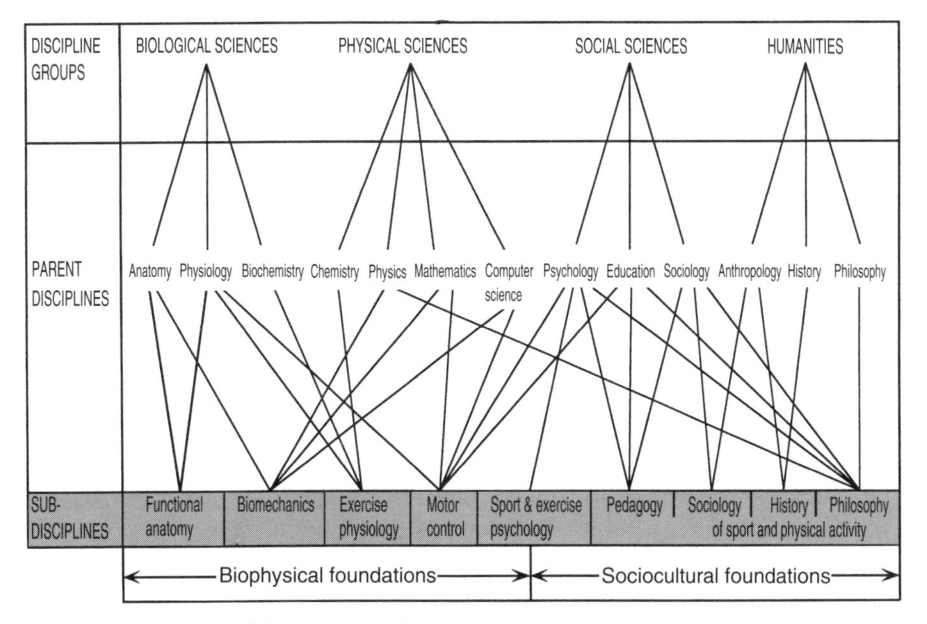

Figure 1.1 One possible conceptualisation of the structure of knowledge about human movement. The discipline of human movement studies is represented by the shaded area

biomechanics, exercise physiology, motor control, sport and exercise psychology and the pedagogy, sociology, history and philosophy of sport and physical activity (as illustrated by the shaded box within Figure 1.1). The subdisciplines of functional anatomy, exercise physiology, biomechanics, motor control and sport and exercise psychology (in so much as it focuses on the individual rather than group or societal behaviour) constitute the biophysical foundations of human movement and are afforded coverage in this text. The social psychology of sport and exercise together with the pedagogical, sociological, historical, philosophical, political and cultural aspects of physical activity and sport constitute the sociocultural foundations of human movement. It is readily apparent from Figure 1.1 that each of the subdisciplines draws theories, methods and knowledge from one or more cognate disciplines. It is also important to recognise, however, that each of the specialist subdisciplines draws on only a subset of the knowledge contained in its cognate discipline(s) and that the subdisciplines can, and frequently do, generate theories, methods and approaches of their own not acquired from the cognate discipline(s).

A number of other points also should be noted with respect to the conceptualisation of the human movement studies discipline presented in Figure 1.1:

- The clustering of disciplines into discipline groups and the selection of the sub-disciplinary groups are necessarily somewhat arbitrary and no two conceptualisations of the field, its inter-relations and the naming of its component parts are likely to be identical.
- The layering and linking of cognitive disciplines and human movement studies sub-disciplines evident in Figure 1.1 are frequently reflected in the scheduling and prerequisite subject structuring of many tertiary courses in human movement studies; basic exposure to the cognate disciplines generally precedes exposure to each of the subdisciplines of human movement studies.
- The disciplines and subdisciplines are so organised in Figure 1.1 as to present a

generally progressive shift from left to right from a focus on the micro phenomena to the more macro phenomena of human movement.

Each of the subdisciplines are presented in Figure 1.1 as essentially insular components, having as much (or greater) contact with the cognate discipline as with the other subdisciplines of human movement studies. To some extent this representation is an accurate reflection of the current state of the field and there is increasing concern in many quarters that the growing differentiation and specialisation within human movement studies may produce fragmentation and an inevitable loss of integrity within the discipline base. This fragmentation is reinforced by the dominant tendency in tertiary programs to teach each of the subdisciplinary fields in a separate, independent subject. To this end the field of human movement studies is probably most accurately depicted as being currently multidisciplinary, whereas the desirable direction is to make it more crossdisciplinary and ultimately interdisciplinary (Figure 1.2). A good indicator of the maturation of a discipline of human movement studies will be the extent to which it becomes more interdisciplinary and advances in knowledge are made by crossing the traditional (but arbitrary) boundaries between the subdisciplines and by synthesising material from the subdisciplines rather than importing ideas from the 'main stream' disciplines.

What should the discipline of human movement studies be called?

Although it will be clear from the previous sections that we favour the use of the term 'human movement studies' to describe the discipline, such a term is far from universally accepted. There is essentially no uniformity in the names selected by the numerous university departments offering courses in human movement studies, either between countries or within the same country, as our examples from Australia (Table 1.1) and the United States (Table 1.2) illustrate. The fact that finding a widely accepted name for the discipline always has been, and still remains, a persistent source of debate perhaps indicates the relative immaturity and diversity of our field of study compared with some others.

The term 'human movement studies' was coined first by the British physical educator and psychologist H.T.A. (John) Whiting in the early 1970s and it has been subsequently adopted in parts of Europe, the United States, and Australia. The term 'human movement studies'

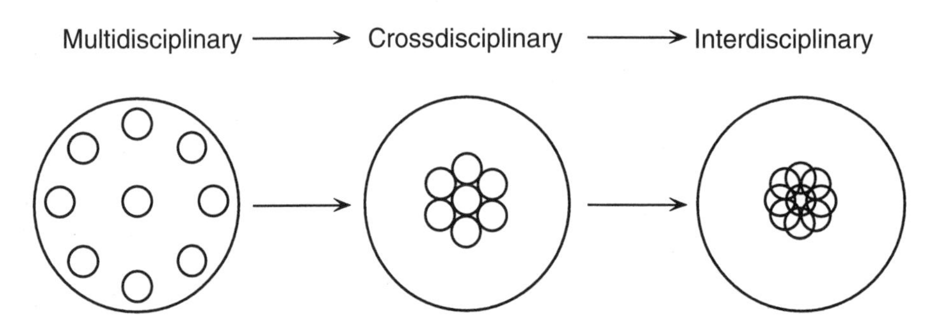

Figure 1.2 A desirable evolutionary progression for a discipline of human movement studies. The components shown by circles represent the subdisciplines identified in Figure 1.1
Source: Zeigler, E.F. (1990), Don't forget the profession when choosing a name!, pp. 67–77, in C.B. Corbin & H.M. Eckert (eds), *The Evolving Undergraduate Major*, Human Kinetics, Champaign, IL. (Copyright 1990 by the American Academy of Physical Education.)

has also become the title for one of the field's journals. The term is useful because it clearly encapsulates the unique and unifying theme within the body of knowledge in our field but is considered by some as being too cumbersome, especially in a public marketing context. The alternative term 'human movement science' has had some following and is probably appropriate for the biophysical aspects of human movement. Where this term is problematic is with aspects of the sociocultural foundations of human movement where knowledge acquisition occurs through methods in addition to those of traditional science. In North America the term 'kinesiology,' literally meaning the study of movement (from the Latin *kinein*, to move and *logos*, a branch of learning), is widely used but it has not been adopted internationally presumably because the term is both poorly understood

Table 1.1 Some names used by Australian university departments offering courses in human movement studies

Exercise and sport science
Human movement
Human movement and health education
Human movement and sports management
Human movement and sports science
Human movement education, research and management
Human movement, physical education and recreation
Human movement science
Human movement studies
Physical education
Physical education and recreation
Physical education, exercise, and sport studies
Sport and leisure studies
Sports science
Sports studies

Table 1.2 Some examples of names considered by university departments in the United States offering courses in human movement studies

Kinesiology	Sport science and movement education
Human performance	Sport science and leisure studies
Sports studies	Sport management
Exercise science	PE and sports programs
Sports science	Human movement studies
HPERD	Human movement
Exercise and sport science	Human performance and sport science
Recreation	Human kinetics and health
Sport, exercise and leisure science	Health and human performance
Sport and exercise science	Human performance and leisure studies
Human performance and health promotion	Human kinetics
Wellness and fitness	Physical culture
Wellness education	Physical education and human movement
Allied health	Exercise and movement sciences
Kinesiology and exercise science	Exercise science
Health and physical education	Exercise science and human movement
PE and exercise science or sport science	Human movement sciences
Recreation and wellness programs	Exercise and sport studies
Human movement studies and PE	Leisure science
Movement and exercise science	Interdisciplinary health studies
Physical education	Movement studies
Physical education and exercise science	Exercise and movement science
Science of human movement	

Source: Razor, J.E & Brassie, P.S. (1990), Trends in the changing titles of departments of physical education in the United States, pp. 82–90, in C.B. Corbin & H.M. Eckert (eds), *The Evolving Undergraduate Major*, Human Kinetics, Champaign, IL. (Copyright by the American Academy of Physical Education.)

in general usage and is often used in a much narrower context to refer simply to the mechanics of human movement. In contrast the terms 'exercise and sport science' and 'physical education' are well understood by the general public but are much narrower in focus than human movement studies. In the case of physical education its current usage is more closely tied to notions of a profession than to one of an all encompassing discipline.

Professions based on the discipline of human movement studies

As we noted earlier, professions, unlike disciplines, have a specific interest in the application of knowledge as a means of solving specific problems, enhancing the quality of life and providing a service to society. Standards of practice in specific professions are typically controlled by professional bodies, who impose minimal training/education requirements, set membership and accreditation criteria and establish codes of professional ethics and the like. There is a wide range of professions that draw on knowledge from the discipline of human movement studies and each of these is represented by one or more professional bodies.

Traditionally the profession most linked to the knowledge base of human movement studies has been physical education, which is the principal historical forebear of the discipline of human movement studies. More recently professions related to exercise management and

Table 1.3 Some major international and national professional organisations relevant to the discipline of human movement studies*

	Exercise science and sports medicine	Physical and health education, recreation and dance
International	Federation International de Medicine Sportive (FIMS)	International Council on Health, Physical Education & Recreation (ICHPER)
	International Council of Sport Science and Physical Education (ICSSPE)	
North American	American College of Sports Medicine (ACSM)	American Alliance for Health, Physical Education, Recreation & Dance (AAHPERD)
		Canadian Association for Health, Physical Education, Recreation & Dance (CAHPERD)
	American Academy of Kinesiology & Physical Education (AAKPE)	
European	European College of Sport Science (ECSS)	Association Internationale des Ecoles Superieures D'Education Physique (AISEP)
	British Association of Sport and Exercise Sciences (BASES)	Physical Education Association (UK) (PEAUK)

*This listing is by no means exhaustive; more specialised bodies are listed in Part 2 of the book.

prescription, sport and exercise science, exercise rehabilitation and therapy, ergonomics, sports medicine, coaching and sport and exercise psychology have emerged, all of which draw in some degree on the discipline of human movement studies. There is now a number of organisations both nationally and internationally that represent key professions whose practices are grounded in human movement studies. Some of the major ones in sports medicine, exercise science and physical and health education, recreation and dance, are displayed together with their acronyms in Table 1.3 as an illustration of the plethora of organisations that exist. Additional details on the functions and goals of three of the major international organisations listed in Table 1.3 are provided in the accompanying boxes, and details of the professional groups representing the specific interests of each of the biophysical subdisciplines of human movement studies are provided in the sectional introductions within Part 2 of this book.

Relationships between the discipline and professions

Thus far our discussions of discipline–profession relations may well have created the impression that the flow of information between the two is unidirectional with the role of the discipline being to generate knowledge that can then be used by the profession(s) as a basis for practice. Although this information flow is important it should be recognised that the ideal relationship between discipline and profession should be one of mutual benefit (i.e. symbiosis) such that information flows as much from the profession to the discipline as in the reverse direction. The professions, in particular, are well positioned to provide questions, problems, observations and issues that can function as a valuable guide and focus for the knowledge-seeking, research-based activities of the discipline (see Figure 1.3). Observations made in practice frequently form the initial basis of hypothesis testing within the discipline.

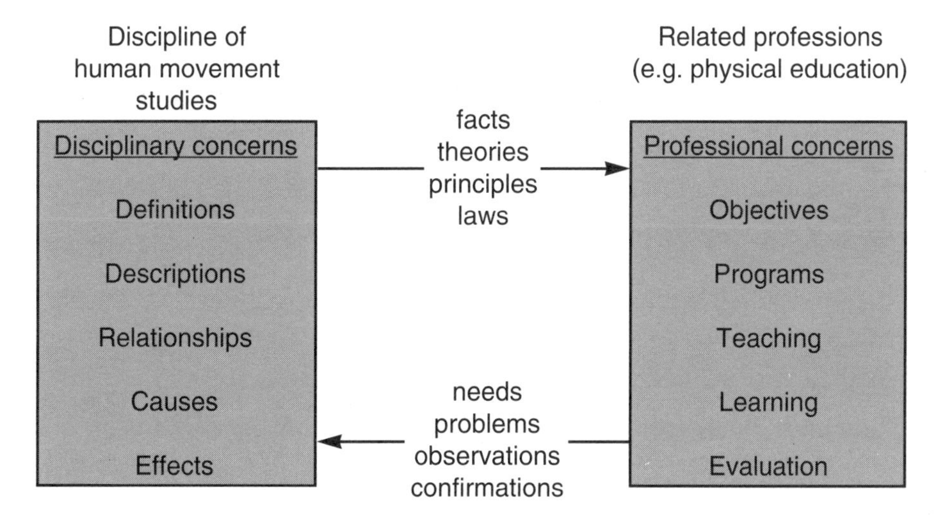

Figure 1.3 A mutually beneficial working relationship between the discipline of human movement studies and a related profession such as physical education
Source: Rivenes, R.S. (1978), *Foundations of Physical Education*, Houghton Mifflin, Boston, p. 17, Figures 1–4. (Copyright 1978 by Houghton Mifflin.)

Box 1.1 International Council of Sport Science and Physical Education

The International Council of Sport Science and Physical Education (ICSSPE) was founded in Paris, France, in 1958, originally under the name of 'International Council of Sport and Physical Education'. The ICSSPE's main purpose is to serve as an international umbrella organisation concerned with the promotion and dissemination of results and findings in the field of sport science, as well as the practical application of these findings on culture and educational contexts. The ICSSPE's aims are to contribute to the awareness of human values inherent in sport; to improve health and physical efficiency; and to develop physical education and sport in all countries to a high level. As a world organisation, the ICSSPE endeavours to bridge the gap between developed and developing countries, and to promote cooperation between scientists and organisations from countries with different political systems.

The ICSSPE has four stated fundamental objectives. These are to:

- encourage international cooperation in the field of sport science;
- promote, stimulate, and coordinate scientific research in the field of physical education and sport throughout the world and to support the application of its results in the many practical areas of sport;
- make scientific knowledge of sport and practical experiences available to all interested national and international organisations and institutions, especially to those in developing countries; and
- facilitate differentiation in sport science whilst promoting the integration of the various branches.

The ICSSPE conducts its scientific work in three main areas, namely sport sciences, physical education and sport, and scientific services (information dissemination). It disseminates information through a variety of publications, of which *Sport Science Review*, an international journal dedicated to thematic overviews of research in sport science, is the best known. ICSSPE serves as a permanent advisory body to the United Nations Educational, Scientific and Cultural Organisation (UNESCO) and regularly conducts research projects on behalf of UNESCO and the International Olympic Committee (IOC). It has eight regional bureaux throughout the world plus close links to a number of other international organisations of physical education and sport science. Membership of ICSSPE is open to organisations and institutions rather than individual subscribers and there are currently some 190 organisations and institutions affiliated with ICSSPE.

More information about the ICSSPE can be obtained from its World Wide Web site (see Box 2.5 for details).

Box 1.2 International Federation of Sports Medicine

The Federation Internationale de Medecine Sportive (FIMS) is an organisation made up of the national sports medicine associations of over 100 countries. It was founded in 1928 during a meeting of Olympic medical doctors in St. Moritz, Switzerland, with the principal purpose of promoting the study and development of sports medicine throughout the world.

The specific objectives of the FIMS are to:

- stimulate and promote the development of knowledge about sports medicine, studying

scientifically the natural and pathological consequences of physical training and sports participation;
- organise and/or sponsor scientific meetings, courses, congresses, and exhibits of a scientific nature on an international basis; and
- cooperate with national and international organisations of sports medicine and related fields in the promotion of sports medicine and human health.

The FIMS hosts a major international conference in sports medicine every three years. In addition, the FIMS is active in the production of position statements on aspects of health, physical activity, and sports medicine. The most influential of these position statements is probably its joint 1995 statement with the World Health Organisation (WHO), entitled *Physical Activity for Health.* In addition to its member national associations, the FIMS also makes available individual membership for medical-trained graduates active in sports medicine and affiliated with one of the member national bodies in sports medicine. Associate membership for non-medical practitioners active in sports medicine is also available.

More information is available about the FIMS through its World Wide Web site (see Box 2.5).

Box 1.3 American College of Sports Medicine

The mission of the American College of Sports Medicine (ACSM) is to promote and integrate scientific research, education, and practical applications of sports medicine and exercise science to maintain and enhance physical performance, fitness, health, and quality of life. The ACSM was founded in 1954 by a group of 11 physicians, physiologists, and educators. The organisation has grown rapidly since its formation. Presently, with over 15,000 members from North America and more than 50 other countries, the ACSM is easily the largest sports medicine and exercise science organisation in the world.

The ACSM's research and educational programs are broad ranging. Its annual four-day meeting, held in May, is one of the major international conferences for the presentation and discussion of new sports medicine and exercise science research. Similarly, the ACSM's official journal, *Medicine and Science in Sport and Exercise* (first published in 1969), is one of the principal international journals in the field for the publication of original research. The College's other publications, such as *Exercise and Sport Science Reviews,* are valuable sources of state-of-the-art reviews on key sports medicine and exercise science topics. In addition to its role in advancing the discipline through disseminating basic and applied scientific research on physical activity, the ACSM is also the peak professional body in North America for the accreditation of individuals seeking to work in clinical sports medicine, exercise science, and the health and fitness industry. This role has been facilitated by the development of key guidelines and policy documents—such as its *Guidelines for Exercise Testing and Prescription*—which have internationally become the 'gold standard' for professional practice.

The ACSM offers five types of membership (Professional, Professional-in-Training, Graduate Student, Undergraduate Student, and Associate), details of which can be found on the College's World Wide Web page (see Box 2.5).

CHAPTER 2

Historical origins of the academic study of human movement

- Scholarly writings on human movement from ancient civilisations (ca. 1000 BC–350 AD)
- The Middle Ages as a period of suppression for the study of human movement (ca. 350–1350 AD)
- Scholarly works on human movement from the Renaissance and Reformation periods (ca. 1350–1650 AD)
- Scholarly works on human movement in the period 1650–1885
- Professionalisation of physical education in the period 1885–1929
- The organisation of research efforts in physical education 1930–1959
- The beginnings of a discipline of human movement studies 1960–1970
- The emergence of subdisciplines and specialisations 1970–
- Future directions, challenges, and opportunities

In order to understand the current structure and status of the discipline of human movement studies and to try to predict, from a factual base, the future direction for the field it is necessary to have some appreciation of the historical antecedents of the contemporary field. The purpose of this chapter, therefore, is to:

- provide an overview of the ancient origins of interest in physical activity
- trace the professionalisation of physical education throughout the late 19th century and early 20th century
- chronicle the international emergence of a discipline of human movement studies from within the profession of physical education
- outline the emergence of specialist subdisciplines within human movement studies
- speculate on future directions, challenges, and opportunities for the field.

This chapter focuses only on the historical development of the broad field of human movement studies. The more specific developments within each of the five major biophysical subdisciplines of human movement studies are discussed separately in each of the sectional introductions in Part 2. Our coverage and categorisation of key periods in the development of a discipline of human movement studies draw heavily on the excellent 1981 review paper (detailed in the Further Reading section) by the American sport historian, Roberta Park.

Scholarly writings on human movement from ancient civilisations (ca. 1000 BC–350 AD)

Athough there is evidence of organised physical activity in both China and Egypt before 1000 BC (in the form of wrestling, tumbling, swimming and ball games in Egypt and ritualised physical training for the enhancement of physical prowess and moral strength in the Chou dynasty in China), the first clear indications of the beginnings of a scholarly focus on physical activity exist in ancient Greece in the period beginning around 450 BC. The framework for both a profession of physical education and a discipline of human movement studies can be claimed to be founded in this classical antiquity period. Organised physical training in the form of wrestling, boxing, gymnastics, swimming, running, discus and javelin throwing was used in the principal city-states of Sparta and Athens from as early as 900 BC as a means of conditioning boys for military service. Regular supervised physical training was provided to Spartan girls as part of their preparation for motherhood. This long tradition of physical activity as an integral part of ancient Greek society provided the platform for a number of the famous scholarly developments of relevance to the understanding of human movement.

Hippocrates (ca. 460–360 BC), the founder of modern medicine, argued for the study of the body (*physis*) as a natural object (rather than as a spiritual or mystical object) and made a number of observations on the relationship between diet, exercise and fatigue that remain equally topical today. The philosopher Plato (427–347 BC) criticised the mind–body dualism implicit in Hippocrates' work, arguing rather for a holistic view of the individual. Plato's many writings contained numerous opinions and observations on the beneficial effects of physical education, including reference to the important roles played by exercise in growth and development and by play and physical activity in the genesis of social skills and a socially responsible community. Plato's pupil Aristotle (384–322 BC) promulgated a positive view of regular physical activity in his writings and considered *gymnastics* (interpreted much more broadly than in its current narrow meaning) as a complete science, which had as its goal the discovery of those exercises and types of training that were of most benefit to health. Two of Aristotle's books, *Movement of Animals* and *Progression of Animals*, while lacking experimental evidence, described many of the phenomena (such as flexion–extension, action–reaction, centre of gravity–base of support) that are still central to modern functional anatomy and biomechanics.

Plato, Aristotle and Socrates were all critical of *athletics*, which placed adverse emphasis on winning as opposed to all round harmonious development of the individual; thus pre-empting by over 2000 years the concerns of modern sports psychologists with the potentially detrimental effects to personal development and co-operation that may arise from an excessive emphasis on competition. Of particular relevance to our discussions in the previous chapter about disciplines and professions was the distinction recognised by Aristotle, in his writings, between the systematic, scholarly understanding of exercise (*gymnasts*) and the practical applications of training techniques (*paedotribes*), which were often based on premises of questionable validity. This distinction (a precursor to the modern discipline–profession division) was also apparent in the writings of the physician Claudius Galen (ca. 129–200 AD), whose contribution to the development of the subdiscipline of functional anatomy is examined in more detail later in this book.

The Middle Ages as a period of suppression for the study of human movement (ca. 350–1350 AD)

In the Middle Ages (the 1000 year period from approximately 350 AD to 1350 AD) scholarly attention to matters physical was essentially non-existent. The Middle Ages was dominated by two philosophical beliefs (asceticism and scholasticism) both of which focused exclusive attention on matters of the spirit and the mind to the complete neglect, and indeed active suppression, of scholarly attempts to understand anything related to the body or human movement. Asceticism was a religious preoccupation with extreme self-denial of the physical in order to exclusively focus on spiritual matters as presented within the evolving Christian faiths. Scholasticism placed a premium on the development of the mind, largely through disciplined study of early Christian writings and the writings of some of the early Greek philosophers, especially Aristotle. In both instances the religious underpinnings of all scholarly work in the Middle Ages actively excluded any systematic study and investigation with respect to the biophysical properties of the human body and its capacity for movement.

Scholarly works on human movement from the Renaissance and Reformation periods (ca. 1350 AD–1650 AD)

The Renaissance (or 'rebirth') was that period from the late 14th century through to the mid-16th century, in which there was a great revival in learning, literature and the arts in Europe. This period, which marks the transition from the medieval to the modern world, was one of renewed scholarly interest in the natural world, including the human body and its capability for movement. The Reformation was the great religious movement of the 16th century that had as its objective the reform of the Roman Catholic Church, which ultimately led to the establishment of the various Protestant churches. The Renaissance and Reformation periods were especially important for the overthrow of much existing dogma about the human body through a number of major advances in knowledge about human anatomy and physiology. In the context of a discipline of human movement studies these advances were especially significant for the subdisciplines of functional anatomy and exercise physiology.

During the Renaissance and Reformation periods the drawings and paintings of Leonardo da Vinci (1452–1519) and Michelangelo (1475–1564) were anatomically precise through the use of knowledge acquired from cadaveric dissections. Similarly the definitive anatomical work (Vasalius' *De Humani Corpus Fabrica* (The Structure of the Human Body)) also arose from extensive meticulous observations from cadaveric dissections. Galileo's (1564–1642) discoveries in physics at this time were central to the establishment of a foundation for the study of the mechanics of human motion (biomechanics) and William Harvey's treatise *De Motu Cord's* (On the Motion of the Heart), which introduced the proposition of blood circulating through the body, was a cornerstone for modern physiology.

Scholarly works on human movement in the period 1650–1885

The late 17th century and early 18th century saw some of the ideals for movement, physical activity and exercise espoused earlier by the Greek philosophers re-emerging in the writings of prominent philosophers such as John Milton and John Locke in England and Jacques Rousseau in France. Locke's (1693) *Some Thoughts Concerning Education* contained the famous dictum *mens sana en corpore sano* (a sound mind in a sound body) and this clearly represented a complete reversal of the prevailing philosophy of the Middle Ages. It followed implicitly from Locke's dictum that a dualism of mind and body was inappropriate and that, logically, the study and nurture of one necessitated also the understanding and development of the other. Rousseau's *Emile* in 1762 advanced the view that movement, in the form of free play, was critical to cognitive, perceptual and motor development, thus pre-empting one of the predominant themes of modern research in motor control and pedagogy.

The 19th century was a period of great scientific discoveries, many of which laid the foundations for a modern discipline of human movement studies. Among the influential physiological discoveries during this period were the electrical properties of muscle, the heat and force properties of muscle fibres, the role of oxygenated blood in respiration, the transfer of oxygen across cell membranes, the function of the liver in carbohydrate metabolism, the role of the cell in energy exchange and the chemical basis of physiological processes. Major contributors to original knowledge were Duchenne, with his *Physiologie des Mouvements* in 1865, Du Bois-Reymond, with his *Physiology of Exercise* published in Berlin in 1885, and Marey, Mayer and Pflüger. Understanding of the neural bases of movement were significantly advanced by Bell's discovery of the respective sensory and motor functions of dorsal and ventral root ganglia in the spinal cord (articulated in his 1830 book *The Nervous Systems of the Human Body*) and the studies by the German psychologist Hermann von Helmholtz and others on nerve conduction velocity. Similarly the subdiscipline of biomechanics was ushered into a new period of measurement precision by the release in 1887 of Eadweard Muybridge's monumental, 11-volume *Animal Locomotion*, containing, for the first time, techniques for the high-speed, sequential photographic analysis of human and animal gait.

At the same time as these major advances in knowledge relevant to the understanding of human movement were taking place the field of physical education was beginning to take shape in western Europe. In Germany, Sweden and England formalised school-based programs in physical education had either emerged or were in the process of emerging, although the programs were of somewhat different form and designed for different purposes in each of the countries. These programs had a profound effect on the emerging shape of the physical education profession, not only in Europe but worldwide. The European impact on the development of physical education in North America is of particular significance because it was ultimately from within this framework that a discipline of human movement studies—as we currently know it in the western world—ultimately emerged. To this end, tracing the development of the profession of physical education and the discipline of human movement studies in North America from the late 19th century to the current day is important and enlightening in terms of understanding the current international form of the discipline as well as some of its professions.

Professionalisation of physical education in the period 1885–1929

Although physical education and sport were well established in the curricula of many European schools (such as the famed Rugby and Eton schools of England) by the early 19th century and in a number of North American public schools and colleges by the 1860s, it was not until 1885 that the first professional organisation for physical education, the American Association for the Advancement of Physical Education (AAAPE), was founded. This body was the forerunner of the current American Alliance for Health, Physical Education, Recreation and Dance (AAHPERD) (see Table 1.3). Prime movers in the formation of AAAPE were Edward Hitchcock (a medical doctor from the Harvard Medical School, who was Director of the Department of Hygiene and Physical Education at Amherst College), Dudley Sargent (a medical doctor from Yale, who was Director of the Hemenway Gymnasium at Harvard) and Edward Hartwell (a medical graduate from Ohio and a PhD graduate from the Johns Hopkins University, who was Director of the gymnasium at the Johns Hopkins University).

Hitchcock collected extensive systematic anthropometric data, culminating in the publication of the *Anthropometric Manual*, which first appeared in 1887. Sargent also collected extensive anthropometric data although his principal legacy is in the area of strength development and the vertical power test that bears his name. Hartwell contributed survey research

Figure 2.1 Dr Dudley Sargent and exercise stations at The Hemenway Gymnasium at Harvard University in 1885
Source: Schrader, C.L. (1948), The old Hemenway Gymnasium *Journal of Health and Physical Education,* 19, p. 476. (Reprinted courtesy of Harvard University Archives.)

on physical education. The work of people such as Hitchcock, Sargent, and Hartwell, collected under the rubric of the annual AAAPE Conferences, plus the work of others such as Norman Triplett, who made early (1898) observations on social facilitation effects in competitive track cyclists, provided the framework for the early development of a research basis for physical education practice. Some of this research work was disseminated to practitioners through the AAAPE's own professional journal, the *American Physical Education Review*, first published in 1896.

The first quarter of the 20th century saw a marked increase in the publication of books related to the profession and especially the discipline of physical education. Noteworthy among these were *Principles of Physiology and Hygiene* by George Fitz in 1908, *Exercise in Education and Medicine*, by the famed Canadian doctor and physical educator R. Tait McKenzie (see Box 2.1) first published in 1909, *Gymnastic Kinesiology* by William Skarstrom in the same year, *Physiology of Muscular Exercise* by F.A. Bainbridge in 1923 and *Massage and Therapeutic Exercise* by Mary McMillan in 1921—the last being one of the earliest texts for the emerging profession of physiotherapy. Perhaps even more importantly for the ultimate establishment of a credible discipline, original research on aspects of human movement began to appear in prestigious medical journals of the day, Schneider's 1920 paper on cardiovascular ratings of physical fatigue and efficiency in the *Journal of the American Medical Association* and A.V. Hill's study on 'oxygen debt' in the *Quarterly Journal of Medicine* being classical examples. There was a clear concern within the professional body in the 1920s about the need to increase research activity to make physical education more scientific.

The organisation of research efforts in physical education 1930–1959

The year 1930 was noteworthy in the development of both the discipline and the profession as the professional body, by then known as the American Physical Education Association, launched two new journals. These were the *Journal of Health and Physical Education* (the predecessor to the current *Journal of Health, Physical Education, Recreation and Dance*, which was to serve as a forum for the discussion of professional issues, and the *Research Quarterly* (the predecessor to the current *Research Quarterly for Exercise and Sport*), which heralded a growing commitment toward research. Research work in the 1930s was primarily in the area of tests and measurement, with the development (particularly by Brace, McCloy and others) of physical fitness and motor ability tests for use in the public school, college and university education systems. These tests were initially general in nature but, with refinement, became more multidimensional in design. The main biophysical subdiscipline of human movement studies to progress its knowledge base in the 1930s was exercise physiology, where the excellent work of Dill, Margaria and associates at the Harvard Fatigue Laboratory added considerable new understanding to the mechanisms limiting exercise performance and underlying recovery from exercise.

The outbreak of the Second World War in 1939 brought to a temporary halt much of the basic research work of relevance to human movement but provided an immediate incentive for applied, problem-driven research and this benefited understanding in some spheres of human movement studies. In particular, problems created by the discovery that large numbers of draftees were failing to meet minimum physical fitness standards (in the US about one-third of the draftees examined were found to be unfit for service) led to renewed interest in population-based physical fitness and a demand for professionals with a knowledge of, and training in, the prescription of exercise. The instrumental requirements of war also resulted in great interest developing in the use of ability tests to aid in the selection of people for specific tasks (what became known as 'man–machine matching') and in optimal procedures for the rapid acquisition of new movement skills. Both these concerns provided much needed research impetus to the motor control field. The dominant research figure

Box 2.1 R. Tait McKenzie (1867–1938)

Robert Tait McKenzie was born in eastern Ontario, Canada on 26 May 1867 and throughout his 71 years of life made unparalleled contributions not only to the profession of physical education but also to the broad understanding and appreciation of human movement in general. McKenzie, in a manner reminiscent of the Renaissance man Leonardo da Vinci, made outstanding international contributions in not one but three fields, namely medicine, physical education and the arts. He gained a medical degree from McGill University in 1892 and from there held positions as the Medical Director of Physical Training at McGill before moving to the United States in 1904 to a new position as full professor in physical education at the University of Pennsylvania. In his joint roles as physician and physical educator McKenzie made outstanding original contributions to both the discipline and the profession. He published some 24 professional articles on various aspects of physical activity and health at a time when scholarly publishing was an unusual rather than expected activity. His first text *Exercise in Education and Medicine* was extremely influential in the medical profession and ultimately also in physical education, a field he guided to professionalisation through his proactive role in the leadership of the American Physical Education Association and in the formation of the American Academy of Physical Educators. A later text, *Reclaiming the Maimed—A Handbook of Physical Therapy*, based on his experiences as an army physician during the First World War, laid the foundations for the practice of physical therapy. For all these pivotal contributions to the field of human movement studies, R. Tait McKenzie's most enduring contributions are through his many, world-renowned, sculptures of athletes and of the beauty of the human body in motion. His sculptures reflected not only a rare skill in and appreciation of art, but also the extent of the precision of his observations and anthropometric measurements of athletes, undertaken as part of his scholastic work. Among other tributes following his death on 28 April 1938, the American Alliance for Health, Physical Education and Recreation (AAHPER) established in 1955 the R. Tait McKenzie Memorial Address lecture series.

Figure 2.2 Dr R. Tait McKenzie
Source: Davidson S.A. & Blackstock, P. (eds) (1980), *The R. Tait McKenzie Memorial Addresses*. (Copyright 1980 by the Canadian Association for Health, Physical Education and Recreation.)

Sources: Day, J. (1967), Robert Tait McKenzie: Physical education's man of the century, *Journal of the Canadian Association for Health, Physical Education and Recreation*, 33(4), 4–17; Davidson, S.A. & Blackstock, P. (eds) (1980), *The R. Tait McKenzie Memorial Addresses*, CAHPER, Canada; Kozar, A.J. (1975), *R. Tait McKenzie: The Sculptor of Athletes*, The University of Tennessee Press, Knoxville.

Figure 2.3 'The sprinter'—R. Tait McKenzie's first attempt at sculpture in the round
Source: Davidson, S.A. & Blackstock, P. (eds) (1980). *The R. Tait McKenzie Memorial Addresses*. (Copyright 1980 by the Canadian Association for Health, Physical Education and Recreation.)

in the 1940s and especially 1950s was Franklin Henry, working at the University of California, who contributed original research both on the physiology of exercise and on the acquisition and control of motor skills (Box 2.2). His work demonstrated, among other things, the highly specific nature of movement skills and, hence, the inappropriateness of general motor ability testing.

The 1940s also saw the initiation of a number of ambitious longitudinal research programs aimed at understanding normative motor development. Of these the University of California Adolescent Growth Study launched by Harold Jones, Nancy Bayley and Anna Espenschade was probably the first but it was followed soon after by comparable programs at the University of Oregon (the Medford Boys' Growth Study), at the University of Wisconsin–Madison, at Michigan State University, and in Canada at the University of Saskatchewan. Although these research activities continued through the 1950s into the 1960s and beyond, the profession of physical education grew somewhat away in focus from its emerging research arm, becoming preoccupied in the 1950s with professional practice issues related to teaching and coaching.

The beginnings of a discipline of human movement studies 1960–1970

The foundations for a discipline of human movement studies were well and truly laid by the early 1960s. Research activity, which had been sporadic and had certainly taken second place in importance to practical issues related to the profession of physical education throughout the first half of the century, had developed in both quantity and quality by the 1960s to the extent of demanding greater recognition and identity. The influential papers, described in Chapter 1, by Henry in 1964 and Rarick in 1967, which tried to define a unique and unifying theme for the growing but diverse research work conducted under the banner of physical education, provided an important catalyst for the beginnings of a discipline of human movement studies and for the encouragement of research in physical education not directly anchored to the solving of practical problems.

Henry's writings in the United States were paralleled by similar efforts by John Whiting

Box 2.2 Franklin Henry (1904–1993)

Franklin M. Henry came to the discipline of human movement studies in an unusual fashion, first completing formal training (a BA in 1935 and a PhD in 1938) in psychology at the University of California, Berkeley, before joining the staff of the then Department of Physical Education in the same university in 1938. He remained at that same institution until his retirement in 1971, contributing, in the interim, landmark studies in both the motor control and exercise physiology fields in addition to his contribution, discussed elsewhere, to the defining of the discipline. His principal scholarly contributions were to the understanding of metabolism and cardiovascular functioning during exercise, the specificity of training and the pre-planning of the control of rapid movements. Throughout his career he published over 120 original research papers in prestigious journals such as *Science, Journal of Applied Physiology, Journal of Experimental Psychology* and *Research Quarterly* as well as supervising over 80 research postgraduate students who, in turn, went on to contribute substantially to original knowledge in a number of the subdisciplines of human movement studies. Like many scientists of his era Franklin Henry had to construct his own equipment in order to undertake experimental work and in this regard his early training in electronics stood him in good stead. In addition to arguing strongly for an academic discipline of physical education Franklin Henry also contributed professionally through taking a lead role in the 1966 formation of the North American Society for the Psychology of Sport and Physical Activity. Professor Henry passed away on 13 September 1993.

Figure 2.4 Professor Franklin M. Henry
Source: The Creative Side of Experimentation (p. 37) by Conrad Wesley Snyder, Jr & Bruce Abernethy (eds), Champaign IL: Human Kinetics Publishers. (Copyright 1992 by Human Kinetics Publishers; reprinted with permission.)

Sources: Henry, F.M. (1992), Autobiography, pp. 37–49, in C.W. Snyder, Jr & B. Abernethy (eds), *The Creative Side of Experimentation*, Human Kinetics, Champaign, IL; Park, R.J. (1994), A long and productive career: Franklin M Henry—Scientist, mentor, pioneer, *Research Quarterly for Exercise and Sport*, 65, 295–307.

and others in England who went to great lengths to describe and detail human movement studies as a legitimate field of study (see Box 2.3). Whiting was responsible, in 1975, for the formation of a new journal bearing the disciplinary title of *Journal of Human Movement Studies*. Fuelling the emergence of a discipline worldwide was a search for academic credibility and worthy recognition by those undertaking fundamental research on human movement. With

Box 2.3 H.T.A. (John) Whiting (1929–

John Whiting was born in London in 1929 and gained employment as a school physical education teacher for a number of years following his initial physical education teacher training at Loughborough College in 1953. Whiting joined the Department of Physical Education at the University of Leeds in 1960 and from this position pursued higher qualifications, completing a PhD in experimental psychology in 1967. His doctoral research on the acquisition of ball skills was a complete break from the laboratory-based research that dominated the subdiscipline of motor control at the time. The 1969 textbook that stemmed from his PhD, *Acquiring Ball Skill. A Psychological Interpretation*, became standard reading for students of motor control and skill acquisition for generations. At the level of discipline, John Whiting initiated, in the 1970s, a blueprint for the foundation of an academic discipline of human movement studies plus a series of state-of-the-art-reviews on aspects of the field, including the emerging field of sport psychology. In 1975 he launched the *Journal of Human Movement Studies* and, in 1982, after moving to the Faculty of Human Movement Sciences at Free University, Amsterdam, the journal *Human Movement Science*. Both of these journals provided much needed outlets for the publication of original research in the discipline. John Whiting retired back to England in 1989 to an honorary position in the Department of Psychology at the University of York, while continuing active involvement in the field as ongoing editor of *Human Movement Science*.

Figure 2.5 Professor H.T.A. Whiting
Source: The Creative Side of Experimentation (p. 79) by Conrad Wesley Snyder, Jr & Bruce Abernethy (eds), Champaign, IL: Human Kinetics Publishers. (Copyright 1992 by Human Kinetics Publishers; reprinted by permission.)

Sources: van Wieringen, P.C.W. & Bootsma, R.J. (1989), *Catching Up: Selected Essays of H.T.A. Whiting*, Free University Press, Amsterdam; Whiting, H.T.A. (1992), Autobiography, pp. 79–97, in C.W. Snyder, Jr & B. Abernethy (eds), *The Creative Side of Experimentation*, Human Kinetics, Champaign, IL.

the initial attempts of Henry, Rarick, Whiting and others to define a discipline came a transition in the course offerings and departmental names of tertiary institutions offering studies in the field away from physical education to alternatives, as discussed in Chapter 1, such as human movement studies or kinesiology.

The emergence of subdisciplines and specialisations 1970–

Following the initial efforts at defining the scope and unifying nature of the discipline in the mid-1960s and beyond, the decades since that time, and most especially the 1970s, have been characterised by the emergence of specialised subdisciplines (as depicted in Figure 1.1), each with its own professional bodies, meetings and research journals. Examples of the proliferation of specialist journals emerging in this time include the *International Journal of Sport Biomechanics*; the *Journal of Motor Behavior*, *Medicine and Science in Sport and Exercise*; the *Journal of Sport and Exercise Psychology*; the *Journal of Sport History*; the *International Review of Sport Sociology* and the *Journal of Teaching in Physical Education*. Whereas on the one hand the emergence of specialist subdisciplines is obviously a positive sign for the discipline of human movement studies in terms of adding extensive depth to the knowledge base about human movement, it does create the potential for fragmentation (as noted in Chapter 1).

Coupled with the maturation of the discipline has been a proliferation of university programs, many of which offer courses in the discipline in addition to (or sometimes as an alternative to) traditional vocational courses in the profession. The relaxation of the rigid coupling of research on human movement studies to the profession of physical education has also created the potential for the knowledge base of human movement studies to be more actively incorporated into the practice of professional groups other than physical educationists. In a period of growing worldwide awareness of the importance of health and preventive medicine, and of the central role of regular physical activity in promoting health, new and exciting links between the discipline of human movement studies and a variety of health professions are clearly both possible and desirable.

Future directions, challenges, and opportunities

The discipline of human movement studies is at an exciting but critical stage in its development. An awareness of the historical antecedents to the modern field of human movement studies provides an invaluable background from which to speculate upon the challenges and opportunities that face the field both now and in the future.

Like other scientific disciplines, human movement studies has had the opportunity, at various times in its development, to not only strengthen existing links with other fields of science and its professions, but to also forge new partnerships. As we have seen in the preceding sections, in the early stages of development of basic research about human movement, issues related to physical activity and health were very much to the foreground and close interactions existed between early physical educators, physiotherapists, and medical doctors.

Over the past few decades, the major focus within human movement studies research has been more on issues related to sport science and performance enhancement than on issues of health, especially the health of the general population. However, there are clear signs of cyclic change in this regard improving international awareness of the central place of regular physical activity in disease prevention and health promotion (a topic we will cover in some detail in Chapters 15 and 24). This awareness and recognition provides an opportunity for human movement studies as a discipline to make new and reinforce existing links to a variety of health professions.

The next decade offers researchers and practitioners from human movement studies the opportunity to not only cement links with traditional areas of physical education and sport but to also revitalise old links with medicine and the therapies and to form new links with emerging fields such as human-computer interaction, robotics, bio-medical engineering, ergonomics, and music and the performing arts.

In addition to challenges and opportunities created by linking the discipline of human movement studies to other disciplines and professions, the future also poses challenges and opportunities within the field itself through the opposing forces of knowledge specialisation and fragmentation. At one level the expansion of specialist knowledge within each of the field's sub-disciplines adds immeasurably to the depth of our understanding about human movement. However, such specialisation also has the potential to work against the essential integrative understanding of human movement which is ultimately sought by the field.

Advances in molecular biology have so dominated biology as a whole over the past decade that studies of whole systems and integration between different sub-systems (such as the anatomical, mechanical, physiological, neural, and psychological systems that collectively comprise the biophysical bases of human movement) have been forced somewhat to the background. The growth of specialisation needs to be tempered by continued attempts to integrate knowledge across the different specialist sub-disciplines of human movement studies in order to continue to advance our knowledge in a consolidated manner. (This is an issue that we will return to again in Chapter 24.)

Undoubtedly one of the greatest challenges of attempting to integrate understanding of human movement comes from the incredible rate of information expansion experienced over the past decade. This explosion of information has occurred across all sciences but is especially pronounced within human biology. For example, new partial human gene sequences are currently being added to the molecular biology databases at the previously inconceivable rate of about 8000 per week.

With more people now working in science than ever before, the rate of publication of new research has well exceeded the capacity of scientists to keep abreast in their own field of specialisation, let alone remain up-to-date with developments in other sub-disciplines. Fortunately the past decade has also seen quantum changes in the technologies available to store, retrieve, and search for information. A number of these technologies are now becoming increasingly central to managing the information explosion in science.

Two major developments in information management are particularly noteworthy. The first of these has been the development of the *CD-ROM* (Compact Disc-Read Only Memory), which has provided the technology to support the development of major databases of information that are transportable and easily searchable via personal computer. CD-ROM technology has been especially valuable in the enhancement of bibliographic resources, allowing researchers to rapidly search for published studies on specific topics.

Literature searching in human movement studies has been greatly facilitated through CD-ROM databases such as *MEDLINE* (which contains all the reference material on health from Index Medicus, Index to Dental Literature, and the International Nursing Index); *PSYCLIT* (produced by the American Psychological Association and covering published psychology research back to 1974); and *SPORT DISCUS* (produced by the Canadian Sport and Information Resource Centre and containing nearly half a million bibliographic references to sport and sport science). These and many other CD-ROM databases are now readily available in most University libraries.

The second major development which is changing not only aspects of scientific communication and information transmission, but communications in general, has been the establishment of the *internet*, a network of networks allowing electronic communication between computer systems all around the world.

Direct communication on the internet via *e-mail* (electronic mail) provides a type of rapid postal service allowing messages and information files to be interchanged quickly between researchers at different locations throughout the world. Electronic discussion groups, built out of the e-mail technology, permit the sharing of information, ideas, and discussion between scientists with common interests.

A large number of *electronic discussion groups*, many with thousands of participants, already exist in a range of areas within the broad field of human movement studies (Box 2.4). The internet provides the potential for researchers to not only make direct electronic communication with other researchers with similar interests but to also browse information of all types stored on single specific computers—information made accessible in the public domain through computer networking. The *World Wide Web (WWW)*, through which one may 'visit' these different internet sites, is something of an electronic encyclopedia through which both text and images may be retrieved on an almost infinite array of topics.

The WWW already contains a large array of sites on organisations and topics of direct interest to students of human movement and provides the technology to even make the CD-ROM obsolete in the future (Box 2.5). Without doubt, the next generation of researchers, practitioners, and students of human movement will be more dependent than their predecessors upon multi-media information available electronically and less dependent upon traditional printed material.

Box 2.4 Electronic Discussion Groups

Electronic networks such as the internet have brought about a revolution in scholarly communication. One of the many powerful features of the internet is the capacity to establish electronic discussion groups or bulletin boards in which multiple (often thousands of) users can receive and reply to the same message. Using *listservers* based at a particular location (address) on the internet, electronic discussion groups may be established for the discussion of current topics, announcement of conferences and position vacancies, requests for information, etc., in specific fields of interest. Anyone with internet connection and an e-mail account can join these electronic discussion groups by sending a simple message generally of the form 'subscribe <list name> <your name>', to the e-mail address at which the specific listserver of interest is located. Table 2.1 lists some of the many electronic discussion groups of relevance to human movement studies; it also provides details of how and where to join these lists. A complete list of listservers may be obtained by sending the simple message 'list' to listserv@hearn.bitnet.

Box 2.5 Human Movement Studies on the World Wide Web

The World Wide Web (WWW) is perhaps best thought of as an enormous electronic encyclopedia, which, through the seamless connection of large band networks, provides access to an enormous array of information on a near infinite array of topics stored at specific sites on individual computers around the world. Through the WWW one can rapidly access information about organisations, information in the form of text and/or high quality graphic images, or information in the form of stored data files. The information on the WWW is constantly changing as new sites are added and information at existing sites is updated. One of the biggest difficulties, given the size of the WWW and the amount of information on it, is how to readily locate relevant information. Most WWW browsers (such as Netscape©) provide search 'engines' to assist in this regard, while an alternative approach is to use the many links between different sites (hence the term 'web') to 'surf' the internet to find sites of interest. Table 2.2 provides a sample of some of the many sites of interest to students of human movement, along with their WWW addresses (known as Universal Resource Locators, or URLs). You may also wish to commence your search by visiting the University of Queensland's WWW site (http://www.uq.edu.au/) or the site of Human Kinetics, our publisher (http://www.humankinetics.com/).

Table 2.1 Some internet discussion groups relevant to human movement studies

Listserver description	Message to join list	Where to send the message
Biomechanics	sub biomch-1 *your name*	listserv@nic.surfnet.nl
Clinical gait	subscribe cga *your name*	cga@info.curtin.edu.au
	suscribe clgait-1 *your name*	mailserv@shcc.org
Exercise physiology	subscribe sportsci	majordomo@stonebow.otago.ac.nz
Neuromuscular control	subscribe neuromus *your first and last names*	listserv@sjuvm.stjohns.edu
	subscribe neuromotor-control	listserv@ai.mit.edu
Motor development	*send message to jc60@umail.umd.edu*	listserv@umdd.umd.edu
Sport psychology	subscribe sportpsy *your name*	listserv@vm.temple.edu
Sports medicine	subscribe amssmnet *your name*	listserv@msu.edu
	join sport-med *your name*	mailbase@mailbase.ac.uk
	sub sptsinj-1 *your name*	listproc@anthrax.ecst.csuchico.edu
	sub athtrn-1 *your name*	listserv@iubvm.ucs.indiana.edu

Table 2.2 World Wide Web addresses for some sites relevant to human movement studies

Topic area	URL
Major Professional Organisations	
International Council of Sport Science and Physical Education (ICSSPE)	http://www.jyu.fi/~ icsspe/
International Federation of Sports Medicine (FIMS)	http://cac.psu.edu/~ hgk2/fims/
American College of Sports Medicine (ACSM)	http://www.acsm.org/sportsmed/
Major Link Sites	
Biomechanics World Wide	http://dragon.acadiau.ca/~ pbaudin/biomch.html
Sports medicine: Medweb	http://www.gen.emory.edu/medweb/medweb.sportsmed.html
Sport psychology internet links	http://www.psyc.unt.edu/apadiv47/links.htm
Physical therapy-related web sites	http://indy.radiology.uiowa.edu/Providers/Physical Therapy/CORE/PTREL/www.html
Specialised Topics	
Anatomy	http://www.mic.ki.se/Anatomy.html
Electromyography	http://nmrc.bu.edu/nmrc/detect/emg.htm
Exercise and health	http://www.ksu.edu/~ kines/THB.html
Heart disease	http://www.ksu.edu/~ kines/heart_page.html
Nutrition in sport	http://www.ausport.gov.au/nutrit.html
Information Retrieval Sources	
Institute for Scientific Information	http://www.isinet.com/
PsycINFO	http://www.healthgate.com/
Sport Information Resource Centre	http://www.SPORTquest.com/
Coaching Science Abstracts	http://www-rohan.sdsu.edu/dept/coachsci/intro.html
Australian Sports Commission National Sports Information Centre	http://www.ausport.gov.au/nsicmain.html

Further reading

Chapter 1: Human movement studies as a discipline and a profession

Bouchard, C., McPherson, B.D. & Taylor, A.W. (eds) (1992), *Physical Activity Sciences*, Human Kinetics, Champaign, IL.

Brooke, J.D. & Whiting, H.T.A. (eds) (1973), *Human Movement: A Field of Study*, Henry Kimpton, London.

Brooks, G.A. (ed.) (1981), *Perspectives on the Academic Discipline of Physical Education*, Human Kinetics, Champaign, IL.

Corbin, C.B. & Eckert, H.M. (eds) (1990), The evolving undergraduate major, *American Academy of Physical Education Papers*, No. 23, Human Kinetics, Champaign, IL.

Henry, F.M. (1964), Physical education: An academic discipline, *Proceedings of the 67th Annual Meeting of the National College Physical Education Association for Men Washington, DC*, AAHPERD, Washington, pp. 6–9. (Reprinted in Brooks (1981) pp. 10–15).

Janz, K.F., Cottle, S.L., Mahaffey, C.R. & Phillips, D.A. (1989), Current name trends in physical education, *Journal of Physical Education, Recreation and Dance*, 60(5), 85–92.

Rarick, G.L. (1967), The domain of physical education, *Quest*, 9, 49–52. (Reprinted in Brooks (1981) pp. 16–19.)

Rivenes, R.S. (ed.) (1978), *Foundations of Physical Education*, Houghton-Mifflin, Boston.

Chapter 2: Historical origins of the academic study of human movement

Adams, W.C. (1991), *Foundations of Physical Education, Exercise, and Sport Sciences*, Lea & Febiger, Philadelphia, Chapter 1.

Dill, D.B. (1974), Historical review of exercise physiology science, pp. 37–41, in W.R. Johnson & E.R. Buskirk (eds), *Science and Medicine of Exercise and Sport* (2nd edn), Harper & Row, New York.

Massengale, J.D. & Swanson, R.A. (1997), *History of Exercise and Sport Science*, Human Kinetics, Champaign, IL.

Park, R.J. (1981), The emergence of the academic discipline of physical education in the United States, pp. 20–45, in G.A. Brooks (ed.), *Perspectives on the Academic Discipline of Physical Education*, Human Kinetics, Champaign, IL.

Rasch, P.J. & Burke, R.K. (1974), *Kinesiology and Applied Anatomy: The Science of Human Movement* (5th edn), Lea & Febiger, Philadelphia, Chapter 1.

Ryan, A.J. (1974), History of sports medicine, pp. 13–29, in A.J. Ryan & F.L. Allman, Jr. (eds), *Sports Medicine*, Academic Press, New York.

Part 2

The biophysical subdisciplines of human movement

Section 1
Anatomical bases of human movement: The subdiscipline of functional anatomy

Introduction

What is functional anatomy?

At a simplistic level, human anatomy is the study of the structure of the human body, in contrast to physiology, which is the study of its function. The term functional anatomy may therefore appear to be synonymous with physiology, but it needs to be recognised that the disciplines of anatomy and physiology cannot be so easily separated. When applied to the musculoskeletal system, functional anatomy is the study of movement and the effects of physical activity. Functional anatomy therefore is dynamic anatomy, which considers both the short-term and long-term effects of activity on the musculoskeletal system. It is obvious that functional anatomy overlaps with both physiology, because of its functional approach, and biomechanics, because functional anatomists consider the musculoskeletal system to be a mechanical system. The boundaries between the three areas are ill-defined as evidenced by the titles of two good functional anatomy books: *Physiology of the Joints* and *Biomechanics of the Musculoskeletal System.*

Typical questions posed and problems addressed

The subdiscipline of functional anatomy concerns itself with answering a range of questions related to physical activity and the musculoskeletal system. Some typical questions include these:

- What functions do bones perform?
- How strong are bones?
- How do muscles produce movement?
- What prevents dislocation of joints during movement?
- How can the size and shape of a person be described?
- Do children or adults have more bones?
- Are children merely scaled-down adults?
- Why do older women in particular suffer more fractures than members of other groups?
- What adaptations occur when a person begins a regular exercise program?
- Is there an optimal level of exercise for the integrity of the musculoskeletal system?

Levels of analysis

Approaches to the study of anatomy

Anatomy is a very visual science. In gross or macroscopic anatomy classes, the unaided eyes are used to study the structure of the human body, whether directly from a cadaver, a model, a chart, or an anatomical atlas that includes many illustrations or photos of different parts of the body. In the study of human movement, the observation of surface features is also important. Therefore surface anatomy, which requires the skills of observation (using the sense of vision) and palpation (using the senses of touch and pressure), is an important precursor to the analysis of human movement.

In histology, light microscopes are used to aid visualisation of structures, tissues, and the cells that form tissues, and electron microscopes are used to aid visualisation of cells and structures within cells. Microscopes are used in functional anatomy to define the responses of tissues and cells to physical activity.

There are many areas of study within the discipline of anatomy, only a few of which will be covered here. The chapters that follow take both a systematic and functional approach to anatomy by discussing the function and properties of structures that form the musculoskeletal system. The three main systems to be discussed are the skeletal system (osteology), the joint system (arthrology) and the muscular system (myology).

Historical perspectives

Key developments

Although an understanding of history is not a prerequisite for the study of anatomy, an overview of its development is valuable as a means of gaining a better appreciation of the discipline. The term 'anatomy' is usually claimed to derive from the Greek word, *temnien*, meaning 'to cut'. However, manuscripts on the anatomy of the pig produced in Salerno during the period 1000–1050 AD explain that the Greek term 'anatomy' derives from *ana*, which means 'straight' and *thomos*, which means 'division'. The combination of the two creates a word that means 'correct division' so that in anatomy the dissection must be performed according to set rules.

Anatomy is still learnt by many students who dissect human cadavers as part of their course, following instructions written in dissection manuals. For many other students the material has already been dissected by others, ready for inspection of the important structures. Even if you never see a cadaver, you will see a model or a chart and these are derived from the investigation of human bodies. Possibly the first anatomical illustrations were drawn in 1522 AD by Bereugario de Carpi.

The well-known medical historian Charles Singer has dubbed Herophilus from Alexandria, in Egypt, as the 'Father of Anatomy' because he was probably the first to dissect both human and animal bodies for the purposes of anatomical instruction. At that time, however, anatomy was really only an investigative technique used by physiologists and surgeons. After a brief period of prominence by the Alexandrian school (300–250 BC) during the Hellenic period, there followed a period of Roman rule during which anatomy was almost exclusively based on animal dissection, although there is evidence of the occasional use of cadavers. The most famous anatomist of this period is Galen (129–200 AD), a Greek who was one of the greatest physicians of all time.

Galen wrote prolifically and his writings were based on both dissection and experiment. It was Galen who demonstrated that the arteries (meaning 'air carriers') actually carried blood. Many terms first used by Galen in relation to bones, joints, and muscles have survived to be part of the modern nomenclature. In fact, most anatomical terms are derived from Latin or Greek but are now often anglicised for ease of use. After this glorious period from 50 AD to 200 AD, anatomy, and medical science in general, waned with the dominance of the 'practical outlook' required by the Roman Empire during the period of its famous decline. This demise provides a classic illustration of how science is dependent on theory-driven and curiosity-driven research, both of which may disappear when excessive demands for studies with immediately applicable results are made.

The Dark Ages lasted until the beginning of the second millennium, when, during the 13th century AD, there was a great intellectual awakening in medicine led by scientists at the University of Bologna. A professor from Bologna, Mondino (1270–1326 AD), restored the techniques of anatomy, including anatomical dissection, and gave anatomy the theoretical foundation required to be considered as a scientific discipline, with a conceptual basis separate from surgery and physiology. From 1300 to 1325 Mondino personally dissected cadavers as part of the medical curriculum.

It was not only scientific anatomists who were interested in the structure of the human body. Artists also used scalpels to gain a better appreciation of the human form. Results of this understanding include the anatomical studies of Leonardo da Vinci (1452–1519 AD) and the anatomically correct paintings and sculptures of Renaissance artists such as Michelangelo (1475–1564 AD). The medical historian, Charles Singer, has stated that Leonardo da Vinci was not only 'one of the greatest biological investigators of all time' but also possibly the greatest genius in human history. He was probably the first to question Galen's authoritative teachings. Today, his drawing illustrating the proportions of the human body (Figure I.1) is arguably the most popular and widely recognised symbol of humans as physical and spiritual entities.

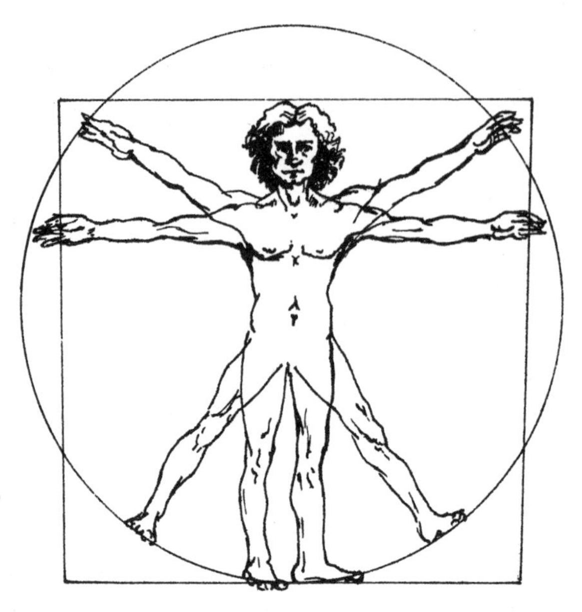

Figure I.1 A drawing by Leonardo da Vinci entitled 'The Proportions of the Human Figure'. (Redrawn by L. Blood, 1994)

Galen's prolific writings survived to the Middle Ages and his teachings influenced anatomical thinking to such an extent that dissections were performed merely to demonstrate his ideas. If the cadaver did not confirm Galen's teachings, then the cadaver was thought to be wrong, but, of course, it did not take long for this approach to be questioned.

It was left to Andreas Vesalius of Brussels (1514–1564 AD) to reform the science of anatomy. He did this as a professor of anatomy at the University of Padua, when he published *On the Fabric of the Human Body*, a work of seven books, in 1543. To Vesalius 'fabric' meant 'workings' and the first two books on bones and muscles indicate that he regarded the human body as a machine designed to perform work. Vesalius did not completely reject the teachings of Galen but he thought that dissection provided the opportunity to observe the human body and make comparisons with textbook descriptions. In this way previous knowledge could be confirmed or questioned. It is interesting to note that this major advance in the biological sciences was paralleled by a major publication in the physical sciences. It was in the same year that Nicholas Copernicus published *On the Revolutions of the Celestial Spheres*, in which he argued that the Earth was not the centre of the universe.

In the history of anatomy, teaching has had equal prominence with discovering new structures or techniques. Possibly two of the greatest teachers of human anatomy in the English language were brothers William Hunter (1718–1783 AD), an obstetrician, and John Hunter (1728–1793 AD), a surgeon. The Hunterian Museum and Library at Glasgow University houses books, pictures, and specimens donated by William, and the Hunterian Museum at the Royal College of Surgeons in London still includes some of the 13 000 specimens of human and animal material donated by John. Possibly the best known modern anatomy textbook in the English language is *Gray's Anatomy*, which is now in its 37th edition. Henry Gray (1827–1861 AD), who was a lecturer in anatomy and curator of the anatomy museum at St George's Hospital, London, wrote the first two editions starting in 1858.

In earlier textbooks, such as the first 20 or more editions of *Gray's Anatomy*, many structures derived their names from the person who first described it. Anatomical terminology therefore provided a history lesson, but it was not a very accurate one. Insignificant people were sometimes remembered by important structures whereas truly great anatomists had insignificant structures named after them. As a result of a series of international meetings since the Second World War, terms derived from the name of an anatomist (eponyms) are now no longer used in anatomy.

Anatomical terms are now descriptive and logical but there are still many named structures within the human body that make study difficult for all students of anatomy. Present students should be grateful that at least there is usually only one official name for each structure and not several as there were earlier this century. Two examples of eponyms that have lingered on in clinical and common usage are the Eustachian tubes (now called the auditory tubes) named after a 16th century Italian anatomist, Eustachius, and the Fallopian tubes (now called the uterine tubes) named after another 16th century Italian anatomist, Fallopius.

Key people

Most histories of anatomy concentrate on gross anatomy and, later, microscopic anatomy, and tend to overlook many of the key innovators in the field of functional anatomy. Geralamo Mercuriale published a book in 1577 AD entitled *The Practice of Gymnastics* in which he outlined the medical, athletic, and military aspects of exercise. Instructions on correct exercise technique were provided and it was claimed that exercise could improve health. He discussed the effects of physical activity on healthy and diseased individuals and stressed that exaggerated exercise can cause damage, especially in competitive athletics. These are all issues that are still topical today and that will be addressed in the chapters that follow. The great anatomist of the 16th century, Vesalius, had a dynamic approach to the musculoskeletal system, and Galileo noted in 1638 that there was a direct relationship between body mass, physical activity and bone size.

This idea of the relationship between form and function was extended by a number of 19th century German scientists, culminating in the publication in 1892 of a book entitled *The Law of Bone Transformation*, written by a German anatomist, Julius Wolff. This book included what is now known as Wolff's law, which was based on engineering analyses of the small bony rods (trabeculae) that form spongy bone. He summarised the work of a number of researchers by concluding that there was a strong correlation between the direction of the trabeculae and the lines of force acting through the bone. The law related to the architecture of different bones but others realised that there was also an adaptive relationship that is still of major interest. Research still continues today on the mechanisms responsible for the biological adaptation of bone to various levels of exercise (see Chapter 6).

Most actions of muscles had been determined from knowledge of muscle attachment sites, and lines of pull of the muscle related to the joint axes of rotation. Guillaume Duchenne (1806–1875 AD), a French physician, attempted to confirm these hypotheses and his experiments and observations over many years culminated in the publication in 1865 of his influential text *Physiology of Motion: Demonstrated by Means of Electrical Stimulation and Clinical Observation and*

Applied to the Study of Paralysis and Deformities. Of major interest to modern functional anatomists is the recording of the actions of almost all of the skeletal muscles in the human body.

Particularly since the Second World War, the converse of Duchenne's technique has been used. When a muscle is stimulated by its nerve it responds in two ways—by producing an electrical signal and a mechanical force. Nobody has yet directly measured the force within the muscle tissue of a living human, although tendon forces have been measured. It is possible to detect the electrical signal produced by muscle tissue using a variety of electrodes. This technique, known as electromyography, is used to describe muscle activity during human movement. The best known name in this field for the past four decades has been John Basmajian (1921–) from North America, who has held professorships in both Canada and the United States. He is a former president of the American Association of Anatomists. Not only has Basmajian contributed greatly to the elucidation of dynamic human muscle function in a variety of situations, but he has been very involved also in the teaching of gross and functional anatomy through his authorship of many books, including the well-known *Muscles Alive: Their Functions Revealed by Electromyography*.

Although the anatomy of adult humans is of major interest to anatomists it is not their sole interest. The structural changes that occur during childhood, and later, have also been studied extensively by auxologists. The field of auxology is defined as the science of biological growth, and key concepts from this field will be examined in some detail in Chapter 5. As was observed in Chapter 2, many human growth studies have been performed throughout the world. Many of the recent researchers in this field have been trained at the Institute of Child Health at the University of London, in a program headed by the noted auxologist, Jim Tanner. Among many books published by Tanner is one on the history of human growth studies and listed as a further reading source to supplement material in Chapter 5.

Professional training and organisations

Anatomy is a foundation subject for many professions including medicine and all the allied health professions. Many qualified professionals often find it useful to revise their anatomy. For example, people interested in sports medicine may attend a session in an anatomy laboratory as part of their continuing education. Likewise, surgeons, nurses, sports coaches, ergonomists, and others also find it valuable to continually update their knowledge of anatomy.

Most countries or regions have an association comprised of practising anatomists which holds regular conferences to present the latest research of members and to discuss new directions in the teaching of anatomy. The International Federation of Associations of Anatomy (FIAA) also holds a conference every 5 years. Members of the International Society for the Advancement of Kinanthropometry (ISAK) have a particular interest in the relationships between body dimensions and human performance. Functional anatomists frequently find more interest in the meetings of professional societies in biomechanics than in the meetings of general anatomy societies.

CHAPTER 3

Basic concepts of the musculoskeletal system

- Tools for measurement
- Structure and function of the skeletal system
- Structure and function of the articular system
- Structure and function of the muscular system

The purpose of this chapter is to introduce key concepts related to the structure and function of the skeletal system, the system of joints (the articular system) and the muscular system, and to describe the tools for measurement of these systems.

Tools for measurement

Language is very important for basic gross anatomy, because of the descriptions required to explain the position, relative size, and relationships of each anatomical feature. However, learning anatomy from a textbook can be very difficult. Learning is improved with the aid of atlases, which may have many artistic impressions, or photographs, of the different structures. Nevertheless anatomy is best learnt in a fully equipped anatomy laboratory with human cadaver specimens. Basic anatomy involves the identification of many structures and can be learnt without the use of numbers because it has traditionally been a qualitative subject. Anatomy as part of the modern biological sciences now employs measurement so it is also quantitative in its approach. Many of the concepts introduced in this chapter will be presented qualitatively but it should be realised that these concepts have been verified experimentally using quantitative methods.

Bone density in living humans can be determined using radiological and other more direct techniques. Bone structure can be visualised under a microscope but special preparation techniques are required because it is a hard tissue. Chemical analyses can also be performed to determine the composition of bone. Movement relies on the integrity of joints, and goniometers are examples of instruments used to measure ranges of joint motion. Muscles produce forces that have an external effect and these effects can be measured by different types of dynamometers, which basically measure muscle 'strength'. The electrical signal generated as a muscle contracts can also be detected, recorded and analysed using electronic equipment and computers.

In mechanical testing of biological tissues biologists use equipment that is similar to that used by civil engineers to test the properties of concrete, and by mechanical engineers and metallurgists to determine the properties of metals. In fact, with knowledge of artificial materials, a better understanding of biological tissues can be developed. For example, the composition and mechanical properties of fibreglass are often used as an analogy to help to understand the structural properties of bone.

Structure and function of the skeletal system

The skeletal system is the framework for the human body

Humans are vertebrate animals. Attached to our 'backbone', or vertebral column, are other bones that form the framework of the human body, much like the beams and studs of a house, or the chassis of a car. These analogies are imperfect, however, because they imply that the skeletal system has only mechanical functions. As we will see, the skeletal system also performs important biological functions.

The skeletal system has mechanical and physiological functions

The most obvious mechanical function of the skeletal system is support for weight-bearing. The skeletal system also protects internal organs, for example protection of the brain by some bones of the skull and protection of the heart and lungs by the ribs. The major bones of the limbs in particular provide 'rigid' links between joints, and these bones also provide sites for muscle attachment. These functions of the skeleton in providing linkages and sites for muscle attachment facilitate human movement and form the basis of the mechanical models of the human body used by many biomechanists.

In addition to its mechanical functions, the skeletal system also has important physiological functions. Bone tissue is involved in the storage of essential minerals such as calcium and phosphorus. Bone marrow is also involved in the production of blood cells and is part of the body's immune system.

It must be remembered that bone is a living dynamic tissue. The framework of a house is often made of timber, which was once a living tree. Similarly the bone as seen in an anatomy laboratory is changed from its condition in the living body. When subjected to large forces over a long period, the framework of the house may start to fail, just as the metal in a car may start to fatigue or rust. Living bone has the advantage that it may heal when broken and even carry out maintenance to prevent failure. Some bone researchers believe that the stimuli to bone adaptation include the microcracks that form during increased levels of physical activity. (Adaptation in response to physical activity will be explored further in Chapter 6.)

Designers of cars know that vehicles must be strong enough to withstand normal road use but that they should also be as light as possible for efficiency. Likewise, bone needs to be strong yet light so that the muscle strength required to move the body is not excessive. Motoring magazines report this as the power-to-weight ratio of cars. In the human skeleton there is a requirement for minimum mass to aid human movement, but there must be sufficient mass to withstand the forces of gravity, especially during weight-bearing exercises such as walking, running, and jumping. In addition, there must be enough bone material to ensure low incidence of fractures when the bones are subjected to large forces such as those involved in jumping, falling, football tackling and other direct blows to the body. Bone has to be rigid but it must also adequately cushion forces to function optimally.

In a car, the metal in the engine tends to be thick and rigid because it must resist deformation but the body panels are more cosmetic and help to absorb the energy of impact during a collision. As with the metal in the car, the composition of bone in the skeleton is also not uniform. There is compact bone, which is dense like ivory, and spongy bone, which is a lattice meshwork of bony rods. This organisational structure can be likened to a bicycle wheel with spokes, where the spokes maintain the shape of the rim but also help to cushion forces from the ground. (A solid bicycle wheel would be too stiff for general purposes.) Metallurgists and engineers have at their disposal different building materials, whereas the two types of bone within the human skeleton are made of the same material. Compact and spongy bone differ mainly in their porosity although compact (also called cortical or dense) bone is also more organised than spongy (also termed cancellous or trabecular) bone.

In spongy bone, each bone cell is close to a nutrient supply because the bony tissue is surrounded by blood vessels and associated material. When bone is remodelled to become compact, bone cells may be removed from their nutrient supply because now the blood vessels are surrounded by bony tissue. This lack of nutritive supply would result in the death of each cell unless compact bone was organised in a specific way. Figure 3.1 illustrates the microscopic organisation of this compact bone, which is contrasted with the organisation of spongy bone.

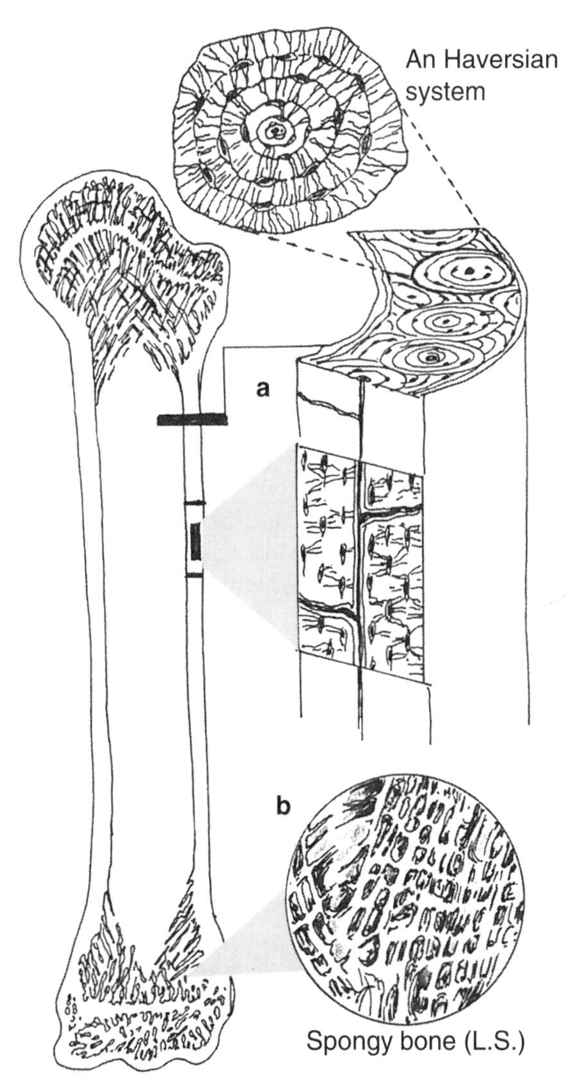

Figure 3.1 Organisation of bone material indicating longitudinal and transverse sections of the compact bone in the shaft (a) and the structure of spongy bone in the expanded ends of a typical long bone (b)

Bone has organic and inorganic components

The bone in the living body contains about one-quarter of its mass as water. In adult bone, after removal from the body and drying, about two-thirds comprises inorganic crystals consisting of calcium and phosphorus. Almost all the body's calcium is stored in bone and is released from there as required. Most of the calcium is in the compact bone but the calcium in the spongy bone is more easily released when needed. One of the important functions of calcium is its role in initiating muscular contraction. Organic collagen fibres comprise most of the remaining one-third of dry bone. Bone cells are another organic component of bone.

There are four types of bone cells

In the matrix of bone there are bone cells called osteocytes. Also associated with bone are bone-forming cells, called osteoblasts, which differentiate from the precursor osteogenic cells. Depending on the local environmental conditions, osteocytes can become osteoblasts and vice versa, so that adaptive responses can occur when there are mechanical stimuli. There are also bone-eroding cells, called osteoclasts, which form from collections of specific types of white blood cells called monocytes. Remodelling of bone, which is a continual process in adults, involves both erosion and deposition of bone tissue by the different cells. A full cycle of remodelling takes about 3 months.

Long bones have hollow shafts with expanded ends

The major forces that must be withstood by the long bones of the limbs, particularly the lower limbs, are compression along the long axes of the bones, bending, and twisting (see Figure 3.2). This loading requires optimal arrangement of the bone material but what is the most efficient architecture? The shafts of long bones are hollow and this has mechanical advantages over a solid rod of the same mass. If both a pipe and a solid rod have the same amount of material the pipe is stronger in relation to bending than the rod so that a hollow shaft is more efficient in terms of the relationship between strength of the whole bone and its weight. Because the bone material is further from the centre in a tube, a hollow shaft also resists twisting better than a solid rod. Look at the steel beams used in building construction and note they are shaped like an 'I' so that most of the steel is concentrated at the top and bottom of the beam where the bending moments are largest.

During movement, especially when landing from a jump, forces are transferred from one bone to the next via the joints. A large contact area between the bones results in less pressure on the ends of the bones, so expanded ends are advantageous. Much of the material forming the expanded ends of bone is spongy, which absorbs energy during impact. Most of us know the difference between jumping on a trampoline and then jumping off the trampoline onto a concrete floor. In fact the difference is so great that we are advised never to do this. Compact bone is much like concrete because it does not deform much but is very strong, whereas spongy bone is more like a trampoline, which cushions the force of landing as it deforms. In fact spongy bone is 10–15 times more deformable than compact bone. At the end of the long bone, a thin outer layer of compact bone protects the overlying cartilage during impact and transfers the forces to the underlying spongy bone, much like the rim of a bicycle wheel (Figure 3.3).

Bone has important mechanical properties

It has been argued that both stiffness and flexibility are important properties of bone. In a car the axles must be stiff to resist deformation as they transfer the forces from the engine to the wheels, but springs are designed to deform and shock absorbers to dampen the motion.

Engineers can use different materials to suit particular purposes; they choose steel, aluminium, plastic or other alternatives depending on the purpose of the part. To produce the optimal mechanical properties, metallurgists develop alloys, which consist of different types of metal bonded together. Steel, for example, is manufactured from iron and carbon, and bronze from tin and copper.

Is bone like a homogeneous metal, such as copper or iron, or does it consist of different components? The answer is the latter; consequently, bone is described as being a composite material. One useful analogy for bone is fibreglass, which consists of glass fibres cured in an epoxy resin. The final product, fibreglass, has mechanical properties that are superior to those of its individual components. The main mechanical components of bone are collagen and calcium salts. The collagen provides the toughness and flexibility and contributes to the tensile strength, while bone's hardness and rigidity are due mainly to its calcium salts, which also contribute principally to the compressive strength of bone. Being a composite material, bone is stronger than collagen in tension and stronger than calcium salts in withstanding compression. The bonding between the organic collagen and the inorganic salts is therefore very important in determining the mechanical properties of bone. It has been calculated that the optimal mineralisation of bone, in terms of strength and flexibility

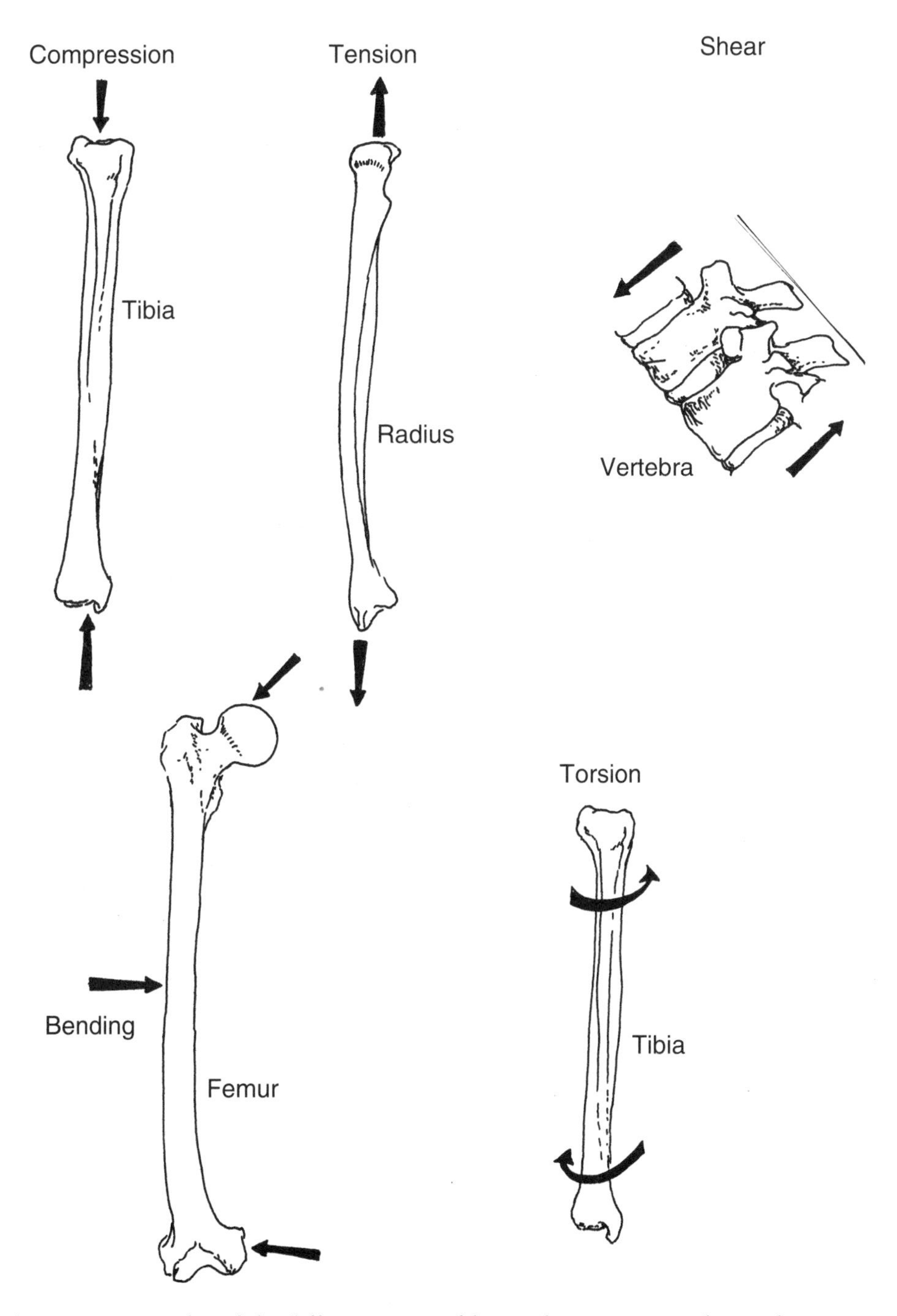

Figure 3.2 Examples of the different types of forces that may act within and between bones. The bones of the lower limb are often under compression during standing, while the bones of the upper limb may be subjected to tension while holding a load in the hand. During forward bending of the trunk there are shear forces between adjacent vertebrae. During walking and sports activities the long bones of the lower limb may be acted on by bending and torsional (twisting) forces

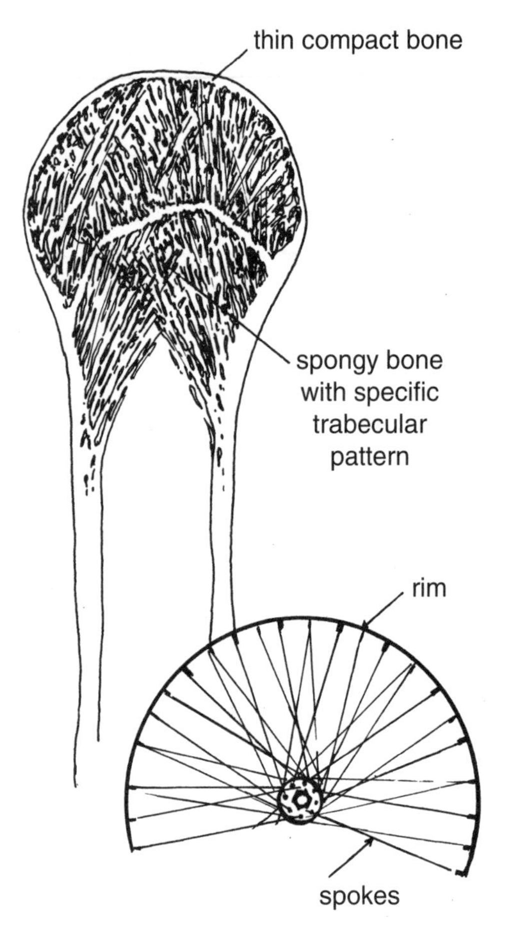

Figure 3.3 Structure of a long bone with compact bone in the shaft and spongy bone in the expanded ends, supporting a thin layer of compact bone. The structures at the end are compared with a bicycle wheel indicating that the thin layer of compact bone is like the rim and the trabeculae forming the spongy bone act like the spokes of the wheel

properties, is two-thirds mineral to one-third organic material. This is precisely the proportion of calcium salts to collagen in healthy adult bone.

In the laboratory calcium can be chemically removed from a long bone in the leg, such as the fibula, which can then be tied in a knot because it loses its stiffness. If the collagen is removed by heat, the bone will easily shatter because it is too brittle. A comparison between the mechanical properties of bone and other materials indicates that bone and wood from the oak tree are similar in their properties of strength and stiffness. Bone is about as flexible as fibreglass although weaker, but is stronger and more flexible than ordinary glass. Ceramics in general are about one-third to one-half as strong in compression as bone. The blocks of older car engines were manufactured from cast iron, which is similar to bone in its tensile strength, but bone is three times lighter (that is, less dense) and much more flexible. Bone is clearly a remarkably engineered material.

Bones have different shapes and organisation

Many different types of materials with different mechanical properties are used in a car, whereas most of the framework of the adult human body is composed of bone, either compact or spongy. When there are different functional requirements there are variations in the arrangements of the two types of bone and there are differently shaped bones. Flat bones such as the bones of the skull protect the brain, long bones in the limbs provide rigid links between major joints, and short bones in the foot help to cushion ground reaction forces during locomotion (see Figure 3.4). You will notice that some of the major functions of bone are repeated, but now there is a specific structure–function relationship defined. All these bones consist of both compact and spongy bone, in differing proportions and structural arrangements. The major difference between compact and spongy bone is the density because of the difference in porosity. The actual bone material has similar composition and mechanical properties.

Amount, density, and distribution of bone material have major effects on the mechanical properties of a whole bone. Compare the resistance of a long plank of wood to bending in the vertical direction when it is supported at both ends and placed either vertically or horizontally. Also compare the deformation of high-density and low-density foam when a person lies on a mattress made of these materials.

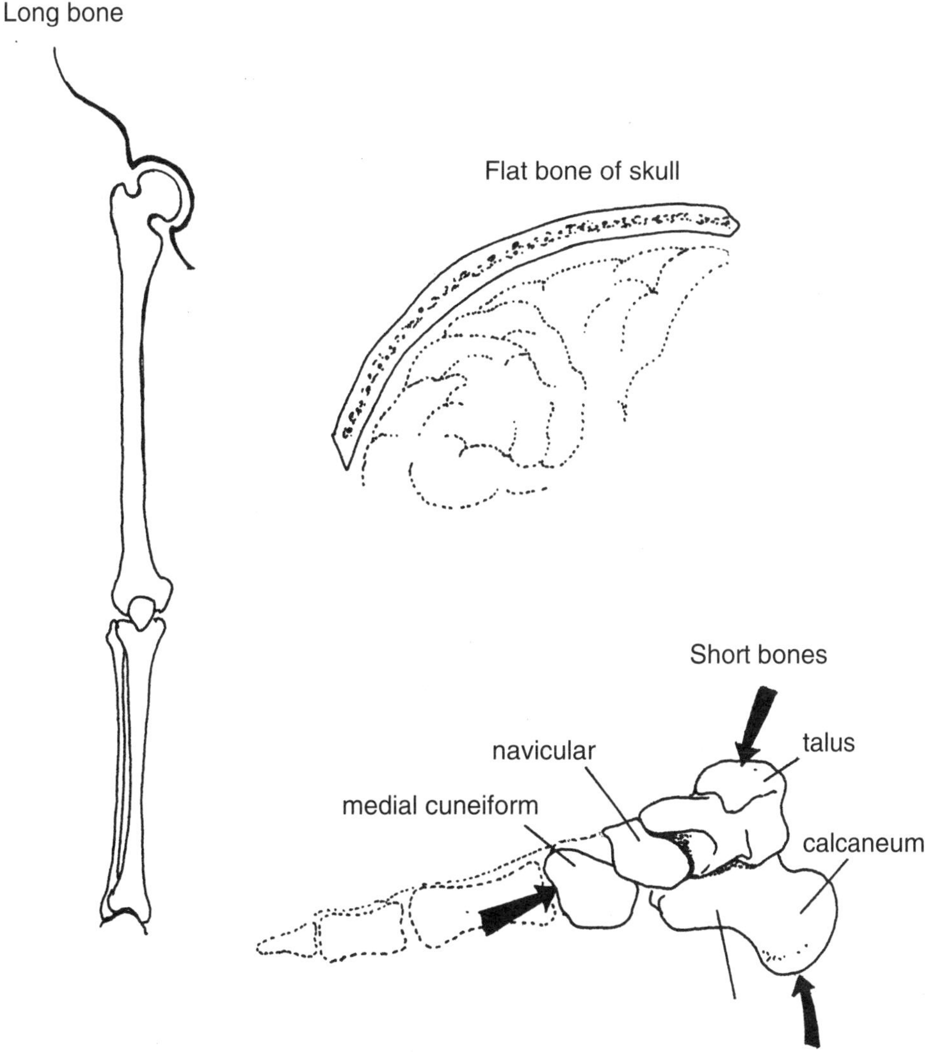

Figure 3.4 Examples of bones with different shapes that serve different functions. The long bones of the lower limb act as levers about the major joints, the flat bones of the skull help to protect the brain, and the short bones in the foot resist the large compressive forces generated during walking and running

Bones perform a variety of functions

To a greater or lesser extent all bones must perform the mechanical functions of support, protection and leverage for soft tissue structures. The dominant function needed is related to the shape of the bone. When all functions are almost equally important the shape is described as irregular. The best example of an irregular bone is the vertebra, which consists of parts, each of which has certain mechanical functions that are more important than others. The different parts of a typical vertebra are illustrated in Figure 3.5. The body of each vertebra has a major weight-bearing function and is like a short bone that cushions compressive forces passing along the length of the vertebral column. The vertebral arch protects the spinal cord and, particularly the lamina of the arch, is like a flat bone. The processes, attached to the vertebral arch, provide leverage to the attached ligaments and muscles, much like long bones in which the shafts resist bending and twisting forces. Note the arrangements of compact and spongy bone illustrated in Figure 3.5.

Structure and function of the articular system

Joints can be classified according to their structure or function

A joint is a union of two or more bones. The material between the bones forming the joint may differ and is the basis for the structural classification of joints. All anatomical joints may be described as fibrous, cartilaginous, or synovial. Most of the major joints with which we are familiar, such as the shoulder, elbow, knee, and ankle, are examples of synovial joints, so we will focus on these exclusively in this section.

The major function of joints is to allow movement but at the same time to remain stable. For each joint there are requirements for both mobility and stability. All joints allow some movement and joints can be classified according to the amount of movement allowed. Synovial joints allow a relatively large amount of movement; consider, for example, the ranges of motion that are possible at both the shoulder and elbow joints.

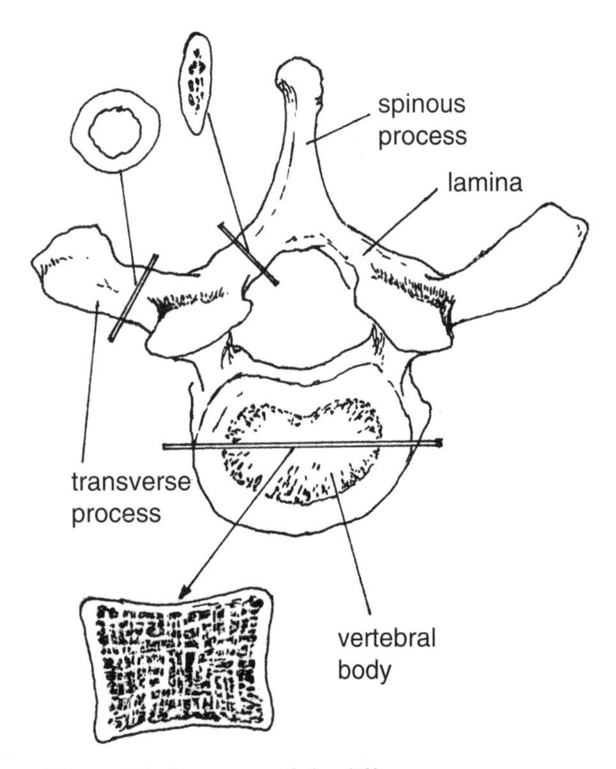

Figure 3.5 Structure of the different components of a typical vertebra. In section, the body is like a short bone, the lamina of the arch is like a flat bone, and the transverse process is like a long bone

Synovial joints have characteristic features

There are a number of characteristic features associated with a typical synovial joint (Figure 3.6). On the ends of the bones forming the joint there is articular cartilage, consisting of collagen fibres in a liquid matrix. Articular cartilage has a high water component of about 80 per cent and has been likened to a sponge in which water can be squeezed out. It will absorb water when not being deformed. Articular cartilage forms a relatively smooth bearing surface and also acts, through its capacity to deform, to cushion forces.

A joint capsule forms part of the boundary of the joint. The material of the joint capsule contains a high proportion of collagen fibres and may be described as dense connective tissue. The joint capsule provides some intrinsic stability to the joint and provides some resistance to motion. Forming the inner layer of the joint capsule is the synovial membrane, which has a number of important functions. It produces the fluid within the joint, both by filtration of blood and by active secretion. It also removes the cell debris that results from 'wear and tear' within the joint.

A major characteristic of a synovial joint is the cavity bounded by the articular cartilage and the synovial membrane of the joint capsule. This joint cavity contains a small amount of synovial fluid, which, in turn, contains

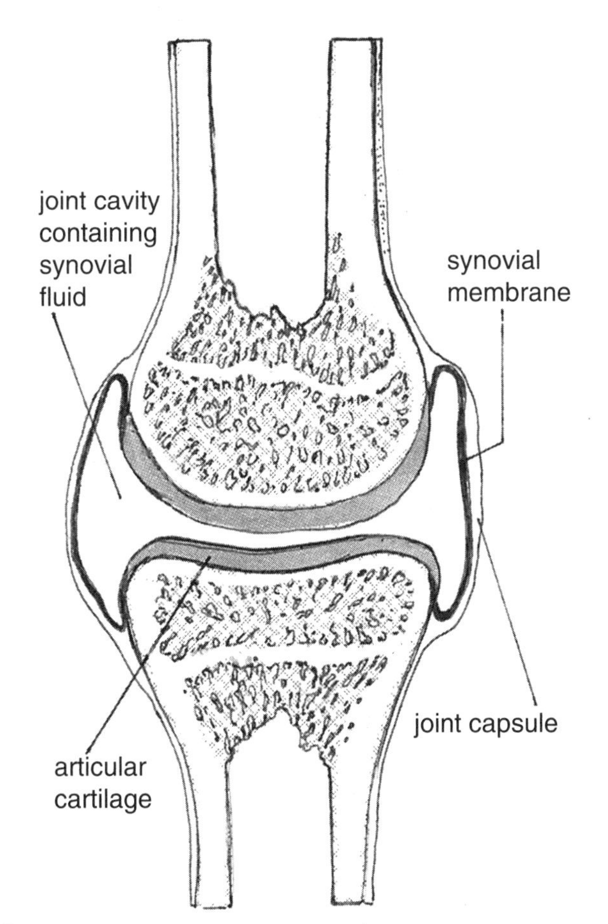

Figure 3.6 Representation of a typical synovial joint indicating its characteristic features. Individual joints may be more complex and specialised. Usually, synovial joints have ligaments associated with them

constituents of blood, substances secreted by the synovial membrane and some products that are the result of abrasive wear within the joint.

The synovial fluid has three important functions: lubrication, protection, and nutrition. The viscosity of the fluid can change according to local environmental conditions so it may be thick and protective, like grease, or thin and lubricating, like oil. Because articular cartilage does not receive an adequate blood supply, nutrients are supplied via the synovial fluid in contact with the cartilage. Pressure during movement helps to squeeze fluid out from within the cartilage, which can then re-enter when the pressure is removed. Physical activity thus promotes the nutritional function of synovial fluid.

To help maintain the integrity of the synovial joint, associated ligaments attach from bone to bone and cross the joint. Ligaments consist of collagen fibres, which are also a constituent of bone. In ligaments the collagen forms about 90 per cent of the structure and the fibres tend to run parallel to each other so that a ligament is an example of dense, regular connective tissue. The ligaments help the joint capsule to provide stability and they also function to guide the movements of the joint. In doing so they provide some resistance to joint motion. Ligaments are basically passive structures that resist external tensile forces, that is, forces that tend to separate the bones forming the joint.

Natural joints are more complex than artificial joints

It is well known that joints allow movement in cars and bicycles. These are relatively simple joints compared with the synovial joints found in the human body and engineers are still trying to explain the functions of biological joints. The design of artificial joints for replacement of human joints such as hips and knees is progressing rapidly but replacement joints still do not work as well as the normal joints. To give some idea of the complexity of natural joints it is better to contrast them with the types of joints in a car. The bearing surfaces in the artificial joints of a car are usually of smooth hard metal whereas the articular cartilage is less smooth but quite deformable. Artificial joint surfaces are very regular, usually part of a circle or sphere, whereas the surfaces in synovial joints are ovoid (egg-shaped) resulting in much more complex movement patterns. Any wear results in rapid deterioration of artificial joints but cell debris can be removed continuously within a synovial joint. To remove debris in the joints of a car it has to be taken to the local garage for a 'grease and oil change'.

Synovial joints can be classified according to different criteria

The classical anatomical view can be used to classify synovial joints on a descriptive basis, or the view of engineers can be utilised to produce a classification system that can explain joint lubrication and wear. This section will not try to provide the details of the alternative classification systems but instead will summarise the criteria on which each system is based. Both structural and functional criteria are used.

Anatomically, each synovial joint can be classified on the basis of the approximate geometric form of its articulating surfaces. For example, the hip looks like a ball in a socket (Figure 3.7*a*), much like the joint between a car and a trailer. Synovial joints can also be classified according to the gross movements permitted. For example, the ankle joint allows movement basically in only one plane and so it acts like a hinge (Figure 3.7*b*). The ankle could also be classified as a uniaxial joint because it allows movement about only one principal axis. The joints forming the knuckles of the fingers are classified as biaxial because they allow motion about two principal axes (Figure 3.7*c*). Note that the fingers can be moved in two directions at right angles: backwards and forwards and from side to side.

The complexity of organisation of the joint structures is another criterion. If a joint consists of two bones, with one pair of articulating surfaces, it can be classified as simple, such as the joints of the fingers. If there is more than one pair of articulating surfaces, such as within the elbow joint capsule where there are three pairs of articulating surfaces, the joint is classified as compound (Figure 3.7*d*). Sometimes

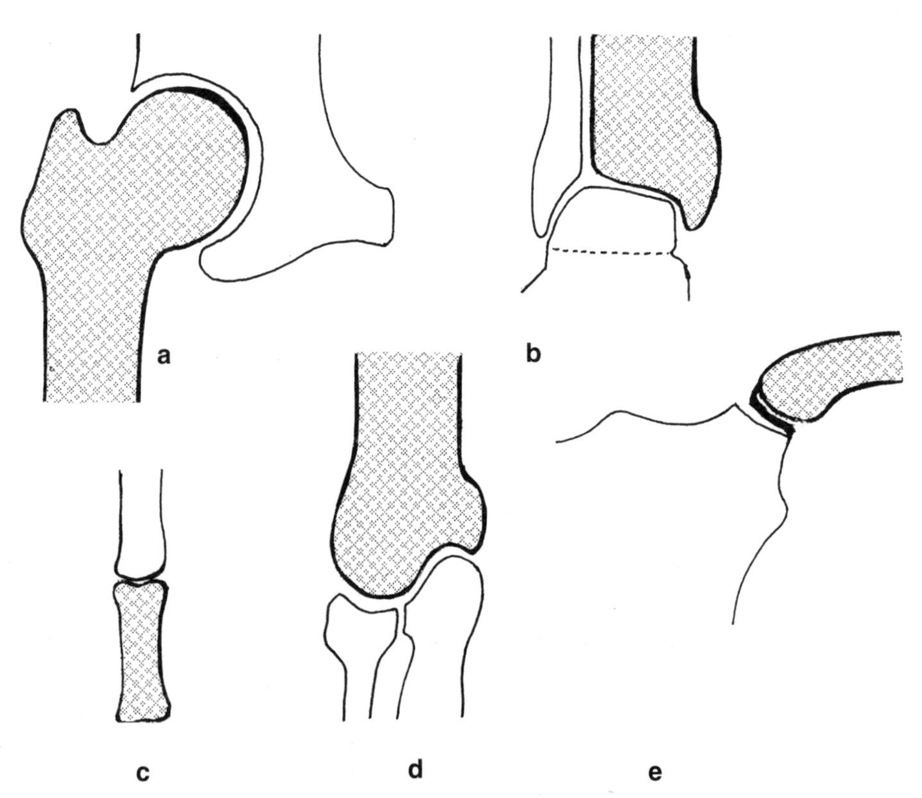

Figure 3.7 Representations of some different types of synovial joints: spheroidal hip joint (a), hinge-like ankle joint (b), simple interphalangeal joint of finger (c), compound elbow joint (d), and complex joint between sternum and clavicle (e)

synovial joints contain intra-articular structures such as a cartilaginous disc or meniscus, and these joints can be classified as complex (Figure 3.7*e*). An articular disc within the joint effectively doubles the number of articulating surfaces.

Synovial joints allow a range of movements

Movements that occur at synovial joints have traditionally been described in terms of the major planes of the body (Figure 3.8). Anatomically, movements that occur in the sagittal plane have been termed 'flexion', when the angle between the limb segments decreases, and 'extension' when the angle increases (Figure 3.8). These gross descriptions of directions of movements of body segments can be contrasted with the more mechanical approach, in which the terminology is related to the movement that occurs between the articular surfaces in contact.

Using the engineering approach, the relative motion between articular surfaces can be described as spin, slide or roll (Figure 3.9). These movements have an effect on the frictional resistance. For example, resistance is much less when the wheels of a car are rolling freely than when the brakes are applied and the tyres slide along the bitumen. Although there is a tendency for rolling to occur, the frictional resistance to all movements at the interface between the articular cartilages is very low. The low coefficient of friction in synovial joints is related to the properties of the articular cartilage and the synovial fluid.

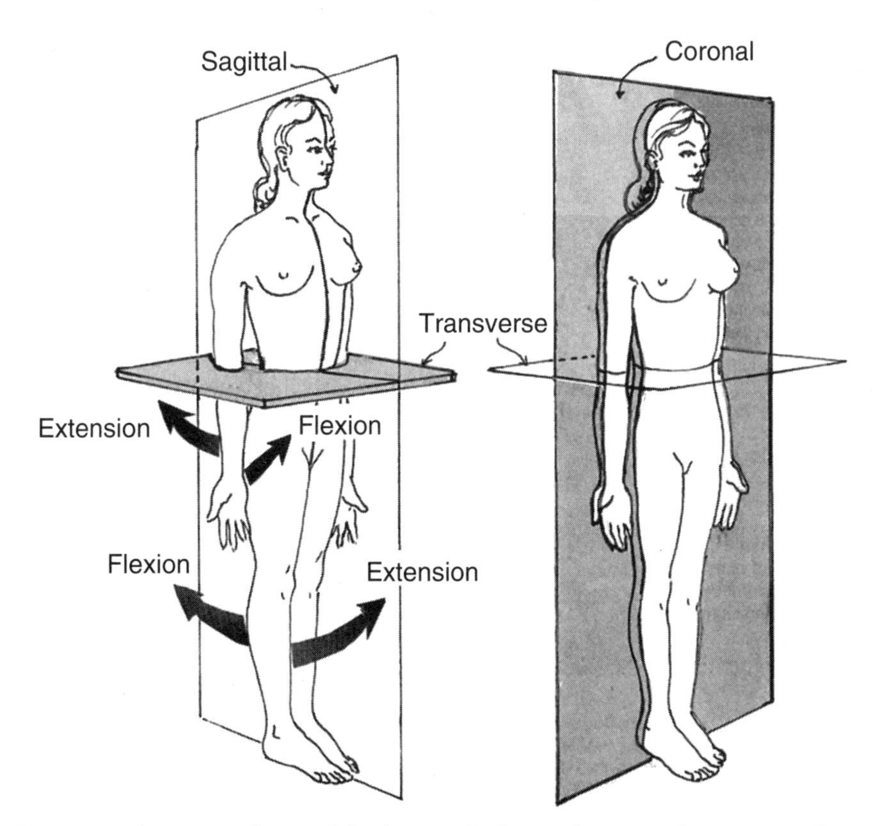

Figure 3.8 The major planes of the human body are shown with respect to the anatomical position. The movements of flexion and extension of the elbow and knee are illustrated. These movements occur in the sagittal plane

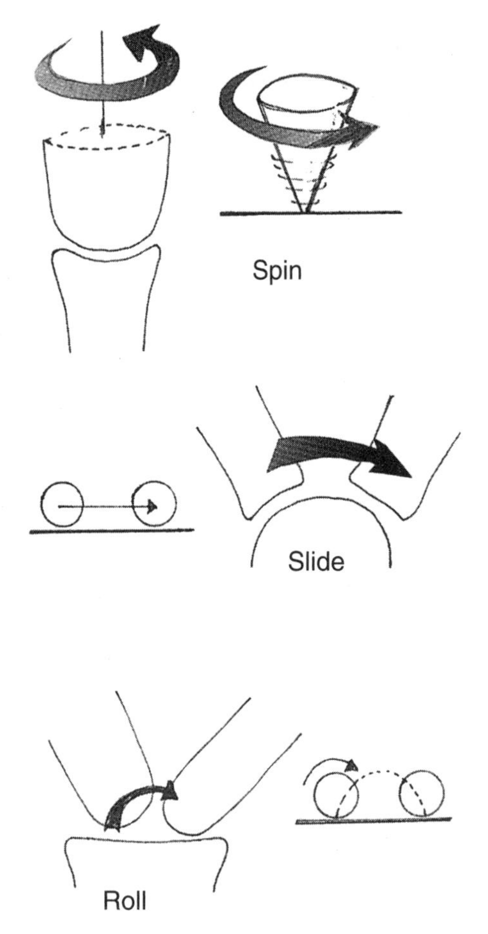

Figure 3.9 The motion of spin, slide, and roll between articular surfaces in a synovial joint. The movements can be likened to the spinning of a top, the skidding of a tyre on a road, and the normal rolling of a car tyre on the road

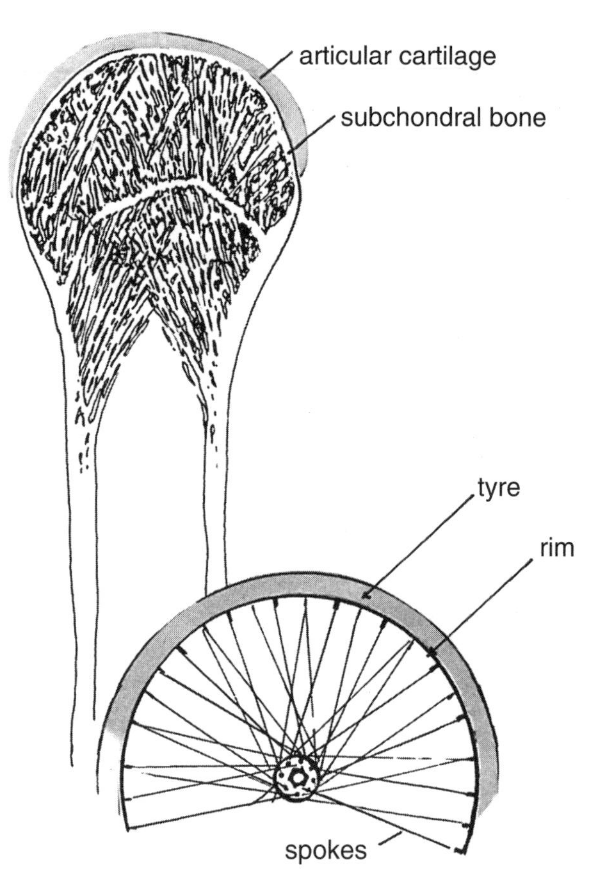

Figure 3.10 The expanded end of a long bone is compared with a bicycle wheel. The deformable articular cartilage acts like the air-filled tyre, which is supported by a light but solid rim, like the thin layer of compact bone. The spokes help to support the rim and help to maintain its shape while also acting to cushion forces from bumps in the road. The trabeculae of the expanded ends line up along the major lines of force to perform a function similar to the spokes of the wheel

Joint protection, lubrication and wear

The structures in contact within a synovial joint are the articular cartilages on the ends of each bone. The cartilage is smooth and deformable so that it can cushion forces applied to its surface. The subchondral bone, which simply means the thin layer of compact bone under the cartilage, provides a solid base and helps to protect the cartilage from damage. This thin layer of compact bone sits on a more deformable network of bony rods forming the spongy bone (Figure 3.10).

The synovial fluid acts as a lubricant; friction between articular cartilage in a synovial joint is less than between two blocks of smooth ice. There are many lubrication models based on the different types of joints that have been designed by engineers. None of these models can yet adequately explain all the features of human synovial joints.

A number of important characteristics of synovial joints make them different from artificial joints. Movement at a synovial joint can best be described as oscillatory but in most load-bearing joints in machines the angular motion is in only one direction. The load-bearing surface in a synovial joint is deformable cartilage, which is both elastic and porous, and the synovial fluid has particular chemical properties that allow it to act as a boundary layer lubricant. This means that it has lubricating properties even when the synovial fluid between the load-bearing surfaces is only a few molecules thick. Synovial fluid also has physical properties that allow it to act as a fluid-film lubricant when there is a finite thickness of fluid between the load-bearing surfaces.

The joint is the functional unit of the musculoskeletal system

In human movement studies and functional anatomy there is an emphasis on movement and therefore the joint is the focus of functional musculoskeletal anatomy. The characteristic features of a synovial joint have been listed (Figure 3.6) but each joint has associated with it a number of other structures. These include bones that act as levers on either side of the joint, and help in force cushioning as explained earlier (Figure 3.10). Skeletal muscles have a role in movement because they cross joints, so muscles initiate and control movement. The forces they produce across the joint also stabilise it so that muscles are secondary stabilisers, in addition to the associated ligaments. If you contract all the muscle groups about a joint simultaneously so that no movement occurs you will realise that the joint is much more stable. It is also stiffer if somebody else attempts to move the joint. Muscle forces are transmitted to bony attachments via tendons so that it is the musculotendinous unit that is of major significance. Nerves are also associated with joints. As we will see in Chapter 16, motor nerves within the central nervous system provide some control over the muscles producing actions at a joint, whereas sensory nerves provide feedback about joint position and movement from a variety of sensors located in the joint capsule and ligaments and in the muscles and tendons.

A human synovial joint is analogous to a tent, which relies on the interaction between a number of structures for its stability. The bones with cartilaginous ends are like the poles of a tent, which resist compressive forces. A tent that relied solely on the poles would be very unstable because it would easily be blown over by wind. For extra stability guy ropes are attached to the poles and these ropes must be flexible (or attached to springs) to perform their functions optimally. Ligaments are flexible structures that stabilise joints. Joints must not be completely stable because they must allow movement. Mobility and stability are two competing requirements of joints. Stabilising features of joints, which tents do not share, are the musculotendinous units that are under neural control and produce and/or control movements at the joints. The time required to damage a ligament is less than the time of a simple muscular reflex response to stretch. Therefore, in terms of injury prevention, the muscle must be activated prior to impact by external deforming forces.

Injury to any intrinsic joint structure or to the associated structures will result in functional impairment of the entire joint and adjacent body segments. Functional impairment of one joint often results in a 'chain reaction' affecting adjacent joints. For example, an ankle or knee injury will affect the whole lower limb, and therefore performance during walking or running will be adversely affected.

Structure and function of the muscular system

The muscular system is associated with other structures

The function of muscles is related to their structure. Muscles cross joints and the major skeletal muscles of the trunk and limbs are attached to bones at both ends. The musculotendinous unit consists of a chain of structures: bone–muscle–tendon–bone. This arrangement is not the only one, but it indicates that muscles can attach to bone 'directly' or indirectly via their tendons.

The attachment sites of muscle and tendon are important because they determine the action of each muscle. Whenever a muscle contracts it tends to pull its two attachment sites closer together and so the relationship between direction of pull of the muscle and the axis of rotation of the joint determines the resulting joint action. This type of analysis is the basis used to define the actions listed in tables in all basic gross anatomy books. Functional anatomists, starting with Duchenne, have used other techniques to try to unravel the complexity of muscle contraction when different muscles act simultaneously, or joint position is changed from the 'normal' anatomical position as illustrated in Figure 3.8.

Muscles have different structural features

So far, the term 'muscle' has been used in our discussion in relation to only one type of muscle, skeletal muscle. But there are other types of muscle tissue also. Smooth muscle is found in the walls of the digestive system and certain blood vessels. Cardiac muscle forms the major part of the walls of the heart. However, in this and subsequent functional anatomy chapters we shall restrict our discussion to skeletal muscle only. Whenever the term 'muscle' is used it will therefore be as shorthand for skeletal muscle.

Not all muscles look the same. The 'typical' muscle has a belly and tapers toward a tendon at each end but there are other muscle shapes. Examples of muscles with different architectural arrangements are illustrated in Figure 3.11. Note that some have 'muscle' attachments while others have tendinous attachments to the bone. In some muscles the fibres align in the direction of pull of the whole musculotendinous unit, while in others the fibres are at an angle to the direction of pull of the whole unit. The architecture of whole muscles has functional implications because of their effects on the total force that can be produced, and the range of motion through which the muscle can operate.

When viewed under a light microscope the most obvious feature of muscle is the striated (striped) appearance of the muscle fibres that form the muscle tissue. This appearance is produced by tens of thousands of repeating units in series, forming each fibre. These repeating units, called sarcomeres, can be observed in a photograph taken by an electron microscope. The sarcomere is the structural and functional unit of muscle (Figure 3.12).

Muscles have five distinguishing properties

Muscle as a tissue has two main properties: its excitability in response to nerve stimulation and its contractility in response to the stimulation. Part of the response to neural stimulation is the production of an electrical signal by the muscle fibres. As these electrical signals travel along the muscle fibres, muscle can be said to have a third property—conductivity. A whole muscle also contains a large proportion of connective tissue, in addition to the tendon, and so the structure of the muscle tissue and the mechanical characteristics of the connective tissue allow a fourth property, namely extensibility. The connective tissue component consists of thin sheets surrounding each fibre, each bundle of fibres, and the whole muscle itself (see Figure 3.12). It is the connective tissue that is mainly responsible for the fifth property, elasticity. Because muscle tissue responds to neural activation by producing a force it is regarded as an active tissue, in contrast to the passive connective tissues, which can only resist applied forces.

When muscles contract they can do so in three different ways

It is the muscular system that provides the power for the human body to perform work. Muscles cross joints so that their contraction produces movement and whenever a muscle

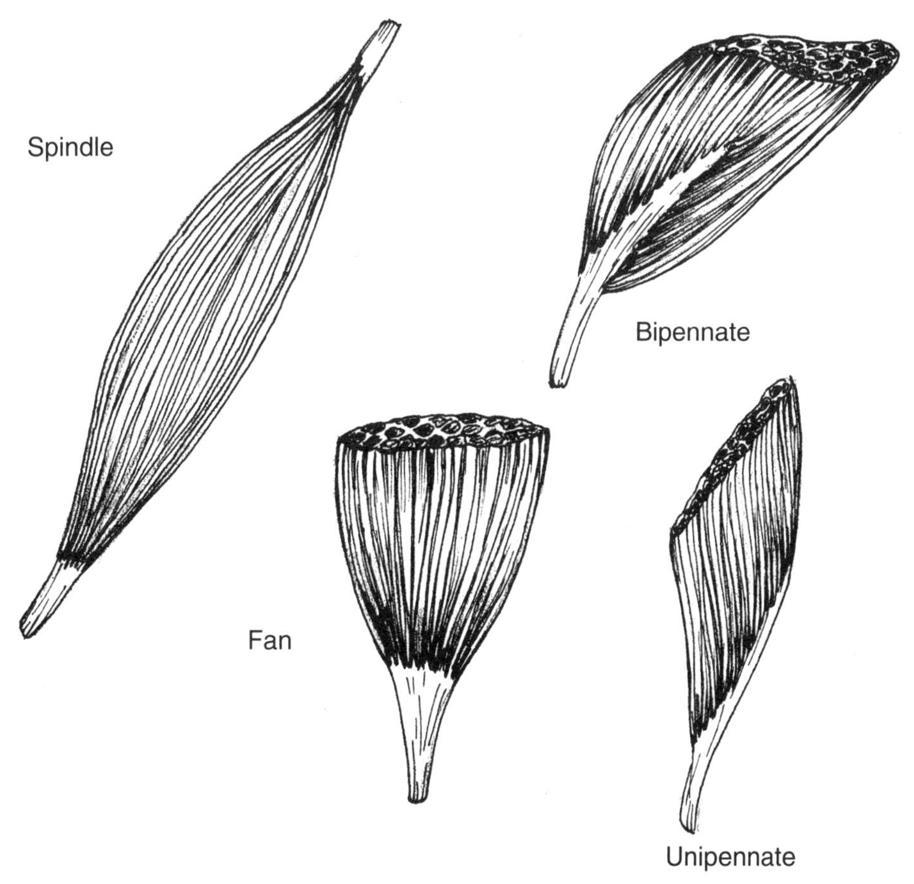

Figure 3.11 Examples of muscles with different shapes and different arrangements of fibres. Muscles attach to bone either directly or via a tendon. Muscle fibres are oriented in the direction of pull of the whole muscle (for example spindle) or at an angle to this direction (for example unipennate)

contracts it tends to shorten and pull the two bony attachments closer together. This action of a muscle is termed concentric, in contrast to an eccentric action, which occurs when a muscle is activated but is lengthening. In this situation other forces prevent the muscle from shortening. During eccentric actions, muscles control the movement produced by the other forces, which include external loads. During an isometric action (the third type of muscular action), the muscle is activated but the overall length of the musculotendinous unit does not change, so this type of action is important for the stabilisation of joints. Muscles may therefore act concentrically to produce movement, eccentrically to control movement, or isometrically to maintain posture and enhance joint stability (Figure 3.13). Joint stability is also a byproduct of the dynamic types of action.

Muscular contraction can be described

Skeletal muscle cells are elongated and contain many nuclei; hence muscle cells are called fibres. There is a basic repeating unit, called a sarcomere, which is the distance between adjacent Z discs. Each skeletal muscle fibre is formed by tens of thousands of sarcomeres. Illustrations of longitudinal skeletal muscle fibres in a lengthened condition and a shortened

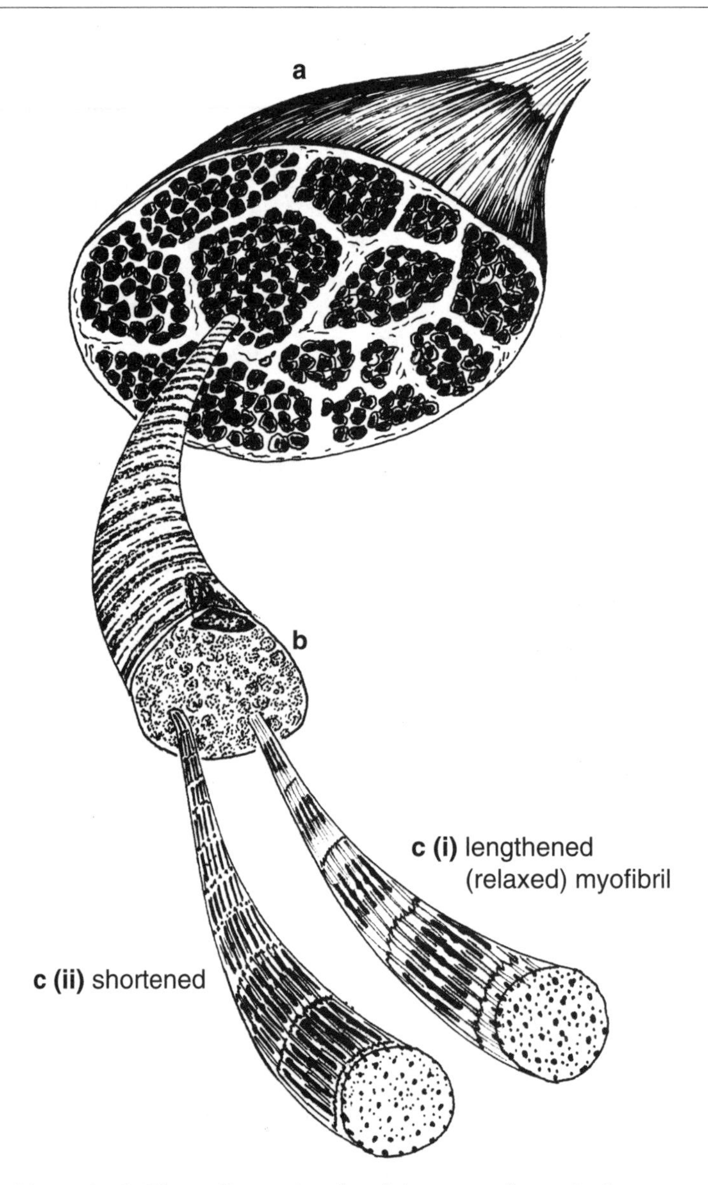

Figure 3.12 Three-dimensional architecture of a typical whole muscle (a). The appearance of muscle as seen under a light microscope (b), as well as the structure of a muscle fibre as seen under an electron microscope (c), are also illustrated in both the lengthened (i) and shortened (ii) positions

condition are shown in Figure 3.12. The alternating thick and thin filaments produce the characteristic striations. In the lengthened position there is little overlap between the protein filaments, but in the shortened position there is much more overlap. These observations led to the 'sliding filament' hypothesis of muscle contraction, which is a descriptive hypothesis. When there is shortening of the sarcomere, there is overlapping of the thick and thin filaments. Further research led to the hypothesis that cross-bridges were formed between the thick myosin filaments and the thin actin filaments attached to the Z disc. This has been termed the 'cross-bridge' or the 'ratchet' hypothesis, which is a mechanistic or explanatory

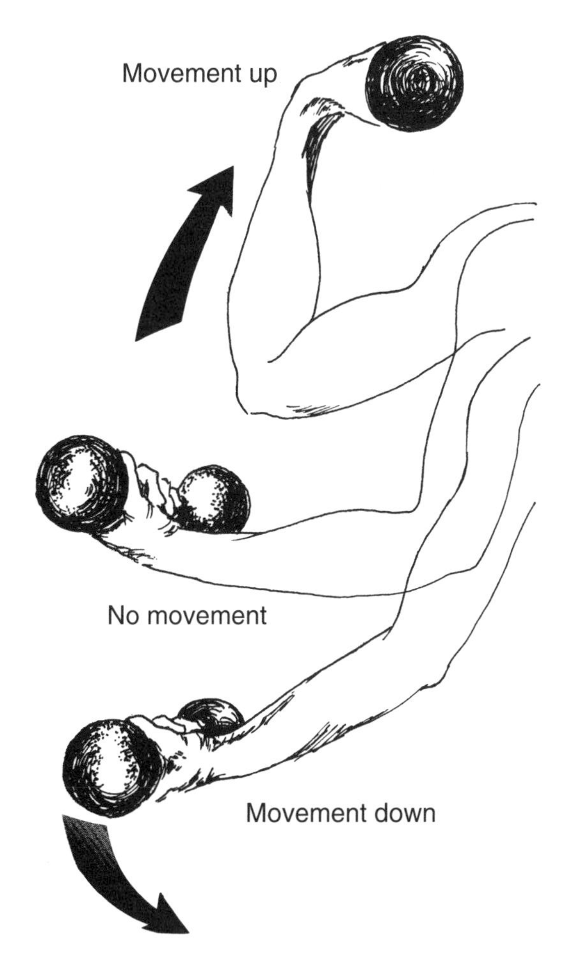

Figure 3.13 The three major types of skeletal muscle action are isometric, concentric, and eccentric. Isometric actions occur in static situations whereas concentric and eccentric muscle actions occur during dynamic tasks. In this case, the elbow flexor muscles act concentrically to move the load up against the resistance of gravity, and then act eccentrically to control the downward movement

hypothesis. Both hypotheses, which are generally accepted, indicate that there are both upper and lower limits to sarcomere, and hence muscle, length.

Joint actions can be explained

The joint actions caused by muscles would be relatively simple to predict and explain if each muscle crossed only one joint, that is, if all muscles were monoarticular. However, in addition to monoarticular muscles, there are also muscles that cross two joints (biarticular), such as the hamstring muscle group at the back of the thigh, and others that cross more than two joints (polyarticular), such as the muscles in the forearm that bend the fingers. A basic rule for predicting joint movement is that when a muscle is activated it shortens and tends to produce all the joint actions of which it is capable. If these actions do not occur it must be the result of external forces, or other muscles, preventing some of the actions. In these situations even a purely mechanical analysis of the muscular system becomes complex.

In a whole muscle the shortest length is determined by overlapping of the actin and myosin filaments in each sarcomere of the muscle tissue, but the upper limit is determined mainly by the extensibility of the connective tissue component of the muscle. When joints are moved through their full range, most muscles in the human body normally operate in a length range that is between the upper and lower limits, but there are some examples where the limits are reached. Generally the monoarticular muscles operate well within their length limits because their length is related to movement at one joint only, but biarticular and polyarticular muscles may reach their limits because their lengths are determined by motion at a series of joints.

We can demonstrate both the upper and lower length limits of muscles crossing more than one joint in the following way (refer to Figure 3.14). Stand on one foot and bend the unsupported limb as far back as possible at the hip (extend the hip). Now try to bend (flex) the knee as far as possible (Figure 3.14*a*). You will notice that the range of knee flexion is less than normal when the hip is extended. (This can be verified by trying the same knee movement with the hip flexed.) Now lie on your back and bend (flex) one hip so that the thigh is vertical, then try to straighten the knee (Figure 3.14*b*). Most people are unable to do this. If you can straighten your knee in this position, flex the hip more and try again. These two examples principally indicate the lower and upper length

Figure 3.14 Exercises used to demonstrate the shortened and lengthened positions of the hamstring muscle group. The hamstring muscles extend the hip and flex the knee and when these motions are combined the muscle becomes very short (a). When the muscle group is lengthened at both ends by the opposite actions (hip flexion plus knee extension) (b), the muscle stretch can be felt

limits of the hamstring muscle group—a group that crosses both the hip and knee joints.

To demonstrate the lower limit of the muscles that bend the fingers, first bend (flex) the wrist so the palm of the hand is brought as close as possible to the forearm and then try to 'make a fist'. Notice this is a very weak grip compared with your strongest power grip. Notice the position of the wrist when the grip is strongest. The muscles that provide most of the strength for bending the fingers have their bellies in the forearm and tendons that cross a number of joints including the wrist and others within the hand.

Muscle activation produces both mechanical and electrical responses

Muscle is an excitable tissue that responds to neural stimulation by contracting. But this mechanical response is preceded by an electrical response from the muscle fibres; in fact the chemical changes within the muscle fibre are dependent on the electrical response. The structural and functional unit of the neuromuscular system is a motor unit that consists of a single motor nerve fibre and all the muscle fibres it innervates (Figure 3.15). Muscles over which we have fine control, such as the small muscles in the hand, have relatively few muscle fibres per motor unit whereas the large muscles of the lower limb, over which we have only relatively coarse control, have many more muscle fibres per motor unit.

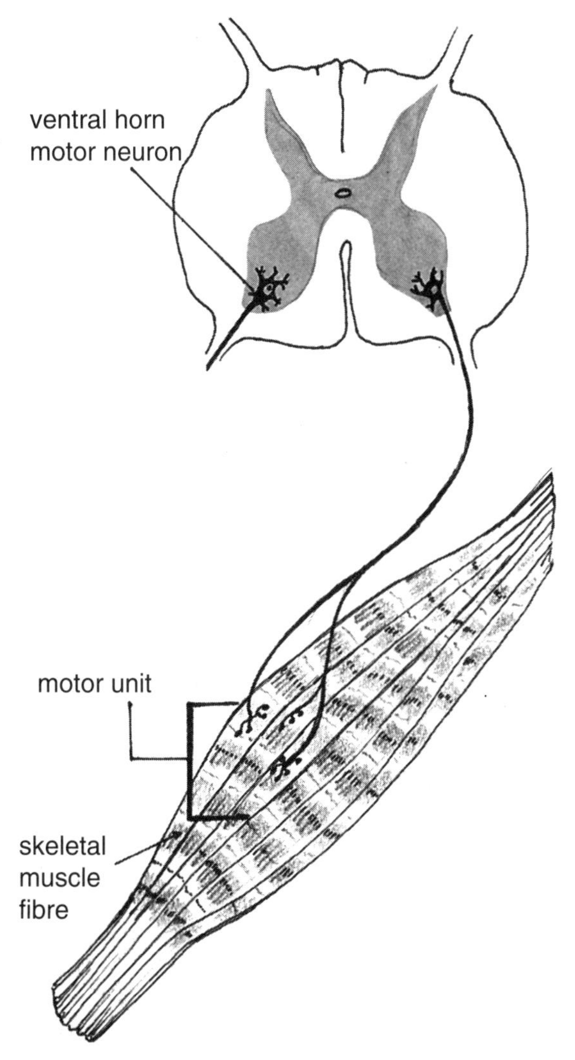

Figure 3.15 Illustration of a motor unit in a muscle of the upper or lower limb where the nerve fibre originates in the spinal cord

Every muscle consists of many motor units and the contractile force produced by the whole muscle is partly determined by the number of motor units that are activated. Neural activation produces a volley of contraction–relaxation cycles. Motor unit activation is overlapped so as to produce and maintain smooth contractions. If the contraction–relaxation cycles occurred simultaneously for all motor units the overall contraction would be jerky. A fundamental principle governing muscle contraction is therefore that there are asynchronous volleys of impulses travelling down the many nerve fibres innervating a single muscle.

Of the electrical and mechanical sequelae that follow neural activation, it is often easier to detect the electrical events. Electrical activity of a muscle can be detected using electrodes as part of a technique called electromyography (EMG). Electromyography is a basic tool in functional anatomy (see Figure 3.16) and has been used to determine muscle roles in movement situations, to provide biological feedback to a person attempting to improve motor performance, and to investigate the effects of strength training.

Strength is determined by muscle size and its relationship to the joint axis

Strength training is popular among people training to improve their motor performance and/or health, but what is strength? Mechanically, strength is determined by both the force generated by muscular contraction and the leverage of the muscle at the joint. This concept, which is termed moment of force, is examined in detail in Chapter 7. Muscle force is proportional to the cross-sectional area of the muscle. This cross-sectional area is measured perpendicular to the longitudinal axes of the muscle fibres

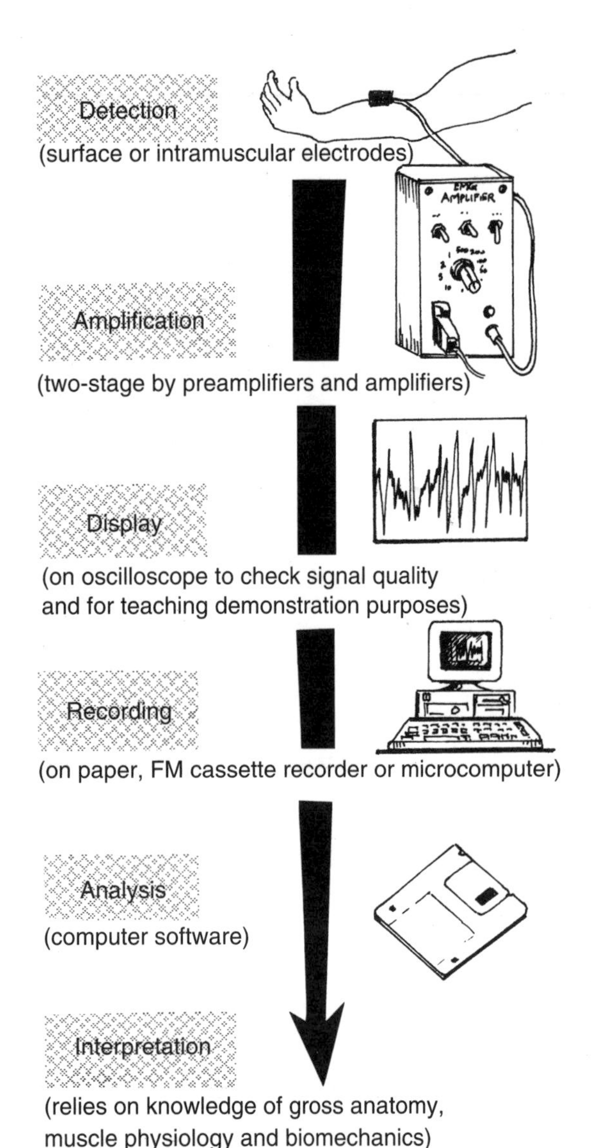

Figure 3.16 The processes involved in electromyography

and so is dependent on muscle architecture (see again Figure 3.11).

Range of joint motion can be limited by various factors

The control of joint motion and joint stabilisation relies on more factors than those illustrated in Figure 3.6. When a joint reaches the end of its range of motion, there are other possible causes of this limitation in addition to the intrinsic features of the joint. When joint motion occurs, there is bending between the body segments and the tissues on one side are compressed whereas those on the opposite side are stretched. Thus, the most obvious limiting factor is the tension in the joint capsule and associated ligaments on one side of the joint. In addition, and most commonly, stretch of the associated musculotendinous tissues also restricts the range of joint motion. Some of the resistance to joint motion through its range is also provided by stretching of the skin. Sometimes apposition of soft tissues also restricts movement and, rarely, apposition of bony parts forming the joint will restrict movement. This bony contact would only occur at the extreme end of the range and is potentially injurious in dynamic situations. In people who habitually use a squatting posture, marks are made on the bones at the front on the ankle. These marks are formed as a result of the static compressive forces on the bones when the ankle is bent more than its normal range. In summary, the structures that limit joint range of motion are usually those soft tissues that are under tension.

CHAPTER 4

Basic concepts of anthropometry

- Defining anthropometry
- Tools used for measurement
- Body size
- Body shape and the relationships between different body segments
- How is body composition estimated?
- Somatotyping is a short-hand way of describing body build
- Human variation

In this chapter we examine the field of anthropometry and discuss the measurement of, and variation in, human body size, shape and composition and the implications of these variations for movement capability.

Defining anthropometry

Anthropometry is the science that deals with the measurement of the size, proportions and composition of the human body. In most cases it is the size that is directly measured, and these direct measurements can be combined to indicate the shape of the whole body or body segments (Figure 4.1). Body composition usually involves using direct anthropometric measurements to predict the relative amount of a particular component in the whole body. The best known example is the use of skinfold thicknesses to predict the percentage of fat in the body.

One specialised branch of anthropometry of particular relevance to the study of human movement is kinanthropometry. Kinanthropometry has been defined by the International Society for the Advancement of Kinanthropometry as 'the scientific specialisation dealing with the measurement of humans in a variety of morphological perspectives, its application to movement, and those factors that influence movement, including:

- components of body build, body measurements, proportions, composition, shape, and maturation
- motor abilities and cardiorespiratory capacities
- physical activity, including recreational activity as well as highly specialised sports performance'

Defined in this way, kinanthropometry is clearly an important scientific specialisation within

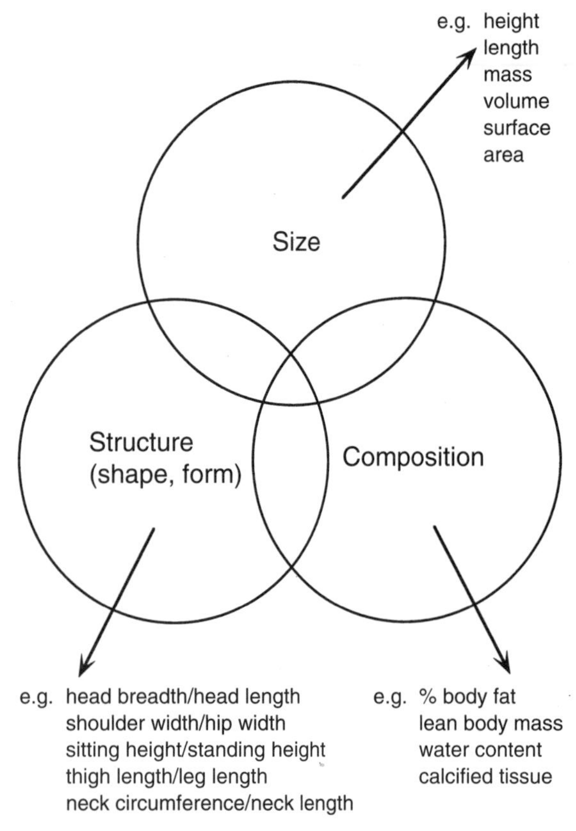

Figure 4.1 Venn diagram indicating that the three components of anthropometry overlap with each other

human movement studies and associated disciplines.

Tools used for measurement

In anthropometry, stadiometers are used to measure height (stature), anthropometers to measure length of body segments, tapes to measure body segment circumference, bicondylar calipers to measure bone diameter, skinfold calipers to measure thickness of skin plus subcutaneous fat, scales to measure mass and so on. In osteometry bone length and diameter are measured on skeletal remains.

Simple or more complex mathematical manipulation can be used to derive indices to describe the shape of a body segment. The cross-sectional shape of the thorax (thoracic index) can be derived by dividing the antero–posterior (from front to back) diameter by the transverse diameter. Statistical procedures are generally used to formulate prediction equations when the direct measurement of a parameter presents certain practical difficulties such as cost, time or technical complexity. For example, there are many equations used to predict percentage of body fat (almost as many equations as there are authors in the field!). Most of these equations have been derived by correlating a battery of anthropometric measurements with body density determined by underwater weighing. Only the best correlating measurements are included in the final formula (see Box 4.1). Other techniques are also available to define body composition and these are outlined later in this chapter.

In kinanthropometry, the basic measurement tools may be anthropometric equipment but there are important skills required by the researchers for interpretation of the results. Required knowledge includes the anatomy and biomechanics of the human musculoskeletal system. Because of the breadth of knowledge required by kinanthropometrists, this aspect will not be emphasised here; however, examples will be provided in Chapter 6.

Body size

The possible number of measurements of the dimensions of the human body is essentially limitless. The major anthropometric manuals recommend a core number of measurements, with a longer list of optional measurements. The choice of options is determined by the objectives of the study. In most cases, techniques to measure the size of the whole body and its different segments are described. Body size includes height and mass, and segment size includes length, circumference, diameter, and so on. Individual segment mass cannot be directly measured so this is predicted from previous research results. As the techniques of anthropometric measurement are critical to measurement validity, predetermined measurement protocols must be followed rigorously. Different manuals may describe different protocols for the same body segment so it is crucial

Box 4.1 How can body density be predicted?

Researchers from the Department of Human Sciences, University of Technology, Loughborough, England, and the Human Nutrition Unit, Istituto Nazionale della Nutrizione, Rome, took 21 anthropometric measurements from 138 male Italian shipyard workers. Examples of the mean values obtained were 1.73 m for height, 76.1 kg for mass, 10.4 mm for triceps skinfold thickness, 37.7 cm for calf girth, and 7.0 cm for bony width across the elbow.

The main aim of the project was to measure body density by underwater weighing with simultaneous estimation of the volume of air in the lungs, and to compare the results with five prediction equations based on anthropometric measurements, which had previously been published. From underwater weighing it was calculated that the average percentage body fat was 22.3. The previous equations from the literature showed correlations with the measured body density of between -0.08, which is very poor, to 0.85, which is quite good. The results tended to confirm the general conclusion that equations developed to predict body density are population-specific.

The authors of this study also developed their own predictive equation using a statistical technique known as stepwise multiple regression. Variables included in the final equation were age and two skinfold thicknesses. The correlation of this specific prediction equation with the measured body density was 0.89.

Source: Norgan, N.G. & Ferro-Luzzi, A. (1985), The estimation of body density in men: are general equations general? *Annals of Human Biology* 12, 1–15.

that the measurer/experimenter choose one particular protocol and adhere to it closely.

Body shape and the relationships between different body segments

Body shape can best be described by the proportions between the size of different body segments. These proportions indicate the structure of the body. The relationship between different measurements may be termed dimensionality. The relationship between height and weight is a common example. In modern nutritional surveys the body mass index (BMI) is often quoted. This is the mass (in kilograms) divided by the height (stature in metres) squared; BMI = mass/height^2. This squared relationship between mass and height was first described by a Belgian mathematician, Quetelet, and it can be contrasted with the ponderal index in which the height is related to the cube root of the mass. The relationship between height and mass will be explored further in the next chapter (Chapter 5) when considering human growth.

Many indices have been used to describe shape, such as the cephalic index for head shape described by the ratio between length and breadth of the skull. Sometimes, when people stand up we can see a large difference in their height but when they sit down the difference is very small. This observation is related to their body proportions. Some people have relatively long lower limbs whereas for others these limbs are relatively short. To describe this relationship the sitting height is divided by the standing height and multiplied by 100, so it is expressed as a percentage. If the lower limbs are investigated, the ratio of the length of the thigh (limb segment between hip and knee) to the leg (limb segment between knee and ankle) can be calculated. Sprinters tend to have relatively short muscular thighs and long skinny legs. The relationship between performance and body proportions relies on mechanical knowledge for interpretation of the results and is consequently a topic of considerable interest for biomechanists.

How is body composition estimated?

The body is composed of different tissues. How is the body composition estimated? Anthropometric models and anatomical models of the composition of the human body can be contrasted. Anatomical dissection has been used to describe the different components of the human body, separating the body into the six masses of skin, muscle, bone, nervous tissue and tissues from other organs, plus fat (Table 4.1). Anthropometric models seek to estimate body composition using non-invasive means. One of the original anthropometric models, involving five components, was proposed by Matiegka in 1921, but this model was simplified by Behnke in 1942 to a two-component model, which is now widely accepted (see Boxes 4.1 and 4.2). This model seeks to separate lean tissue (such as bone and muscle) from fat tissue, because of the realisation by health professionals that excess fat rather than lean tissue is a prime risk factor for premature death. Measurement of skinfold thickness was introduced in an attempt to derive an indirect, non-invasive measure of body fatness.

Table 4.1 Tissue proportions in the human body, as measured by dissection of 25 cadavers

Males	Females
28.1% fat	40.5% fat
skin = 8.5% lean body mass	
muscle = 50% lean body mass	
bone = 20.6% lean body mass	
undifferentiated tissues constitute the rest of the body	

Source: Data taken from Clarys J.P, Martin, A.D. & Drinkwater, D.T. Gross tissue weights in the human body by cadaver dissection, *(1984), Human Biology,* 56, 459–73.

The two-component model of Behnke is used in the calculation of body density, in which lean and fat tissue each have an assumed specific

Box 4.2 Can body composition be estimated by indirect methods?

A group of seven researchers from five different universities in the United States used underwater weighing to determine the body density of 310 subjects aged between 8 and 25 years. The subjects, from Illinois and Arizona, represented both genders and both major racial groups. The age range included all stages of physical maturation. Body fatness was determined by three sequential techniques. The body density was corrected by calculation of the effect of air contained within the body, and in addition the total body water was measured by deuterium oxide dilution. Bone mineral measurements were also made on forearm bones using photon absorptiometry. Nine skinfold measurements were used for the prediction of percent body fat.

As an example, the mean total body mass of the pubescent girls was 43.2 kilograms. Their percentage of body fat, calculated from underwater weighing, was 27.9, which reduced to 24.8 when body water was accounted for. When bone mineral was also taken into account, calculated percentage of body fat was further reduced to 23.7. The authors' interpretation of their results was that chemical maturity has not been reached during puberty so that adult prediction equations lead to an underestimate of body density, and hence an overestimate of percent body fat.

For the 59 pubescent subjects, the sum of three skinfolds (triceps + calf or triceps + subscapular), plus an indication of physical maturation, was able to predict percent body fat reasonably well (correlation coefficients between 0.85 and 0.91).

Source: Slaughter, M.H., Lohman, T.G., Boileau, R.A., Horswill, C.A., Stillman, R.J., van Loan, M.D., & Bemben, D.A. (1988), Skinfold equations for estimation of body fatness in children and youth, *Human Biology* 60, 709–23.

Box 4.3 Can it be assumed that the density of the human body is uniform?

A study on the leg of a single cadaver and the leg of a living patient, performed by a research group consisting of two Australians and a visiting Canadian, provided morphological data that have important implications in biomechanical analyses. As the authors (from the Department of Human Movement, University of Western Australia, the Department of Medical Physics, Royal Perth Hospital, and the College of Physical Education, University of Saskatchewan) point out, in their introduction, existing biomechanical models rely on at least three assumptions, all of which are questionable.

Computerised tomography (CT) cross-sectional scans were taken at six levels between the knee and ankle. A three-component model, consisting of bone, skin plus fat, and muscle plus connective tissue, was used to analyse each cross-section. Mass and volume of the cadaver leg were directly measured as well as being modelled. One model used a constant density assumption and another was based on variable density from the CT scans. Cross-sectional areas at various levels were measured and modelled. The different techniques were used to calculate mass and volume of the patient's leg. Centre of mass and moment of inertia of the cadaver leg were directly measured and estimated from the models. These parameters were also calculated for the patient's leg.

Underwater weighing of the cadaver leg and CT calculations from the cadaver and patient legs indicated that density varied along the length of the leg. The pattern of density changes along the cadaver leg were similar for CT-derived figures and for underwater weighing, but the CT-derived values were consistently from one to three per cent higher. At each level the CT-derived densities were similar for the cadaver and the patient leg (Figure 4.2).

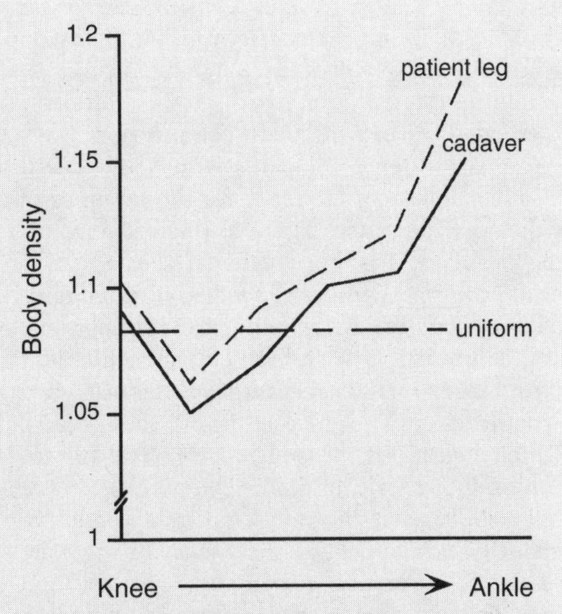

Figure 4.2 CT-derived densities of a cadaver (solid line) and patient leg (broken line) compared with a uniform density assumption
Source: Ackland, T.R., Henson, P.W. & Bailey, D.A. (1988), The uniform density assumption: its effect upon the estimation of body segment inertial parameters, *International Journal of Sport Biomechanics*, 4, 146–55. (Copyright 1988 by Human Kinetics Publishers, adapted with permission.)

Overall, the calculated values of centre of mass and moment of inertia using the uniform density assumption and CT-derived variable density were similar; however, the authors predicted that this similarity may not be the case in the trunk, in which density variations are likely to be larger.

Source: Ackland, T.R., Henson, P.W. & Bailey, D.A. (1988), The uniform density assumption: its effect upon the estimation of body segment inertial parameters, *International Journal of Sport Biomechanics*, 4, 146–55.

gravity. Basically, lean tissue sinks in water because of its higher density and fat floats on water. In the underwater weighing technique, which has often been used as the definitive method of estimating body composition, body weight in water is compared with body weight in air. Allowances are made for the volumes of air in the lungs and digestive tract. Using this technique, it has been calculated that some elite power athletes have a negative percentage of body fat, which is biologically impossible and incompatible with life! This measurement error arises mainly because of the assumed specific gravity of non-fat tissue (about 1.1), which includes bone and muscle. The tissues comprising the lean tissue component of the body are variable in their densities and relative quantities. Also the specific gravity of the body varies between one region and another, and may also vary within the one body segment (see Box 4.3). If underwater weighing can lead to incorrect estimates of body density, and therefore the percentage of body fat, results from this method cannot be used for valid comparison with skinfold measurements. Therefore, many researchers now recommend that skinfold measurements be reported directly and not used in an equation that aims to predict percentage of body fat. Skinfold thickness has traditionally been measured using calipers, but ultrasound and electrical impedance have also been used more recently. The advantages of these more expensive electronic techniques over skinfold calipers are debatable and not universally accepted.

Lean body mass can be estimated by alternative techniques, including gamma-ray spectrometry, which uses a radioisotope of potassium (^{40}K), and hydrometry, which involves an isotope of water. Direct chemical analysis allows the definition of quantities of at least four basic constituents (water, protein, mineral, fat) within a sample of tissue but this technique is not suitable for living subjects. If muscle is of major interest, the concentration of creatinine in urine can be used as an estimate of muscle mass.

Radiographic techniques can be utilised in a three-component model of the human body. In dual energy X-ray absorptiometry (DEXA) a low-dose radiation source allows estimation of the masses of calcified tissue, fat, and non-fat tissue within the body. The modern imaging techniques of computerised tomography (CT—see Box 4.4) and magnetic resonance imaging (MRI) have the potential to define proportions of many different types of tissues within regions of the body. In fact, some of the information provided later in this chapter has been based on the results of these types of studies. Magnetic resonance spectroscopy is also used to investigate chemical components of the body.

Somatotyping is a short-hand way of describing body build

In about 400 BC, the famous Greek physician Hippocrates described different body shapes. He categorised people either as *habitus apoplecticus*, who were short and thick, or *habitus phthisicus*, who were tall and thin. His observation was that the fat people suffered from more circulatory problems and that the thin people had more respiratory problems such as consumption (tuberculosis). Hippocrates' observations were prophetic as obesity is today listed as one of the major risk factors for heart disease.

Physique and personality are sometimes linked, such as in the depiction of 'jolly fat men'. This idea was formalised in 1940 by William Sheldon and his collaborators, S.S. Stevens and W.B. Tucker, who published a book entitled *The Varieties of Human Physique: An Introduction to Constitutional Psychology* in which they proposed a classification of physiques, relating them to temperament and even to mental disorders. Any concept of general relationships between these physical and psychological factors has long been abandoned, but the use of body measurements and expressing them in a short-hand way to describe physique is still current.

Initially it was thought that body types were genetically determined and so did not change throughout life. It is now agreed that body types do have a genetic component but are also a product of environment. Somatotypes were originally described as genotypes but they are now con-

Box 4.4 What is the effect of training on muscle cross-sectional area and strength?

Researchers from the Department of Physical Education and Sports Studies and the Department of Anatomy and Cell Biology at the University of Alberta, Edmonton, Canada, examined thigh cross-sectional images produced by computerised tomography (CT) in 14 males with previous strength training experience. The areas of the quadriceps femoris and hamstring muscle groups were measured at three different times during an intensive 12-week resistance training program consisting of concentric contractions only. Knee extension and flexion strengths were also measured isokinetically, during both concentric and eccentric contractions (see Figure 3.11).

Previous studies by the authors indicated that measurements of muscle cross-sectional area were reliable. Test-retest measurements indicated that the strength measurements were also reliable. Six weeks after the commencement of training, the quadriceps (knee extensors) had significantly increased in size (by 2.6 per cent) but the hamstrings (knee flexors) had not. The strengths of knee extension and flexion during both concentric and eccentric contractions had all increased significantly at this stage (by between 7.8 and 10.2 percent). The results suggested that there may have been a predominance of neural adaptation during this initial period. After a further 6 weeks of high resistance concentric training, all muscle sizes and knee joint moments had significantly increased again. Percentage increases in strength were about 4 times greater than percentage increases in muscle size. The authors concluded that, because concentric training produced increases in eccentric strength, strength training effects are not completely specific to the type of contraction during training.

Source: Petersen, S.R., Bell, G.J., Bagnall, K.M., & Quinney, H.A. (1991), The effects of concentric resistance training on eccentric peak torque and muscle cross-sectional area, *Journal of Orthopaedic and Sports Physical Therapy*, 13, 132–37.

sidered phenotypes because of environmental influences on body shape, and the fact that somatotype can change throughout life. Anthropometric somatotyping uses measurements to describe shape and composition of the body at a particular time. Size is reported separately.

Definitions of somatotypes given in medical dictionaries are based on the preponderance of the derivatives from the three basic embryological tissues (see Chapter 5), but the most commonly used somatotyping techniques bear little resemblance to the basic definitions. The Heath–Carter anthropometric somatotyping technique is typical and its description provides a useful overview of the most commonly used somatotyping methods. Ten anthropometric measurements are recorded for the calculation of the Heath–Carter anthropometric somatotype but, according to the protocol, each measurement must be repeated at least once. Checking the reliability of measurements is a fundamental principle in any scientific experimentation.

In the Heath–Carter technique, the subject is defined by three numbers, each representing one component. Table 4.2 summarises the ten measurements taken to calculate the three components. The sum of three skinfolds from the arm and trunk are used to calculate the first component, which may be termed 'endomorphy'. This component defines the relative fatness of the person, especially when it is related to the person's height. The second component is calculated by determining the sizes of the elbow and knee, and arm and calf, relative to the subject's height. Termed 'mesomorphy', it represents musculoskeletal development. The skeletal development is determined by measuring the diameter of the bones at the elbow and knee. The muscular development is determined by measuring the circumference of the arm and

Table 4.2 Ten measurements used to calculate the three components of the Heath–Carter anthropometric somatotype

Four skinfolds measured on the right-hand side of the body with skinfold calipers, and expressed in mm:

- Triceps on the back of the arm
- Subscapular just below the lowest part of the shoulder blade
- Suprailiac just above the bony projection of the pelvis at the approximate level of the umbilicus
- Medial calf inside the widest part of the calf

Height measured with a stadiometer, and expressed in millimetres (for calculation of mesomorphy) or centimetres (for calculation of ectomorphy).

Two bone breadths measured on both sides of the body with bicondylar calipers, and expressed in centimetres:

- Humeral epicondylar elbow width
- Femoral epicondylar knee width

Two limb segment circumferences measured on both sides of the body with an anthropometric tape, and expressed in centimetres:

- Arm elbow bent with biceps muscle maximally contracted
- Calf largest part of calf when standing normally on both feet

Weight (mass) measured on scales, and expressed in kilograms

calf, after the estimated fat component of these segments has been accounted for. Bone diameters have more of a genetic determination than the segment circumferences, which are very much influenced by the type and quantity of physical activity. The third component is calculated from height and mass, using a cubic relationship. This component is termed 'ectomorphy' and a high rating on this component indicates lightness for a given height.

Every person is rated on each of the three components and the rating is positioned on a special diagram called a somatochart (Figure 4.3). At the centre of this triangular-shaped diagram is the 'unisex phantom', with males and females plotted on the same somatochart. A typical endomorph is the sedentary middle- aged man, 'Norm', made famous by the 'Life Be In It' campaign, or a Sumo wrestler (Figure 4.4*a*). Typical mesomorphs are elite power athletes, especially weight lifters, boxers, and wrestlers (Figure 4.4*b*). Typical ectomorphs are ballet dancers and female gymnasts. High jumpers and marathon runners also tend to fit into this category (Figure 4.4*c*). These examples indicate

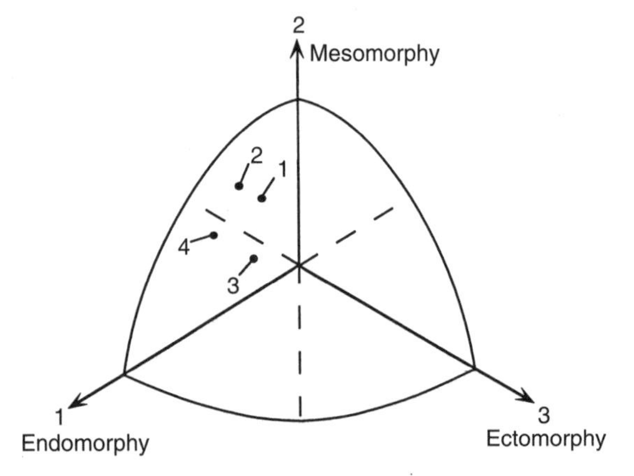

Figure 4.3 Somatochart indicating the somatotypes of four groups within the Canadian adult population. The four groups are (1) males younger than 30 years (3.5–5–2), (2) males older than 30 (4–5.5–2), (3) females younger than 40 years (4.5–4–2.5), and (4) females older than 40 (5–4.5–2)
Source: Bailey, D.A., Carter, J.E.L & Mirwald, R.L. (1982), Somatotypes of Canadian men and women, *Human Biology*, 54(4), 813–28. (Copyright 1982 by Wayne State University Press; adapted with permission.)

Figure 4.4 Typical extreme somatotypes in different sports: endomorph (a), mesomorph (b), and ectomorph (c)

one aspect of kinanthropometry—defining relationships between body size, shape, and composition, and human performance, and then explaining the relationships.

Human variation

It is obvious that there is a wide range of 'normal' in almost all anthropometric measurements. Some individual differences may be explained by more general differences, such as differences between males and females, racial differences, age differences, and differences in types and amounts of activity. This short list implies two main sources of difference: genetic and environmental. Environmental factors include the mechanical environment related to the magnitude and regularity of muscular forces, and metabolic factors, which may include deficiencies of certain nutrients in the diet. Differences related to age (Chapter 5) and activity (Chapter 6), are addressed later.

Human variation in the musculoskeletal system

Bone sizes obviously differ between individuals, but it has also been shown that bone density exhibits racial differences. In North America, Afro-Americans have a higher bone density than White Caucasians. In fact, some Afro-American football players have such high body densities that they have a negative percentage body fat, based on prediction equations.

Peak bone mass in women is about 30 per cent less than in men. This difference between the sexes is an example of sexual dimorphism, which in this case is due mainly to differences in physical size. Any differences in bone density between the sexes are small and are also site specific, rather than general.

Muscle volume differences between males and females are related mainly to differences in body size, but there appear to be differences also between muscle size related to the presence of different sex hormones and their interaction with physical activity. Generally the muscle volume of adult males is 40 per cent more than in adult females, but this is not a large difference when muscle volume is expressed as a proportion of total body volume. Relative muscle mass in the lower limbs of males and females is similar; however, there are larger differences in the upper limbs. Studies that have analysed the tissues of the arm using computerised tomography (which provides an image of the cross-section of the arm) have found the average cross-sectional area of muscle plus bone to be 62 cm^2 in a 30 year old male, compared with 35 cm^2 in a female of the same age.

Human variation in physical dimensions

There are many examples of sexual dimorphism in body dimensions but often there is such a large overlap between the two sexes that it would be difficult to predict the sex of an individual solely on the basis of body dimensions. Males tend to be taller and heavier than females but these are not distinguishing features. At about 20 years of age, the ratios between males and females are 1.08:1 for height, 1.25:1 for weight, and 1.45:1 for lean body mass. These differences have important implications for both nutritional requirements and athletic performance.

The difference in shape between adult males and females is well known and is mainly due to differences in shoulder and pelvic size. By measuring the width of the shoulders in centimetres and multiplying this by 3, then subtracting the width of the pelvis, almost 90 per cent of the adult population can be correctly identified as male or female. This relationship between shoulder and hip width is termed the 'androgyny index' and is a good example of the use of proportions to describe shape and, in this case, emphasise differences. Sex-related differences in shape and composition are also indicated in Figure 4.3. An interesting example of sexual dimorphism relates to the position of a car driver's seat. Usually a male driver has to push the seat back when he sits in the seat after a female has been driving, but the difference in seated height is not so obvious; men not only have longer lower limbs than women but the lower limbs are also slightly longer relative to their trunk length. Of course the difference is not as great as it may appear from the car seat example because car manufacturers slope the runners, on which the seat slides, backward and downward.

Some sports such as basketball favour tall players whereas jockeys are usually short because they must be very light. Figure 4.5 indicates where a typical 200 cm basketballer and a 155 cm jockey are positioned relative to the young adult male population of Queensland. In height comparison, the average North American man is taller than 95 per cent of Asian men, which probably explains the restricted interior space in Japanese cars when they were first exported during the 1960s. It is predicted that the height difference will be smaller in future generations, and the basis of this prediction will be explained in the next chapter on growth and maturation.

There are also racial differences in a number of proportions including the cephalic index and the sitting height to standing height ratio. Asians tend to have the smallest cephalic index indicating their relatively broad skulls. The smallest sitting height to standing height ratio

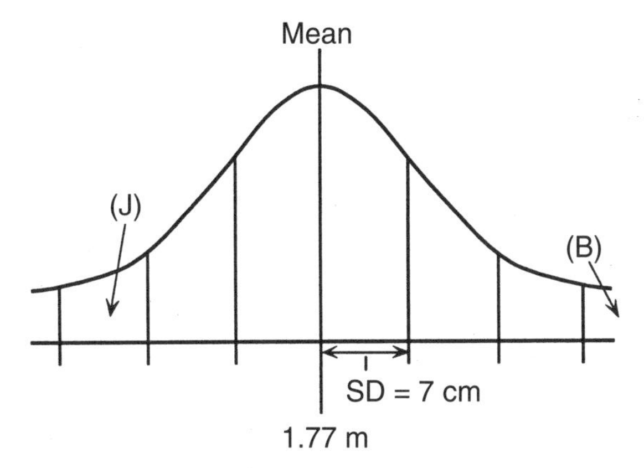

Figure 4.5 A curve indicating the normal distribution of self-reported heights among almost 100 000 young adult males in Queensland. Heights of typical jockeys (J) and professional basketballers (B) are indicated. The normal distribution is drawn from data supplied by the Queensland Department of Transport

is in the Australian Aborigine, who has the longest lower limbs relative to height. On average this figure is under 50 per cent compared with over 53 per cent for people from South East Asia. In a photographic atlas of Olympic athletes, Jim Tanner emphasised proportional differences between Afro-American male athletes and Caucasians. The relatively longer lower limbs of the black athletes and other differences related to the lower limb are thought to partly explain differences in general performance in the 100 m sprint. Although this example is appealing, body dimensions are only one factor in performance and there are many examples of individuals who do not conform to common perceptions of physical requirements. For example, the basketballer 'Spud' Webb was able to win an NBA 'slam dunk' competition even though he is only 170 cm tall.

Statistically there are trends relating body size and shape to athletic performance, and certain generalisations follow from these trends. One statistician has argued that taller athletes tend to perform better in almost all Olympic athletic events. Graphs of the relationship between height and performance certainly provide evidence in support of this concept.

CHAPTER 5

Musculoskeletal changes throughout the lifespan

- Defining auxology and gerontology
- Tools used for measurement
- Physical growth, maturation and ageing
- Physical growth, maturation and ageing of the musculoskeletal system
- Changes in body dimensions throughout the lifespan
- How is age determined?

In this chapter we bring the basic concepts about the musculoskeletal system and about anthropometry introduced in Chapters 3 and 4 respectively to focus on the issue of lifespan changes. After initial definitions and discussion of measurement procedures, principal attention is given in this chapter to three major issues: description of the general process of physical growth, maturation and ageing; examination of age-related changes in the skeletal, articular and muscular system; and examination of changes in body dimensions throughout the lifespan.

Defining auxology and gerontology

The terms 'auxology' and 'gerontology' can be defined as the sciences of growth and ageing respectively. Human auxology (human growth) research began only this century when objective measurement techniques first became available. Of particular interest in auxology are the ages of onset of changes, the magnitudes of these changes, and the duration of the change period. All of these factors are highly variable between individuals but the order in which major changes occur is relatively constant. The knowledge available from auxology provides a basis for determining whether people are older or younger physically than their chronological age. Gerontology is becoming an increasingly important field of study as the percentage of aged individuals within the community increases. One focus in gerontology that is of particular importance to the subdiscipline of functional anatomy is the determination of the relative effects of ageing *per se* and inactivity on the musculoskeletal system. Their relative effects need to be determined so that the probable effects of physical activity programs can be estimated.

Tools used for measurement

Growth involves a measurable change in either quantity of tissue or size of body segments. Therefore, for many large human growth and ageing studies, the tools used are the same as those used for anthropometry (Chapter 4). Most pregnant women in Australia have an ultrasound scan at about 12 weeks of pregnancy and the measurements taken by the obstetrician include the width of the fetal skull and the length of the femur. These measurements are taken to confirm the age of the fetus. During postnatal growth, physiological measurements may be taken also (see Chapter 14).

Researchers are often interested in the skeletal development of children and in the skeletal changes associated with ageing. Both cases require the use of radiological techniques. Skeletal age of children can be determined by hand/wrist radiographs and the bone density of adults can be measured using computerised tomography (CT) scans. Recently, body composition, such as amount and placement of fat, has been determined using magnetic resonance imaging (MRI), which can be used also to determine cross-sectional areas of muscles and bones. CT is better for visualising hard tissues such as bone whereas MRI is better for the imaging of soft tissues such as muscle.

Physical growth, maturation and ageing

> All the world's a stage,
> And all the men and women merely players.
> They have their exits and their entrances,
> And one man in his time plays many parts,
> His acts being seven ages. At first the infant,
> Mewling and puking in the nurse's arms.
> Then, the whining school-boy with his satchel
> And shining morning face, creeping like snail
> Unwillingly to school. And then the lover,
> Sighing like furnace, with a woeful ballad
> Made to his mistress' eyebrow. Then, a soldier,
> Full of strange oaths, and bearded like the pard,
> Jealous in honour, sudden, and quick in quarrel,
> Seeking the bubble reputation
> Even in the cannon's mouth. And then, the justice,
> In fair round belly, with good capon lin'd,
> With eyes severe, and beard of formal cut,
> Full of wise saws, and modern instances,
> And so he plays his part. The sixth age shifts
> Into the lean and slipper'd pantaloon,
> With spectacles on nose, and pouch on side,
> His youthful hose well sav'd, a world too wide
> For his shrunk shank, and his big manly voice,
> Turning again toward childish treble, pipes
> And whistles in his sound. Last scene of all,
> That ends this strange eventful history,
> Is second childishness and mere oblivion,
> Sans teeth, sans eyes, sans taste, sans everything.
>
> *(William Shakespeare,* As You Like It, *Act II, Scene VII, lines 139–66 (speech by Jacques))*

Compare 'the seven ages of man' as defined by William Shakespeare (1564–1616) with the stages of the human lifecycle listed in Table 5.1. About one-quarter of the human lifespan is spent growing. Humans have a long period of infant dependency followed by a unique extended childhood growth period between infancy and puberty. Puberty is the period of rapid growth leading to sexual and physical maturity. Some people claim that ageing begins as soon as the peak in any chosen parameter is reached; this would imply a 40–50 year period of ageing. Ideally this process should be as gradual as possible.

Embryological development

Development of a human begins when the ovum is fertilised by a spermatozoon, forming a zygote, which is a single fertilised cell. The zygote quickly begins to divide and during the first 3 weeks of development, three primary germ layers differentiate. These are the ectoderm, which eventually forms nervous tissue and part

Table 5.1 Stages in human development

Landmark events	Major stage	Period within lifecycle	Approximate age
Conception			
	Prenatal	Preembryonic	Weeks 1–3 after conception
		Embryonic	Weeks 4–8 after conception
		Fetal	Months 3–9 after conception
Birth			
	Postnatal	Infancy	Years 1–2
		Childhood	Years 3–10
Peak height velocity			
		Puberty	Years 11–17
		Young adult	Years 18–40
		Middle aged	Years 41–60
		Elderly adult	Years 61–
Death			

of the skin, the endoderm, which forms the visceral organs of the body, and the mesoderm, which forms the tissues and structures of the musculoskeletal system. Only the mesoderm will be considered further here.

During the embryonic period, there is rapid development of the mesoderm, characterised by a series of orderly and irreversible stages through which each organism passes (Figure 5.1). From relatively undifferentiated cells and tissues, there is a progression of changes leading to highly specialised cells and tissues. By the end of the embryonic period the embryo is about 35 mm long, measured from the top of the head to the buttocks, and all the fingers and toes can be seen. Most tissues differentiate during this period and a number of landmark events occur.

During the fetal period there is rapid growth of the tissues and structures that have already been formed during the preceding embryonic period. Growth of tissues can occur by an increase in the number of cells forming the tissue, called *hyperplasia*, or by an increase in the size of each cell, called *hypertrophy*. Hyperplastic growth is a feature of the fetal period.

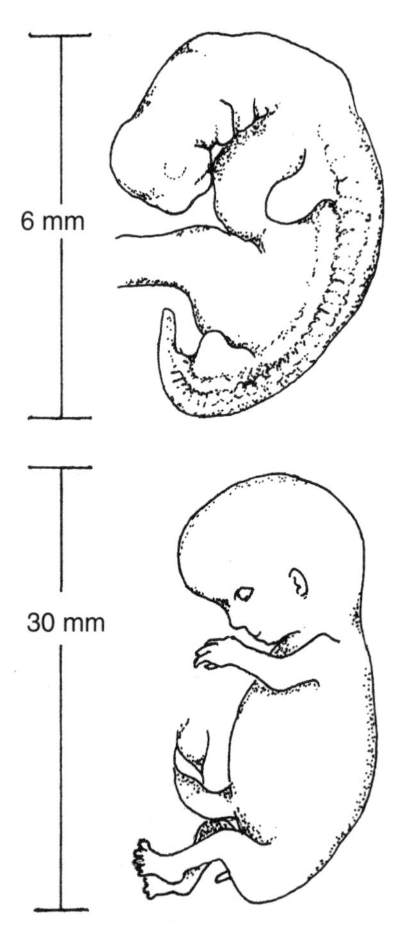

Figure 5.1 Illustrations of an embryo at 30 days and 55 days of development indicating sizes by the crown–rump length

The postnatal years

Birth is a major landmark event and provides a date from which age is calculated for the rest of a person's life. Despite the accuracy of chronological age it is not a good indicator of biological age as we shall see later in this chapter. During infancy the baby learns to sit up, crawl, stand and walk, so it is a period of rapid motor development. During childhood, growth continues at a steady rate. There is another period of rapid growth during puberty, which is associated with the development of the secondary sexual characteristics.

The term *maturation* can be used to describe the physical changes that occur between conception and maturity in all healthy growing children, usually in a similar order as a result of internal processes. These internal processes appear to be genetically determined in contrast to external environmental factors, which affect maturation in the process of adaptation. (The body's accommodation, or adjustment, to the immediate environment will be discussed in Chapters 6 and 14.)

During early adulthood many physical characteristics exhibit their peak. The ageing process begins during early adulthood and continues through the middle years to old age. One aim of physical activity is to delay some of the physical changes associated with ageing. Indeed it has been shown that some of the so-called effects of age are related more to decreased physical activity than to some inbuilt ageing mechanism (see Chapter 14).

Ageing hypotheses

A number of hypotheses have been proposed to explain biological ageing. Some people think that ageing is genetically determined and so is programmed from conception. Others think that constant exposure to background radiation causes mutations or that errors creep into the ribonucleic acid (RNA) within cells so that the cells become less efficient with each generation. There is little current evidence for either of these hypotheses.

With respect to ageing specifically in the musculoskeletal system, a 'wear and tear' theory abounds although this seems to ignore the generally positive adaptations the musculoskeletal system makes in response to physical activity, as will be outlined in Chapters 6 and 14. Another possibility, related to the musculoskeletal system, is the cross-linking theory, which proposes that there is increased cross-linking between collagen and elastin fibres. This cross-linking does seem to have an effect related to joint range of motion but whether it could have major effects on general health is debatable. Other hypotheses that try to explain the ageing phenomenon are related to changes in chemistry and reductions in the response of the immune system.

Physical growth, maturation and ageing of the musculoskeletal system

For the musculoskeletal system the major periods of growth are the fetal and pubertal stages. During the early fetal growth period there is rapid multiplication of cells (hyperplasia) whereas during the pubertal period growth is caused mainly by development and enlargement of existing cells (hypertrophy). Development can also occur via the replacement of one type of tissue by another.

The development of bone involves a number of stages

Adult human bones appear in a variety of shapes and sizes. Their basic shape is genetically determined; however, their final shape is influenced greatly by the environment in which they develop. Environmental influences include mechanical factors, such as muscle forces acting on the developing bone, and metabolic factors, which include the supply of nutrients.

Typical long bones, such as the humerus in the arm and the femur in the thigh, develop via a process of endochondral ossification in which a cartilaginous model precedes the bone formation. The cartilage is eventually replaced by bone. The cartilage model initially is small and its replacement by bone begins at an early stage of development. The primary ossification centres appear near the middle of the shaft of the future long bone at about 8 weeks after fertilisation when the embryo is about 35 mm long (see Figure 5.1).

The cartilage model of the future bone grows before it is replaced by bone. This replacement occurs in a series of stages. Initially, the cartilage model grows by two processes: an increase in the number of cartilage cells and then by an increase in the size of each cell. Next, the gel-like matrix surrounding the cartilage cells is calcified. This hardening of the tissue surrounding the cells effectively cuts off their supply of nutrients as well as prevents the removal of the metabolic waste products. The cells eventually die so that the calcified cartilage has a honeycomb appearance. Invading blood vessels bring nutrients and bone-forming cells (osteoblasts) to the calcified cartilage to lay down new bone. This new bone has a disorganised appearance. If ossification were the final stage, a long bone would not develop its hollow structure. Therefore some of the newly laid down bone is removed from internal sites by special bone-eroding cells. These osteoclasts work with osteoblasts during modelling of the bone, causing the cortex to drift away from the central axis of the shaft during growth. Changes in shaft diameter and compact bone thickness can occur at any age by a remodelling process. Bone can respond to changes in mechanical stimuli by changing its structure. This adaptation is one indication that bone is a dynamic, metabolically active tissue.

One type of remodelling is the replacement of the originally laid down bone by compact bone, which has an organised structure (see Figure 3.1). Most primary centres of ossification (in the developing shaft) appear before birth whereas most secondary centres of ossification (in the developing ends of the bone) appear after birth. The time of appearance of these secondary centres of ossification ranges from the centre in the femur near the knee, which is normally present at birth, to the centre in the clavicle (collar bone) closest to the sternum (breastbone), which does not appear until about 18 years of age.

The epiphyseal plate exhibits the stages of bone development

When both the shaft and ends of a long bone are developing they are separated by a 'growth plate' or epiphyseal plate. Within this region all the stages of bone development can be seen under a microscope (see Figure 5.2). Bones grow in length by the replacement of the growing cartilage by bone, and growth in height of a person stops when the epiphyseal plates 'fuse'. Actually the growth ceases because the cartilage cells no longer respond to hormonal influence. Books on growth and development provide tables indicating time of appearance of the primary and secondary centres of ossification plus the age at which union of the various parts of a bone occurs to form a complete adult bone. The anterior fontanelle, or soft spot on the skull of a newborn baby, closes during the second year but the sternum does not become a single bone until some time in old age. Even the five parts of the femur do not completely unite until the early twenties.

Growth in length and width of bone are two different processes

Although growth in height ceases at a certain age, changes in thickness of the compact bone of the shaft and the density of spongy bone can occur at any time of life (see Chapter 6). Growth in the thickness and diameter of long bones occurs by appositional growth, during which bone is added on the outside of the shaft and removed from the inside of the shaft.

Skeletal composition changes throughout the lifespan

The major constituents of bone are the inorganic salt crystals and the organic collagen. The relative proportions of these change throughout the lifespan. In a child the flexibility of the bone is related to the large proportion of collagen. It is thought that the rapid growth of children's bone and the consequent amount of remodelling result in less than optimal mineralisation. In a young adult the inorganic material is close to the optimal two-thirds mineralisation so the bone is strong and tough. The brittleness of bone in elderly people is only partly related to the increased proportion of the inorganic

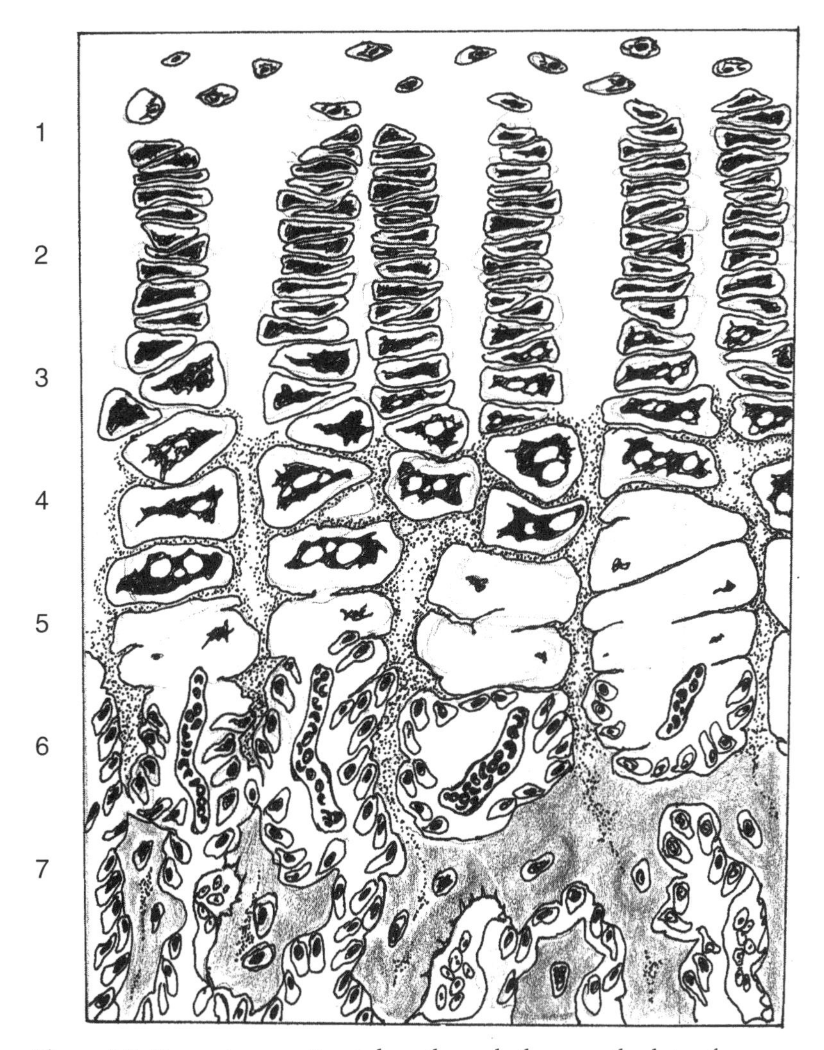

Figure 5.2 Zones in a section taken through the growth plate of a developing bone, as seen under a microscope. The zones, numbered from the end of the bone towards the shaft, may be called: zone of resting cartilage 1; zone of cartilage proliferation 2; zone of cartilage maturation 3; zone of calcification 4; zone of cartilage cell death 5; zone of bone formation 6; zone of bone resorption 7. Each of the zones represent a stage in the replacement of the cartilaginous model by bone so that the zones illustrated do not migrate downward, but rather they maintain their position and change their nature. This effectively increases the distance between the growth plates at each end of the bone

salts. Of more significance in the skeletal system of aged persons is the reduced total mass of bone material in the body. This results from the increased porosity of bone, which, in turn, causes decreased density of bone.

In a newborn baby the skeleton accounts for about 13 per cent of the total body weight, but about two-thirds of the skeleton is cartilaginous. There is a similar proportion of skeletal material in the mature adult but just over one-tenth of the skeleton is cartilaginous. Of the bone, which comprises half of the total skeletal material in the adult, four-fifths is compact and one-fifth is spongy. There is a relatively large amount of bone marrow, which constitutes about one-third of the total skeletal material. There is also joint tissue, such as capsule and ligaments, which constitutes almost one-tenth of the total material forming the skeleton and joints of the adult body.

Structural changes in bone are often associated with ageing

In an 80 year old male the density of bone in the spine is often 55 per cent of that in a 20 year old and in elderly females the spinal bone density is often only 40 per cent of its peak value. Density is mass per unit volume of tissue and it is the mass of the bone that significantly decreases during ageing. Bone mass decreases mainly because the number of trabeculae in spongy bone decreases (see Box 5.1), combined with resorption, and consequent thinning, of the trabeculae. In the compact bone the size of the Haversian canals (see Figure 3.1) increases and the cross-sectional area of bone surrounding each Haversian canal decreases.

Not only is there a large decrease in bone mass but there is also some decrease in bone volume because there is decreased thickness of the compact bone, especially in the shafts of long bones. Particularly in males, there are changes in whole bone architecture that counteract the expected decrease in strength due to decreased bone volume. The major change is an increased diameter of the shaft, particularly of long bones of the lower limb. The actual decrease in strength of the male human femur between the ages of 40 and 70–80 years is of the order of 20 per cent, which is about half the decrease in bone density that occurs over the same period.

With ageing, bone becomes stiffer, partly because of the collagen cross-linking. Eventually, aged bone could be described as being brittle because of the increased inorganic component and the weakening of the bonds between the salt crystals and the collagen. A consequence of these mechanical changes is that aged bone can absorb less energy before it fractures so it is more liable to fail when subjected to large forces.

Osteoporosis

For most tissues in the body their mass at any time is determined by the balance between the simultaneous formation and destruction of the tissue. *Osteopenia* is the name for reduced bone density, which is due mainly to decreased bone synthesis, particularly during the post-menopausal years in women. The dramatic decrease in the female hormone, oestrogen, after menopause allows the rate of synthesis to be lower than the resorption rate. The initial effects are mainly on the spongy bone, but later the compact bone in the shafts of long bones also decreases in thickness.

Osteoporosis literally means increased porosity of the bone and is associated with gross structural changes to bone, particularly fractures. In general the porosity of bone in the human femur doubles from 40 to 80 years of age and there is an even greater change in osteoporotic individuals. The effect on the whole bone is that it is weakened and is able to absorb less energy before it fails.

Even though a chemical analysis of the bone material may indicate slightly elevated levels of calcium, the overall mass of bone decreases markedly and therefore there is less calcium stored in the skeleton. Some authors have promoted the idea of bone banks in the discussion of prevention of osteoporosis. During maturation of bones the density and calcium content increases due to a genetic predisposition aided by the interaction between dietary calcium and exercise. Later in life, after about 40 years of age, withdrawal of bone material is inevitable. The structural integrity of the bones will be maintained longer if a person's lifestyle during maturation and early adulthood promoted maximum deposition of bone material. Financial advisers tell us to start saving as young as possible, and this concept is even more important for prevention of osteoporosis. Both the personal lifestyle changes due

Box 5.1 What causes structural decreases in trunk height with ageing?

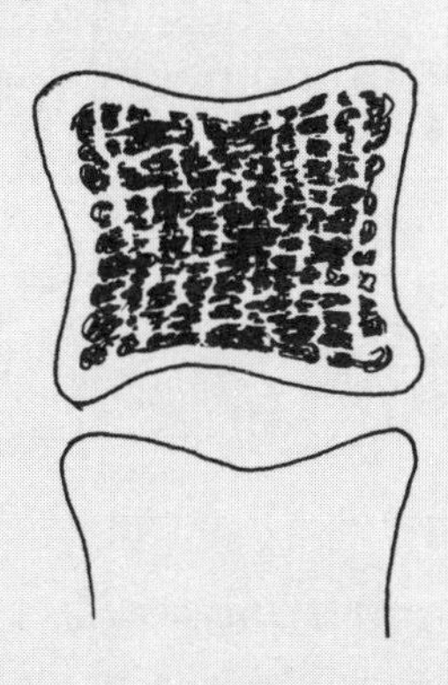

Figure 5.3 Age-related changes in dimensions of the vertebral body and intervertebral discs
Source: Twomey, L., Taylor, J. & Furniss, B. (1983), Age changes in the bone density and structure of the lumbar vertebral column, *Journal of Anatomy*, 136, 15–25. (Copyright 1983 by Cambridge University Press); and Twomey, L.T. & Taylor, J.R, (1987), Age changes in lumbar vertebrae and intervertebral discs, *Clinical Orthopaedics and Related Research*, 224, 97–104. (Copyright 1987 by Lippincott-Raven Publishers; adapted with permission.)

Research workers in the School of Physiotherapy at the Western Australia Institute of Technology (now Curtin University of Technology) and the Department of Anatomy at the University of Western Australia analysed 204 cadaveric lumbar spines from people who had died at ages ranging from 1 day to 97 years. The dimensions of thin sagittal sections of the third lumbar vertebra, removed from 93 adult subjects after their death, were measured. Vertebral body bone density was calculated by dividing the measured dry mass of each section by the calculated volume of each specimen. Another measure of bone density was calculated by analysing the amount of light transmission through each section. Size and orientation of trabeculae were also analysed (see Figure 3.3). The shape of each vertebral body section was defined by dividing the mid-vertebral height of the section by the anterior vertebral height. Average height of each lumbar intervertebral disc was calculated as the mean of three measurements of disc height: anterior, middle, and posterior. Disc shape was described by dividing the middle disc height by the sum of the anterior and posterior disc heights.

All three methods of measuring bone density confirmed that there was a loss of bone density (or increase in porosity) related to ageing. The small decline in the number of vertical trabeculae was not statistically significant but there was a marked decrease in the number of horizontal trabeculae, especially in the females. The thickness of trabeculae was not measured. Shape measurement of the vertebral bodies indicated that there was increasing concavity of the vertebral end plates with age for both sexes. Overall there was no decrease in disc height with ageing, and in fact some individuals exhibited an increase in disc height. In the adult specimens, evidence of disc degeneration was observed more often as age increased but the group with degenerative discs was still a minority in the aged individuals. The authors concluded that the age-related structural changes of the lumbar vertebral column were mainly related to the hard tissues of the vertebral body rather than to the soft tissues of the intervertebral disc (see Figure 5.3).

The authors considered that the results indicated the effects of normal ageing and did not propose a definition of osteoporosis based on the measurements used. They used an engineering concept called Euler's theory to explain that the horizontal trabeculae are very important, as they act as 'cross braces' to support the vertical trabeculae. The loss of horizontal trabeculae with ageing leads to the collapse of vertical trabeculae, which in turn affects the shape of the end plates (at the top and bottom of the vertebral body). These become more concave. The authors cited this example of bone geometry as being more important than mass *per se*. Further, the authors suggested that the increasing concavity of the vertebral bodies explained the loss of trunk height and increased curvature of the vertebral column observed in old age (Figure 5.3). Radiological images of the lumbar spine may give the impression of disc space narrowing because the anterior and posterior disc heights may decrease.

Sources: Twomey, L.T. & Taylor, J.R. (1987), Age changes in lumbar vertebrae and intervertebral discs, *Clinical Orthopaedics and Related Research*, 224, 97–104; Twomey, L., Taylor, J. & Furniss, B. (1983), Age changes in the bone density and structure of the lumbar vertebral column, *Journal of Anatomy*, 136, 15–25.

to, and public health care costs of, osteoporosis are so great that they will be considered further in Chapter 15.

Bones may sometimes fail

Engineers have been unable to design a car that completely protects the passengers from injury although this is one of the principal objectives of the modern automotive industry. A major way of protecting passengers is to enclose them in a cage surrounded by material designed to cushion the forces involved in a collision. The soft tissues of the human body, including fat and muscle, perform this role of a 'crumple zone' to some degree, but sometimes the energy involved in a collision is enough to cause a fracture. A 'greenstick' fracture is so named because the flexible bones of a child tend to splinter in a similar manner to the branch of a young tree. Late in life the fracture is more likely to be of the brittle type because of the increased stiffness of old bones, related to the change in composition during ageing. The bones of elderly adults are also more likely to fail because of the decreased bone mass and density in osteoporotic individuals. Possible patterns of bone loss are illustrated in Figure 5.4.

Specific fractures are related to certain ages; for example, fractures of the neck of the femur (commonly called hip fractures) are common in elderly women. Interestingly, while this injury often occurs during a fall it may be the strong muscular response to the overbalancing rather than the contact with the ground that actually causes the fracture. In other words the fracture may have already occurred before the fall was completed.

There is an increase in incidence of forearm fractures near the peak of the pubescent growth spurt. This is associated with increased porosity of compact bone as the remodelling space increases to provide calcium to the rapid growth regions. In developing children the cartilaginous growth plate may also fail and this most often occurs at the zone of maturation, or hypertrophy (see Figure 5.2).

Developing bone may be vulnerable to repetitive muscular forces

The development of bone has implications for potential injury to the musculoskeletal system. Injury to the cartilaginous growth plates can be produced by single excessive forces causing trauma or by repetitive compressive forces or tensile forces produced by musculotendinous pull on an area of developing bone.

In sports medicine, specific injuries have been attributed to certain activities. One of the

Box 5.2 What is the pattern of trabecular bone loss in adult men and women?

A research team from the Medical Research Council Mineral Metabolism Unit, The General Infirmary, at Leeds, England, including a visiting pathologist from New Delhi, India, investigated the trabeculae of the ilium (part of the hip bone) collected at autopsy. Specimens studied were from 98 men and 86 women, aged 20–90 years. They had been carefully selected over a period of 15 years, to exclude pathology or immobilisation as confounding factors which would contribute to changes in structure of the spongy bone (see Figure 3.1).

Histomorphometry, which is a quantitative microscopic technique, was used to determine trabecular number and width, trabecular bone volume, and other parameters. A loss of trabecular volume, related to ageing, was a general finding and was similar for both sexes. The factors which contributed to this decrease in volume appeared to be sex-related. The number of trabeculae and the total trabecular surface area decreased with age for the women, but this was not very obvious for the men (see Figure 5.4). Trabecular width tended to decrease for the men but this parameter did not change much for the women (see Figure 5.4).

The total removal of some trabeculae was the main parameter associated with loss of trabecular bone volume in women, whereas, in men, the major observation was the general thinning of individual trabeculae. At any stage during bone remodelling, bone volume is the result of the balance between bone resorption and formation. The authors' conclusion from this study was that decreased bone formation was the principal reason for age-related bone loss in men. In contrast, women appeared to exhibit increased resorption of bone.

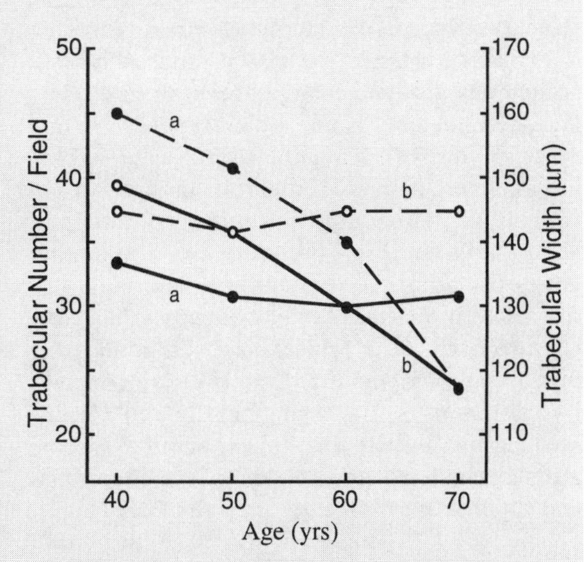

Figure 5.4 Age-related changes over three decades in the number of trabeculae (solid lines) and mean trabecular width (interrupted lines) in the spongy bone of male (a) and female (b) hip bones. Data from Aaron, J.E., Makins, N.B., & Sagreiya, K. (1987), The microanatomy of trabecular bone loss in normal aging men and women, *Clinical Orthopaedics and Related Research*, (215), 260–71.

Source: Aaron, J.E., Makins, N.B., & Sagreiya, K. (1987), The microanatomy of trabecular bone loss in normal aging men and women, *Clinical Orthopaedics and Related Research* (215), 260–71.

best examples is 'little league pitcher's elbow'. The incidence of elbow injuries in immature baseball pitchers was found to be high, and the mechanism appeared to be excessive muscular forces around the elbow. This information led to rule changes in Little League baseball, which restricted the amount of pitching by any one player in both training and matches. The incidence of elbow injuries reduced significantly as a result of these administrative reforms. Other examples of the relationships between activity and types of growth plate problems are wrist injuries in gymnasts and tibial tuberosity avulsions in sports involving sprinting and jumping. (The tibial tuberosity is the point of attachment of the patellar ligament, which transmits the large forces from the quadriceps muscle group.)

Joints develop from the gaps between bones

During early intrauterine development there is a primitive connective tissue, mesenchyme, which has the potential to become cartilage, bone, ligament or some other type of tissue. Mesenchyme is present at between 5 and 6 weeks after conception and then part of this primitive connective tissue differentiates into cartilage about a week later. Between the developing bones the primitive connective tissue may differentiate into a number of different structures characteristic of a typical synovial joint (see Figure 3.6). Peripherally, the primitive connective tissue gives rise to the capsule and associated ligaments that first appear about 7 weeks after conception. Centrally, the tissue disappears and the resulting space becomes the joint cavity during the 8th and 9th weeks. Where the primitive connective tissue lines the capsule and the cartilaginous joint surface, it forms synovial membrane. Movement plays a role in the further development of the joint because it causes the synovial membrane lining the load-bearing surfaces to be sloughed off.

The thickness of articular cartilage is dependent on a number of factors, including age, size of the joint, and whether it is normally a weight-bearing joint. Articular cartilage may be 5–7 mm thick in the hip joints of young healthy adults but it may be 1–2 mm thick in the finger joints of elderly adults.

Range of motion at joints is affected by joint structure and the mechanical properties of the tissues associated with the joints. It is generally perceived that joint range of motion reduces during life, and while this is the general trend the rate of loss is not constant. Ranges of joint motion are very large in a newborn baby. As an example, the range of ankle dorsiflexion is only limited by the contact of the top of the foot against the shin. Try this movement on yourself to indicate how much loss of flexibility has occurred since your birth.

Between the ages of 6 and 15 there is a general trend for joint range of motion to decrease in boys, whereas for girls the effects are variable and joint dependent. Girls are generally more flexible than boys during childhood and adolescence. Interestingly, there are also anthropometric correlates with joint range of motion during this period. Short children tend to be more flexible than tall children, but in adults there is no clear relationship between anthropometric measurement and joint flexibility. Changes in joint ranges of motion between adolescence and young adulthood are variable and, as we shall see in the next chapter, appear to be related to physical activity. Similarly, the general trend for joint flexibility to decrease during ageing may not be completely explained by biological ageing processes.

Synovial joints are usually self-repairing, but ageing people may suffer arthritis, which is literally 'inflammation of joints'. In rheumatoid arthritis there is inflammation of the synovial membrane. When there is degeneration of the articular cartilage the condition is known as osteoarthrosis, which often develops into osteoarthritis. Under 45 years of age more men have osteoarthrosis than women but with increasing age the women overtake the men so that in the total adult population about three times as many women as men have osteoarthrosis. Thus there is some genetic determination of osteoarthrosis to which are added the effects of environmental factors. It appears that the risk factors for rheumatoid arthritis are mainly related to a family history of the condition.

Muscles develop from the longitudinal fusion of cells

Mesoderm that forms towards the end of the 3rd week of development will eventually form the muscle and connective tissues of the body. Some of the mesoderm is segmented so that it is in lumps on either side of the midline and these lumps form the muscles of the trunk.

Other mesoderm will form the muscles of the limbs and this tissue migrates into the limbs when they first appear. Limb muscles migrate toward their final position during further development.

Certain precursor cells differentiate into myoblasts (muscle-forming cells), which fuse longitudinally to form long muscles and muscle fibres. There is still controversy about the relative contribution of hyperplasia and hypertrophy to muscle growth. It appears that most muscle fibres have differentiated by the 7th month of development in utero, but there is continued increase in number for a few months after birth. The number of muscle fibres at this stage appears to be genetically determined and further growth occurs by hypertrophy. Newly formed muscle fibres have about 75 myofibrils per fibre (see Figure 3.12), which increase to over 1000 in early adulthood.

Muscle changes during growth, maturation and ageing

It is thought that the number of muscle fibres is genetically determined, so large people often have more muscle fibres. As muscles grow in length there is an increase in the number of sarcomeres. Although there is very little direct evidence, many believe that the stimulus for growth of muscle is bone growth, which pulls on the attached ends of the muscle. There is developmental growth in both muscle length and cross-sectional area and these factors may also be affected by activity. The effects of physical training on muscle will be summarised in following chapters (Chapters 6, 12 and 13).

In an infant, muscle tissue accounts for about 25 per cent of total body weight but in a young adult the proportion of muscle is more than 40 per cent. In terms of growth, muscle increases from about 850 g at birth to about 30 kg in a young adult male weighing 70 kg.

Strength may be defined as the capacity to produce force against an external resistance and is related to the cross-sectional area of the muscles producing the force. (Muscle strength is discussed further in Chapters 12, 13 and 14.) Muscle volume peaks at about 30 years of age and then gradually reduces. The atrophy of skeletal muscle associated with disuse is much faster than any ageing related atrophy. Muscle elasticity also decreases during ageing so that muscles are stiffer and less extensible. This is one contributing factor to the loss of joint range of motion described earlier.

Changes in body dimensions throughout the lifespan

To examine body dimensional changes throughout the lifespan it would be ideal to perform a study in which individuals were chosen at birth and followed until death, with a battery of measurements and tests conducted at regular intervals during their life. This approach is known as a longitudinal study (see Box 5.3) and even though it may be considered the ideal type of study, there are many potential problems associated with it, most of them of a practical nature. To reduce the time of a study related to growth, maturation, and ageing, a particular period can be chosen for the longitudinal study. Alternatively, groups of individuals of different ages can be measured simultaneously to determine general patterns of change. These cross-sectional studies do not provide information related to any particular individual, but they can provide information about averages of, and variations in, measurements over a large age span (see Box 5.4). Individual data can be plotted on graphs derived from cross-sectional studies (Figure 5.5). A compromise between these two approaches to human growth or ageing studies is the mixed longitudinal design. In this type of study, groups, say 5 years apart in age, are measured at the same time and then followed for 5 years. If there were initially four groups, 20 years worth of data could be collected in 5 years.

Size measurements can be reported directly so there is information on the average heights (Figure 5.6) and weights (Figure 5.7) of Australian and Canadian children. In a cross-sectional study, performed during 1985 by a Queensland research team headed by Professor Tony Parker, 250 junior rugby players in Brisbane

Box 5.3 What is the pattern of growth in Swedish children?

Commencing in 1955, 103 boys and 80 girls were followed from birth to adulthood. Patterns of maturation were analysed by researchers from the Department of Orthodontics, University of Lund, Malmo, and the Department of Pediatrics, University of Goteborg. The subjects were divided into three maturity groups, according to their age at peak height velocity (PHV). On average, the early maturing girls were 10.7 years at PHV, and the boys were 12.6 years. The average maturers were 12.1 years (girls) and 14.0 years (boys) at PHV, and the late maturers were 13.6 years (girls) and 15.6 years (boys). Just over half of the subjects were classified as average maturers, while almost a quarter were early maturers and about a fifth were late maturers.

Between the ages of 5 and 14 years, the early maturing girls were taller than the late maturers (maximum of 13.1 cm difference at 13 years); however, the three groups were very similar in height at 25 years of age. The early maturing boys were taller than the late maturers between the ages of 12 and 15 years, but were actually shorter as adults. The early maturers were 11.8 cm taller than the late maturers at 14 years but were 6.5 cm shorter at 25 years of age. The late maturers were also taller than the average maturers at 25 years, by 4.2 cm.

Source: Hagg, U. & Taranger, J. (1991), Height and height velocity in early, average and late maturers followed to the age of 25: a prospective longitudinal study of Swedish urban children from birth to adulthood, *Annals of Human Biology* 18, 47–56.

Box 5.4 What are the variations in body size and composition in children?

During the 1970s a number of groups of children between the ages of 5 and 12 years were studied anthropometrically in the United States. Three different cross-sectional studies were compared by researchers from the Department of Family and Community Medicine at Bowman Gray School of Medicine, Wake Forest University, Winston-Salem, North Carolina; the results from a total of 3373 children were compared. The North Carolina study, conducted by the authors, included 832 boys and 836 girls from 17 different schools. A sample from Michigan included 127 boys and 98 girls. The National Health and Nutrition Examination Survey I (HANES I, 1971–1974) included 745 boys and 735 girls surveyed from the United States' general population of children. The age distribution in the North Carolina sample was similar to the general population, while whites were slightly under-represented in the sample.

Between the ages of 5 and 12 years, boys grew from about 113 cm tall and 20 kg to 153 cm tall and 45 kg. The variations in both height and mass, as represented by the standard deviations, increased with age. Girls were only very slightly smaller than boys at 5 years (112 cm and 20 kg) but at 12 years they were slightly larger (155 cm and 47 kg), probably because they matured earlier (see Box 5.3). The skinfold thickness at the back of the arm (triceps) was consistently larger in the girls at all ages. Skinfold thickness tended to increase with age, except that for the boys between 11 and 12 years there was a plateau or even a decrease in the triceps skinfold.

In general, the children in the small Michigan sample appeared to be taller, heavier, and fatter than the children surveyed in the other two larger samples. This was an important comparison because the Michigan study had been used to argue for a secular trend (see Box 5.5) toward fatter children. This secular trend was not confirmed by the North Carolina or HANES I studies.

Sources: Diseker, R.A., Michielutte, R., Ureda, J.R., Schey, H.M., & Corbett, W. (1982), A comparison of height, weight, and triceps skinfold thickness of children ages 5-12 in Michigan (1978), Forsyth County North Carolina (1978), and HANES I (1971-1974), *American Journal of Public Health* 72, 730–33.

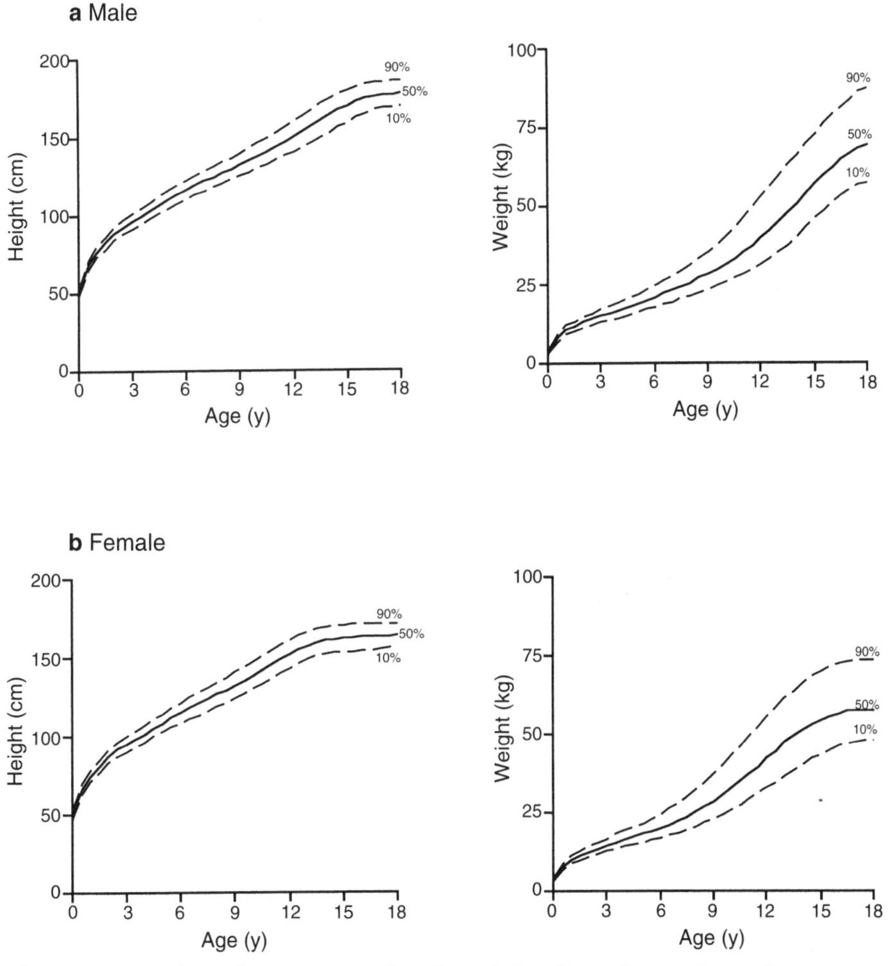

Figure 5.5 Graphs indicating growth in length/height and mass for 50th percentile males (a) and females (b) modified from Hamill, P.V.V. (1978), NCHS growth curves for children, Department of Health, Education and Welfare publication (PHS) 78-1650. Adaptations of these growth curves for Caucasian children are used by family doctors in a number of Western industrialised countries, such as Australia.

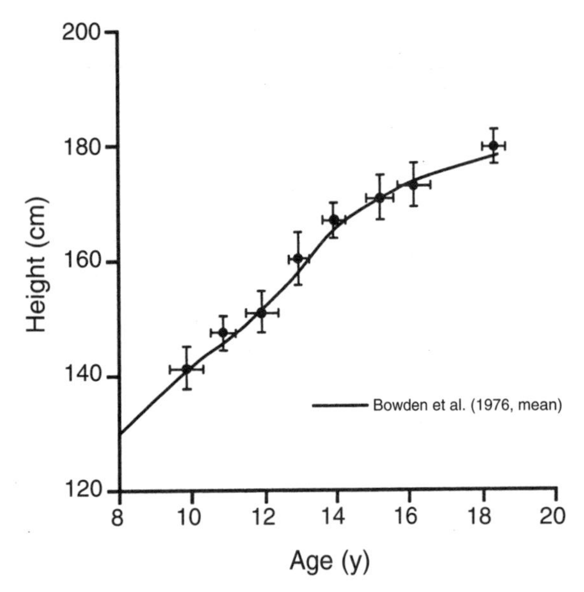

Figure 5.6 Height of Victorian boys, used as a basis for comparison with 250 junior rugby players in Queensland
Source: Parker, A.W. & Kippers, V. (1986), Anthropometric profiles of junior rugby players, unpublished data and Bowden, B.D., Johnson, J., Ray, L.J. & Towns, J. (1976), The height and weight changes of Melbourne children compared with other population groups, *Australian Paediatric Journal*, 12, 281–295.

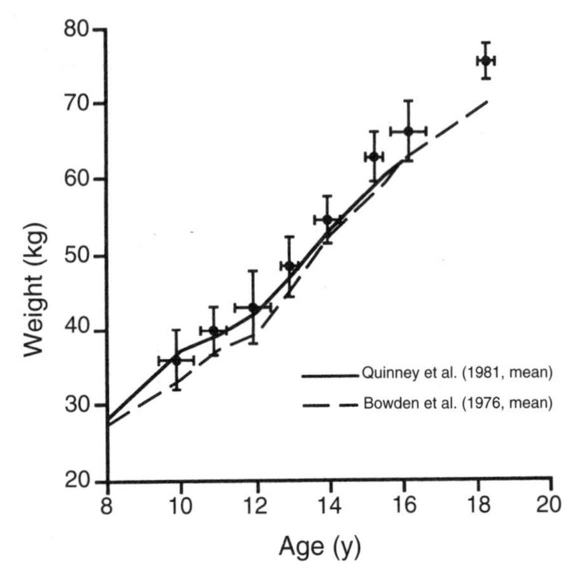

Figure 5.7 Weights of Victorian and Canadian boys, used as basis for comparison with 250 junior rugby players in Queensland
Source: Data from Parker, A.W. & Kippers, V. (1986), Anthropometric profiles of junior rugby players, (unpublished); Bowden, B.D., Johnson, J., Ray, L.J. & Towns, J. (1976), The height and weight changes of Melbourne children compared with other population groups, *Australian Paediatric Journal*,12, 281–95; and Quinney, H.A., Watkinson, E.J., Massicotte, D., Conger, P.R. & Gauthier, R. (1981), The height, weight and height/weight ratio of Canadian children in 1979, *Canadian Medical Association Journal*, 125, 863–65.

were compared with the normal population. Figures 5.6 and 5.7 show that even though the boys were normal in height through all the age ranges, they tended to be heavier in the under 17 and under 19 divisions. Whether this was due to the smaller boys dropping out, or the effects of weight training by the serious competitors, could not be determined.

Height can be plotted against age (as in Figures 5.5 and 5.6) but when height gain is plotted against age for an individual it more obviously illustrates the 'growth spurt' (Figure 5.8). Within this period is the peak height velocity (PHV), which is a major landmark in pubertal growth and is a reference for many other changes that occur. In boys, the proportion of fat in the body may be decreasing during this period of rapid growth. The peak weight velocity is delayed relative to peak height velocity by a few months. Peak strength velocity may be delayed for more than a year. Unfortunately all these peaks in the various parameters can only be determined after the period has past. In other words peak height velocity can only be determined retrospectively; it cannot be easily or accurately predicted.

Size measurements can be combined to provide information about shape

Galileo first described a square–cubic relationship seen in the human body. Assume two objects have the same shape. If one object is twice as tall as the other the cross-sectional area will be four times as large (squared relationship) and

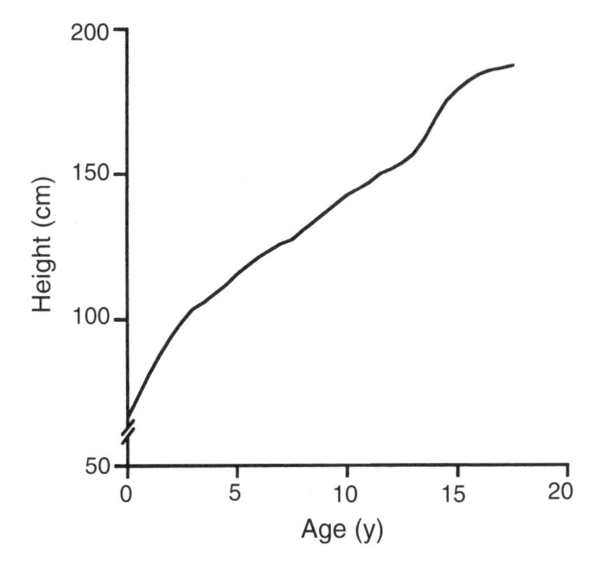

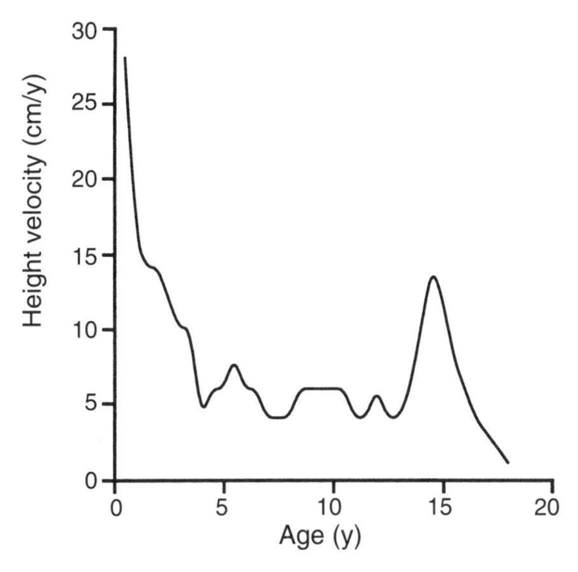

Figure 5.8 Two growth charts for the same individual derived from the same data. When growth in height is expressed as height gain per year the peak height velocity can be determined. These curves are derived from measurements taken by de Montbeillard of his son from the year of his birth in 1759 to 1777

the volume will be eight times as large (cubic relationship). Remember that the area of a circle is πr^2 and the volume of a sphere is $\frac{4}{3}\pi r^3$. Are these relationships likely to describe a normal pattern of growth? Consider a baby who is 50 cm tall and weighs 3 kg when born and assume she grows to 150 cm, that is, three times longer. We shall assume that volume is directly related to mass so we would expect her adult mass to be 3×3^3 kg. The calculated result of 81 kg is very heavy for a 150 cm tall person, so what does this tell us about growth? Basically it demonstrates that the relationship between height and weight during growth is not cubic and so all dimensions are not changing at the same rate. In other words the shape of the human body changes during growth. This change in proportionality can be seen most clearly by considering the relative size of the head at different ages (Figure 5.9). Obviously human growth is not explained by a squared relationship either because the baby girl would have grown to 150 cm and 27 kg in this situation!

In the early 1980s the New South Wales branch of the Australian Sports Medicine Federation started a campaign to counsel boys who were playing either rugby union or rugby league if it was thought that they had what was described as a 'swan neck'. Persons with this type of neck were advised not to play in the front or second row of a scrum and all players were advised to develop their neck muscles by strengthening exercises. At this time there was no objective information related to neck shape so both neck circumference and length were measured on the 250 junior rugby players included in the study mentioned earlier. The neck index (circumference/length) was calculated for each boy and the results are shown in Figure 5.10. These data can be used to determine whether a boy has an abnormally long thin neck.

Changes in body size are not constant

The overall dimensions of the body such as stature and volume (related to body mass) do not increase at a constant rate during growth, nor do they change at a constant rate during ageing. This idea is illustrated in Figures 5.3, 5.5, 5.6 and 5.8 and is recognised by the common expression 'growth spurts'. If there is a specific interest in 'growth spurts' and peak height velocity, mentioned earlier, the height

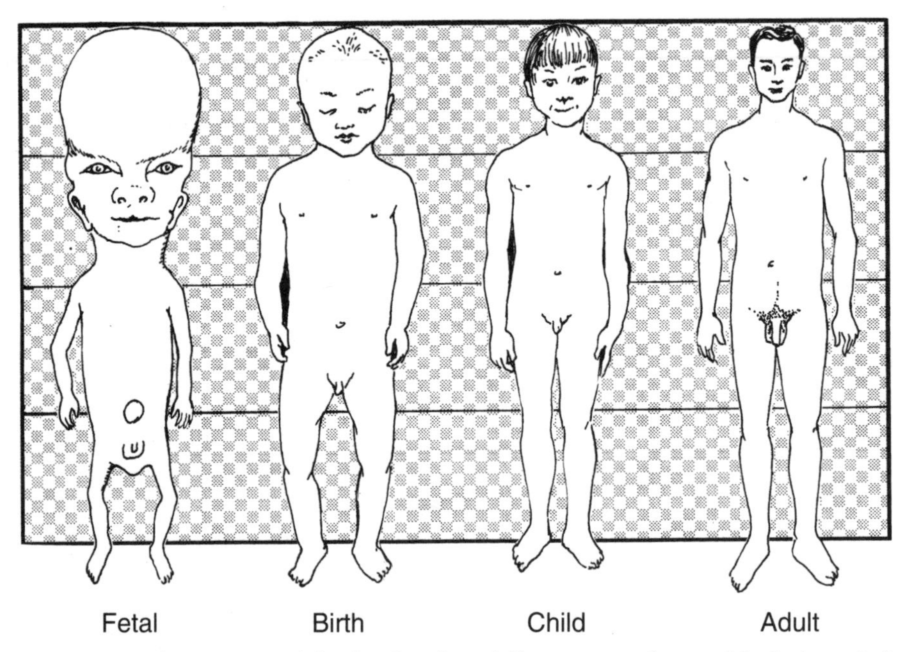

Figure 5.9 Relative sizes of the head at four different ages; the total body is scaled to the same height

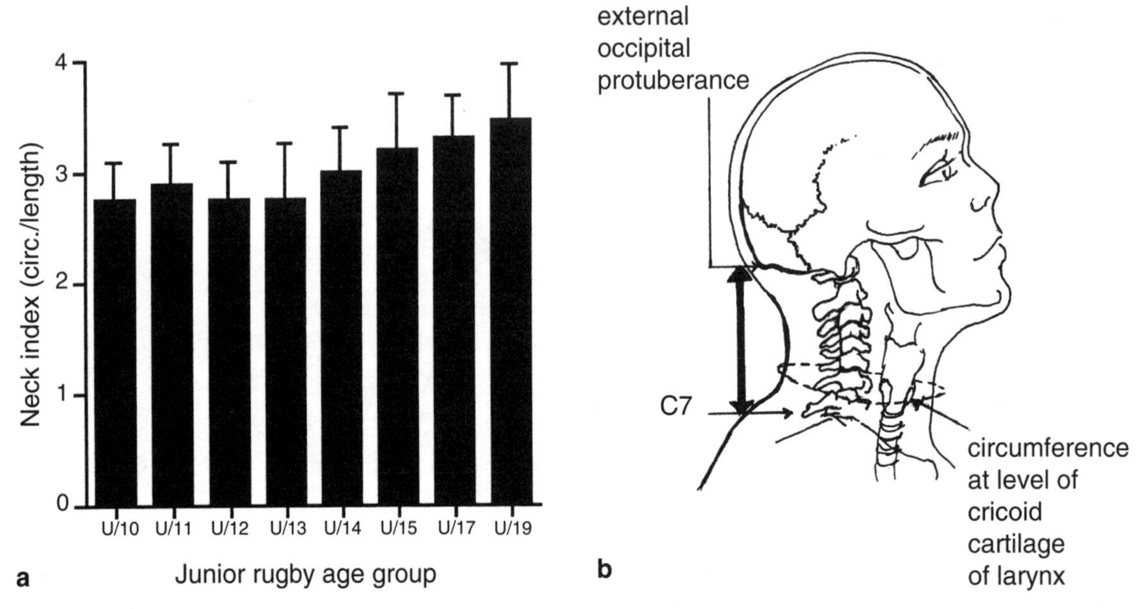

Figure 5.10 Neck indices of 250 junior rugby players (a); the two measurements taken to calculate this index are illustrated in (b)
Source: Data from Parker, A.W. & Kippers, V. (1986), Anthropometric profiles of junior rugby players, (unpublished).

versus age graph could be converted to a height velocity versus age graph by a process of mathematical differentiation. More simply, height velocity can be plotted as height gain per year against age, as illustrated in Figure 5.8.

There is a secular trend in body dimensions

There appear to be differences in size between people of different generations. Certainly it seems that people from the Middle Ages were shorter than modern humans. On a shorter time scale, there may be changes from one generation to the next. Teenagers are often taller than their parents. This generational change is known as a secular trend (see Box 5.5) and it can be illustrated by differences in the heights of Queensland school children from 1911 to 1976 (Figure 5.11). This figure illustrates a number of interesting changes that have occurred in just over two generations. In 1911, 7 and 8 year old

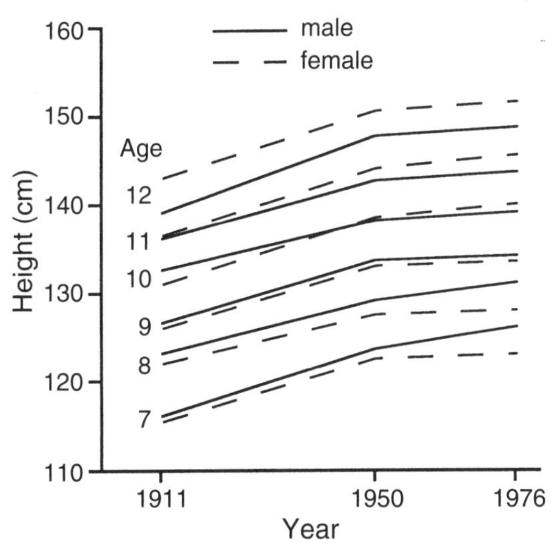

Figure 5.11 Secular trend in height of Queensland schoolchildren from 1911 to 1976
Source: Dugdale, A.E., O'Hara, V. & May, G. (1983), Changes in body size and fatness of Australian children 1911–1976, *Australian Paediatric Journal*, 19, 14–17. (Copyright 1983 by Blackwell Scientific; adapted with permission.)

Box 5.5 Are there differences in body size between children of different generations?

Between 1982 and 1984, 1048 children (539 boys and 509 girls) from 18 separate Hong Kong schools were measured as part of a mixed cross-sectional and longitudinal anthropometric study (see page 79) performed by researchers from the Department of Children's Dentistry and Orthodontics at the University of Hong Kong. The children were aged between 11 years-1 month and 13 years-6 months. Only the results from the 12- and 13-year-olds in this study were compared with subjects from a previous survey (performed between 1961 and 1963) of southern Chinese children living in Hong Kong.

Three measurements each, of both height and mass, were taken to minimise operator variability. Between 1982 and 1984, the 12- and 13-year-old girls were taller and heavier than the boys, but these differences were smaller than they were in the 1961–1963 study. Tabulated results indicated that there were consistent upward secular trends in both height and weight. The largest secular trends were exhibited by the 12-year-old boys who were 6.7 cm taller and 5.7 kg heavier than they were twenty years earlier. Interestingly, the 97th percentile 12-year-old boys were almost 2 cm taller in 1961–63. The 13-year-old girls exhibited the smallest secular trends of 1.4 cm increase in height and 0.9 kg in mass. The authors attributed the positive secular trends to improvements in socioeconomic and sociohygienic conditions, particularly diet and health care.

Source: Ling, J.Y.K. & King, N.M. (1987), Secular trends in stature and weight in southern Chinese children in Hong Kong, *Annals of Human Biology* 14, 187–90.

boys and girls were similar in height but in 1976 the boys were considerably taller. Twelve year old girls are still taller than boys of the same age, therefore there must be an age at which the girls overtake the boys. This was 11 years of age in 1911 but it was already 10 years by 1950. This age of crossover between the male and female heights provides some indirect evidence for another good example of secular trend. Studies from a number of countries show that the age of menarche, the age at which menstruation first occurs, has decreased by up to 4 years during the past century. Girls appear to be maturing earlier relative to the boys than they were in previous generations.

If the heights of 80 year old women were compared with those of 20 year old women, would this provide evidence of a definitive secular trend in height of adult females? Unfortunately the answer is 'not completely' because there are confounding variables. One effect of ageing is decreased stature, due to two main causes. Structurally, the trunk shortens because the vertebrae tend to collapse as a result of decreased bone density (see Box 5.1). Functionally, stature also decreases because of postural changes including exaggerated curvature of the thoracic region of the vertebral column and flexion at the lower limb joints. Some researchers have calculated that structural and postural changes account for a 4.3 cm decrease in stature, with the extra 3.0 cm difference observed between 20 and 80 year old adult females being attributable to the secular trend.

A disturbing secular trend that should be noted is the gradual decrease in bone density. This trend towards lower bone density has been attributed to lifestyle changes, which include inadequate intake of calcium and less than optimal levels of physical activity.

Body segments do not grow at the same rate

An important concept in growth is the differential growth of tissues and body segments. Total body height, usually termed stature, is the total of lower limb length, trunk length and head height. These three components do not grow at

Table 5.2 Relative increase in length of major body segments from birth to maturity

Body segment/s	**Increase in length**
Head	2 ×
Trunk	3 ×
Upper limbs	4 ×
Lower limbs	5 ×

Source: Information kindly supplied by Professor Don Bailey, College of Physical Education, University of Saskatchewan, Canada.

the same rate. The differential changes in size of different body segments during growth are summarised, by rule of thumb, in Table 5.2.

Differential changes in the size of body segments mean changes in body shape because body shape is defined by proportions. The sitting height to standing height ratio changes from birth to adulthood but the rate of change is not constant, with even the sign (direction) of the ratio changing (Figure 5.12). At 2 years of age the ratio is close to 60 per cent. It then decreases to a minimum near puberty, indicating the relatively rapid growth of the lower limbs, after which it increases slightly because of some 'catch up' growth by the trunk. At the age of 9–10 years the sitting height to standing height ratio is similar to the adult ratio of about 52 per cent.

Sexual dimorphism in the sitting height ratio is also illustrated in Figure 5.12. One explanation is that during the period of rapid growth the lower limbs of boys grow relatively faster than those of girls. During ageing the ratio decreases again because of the shortening of the vertebral column, related to both structural and postural changes mentioned earlier. Very often the trunk length of females decreases at a faster rate than in males, related to the higher incidence of osteoporosis and its effect on vertebral height (see Box 5.1).

Body tissues do not grow at the same rate

The various components of the human body were listed in Table 4.1. It was mentioned earlier in this chapter that the relative amount of

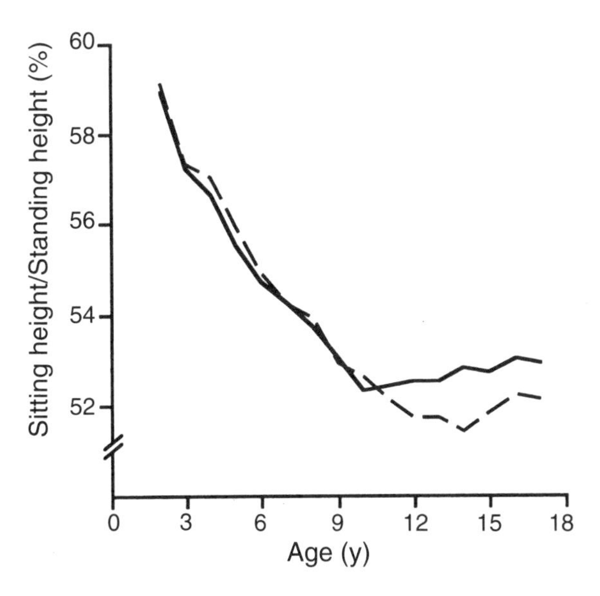

Figure 5.12 Changes in the sitting height–standing height ratio from 2 to 17 years of age for non-Hispanic white American males (broken line) and females (solid line)
Source: Martorell, R., Malina, R.M., Castillo, R.O., Mendoza, F.S. & Pawson, I.G. (1988), Body proportions in three ethnic groups: children and youths 2–17 years in NHANES II and HHANES, *Human Biology*, 60, 205–22. (Copyright 1988 by Wayne State University Press; adapted with permission.)

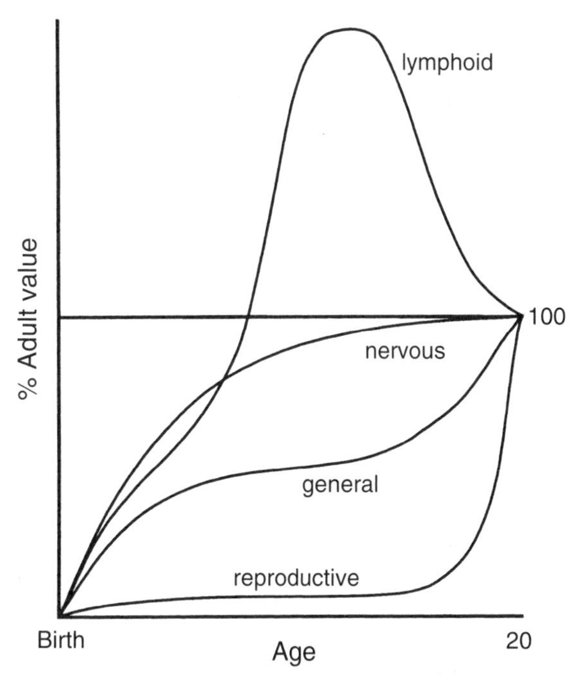

Figure 5.13 Differential growth in nervous, reproductive, and lymphoid tissue compared with the general growth of all tissues

muscle tissue increases during growth and maturation. Consequently, the relative amounts of other tissues must decrease, thus indicating that the various tissues must grow at different rates. The differential growth in some body tissues is illustrated in Figure 5.13. As well as changes in muscle and bone there are changes in major organs such as the brain and liver. In the newborn baby, 13 per cent of the body weight is brain and 5 per cent is liver. In the adult these relative weights have decreased to 2 per cent and 2.5 per cent respectively. Obviously there is a good correlation between the size of the skull and the size of the brain. The protruding abdomen observed during infancy and early childhood is indicative of the relatively large digestive organs, including the liver.

Height growth and changes in body proportions and composition—sexual dimorphism

There is sexual dimorphism in growth indicated by Figures 5.5, 5.6, 5.7 and 5.11, which show that girls are always closer to their adult height than boys. This phenomenon is particularly noticeable during the upper primary and lower secondary school years, when girls are often taller than their male class mates because the female growth spurt usually occurs 2 years earlier than in the male. A rule of thumb is that girls reach half their adult height at about 2 years of age whereas for boys this age is 2.5 years.

There are three main reasons for the tendency of adult males to be taller than adult females. Firstly, because the pubescent growth spurt starts later in males, they are already taller when they enter the period of rapid growth. Secondly, the male growth spurt is more intense, meaning that the peak height velocity is greater. Thirdly, the growth spurt lasts longer in males.

Relative amounts of fat tend to be greater in females at all ages but the major divergence between males and females occurs when girls enter puberty (Figure 5.14). The size of the triceps skinfold between the ages of 1 and 20 years of age exhibit different time courses for males and females. At 20 the skinfold thickness is greater than during infancy in females, but it is less than at 1 year of age in males. The minimum skinfold thickness is at 7 years of age for females, and it then gradually increases, with almost a plateau for about 2 years during puberty. For males there are two minima, at 7 and at 15–16 years of age. The second minimum occurs because there is a decrease in skinfold thickness for about 4 years during puberty. Especially for females, the term 'puppy fat' is probably inappropriate because there is no period during maturation when the skinfold thicknesses are decreasing.

Muscle and bone masses tend to be similar in males and females until about 13 years of age. At this age the hormonal influence on males is such that they have a greater increase in bone and, particularly, muscle mass.

Somatotype changes during growth, maturation, and ageing

A longitudinal study of somatotypes of European males between the ages of 11 and 24 years indicated that there are two stages of puberty. In the first stage between about 11 and 15 years there is a decrease in endomorphy because there is a decrease in skinfold thickness at the same time as there is an increase in height. This increase in height is rapid, so the growth tends to be 'linear', resulting in an increase in ectomorphy. The second stage of puberty is marked by an increase in mesomorphy, which is the result of relative increases in transverse diameter of the skeleton and increases in muscle volume. Ectomorphy decreases between 15 and 18 years of age because of the increased mass of the musculoskeletal system. The general changes in somatotype between 11 and 24 years of age are illustrated in Figure 5.15.

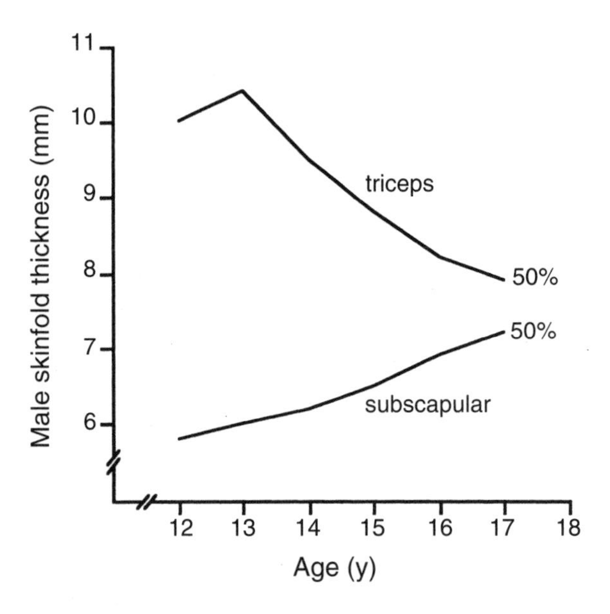

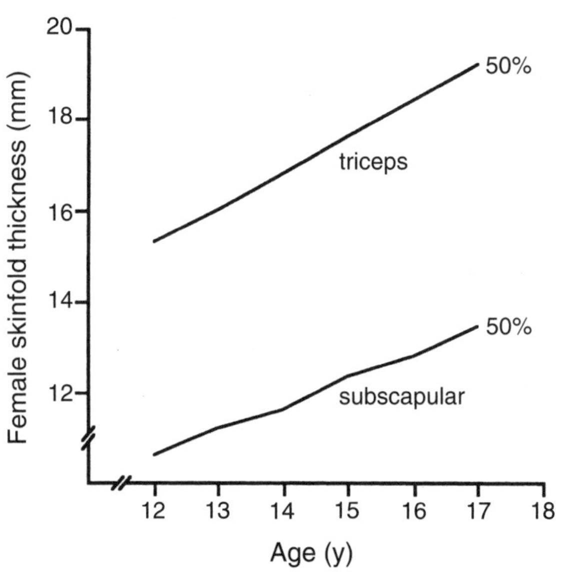

Figure 5.14 Changes in skinfold thickness during growth and maturation for 50th percentile Australian males and females
Source: Court, J.M., Dunlop, M., Reynolds, M., Russell, J., & Griffiths, L. (1976), Growth and development of fat in adolescent school children in Victoria. Part 1. Normal growth values and prevalence of obesity, *Australian Paediatric Journal*, 12, 296–304. (Copyright 1976 by Blackwell Scientific; modified with permission.)

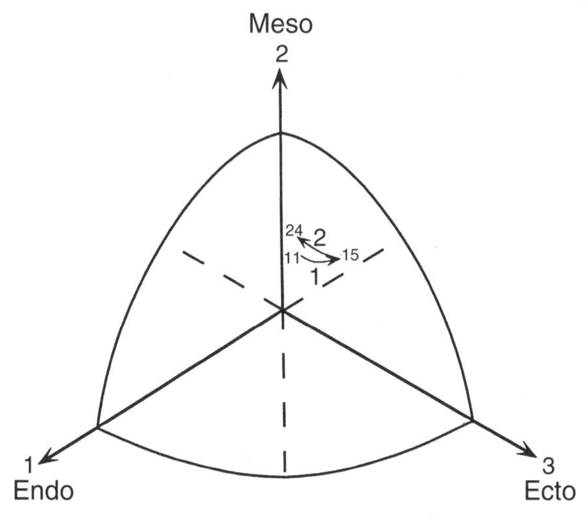

Figure 5.15 Trends in somatotype changes of 14 boys tested between the ages of 11 and 24 indicating two stages (1 & 2) of maturation during this period
Source: Carter, J.E.L & Parizkova, J. (1978), Changes in somatotypes of European males between 17 and 24 years, *American Journal of Physical Anthropology*, 48, 251–254. (Copyright 1978 by John Wiley & Sons; adapted with permission.)

A study of almost 14 000 Canadian adults showed that males over 30 years of age are slightly more endo-mesomorphic than those under 30, and that there are similar differences between women over and under the ages of 40 years (see Figure 4.3).

How is age determined?

Of some interest is the concept of chronological age versus biological age. We all know children who are early maturers and teenagers who are late maturers. A master athlete is very pleased when told by a doctor that he or she has the body of a person 10 years younger. Obviously then, the time since birth is not the only method of representing age.

Examples of maturation include deciduous dentition ('baby teeth') followed by the permanent dentition, the gradual replacement of cartilage by bone, the onset of menarche (beginning of menstruation in girls), and the appearance of secondary sexual characteristics. These changes tend to occur in a similar order but the age of onset of the changes and the period of time required to complete the changes are more variable. Therefore stages in development can be used as a basis for determining biological age during maturation.

In young people, their skeletal or bone age can be calculated by inspection of hand/wrist X-ray films. This site is the most suitable because the appearance of ossification centres and fusion of bony parts take place over many years in a fairly regular pattern and the exposure to X-radiation is less of a hazard than if other parts of the body were X-rayed. Bone age can be determined to the nearest month and then compared with chronological age to determine whether the person is ahead or behind the normal developmental pattern. Clinically, bone age can be used, in conjunction with other measurements, to predict adult height.

Radiographic techniques are not the most suitable to use in large studies because of the ethical issues involved with the exposure to X-radiation. It would be preferable to use less potentially hazardous techniques. An alternative in common use for many years was dental age. We know that there are both 'baby' (deciduous) and 'adult' (permanent) teeth and that the deciduous teeth appear and are lost, and then permanent teeth erupt over many years from infancy to young adulthood. The so-called dental age can therefore be determined but its correlation with skeletal age is relatively poor. Dental age was used over a century ago when the *Factory Act* in Great Britain stated that children under the age of 12 could not work in industry. Because of a lack of adequate birth records, a child was allowed to work when the '12 year old molar' had erupted.

Much of the musculoskeletal development that occurs around puberty is related to hormonal changes associated with sexual maturation. Therefore sexual maturity itself can be determined using stages proposed by Tanner. During the development of secondary sexual characteristics all individuals pass through a common series of body changes, which, of course, are different for males and females. The order of development is relatively constant, but wide variation is exhibited in the age of onset of the first stage of maturation and in the duration of progress through all stages. Genital and associated development is classified to determine which of the five Tanner stages best describes an individual.

Sexual maturity rating has been used in the control of male junior sports in North America because it relates to the musculoskeletal development induced by the increase in levels of testosterone. In a collision sport such as American football, musculoskeletal development is important, and the school administrators in New York State during the early 1980s decided that no high school boy could play football until he had entered stage 3 on the sexual maturity rating. This recommendation was seen as an important sports medicine intervention, protecting boys who were perceived as being most likely to suffer injury during participation in the sport. Because sexual maturity rating involves estimation of genital development it is psychologically invasive and other less embarrassing methods are being developed that it is hoped will be equally valid.

During ageing, such parameters as bone density, range of joint motion, and muscle strength decrease; if these values are more than expected for people of a given age these people could be described as being biologically young for their chronological age. There has been no formalised equation grouping parameters for the elderly as there has been for children.

CHAPTER 6

Musculoskeletal adaptations to training

- Bone is affected by different levels of activity
- Bone is a dynamic living organ that can repair after injury
- Joint structure and range of motion are affected by different levels of activity
- Muscle structure and function are affected by different levels of activity
- Body size, shape and composition can be altered by training
- Lifestyle factors play a large part in determining physique
- Body size and type are related to performance in sports and different events within the one sport

The purpose of this chapter is to examine the changes in the skeletal, articular and muscular systems and in overall body shape, size and composition that occur as an adaptation in response to physical activity.

Bone is affected by different levels of activity

Charles Darwin (1809–1882) recognised the relationship between physical activity and bone mass in his ground-breaking book *The Origin of Species* first published in 1859. In a modern version abridged by the physical anthropologist, Richard Leakey, it is stated in the first chapter, entitled 'Variation under domestication' that:

> ... with animals the increased use or disuse of parts has a marked influence; thus in the domestic duck the bones of the wing weigh less and the bones of the leg more, in proportion to the whole skeleton, than do the same bones in the wild-duck; and this may be safely attributed to the domestic duck flying much less, and walking more, than its wild parents'. (Darwin C. (1979), *The Illustrated Origin of Species*, Book Club Associates, London, p. 50.)

Both human clinical cases and animal experimentation have shown similar adaptability of bone. The major weight-bearing bone in the human leg is the tibia, which is in contact with the femur at the knee. In cases where the tibia is congenitally missing, successful replacement by the fibula has been performed. The increases in size and strength of the fibula in this situation are quite spectacular.

In describing earlier the historical origins of the subdiscipline of functional anatomy, mention was made of the pioneering role of the German anatomists of the 19th century, who demonstrated that the inner architecture of bone reflects the stresses produced by external mechanical forces. Wolff's law was first published in German over a century ago and there have been many translations or interpretations of the law since then. Many later versions ascribe dynamic responses to Wolff although he really only wrote that the directions of trabeculae in spongy bone reflect the main directions of forces acting through the bone. The concept of bone adaptation was introduced by others, who gave much broader meaning to Wolff's law than its original conception.

One version proposed by Sir Arthur Keith in a lecture to the Royal Society in London in 1921 states that:

> Every change in the form and function of a bone or of their function alone, is followed by certain definite changes in their internal architecture, and equally definite secondary alterations in their external conformation, in accordance with mathematical laws. (W.J. Tobin (1955), The internal architecture of the femur and its clinical significance: The upper end, *Journal of Bone and Joint Surgery*, 37A, 57–72).

This translation of the law implies that bones sustain a maximum of stress with a minimum of bone tissue and that bone reorganises to resist forces most economically. The mechanisms of this adaptive process are beyond the scope of this chapter but a number of hypotheses have been investigated and research is still going on in an attempt to find the biological transducer that translates the mechanical message into a biological result. Not all adaptation of bone is positive; under certain circumstances there may be maladaptation (see Box 6.1).

Box 6.1 How does intensive exercise affect the bones of growing animals?

Research workers from the Department of Anatomy at the University of Queensland studied the effects of a 1 month intensive exercise program on the structural and functional properties of 17 pubescent male rats. An equal number of rats that were not specifically trained were used as controls.

For 5 days of each of the 4 weeks of the experiment, the experimental rats ran on a treadmill for 1 h each day and also swam for an equivalent period of time. It was estimated that the rats were working at about 80 per cent of their maximal oxygen consumption. The structure and function of the major lower limb bones were studied. Bone length and width were measured and bone slices were studied under a light microscope. Whole bones were also mounted in a specially manufactured torsion-testing machine to measure the strength of the bones when twisted at a physiological rate.

During the month, the experimental rats ate more than the controls but were lighter at the end of the training period. The long bones of the rat hindlimb were shorter and lighter in the experimental animals. Also the epiphysial plates were thinner. As a result of the intense exercise program the femur was not significantly affected but the tibia exhibited a significant decrease in the amount of energy it could absorb before it fractured.

It was postulated by the authors that the repetitive cyclical loading caused an accumulation of micro-cracks in the bone, which resulted in maladaptation. The proposal that exercise is always beneficial for young animals was therefore questioned by the authors, who indicated that the literature related to stress fractures in humans showed that it was a potential problem.

Source: Forwood, M.R. & Parker, A.W. (1987), Effects of exercise on bone growth: mechanical and physical properties studied in the rat, *Clinical Biomechanics*, 2, 185–90.

There are a number of examples of bone adaptation in response to external forces. It was soon realised after animals and humans were sent into space that there was a rapid reduction in bone mass, probably related to the decreased muscle activity in a zero gravity environment. American and Russian scientists are still trying to develop the optimal exercise regimen to prevent loss of bone in space. Likewise when a limb is immobilised in a plaster cast, evidence of disuse osteopaenia can be observed on X-ray film. Osteopaenia is the mechanism by which bone mass decreases and is due to the imbalance between bone resorption and deposition. As a result, porosity increases and therefore the bone is less dense and appears less radio-opaque on an X-ray film (bone absorbs more X-radiation than any other tissue, except teeth). The bone loss is related to the absence of external forces on the bone, which are produced mainly by muscle activity. It would appear that for bone there is a genetically determined baseline mass of bone that is less than is required for normal functioning. Therefore a certain level of physical activity is necessary. Complete rest will result in rapid adverse responses in the musculoskeletal system.

Generally, lack of physical activity results in a loss of bone mass and with increased activity the opposite occurs. Some of the best examples of the relationship between physical activity and bone mass relate to comparisons made between the humerus of the playing and non-playing arms of professional tennis players. For example, radiological studies have demonstrated that not only was the bone denser in the playing arm, but also that the diameter of the shaft was larger and the compact bone forming the shaft was thicker than the non-playing arm. Recent reanalyses of the original data have shown that the greatest observed increases occurred in individuals who began training at a young age, that is, before puberty. Radiological studies of lower limb and vertebral column bones of runners have shown similar positive effects of chronic activity.

These results have implications for young adults who are advised to optimise their bone density. Cross-sectional studies indicate that adults with active lifestyles have greater bone mass than sedentary adults, and better controlled longitudinal studies indicate that exercise can cause an increase in bone mass. Even in post-menopausal women, exercise can add bone, and exercise is certainly important in maintaining bone mass that would otherwise be lost through ageing. 'Modest' additions of bone can occur in ageing individuals but the biggest contribution of exercise is to reduce the bone loss that would occur if the level of physical activity was inadequate.

The type of exercise is important and this will be discussed further in Chapter 15. Weight-bearing activities, such as walking and running, are associated with the addition of bone but this effect may not be so great in other activities such as swimming and cycling. Elite swimmers appear to have lower bone densities than other elite athletes, and weight lifters have the highest bone density. As indicated in Figure 6.1 the

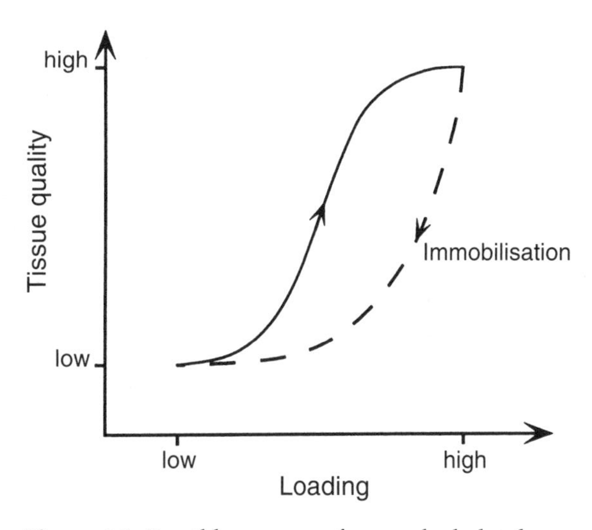

Figure 6.1 Possible courses of musculoskeletal tissue adaptation related to training (solid line) and immobilisation (broken line)
Source: Kannus, P., Jozsa, L., Renstrom, P., Järvinen, M., Kvist, M., Lehto, M., Oja, P. & Vuori, I. (1992) The effects of training, immobilization and remobilization on musculoskeletal tissue. 1. Training and immobilization, *Scandinavian Journal of Medicine and Science in Sports*, 2, 100–18. (Copyright 1992 by Munksgaard International Publishers; modified with permission.)

initial positive changes occur slowly whereas the initial response to decreased activity is rapid, so an exercise program must be regular and must be maintained over a long period of time to be beneficial. The period of a remodelling cycle is about 3 months and about three or four remodelling cycles are required to reach a new steady state in response to increased physical activity.

So far, the positive effects of exercise have been emphasised but the term 'exercise prescription', which is often used, implies that there is an optimal dosage. What is the effect if the optimal dosage is exceeded? Animal studies (see Box 6.1) and clinical evidence from elite junior athletes (see Chapter 5) suggest that intense physical activity can produce maladaptation in growing bones. Stress fractures may occur in young adults, especially when the onset of intense physical activity is rapid as it is among recruits in the armed forces.

These examples emphasise a concept proposed by a number of authors, and summarised in Figure 6.1. From an initial baseline level of activity there is a gradual increase, which results in positive adaptation. During intense activity there may be negative changes, which may lead to injuries such as stress fractures. The following period of immobilisation will result in loss of bone and decreased density, so that the bones are weaker. During rehabilitation, after repair of the injured site, bone will regain its former properties. It is very important for coaches and sports medicine personnel to ensure that this scenario does not become a recurring cycle for an elite athlete because the loss of strength is rapid by comparison with its return. This concept can be applied to other structures of the musculoskeletal system and related to ligament sprains and muscle strains. Figure 6.1 indicates that there is an optimal level

Box 6.2 Does exercise have a positive or negative effect on the bones of adult female runners?

This research was a collaborative effort between the Center for Clinical and Basic Research, Ballerup, Denmark, and the Department of Gynaecology, University of Copenhagen. The aim was to determine the prevalence of menstrual and sex hormonal disturbances in female runners, and to investigate the effects of exercise on bone mass and metabolism. Measures of fitness, gynecologic status, bone mass, and bone metabolism were taken from 205 premenopausal runners who were divided into three groups; 'normally active', 'recreational runners', and 'elite runners'.

Irregular menstruation was most prevalent in the elite runner group, who ran an average of 67 km per week and trained most intensely. Total body bone mineral content (BMC) and regional bone mineral densities (BMD) were measured using dual exposure X-ray absorptiometry (DEXA). Total body BMC and regional BMCs and BMDs were similar in all groups. Lumbar spine BMD was significantly less in the elite runners than in the normally active women; however, this difference did not remain when adjustments were made for differences in age and body mass index between the two groups. Biochemical tests of blood and urine indicated similar bone turnover in the three groups of runners. The amenorrheic runners, on average, exhibited spinal bone densities about 10 percent less than the normally menstruating runners.

The authors concluded that the negative effects of long distance running for women have been overemphasised, except in the rare situation where menstruation ceases for a long period of time.

Source: Hetland, M.L., Haarbo, J., Christiansen, C., & Larsen, T. (1993), Running induces menstrual disturbances but bone mass is unaffected, except in amenorrheic women, The *American Journal of Medicine*, 95, 53–60.

of activity for each person and the challenge for prescribers of exercise is to determine this level on an individual basis.

Bone is a dynamic living organ that can repair after injury

Remember that bone is more like a tree than a piece of wood because it is a living, dynamic, metabolically active tissue with a good blood supply. It undergoes continual remodelling and can therefore recover from damage, unlike most parts of a car, which often have to be replaced when they wear out or are damaged. The major stages during fracture healing are illustrated in Figure 6.2. During a fracture the bone and the tissues surrounding it bleed when the blood vessels are ruptured. Subsequently a blood clot is formed and, later, capillaries invade the clot bringing cells and nutrients. A fibrous callus fills the space between the ends of broken bone and then a bony callus replaces the fibrous callus. Appropriate physical rehabilitation is important

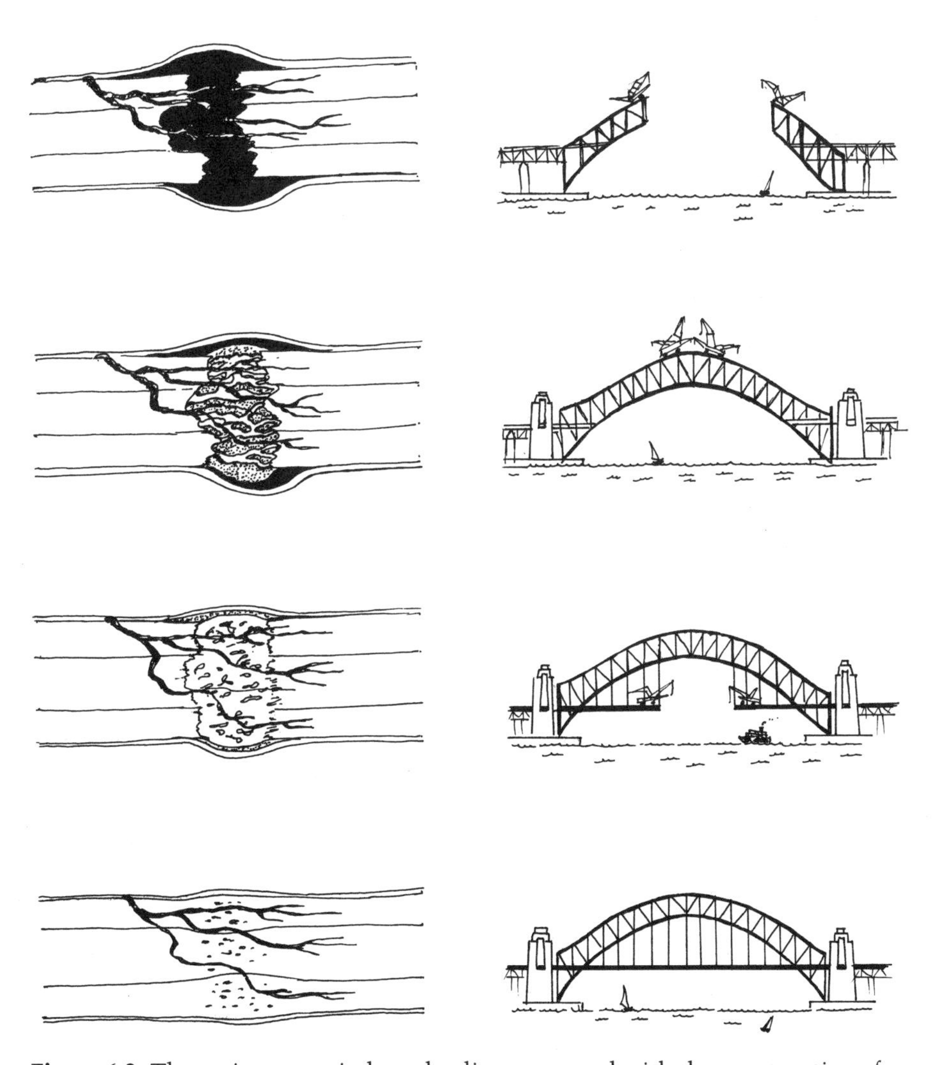

Figure 6.2 The major stages in bone healing compared with the construction of the Sydney Harbour Bridge, which was completed in 1932

because the final strength of the bone will depend on remodelling of the bony callus and this remodelling is affected by physical activity.

Joint structure and range of motion are affected by different levels of activity

Warm-up is usually recommended before any physical activity. Many people believe that a warm-up is performed only for cardiovascular reasons, including increasing the heart rate, but there are also positive effects on the musculoskeletal system. During cyclical exercise, such as jogging, the lower limb joints are continually moving within their range of motion. The short-term effect of this activity on articular cartilage is that it becomes thickened, presumably because it absorbs synovial fluid. Synovial fluid flow into and out of the cartilage improves the supply of nutrients and the removal of waste products of metabolism. The long-term result of chronic exercise is that articular cartilage becomes thickened, except when there are excessive compressive forces on the cartilage, which may occur during repeated downhill running or during heavy workouts involving a lot of jumping.

Indications of degenerative joint disease, or osteoarthrosis, are thinning of the articular cartilage and thickening of the thin layer of compact bone under the articular cartilage. It has been hypothesised that changes in this subchondral bone precede the cartilage thinning. Animal studies have shown that repetitive impulsive loading causes stiffening of the thin endochondral bone so that the cartilage may be more susceptible to damage. The human knee is one of the joints commonly affected by osteoarthrosis and its incidence increases with age. Overall, the incidence is three times greater in females. To these possible genetic and ageing factors may be added environmental factors such as obesity, which is associated with a greatly increased risk of osteoarthrosis.

Many clinicians report that the long-term effects of jogging include osteoarthrosis; however, epidemiological studies that have investigated the relationship between exercise and degeneration of the articular cartilage have found that regular runners do not suffer an increased incidence of osteoarthrosis. Recent studies report that the connection between exercise and later degenerative changes in the articular cartilage is related to previous synovial joint injury such as ligament sprain. Ligament damage may allow abnormal motion within the synovial joint, resulting in excessive localised loading. In a radiological study of almost 400 older physical education teachers, it was found that they exhibited a lower incidence of osteoarthrosis than the general population. The results of this study led the authors to conclude that the increased muscle strength associated with regular physical activity may protect lax joints from developing osteoarthrosis.

Normally the quantity of synovial fluid in a joint is very small. The knee has the largest joint cavity and it contains only about 0.2–0.5 ml of synovial fluid at rest. After a 1 or 2 km run the volume may have increased by up to two or three times. It may be argued that the synovial fluid then can better perform its functions of lubrication and nutrition. Certainly, short-term exercise makes synovial fluid less viscous, and therefore more like oil than grease. It may therefore be a better lubricant, but does it sacrifice this property for the role of protection? The answer is 'yes' but because of the thickening of the articular cartilage, the joint is probably better protected after a warm-up period.

Ligaments are passive stabilising structures of joints and the loads applied to them vary when a joint is moved through its full range of motion. The oscillating increase and decrease in tensile forces on ligaments during physical activity result in adaptations. The size of the ligament increases because of an increase in collagen synthesis so the ligament becomes stronger. There is also an increase in cross-linking within the ligament so it also becomes stiffer. The junctions between the ligament and the bones at either end also adapt to become stronger (see Box 6.3). It has been suggested that endurance exercise has more positive effects on ligament strength than sprint training.

Box 6.3 How does physical rehabilitation affect knee ligaments?

Research workers at the School of Applied Science at the then Tasmanian State Institute of Technology and the Department of Anatomy at the University of Queensland used an animal model to investigate an important biological question related to rehabilitation of the musculoskeletal system.

Five groups of rats, with 10 rats in each group, were studied to investigate the effects of immobilisation, swimming after immobilisation, and swimming training, on the strength of ligaments. Each training session was 1 h and the intensity was gradually increased. Every 4th day was a rest day and the training period was 6 weeks. At the end of the experimental period both the anterior and posterior cruciate ligaments were mechanically tested in an engineering tensile testing machine. Strength and stiffness were measured and the mode of failure of each ligament was microscopically determined. When ligaments fail, their mid-substance may be ruptured, the tear may occur towards one end, or the bone near the attachment site may separate from the surrounding bone (Figure 6.3).

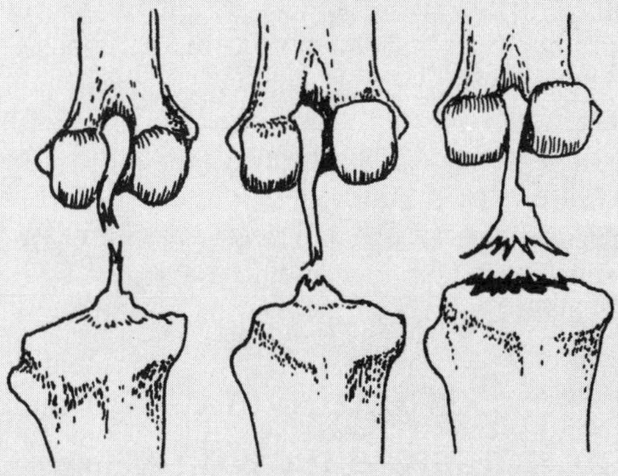

Figure 6.3 Different types of ligament/bone failure produced by a mechanical tension test
Source: Larsen, N.P., Forwood, M.R. & Parker, A.W. (1987), Immobilization and retraining of cruciate ligaments in the rat, *Acta Orthopaedica Scandinavica*, 58, 260–64. (Copyright 1987 by Scandinavian University Press; reprinted with permission.)

Both strength and stiffness of the anterior cruciate ligament were reduced after 4 weeks of cast immobilisation but the strength of the posterior cruciate ligament was not significantly affected. It was proposed that this finding was due to the greater changes in forces applied to the anterior cruciate ligament during normal activity. The 6-week swimming training program was not sufficient to increase ligament strength and stiffness above normal values, but it was sufficient to return the ligaments to their normal values after 4 weeks of immobilisation.

In the immobilised animals, the bone at the ligament insertion sites tended to fail whereas in the remobilised animals the ligament itself tended to fail, indicating that the swimming rehabilitation preferentially improved the mechanical properties of the bone tissue. The authors warn the readers that the required period of remobilisation after ligament surgery is likely to be longer than 6 weeks because of the combined effects of trauma and immobilisation on the ligament.

Source: Larsen, N.P., Forwood, M.R. & Parker, A.W. (1987), Immobilization and retraining of cruciate ligaments in the rat, *Acta Orthopaedica Scandinavica*, 58, 260–64.

Flexibility is a common term used to indicate range of joint motion. It is often listed as one of the factors included in physical fitness, which gives the impression that flexibility is a general characteristic of an individual. In reality, flexibility is very joint specific because it is related to normal activity patterns. One of the best examples concerns the flexibility of ballet dancers. A recent range-of-motion study performed on the hip found that the range of external rotation of ballerinas was much greater than in the normal population; however, their range of internal hip rotation was less than normal. This variability in the ranges of hip motion suggests that their activity patterns, involving externally rotated postures of the hip, produce structural changes that adapt to the forces but then restrict motion in the opposite direction.

A general warm-up, even when combined with massage, seems to have little effect on joint flexibility. Therefore specific joint flexibility exercises must be a separate part of the warm-up routine.

Muscle structure and function are affected by different levels of activity

The terms 'flexibility' and 'joint range of motion' indicate that some of the specific structures of synovial joints, illustrated in Figure 3.6, restrict motion. In most cases the restriction relates to stretched musculotendinous units crossing joints. In fact most flexibility exercises stretch muscle–tendon units and not joint capsules or ligaments. Interestingly, flexibility that relates to muscle–tendon length is perceived as 'good' but when ligaments are stretched the joint may be classified as 'loose', which is perceived as bad because it is associated with increased risk of injury. Joint laxity may in fact require physical therapy or orthopaedic surgery.

Studies on teenage gymnasts and aged persons indicate that when flexibility exercises are performed regularly, the normal decreases in ranges of joint motion can be prevented or even reversed. Certainly stretching exercises performed regularly for a period of weeks, months, or years can increase joint flexibility. When this increase occurs, how does the whole muscle respond to regular static stretching? Remember that whole muscles consist of both muscle and connective tissue. Does the number of sarcomeres in series forming each muscle fibre increase, or does the extensibility and elasticity of the connective tissue component increase? In other words, does a muscle grow longer or become more extensible as a result of flexibility exercises? The answer is probably that increased extensibility of the connective tissue is the predominant factor although sarcomere number will adapt rapidly to any habitual length change.

Weight trainers are often perceived as being 'muscle bound' indicating that their muscle development produces a consequent decrease in the range of joint motion. This general conception appears to be ill-founded. With correct weight training techniques, including flexibility exercises, range of motion can be maintained and even enhanced.

When a weight training program is started there is often a rapid increase in measured strength, followed by a plateau effect, and then further gradual increases. Electromyographic studies have indicated that during the initial 6–8 weeks the muscular activation pattern becomes more efficient so the strength gains are related to improved control of the neuromuscular system. In the next phase there are physical increases in the size of the muscle fibres due to an increase in the number of myofibrils within each fibre. That is, initially there is a phase of neural adaptation (neurotrophic phase) followed by a longer period of muscle hypertrophy (hypertrophic phase) in which the cross-sectional area of each muscle fibre increases. The major effect of weight training is hypertrophy of the muscle fibres—an effect that is discussed further in Chapter 13. It appears that increases in the connective tissue component of whole muscle parallel those of the muscle tissue during strength training programs. Long-term effects of endurance exercise are related mainly to changes within the muscle fibre to make the muscle less fatigable, but there is an increase in the relative amount of connective tissue within the whole muscle.

Exercise has a positive effect on tendon but, in common with other connective tissue structures, the rate of adaptation is much slower than that of muscle. In tendon there is an increase in collagen synthesis and the collagen fibres line up more regularly in a longitudinal direction. In common with other tissues, training that is too strenuous may be injurious (Figure 6.1). When muscle strains occur they often occur at the junction between the muscle and its tendon.

Body size, shape, and composition can be altered by training

It was emphasised in both Chapters 4 and 5 that a person's somatotype is more of a phenotype than a genotype, implying that the somatotype can be modified by factors such as training. Although there is a certain amount of genetic control of the amount of body fat and its distribution, a combination of increased activity and reduced energy intake via the diet will decrease the fat content of the body. Following such a regimen means that a person will become less endomorphic. All studies that have compared athletes with the sedentary population have found that athletes have less body fat and a larger lean body mass, partly because of an increased skeletal mass. Thus, exercise may not alter a person's ectomorphy rating because the decrease in endomorphy may be balanced by an increase in mesomorphy. In other words, the skinfolds will decrease but the muscular circumferences of the arm and leg will increase.

Lifestyle factors play a large part in determining physique

Certain athletes such as body builders, weight lifters and ballet dancers appear to have unusual body shapes that relate more to their adaptation to training rather than to differences that were determined before birth. When measurements of weight lifters and ballet dancers are compared with those of the normal population, interesting similarities and differences emerge (Figure 6.4). Bone size measurements, such as circumference of the wrist, knee, and ankle, are relatively normal. Remember that the bone in the shafts of weight lifters is likely to be denser and heavier (as reported earlier) but these parameters are not measured in normal anthropometric protocols. Width of the expanded ends of long bones is less susceptible to training. Notice particularly that the muscular circumferences, such as biceps, are relatively large in the weight lifters, as a result of muscle hypertrophy, and relatively small in the ballet dancers, related to the demands of their training including physical activity and dietary intake.

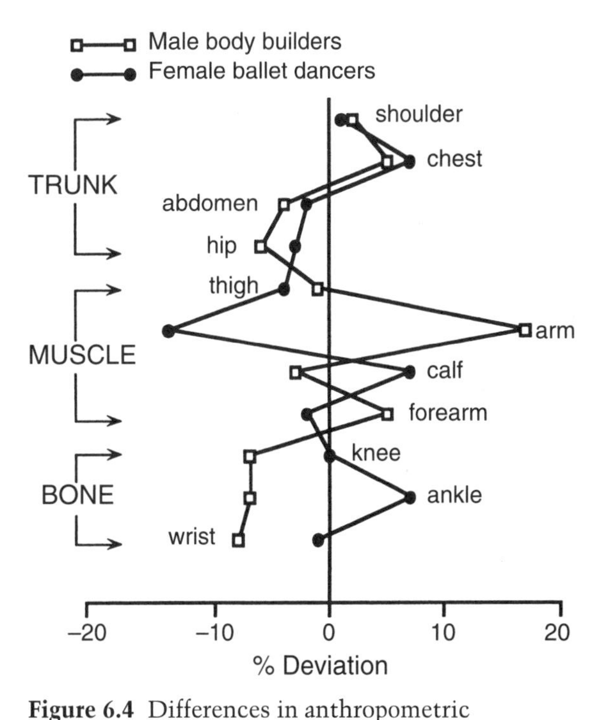

Figure 6.4 Differences in anthropometric measurements between male body builders and female ballet dancers and their respective reference populations. The numbers on the *X*-axis represent the percentage deviation from the respective reference male and female, represented by the solid vertical line
Source: Katch, V.L., Katch, F.L., Moffatt, R. & Gittleson, M. (1980), Muscular development and lean body weight in body builders and weight lifters, *Medicine and Science in Sports and Exercise*, 12, 340–44; and Dolgener, F.A., Spasoff, T.C. & St John, W.E. (1980), Body build and body composition of high ability female dancers, *Research Quarterly for Exercise and Sport*, 51, 599–607.

Training can affect physical and physiological factors at any time of life and this aspect is a major focus of Chapter 14.

Body size and type are related to performance in sports and different events within the one sport

One good example of kinanthropometry is provided by the following quote: 'Javelin throwers and gymnasts have practically identical somatotypes, although the javelin throwers, at 179.5 cm and 76.7 kg, are much bigger than the gymnasts, at 167.4 cm and 67.1 kg.' (M. Hebbelinck & W.D. Ross (1974), Kinanthropometry and biomechanics, p. 545, in R.C. Nelson & C.A. Morehouse (eds), *Biomechanics IV*, Baltimore University Park Press) After reading the next section of the book on biomechanics you might like to explain these similarities and differences. One important aspect is the large difference in the size measurements, such as height and weight. Height is predominantly genetically determined and is not greatly affected by environmental factors under normal conditions. Therefore sports performance is affected by genetically determined factors as well as training and the physical adaptations to training. The physical dimensions that cannot be changed are often used in talent identification, a topic explored more fully in Part 3 of this book.

In a single sport, different events may require mainly either strength or endurance. This continum is seen in running events, where the 100 m sprint and the marathon are at two ends of a strength–endurance continuum. In running, the athlete's own body weight must be transported and runners tend to be lighter than the average person; they get progressively lighter as the distance of the event increases. Sprint and middle-distance runners tend to be taller than normal but long-distance runners tend to be shorter than the average person.

Events such as the shotput require even greater strength than sprinting, and height is also an advantage because it means that the shot can be released from a greater height (and hence travel further for any particular velocity and angle of release). Shotputters and discus throwers tend to be taller and heavier than other track and field athletes. Body weight is not such a limiting factor for these athletes because they do not have to transport their own bodies over a long distance, or lift their centres of mass very far, so they tend to be very heavy. Also a large lean body mass would be advantageous for development of the power required in explosive movements.

Somatotypes also change as the requirements for strength and endurance change. When there is a high strength component, such as in the shotput and discus, the athletes are very mesomorphic and they tend to move from this position on the somatochart downwards and to the right as the endurance component increases (Figure 6.5; see Box 6.4).

Box 6.4 Are there physical differences between players in different sports?

Kinanthropometric studies were performed by a research group in the School of Education at the Flinders University of South Australia led by Bob Withers. The Heath–Carter anthropometric somatotype method (see Chapter 4) was used to characterise 206 national standard sportsmen in 17 sports and 127 female representative squad members in 10 sports. Endomorphy was corrected for height. Age, height and mass were reported separately.

The male athletes were described generally as being balanced mesomorphs. Their mesomorphy ratings were above '4' and the ratings for endomorphy and ectomorphy were both below '4'. When compared with an age and sex matched group of untrained Canadian males, the South Australian athletes were taller, lighter, less endomorphic, more mesomorphic, and more ectomorphic. The overall mean for relative body fat of the athletes was estimated to be 10 per cent.

The most mesomorphic sportsmen were the powerlifters and the least mesomorphic were the long-distance runners. The field lacrosse players were more endomorphic than the Australian Rules football players and the gymnasts. The gymnasts were more mesomorphic than the track and field athletes when considered as a single group. The Rugby Union footballers and swimmers differed on all three components of the somatotype. The footballers were more endomorphic, more mesomorphic and less ectomorphic. The tallest sportsmen were the basketballers and rowers and the shortest groups were the powerlifters and the gymnasts. The heaviest group comprised the Rugby Union footballers and the lightest group comprised the gymnasts.

The authors compared their results with previous studies and provided a good kinanthropometric explanation of the differences between groups. The differences confirm the relationship between somatotype and relative requirements for strength and endurance (see Figures 6.5 and 6.6).

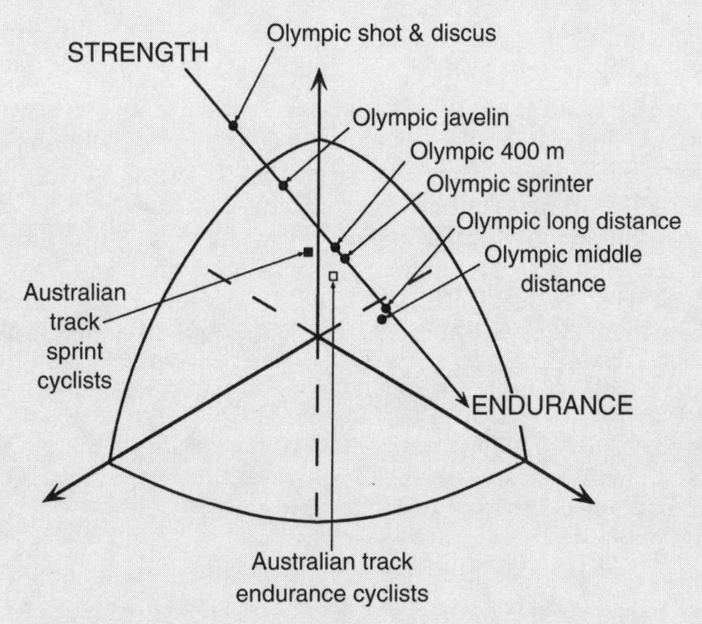

Figure 6.5 Somatochart illustrating the somatotypes of some Olympic track and field athletes and elite Australian cyclists. Notice the strength/endurance continuum
Source: Data from Carter, J.E.L., Aubry, S.P. & Sleet, D.A. (1982), Somatotypes of Montreal Olympic athletes in J.E.L. Carter (ed.), *Physical Structure of Olympic Athletes. Part I. The Montreal Olympic Games Anthropological Project. Medicine and Sport*, vol. 16. S Karger, Basel, 53–80; and McLean, B.D. & Parker, A.W., (1989), An anthropometric analysis of elite Australian track cyclists, *Journal of Sports Sciences*, 7, 247–55.

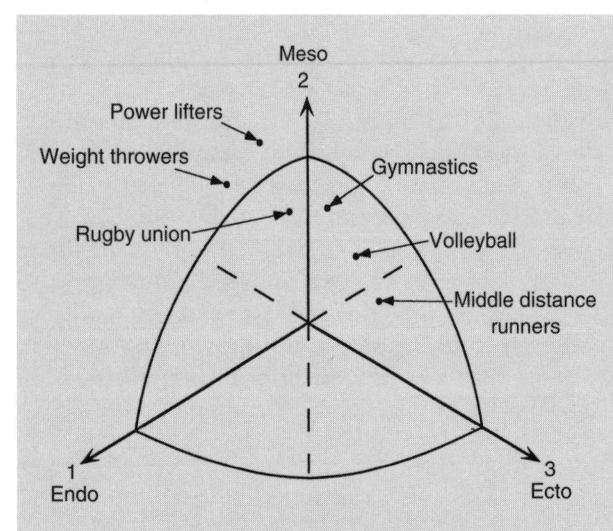

Figure 6.6 Somatochart illustrating the somatotypes of selected elite male South Australian sports people
Source: Withers, R.T., Craig, N.P. & Norton, K.I., (1986), Somatotypes of South Australian male athletes, *Human Biology*, 58, 337–56. (Copyright 1986 by Wayne State University Press; adapted with permission.)

Compared with a reference sample of 135 women, the female athletes were less endomorphic, slightly more mesomorphic, and more ectomorphic. Other studies have found a tendency for endomorphy to decrease as the level of physical activity increases. The basketballers and netballers were taller than those playing soccer, lacrosse, cricket, hockey, and softball and the softballers were lighter than the basketballers. There was a tendency for the netballers, squash players and volleyballers to be less endomorphic and more ectomorphic than those playing badminton, basketball, cricket, hockey, lacrosse, soccer, and softball.

The authors emphasise that there are many factors other than body size and shape that determine successful sports performance. This caveat is particularly important in the case of field games, which require a range of skills, especially perception–action coupling, as well as various components of fitness.

Sources: Withers, R.T., Craig, N.P. & Norton, K.I. (1986), Somatotypes of South Australian male athletes, *Human Biology*, 58, 337–56; Withers, R.T., Whittingham, N.O., Norton, K.I. & Dutton, M. (1987), Somatotypes of South Australian female games players, *Human Biology*, 59, 575–84.

Further reading

Section 1: Anatomical bases of human movement: The subdiscipline of functional anatomy

Singer, C. (1957), *A Short History of Anatomy and Physiology from the Greeks to Harvey (The Evolution of Anatomy)* (2nd edn), Dover, New York.

Chapter 3: Basic concepts of the musculoskeletal system

Van de Graaff, K.M. (1992), *Human Anatomy* (3rd edn), Wm. C. Brown, Dubuque, IA.
Nordin, M. & Frankel, V.H. (1989), *Basic Biomechanics of the Musculoskeletal System* (2nd edn), Lea & Febiger, Philadelphia.
Basmajian, J.V. & de Luca, C.J. (1985), *Muscles Alive: Their Functions Revealed by Electromyography* (5th edn), Williams & Wilkins, Baltimore.

Chapter 4: Basic concepts of anthropometry

Lohman, T.G., Roche, A.F. & Martorell, R. (eds) (1988), *Anthropometric Standardisation Reference Manual*, Human Kinetics, Champaign, IL.
Norton, K.I. & Olds, T.S. (eds) (1994), *Anthropometry and Anthropometric Profiling*, Nolds Sports Scientific, Sydney.

Chapter 5: Musculoskeletal changes throughout the lifespan

Tanner, J.M. (1989), *Foetus into Man: Physical Growth from Conception to Maturity* (2nd edn), Castlemead, Ware, England.
Shephard, R.J. (1987), *Physical Activity and Aging* (2nd edn), Aspen, Rockville, ML.

Chapter 6: Musculoskeletal adaptations to training

Bloomfield, J., Ackland, T.R. & Elliott, B.C. (1994), *Applied Anatomy and Biomechanics in Sport*, Blackwell, Melbourne.

Section 2
Mechanical bases of human movement: The subdiscipline of biomechanics

Introduction

What is biomechanics?

Biomechanics is the study of biological phenomena (processes, function and structure) using the methods of mechanics. Mechanics, the traditional domain of engineers and physicists, was developed to allow people to describe the motion of machines and the movement of the planets and stars. These same processes and theories of motion are equally applicable to living material. Biological phenomena include, at one end of the spectrum, the way in which human and other species move their limbs and bodies in space through to cellular processes such as the transport of nutrients across cell membranes.

Typical questions posed and problems addressed

The principles that underpin biomechanics are those of Newtonian mechanics and mathematics. To study biomechanics, a sound understanding of mechanics, and how to apply these principles to human movement, is necessary. Secondly, a thorough grounding in anatomy is necessary because the biological tissues (bones, ligaments, tendons and muscles) comprise the media of interest. Biomechanists, using either physical or mathematical methods, try to model these tissues and then study how these tissues change across the lifespan and in response to loading, such as that provided by physical activity.

Some of the fundamental questions that drive the scientists who are interested in biomechanics include the following:

- What are the load limits of various structures of the human body?
- How do we measure the loads on the body?
- How can performance be optimised?
- How do body segments interact?
- Through what mechanics is co-ordinated movement produced?
- How efficient is the human body when it moves (e.g. running, cycling) and can this efficiency be altered?

This list is but a sample of key issues in the sub-discipline of biomechanics and by no means exhaustive.

Levels of analysis

Biomechanics research studies phenomena in the full range from the macro level (such as the study of whole-body movement or comparisons among different animal species and limb movements) through to the micro level (such as osmotic forces across cell membranes, propagation of nerve and muscle action potentials, and transportation of nutrients and waste products in the body). The field of biomechanics is thus very diverse and includes the study of people in not only sporting situations, but also in work environments (occupational biomechanics), and rehabilitative settings. For example, some biomechanists study the muscle function and gait of patients suffering from cerebral palsy whereas others examine the design of prostheses that realistically simulate natural motion of the human body.

Movement of fluids in the body, particularly the functioning of the heart and blood flow in the arteries, is another area of study and requires considerable mathematical and hydrodynamic expertise. People who are interested in this area of biomechanics are also interested in developing suitable heart and valve replacements.

Technique analysis in sport and work environments has been a very fruitful area of research of particular interest to students of human movement and has led to guidelines for safe lifting, optimised movement sequences and development of theories on how limb movements are controlled. Optimisation of criteria such as minimal muscle involvement, minimum energy expenditure or maximum efficiency and maximum performance has provided the impetus for much research.

Historical perspectives

Biomechanics has a history that can be traced back to the days of the Greek philosophers. Aristotle, who has been honoured by the title of 'Father of Kinesiology', was the first person to describe the actions of muscles and subject them to geometrical analysis. Archimedes has a principle named in his honour, signifying his contributions to hydrostatics, the study of bodies at rest in fluids, and was the first person to write about the centre of gravity of the human and the laws of levers. Following a dearth of scientific progress over the Dark Ages, da Vinci emerged and provided experiments and theories on motion, structure of the human body, levers, and balance. Galileo further developed the notions of scientific experimentation and also discovered the constancy of the acceleration of gravity. He also showed how the size of bones needs to be scaled in order to withstand the force of gravity.

Borelli (Box II.1), praised by some as the 'Father of Modern Biomechanics', conducted experiments on electrical stimulation of muscles that were well ahead of his time. In the same era, Newton wrote his *Philosophiae Naturalis Principia Mathematica*, and while it would be untrue to state that Newton was directly interested in biomechanics, the influence of his theories of motion and calculus are so far reaching that they impinge on virtually all aspects of biomechanics (Box II.2).

Many other well-known scientists made contributions over the next two centuries, including Young (on eye lenses and sight), Poiseuille (on blood flow), Fick (on joints of the body) and Fenn (on muscle mechanics) to name but a few. These scientists developed the platforms on which modern day biomechanics is based and it is only relatively recently that many of the theories proposed by these pioneers have been able to be tested empirically. As is so often the case, the scientific thought preceded the technology required to test the theory.

Biomechanics has seen a remarkable growth over the past 20 years, particularly with the development of fast but inexpensive digital computers. The computer has allowed biomechanists to study complex movements that, in the early days, could not be studied. The improvement in computing power has also permitted new questions to be asked. For example, early research involved merely describing planar motion whereas, today, measuring and estimating the forces that control movement and the consequences of loads on the body are possible.

Where is the field heading? This is a difficult question to answer. There has been a continued emphasis on development of tools and devices to measure motion or forces. Although this focus has been necessary and important, there is evidence now that the emphasis is changing, that new and important questions are being investigated. That is, biomechanists are becoming less concerned with the means of collecting data and more concerned with questions relating to

Box II.1 Giovanni Alfonso Borelli (1608–1679)

Born in Naples in 1608, Borelli came from humble beginnings. Very little is known of his early days and childhood except that he studied mathematics in Rome. He was obviously very talented in this field and was invited to the chair (similar to a professorial position in an Australian university) of mathematics at the University of Messina. As his reputation grew, he was asked to take up a position at the University of Pisa, a university famous for its great scientists including Galileo. Borelli in fact was a student of Galileo and it was this association that allowed him to develop his mathematical skills further and, more importantly, learn about physics.

Many of Borelli's friends and colleagues were anatomists and physiologists and, through their contact, he applied many of the techniques of physics and mathematics to the living human body. His most famous book *De Motu Animalium* contained many examples of the application of physics and physical principles to the study of animal movement. Borelli has been titled the 'Father of Kinesiology' and is honoured by the American Society of Biomechanics with a memorial lecture (the Borelli lecture) presented at each meeting of the society by an eminent scientist who has made a substantial contribution to biomechanics.

In *De Motu Animalium* Borelli treats two particular aspects of animal dynamics, the movement of the body segments under the action of the forces produced by the muscles, and that field of endeavour that has now become known as muscle mechanics (the study of how a muscle contracts and produces force). In a chapter of his book, he also discusses flexion and extension of joints and how the muscle forces produce these movements and how the combined motion of a series of joints allows the human to walk, run and move in other ways. He also discusses the resistive forces that we experience, namely air resistance and the drag exerted on us when we move in water.

Apparently Borelli, while an intellectual giant, was a very jealous man who despised the success of others. His less than friendly personality led to many fights with colleagues and 5 years before his death, under the threat of arrest for political conspiracy, he fled to Rome. It was here he met Christina of Sweden who protected him and eventually paid for the publication of *De Motu Animalium* after Borelli's death.

how loads alter the structure and function of the body and how these forces can be altered to optimise movement. This positions biomechanics as a subdiscipline with its own contributions to knowledge about human movement rather than simply providing a set of tools or methods that can be used by scientists in the other subdisciplines to describe and measure movement.

Professional organisations and training

For those people wishing to specialise in biomechanics, a strong mathematics/physics/ engineering background is important. If the methods used by physicists and engineers are well understood and a sound understanding of the anatomy and physiology of the system of interest is developed, an individual can expect to enjoy and do well in biomechanics. Computing has become an integral part of most biomechanics research and so an interest in computers can complement the theoretical aspects.

Many professional organisations exist for biomechanics including the International Society of Biomechanics and its affiliated societies in Canada, the United States, Europe, Japan, and some of the new Eastern European countries. These societies hold annual and biannual meetings at which current research is

Box II.2 Sir Isaac Newton (1642–1727)

Sir Isaac Newton was one of the most brilliant scientists in history, making contributions in mechanics, mathematics and optics. His father died when he was a baby and he was raised by his grandmother after his mother remarried when he was 2 years old. He had an extremely tumultuous childhood and it is believed that this unstable time may have led to the psychotic tendencies that were prevalent in his adult life. Apparently he was very anxious when his work was published and became extremely violent and irrational when his ideas were challenged. Thankfully, most of his theories were correct!

Newton was a brilliant mathematician and when studying at Cambridge he voraciously read through most of the published science literature. It was here that he developed the binomial theorem, the basic concepts of calculus and deduced that the force on a planet due to the sun must vary inversely with the squared distance between the sun and the planet. About 20 years later he extended this idea, which became known as his universal law of gravitation.

He was appointed a professor at Cambridge 4 years after graduating from an undergraduate program and in this position he became interested in optics. At this time he developed the first reflecting telescope and was encouraged to present his ideas on other topics in optics. These notions were challenged by Robert Hooke and many bitter disputes followed. Newton chose to withdraw as a result of these disagreements and rather turned his attention to other matters.

After much encouragement from Edmond Halley (the astronomer who has a comet named after him in recognition of his work) Newton formalised his ideas on mechanics. In 1687 he published *Principia Mathematica*, which contained his three laws of motion, the universal law of gravitation, and of course the calculus necessary for these laws. His laws of motion are still used today in physics and mechanics and constitute one of the foundations of modern science. They are used and studied by all biomechanists.

Newton was not active scientifically over his final years, being driven instead by political and theological issues and disputes over scientific credit. His greatest dispute was with the German, Leibnitz, over credit for the development of the calculus. This long standing battle is believed to have precipitated his death in 1727. Posthumously it was agreed that Newton developed calculus prior to Leibnitz but since it was not published for many years, Leibnitz developed it independently of Newton.

presented both orally and via posters. These meetings allow biomechanists to assemble and discuss problems and possible solutions. The most recent developments in communications have been electronic with the introduction of World Wide Web sites, establishment of file transfer protocol (FTP) sites and the continued improvement of discussions on the electronic bulletin board for biomechanics discussion, the BIOMCH-L listserver.

Specialist biomechanists work in numerous environments including tertiary institutions, academies or institutes of sport, government departments (as ergonomists), hospitals (as bioengineers), sporting goods manufacturers such as NIKE, Adidas and Puma and research science institutes, such as CSIRO. Some find work in private industry offering specialist consulting in work place accidents and injuries and in sports technique modification.

CHAPTER 7

Basic concepts of kinetics

- Forces
- Bodies
- Inertia
- Torques
- Couples
- Equilibrium
- Summary

Mechanics is often subdivided into two broad categories: kinematics and kinetics. Kinematics is the science that concerns the description of motion and attempts to answer such questions as 'How far did an object move?', 'How fast did it travel?', 'Where did it go?', and 'How rapidly did it accelerate?'. Kinetics, on the other hand, investigates the causes of motion and involves the study and measurement of the underlying forces and torques that cause motion. Thus kinematics tends to be descriptive whereas kinetics is concerned with mechanisms. In this chapter we examine basic concepts of kinetics related to forces, bodies, inertia, torques, couples and equilibrium. Chapter 8 focuses on basic concepts of kinematics.

Forces

Introductory comments

The word 'force' is often used by lay people to describe the action of pushing or coercing others to act. For example, parents might *force* their children to do their homework or a driver was *forced* off the road by an oncoming motor car. The concept of force in physics or mechanics is not very different from this notion and a force is often described as a *push* or a *pull*. Thus, when walking, forces are applied to the ground; in tennis, forces are applied to the ball by the racquet to change the ball's motion; a dropped object falls under the action of the gravitational force; a rowing shell is propelled as a result of the net forces applied by the rowers and the resistance encountered through the water.

In some situations, the effects of forces are more obvious than others. When a golf ball is struck by a club its velocity is nearly instantly changed from zero to approximately 70 m/s and it is clear that such a change occurs as a result of a very large force. In other situations the visualisation of forces and their effects is not so apparent. Without prior knowledge, understanding and imagination it is difficult to believe that there are forces being applied to stationary objects. For example, a computer resting on a desk is stationary because balanced forces are being applied to it. The gravitational force pushes it toward the ground whereas the desk pushes back with an equal force, preventing it from moving. Only if there was an imbalance of forces would the computer move, for example if it was bumped!

An even more difficult situation to conceptualise is the rare case in which an object is in motion with constant velocity but the net force acting is zero. This situation occurs when a bicycle is rolling with constant speed down a slight incline. The frictional forces applied at the road surface, at the bearing surfaces and by the fluid or air resistance are exactly equal to the component of the weight force acting down the incline, causing the machine to move.

Forces are sometimes referred to as loads and one of their effects is to cause motion. Newton was the first person to formally describe the relationship between force and motion and these relationships are covered in following chapters. A common mistake made by novices to the fields of physics, mechanics and biomechanics is to think that if one can observe motion of an object then it is experiencing a net force. In elementary projectile problems it is assumed that, once released, an object moves with constant horizontal velocity and therefore does not experience a force in that direction. In other words, although the object experiences a net force in the vertical direction, due to gravitational attraction, the horizontal speed remains constant.

Whenever a force acts and causes motion, the change in motion (the acceleration) will be in the direction of the applied net force. That is, when a ball is kicked, it accelerates in the direction of the net force that the foot and all other external forces applied to it. In other words, when a force is applied to a body and a change in the state of its motion occurs, the body will experience a linear (straight-line) acceleration in the direction of the applied force.

In many situations biomechanists use simplifying models in which parts of the human body are assumed to be rigid. At other times the human body or an object being propelled by an athlete can be represented by a single particle. This situation is the simplest case from a dynamics perspective. There are many other circumstances when these two assumptions (rigid body or particle) are invalid. When loads are applied to soft tissues (deformable bodies), the body will stretch and change shape in accordance with its particular compliance (yielding) properties and the applied force(s). For example, when a gymnast lands after performing a gymnastic manoeuvre on to a foam mat, the mat does not experience a uniform acceleration; it deforms under the influence of the loads applied by the gymnast's feet. Thus forces can cause acceleration of a particle and/or they may cause a body to deform.

One of the most critical and interesting aspects in biomechanics is the study of the ways in which loads cause adaptation of biological tissues. This process has been observed in many situations, some of which have been described in the preceding chapters. When people engage in strength training their muscles become stronger and larger than when they began training. Similarly, when a person breaks a bone and is forced to wear a cast while the bone repairs, the muscles atrophy (get weaker and smaller) when they are not required to bear loads. Whereas both positive changes (hypertrophy) and atrophy are rapid and easily observed in muscle tissue, similar adaptations occur in other biological tissues. Without regular weight-bearing activity, bone loses its strength (demineralisation). Similarly, tendons and ligaments also lose strength if loads are not applied to them and they increase their mechanical properties (strength, size, elasticity) if they are subjected to systematic loads.

To summarise, forces can be described as pushes or pulls or, more generally, interactions between two objects. Forces can cause a body to accelerate and/or change shape (deform) and the manifestation of applying loads to biological tissues is that such loads cause the tissues to adapt. If appropriate levels of loading and recovery are provided, the adaptation is positive; that is, the tissues become stronger and typically larger than before training started. If the loads are beyond the limits of the tissue or are applied excessively with insufficient recovery time, the adaptation of biological tissue will be negative. That is, it will lose its mechanical properties and become susceptible to failure.

Characterising a force

Like many quantities in mechanics, force is represented as a vector because two characteristics need to be described in order to ascertain the influence of a force. These two quantities are known as the *magnitude* and the *direction* of the force. If several forces act at any one time, the resultant or net force is the load that will cause the object to move or deform. The resultant force is the vector sum of all the individual forces that act. The unit of measurement of force in the Système International (SI) units is the newton (N). The newton is a relatively small unit of measurement; typical forces experienced on the under surface of the foot during running are of the order of 2000 N. Unfortunately in common language a person's weight is measured in kilograms whereas in physics and biomechanics (as well as all other sciences) weight is a force and is measured in newtons. There is a simple relationship between mass and weight: multiply the mass in kilograms by the gravitational acceleration (9.8 m/s^2, approximately 10). Thus, if one person has a mass of 55 kg then his or her weight would be about 550 N.

There are three fundamental forces of nature: the gravitational, electromagnetic strong and electroweak forces. Almost without exception, the forces that are of concern to biomechanists are the gravitational and the electromagnetic strong forces. Gravitational force emerges as a consequence of the attraction that two objects have for one another and comes about merely because they possess mass. Whenever an object is propelled into the air, including the common situation when people jump up, they are drawn back to Earth by gravitational force. The gravitational attraction toward the Earth is quite strong because of the tremendous mass of the Earth (approximately 6×10^{24} kg) compared with the mass of a human (70 kg).

The large-scale manifestations of the electromagnetic strong force, the force that holds atoms and molecules together, result in what are often called contact or reaction forces. As the name implies, these forces arise when two objects are in contact. Thus when attempting to identify the forces that are acting on a system (for example a person lifting a box or throwing a ball), an excellent rule of thumb is to note that wherever the system (person) is in contact with another object, person or the ground, there will be a contact or reaction force. During walking there are reaction forces at both feet but only when they are in contact with the ground. Similarly, if a person is using a walking cane, then there are reaction forces at both the feet as well as where the cane and hand are in contact. When analysing human movement other non-contact forces (with the exception of gravity) can usually be ignored.

To fully describe a force, three characteristics need to be defined:

- the magnitude of the force (for example 1500 N)
- the line of action of the force (for example along the horizontal, vertically or 30° above the horizontal)
- the point of application of the force (for example at the ankle joint, at the centre of percussion or 5 cm from the joint centre).

Figure 7.1 illustrates this concept for a number of sporting and rehabilitation situations.

In most situations involving human movement, many forces are applied. Even for seemingly simple motion, such as the movement of a ball in flight, a number of forces are applied. These are shown in Figure 7.2. To describe the motion of the ball, we need to know the magnitude, direction and point of application of the *net* or *resultant* force applied. A knowledge of vector addition is necessary to complete this task. Two methods will be described: a graphical

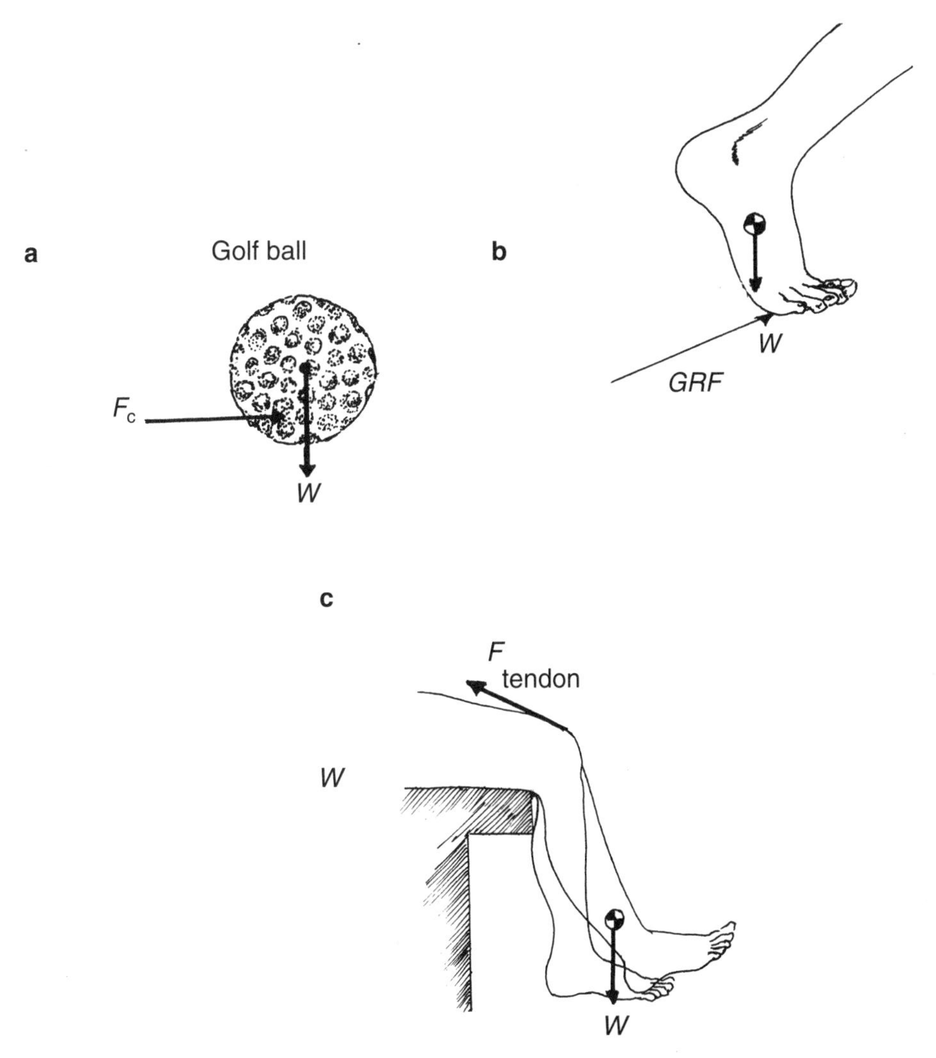

Figure 7.1 Illustrations of forces applied to objects. A golf ball (a) experiences the contact force of the club (F_c) and its weight (W), whereas the foot of a runner (b) has the ground reaction force (GRF) and its weight (W) applied to it. In the last example (c), the quadriceps muscles apply a force (F_{tendon}) to the leg but the weight force (W) provides resistance.

method, which is particularly useful to gain a conceptual understanding of the task, and a mathematical method based on determining the orthogonal components of a vector (that is, those force components acting at right-angles to each other). In the examples described, the forces are *concurrent*, which means that they are all applied at the same point.

Consider the discus shown in Figure 7.3*a* for which a diagram representing the forces applied to it is provided in Figure 7.3*b*. Two forces are applied to it, the lift force, ***L***, which acts in the vertical direction and the drag force, ***D***, which acts in the horizontal direction. What is the *resultant* force applied? To answer this question, the individual force vectors must be added. To add a second vector to the first one, a parallel vector (the same size and direction as the second vector) is drawn emanating from the tip of the first vector. The resultant vector (***R***) is then

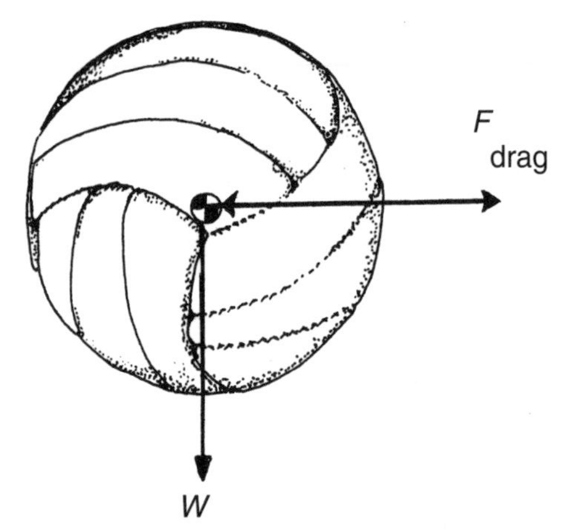

Figure 7.2 The soccer ball in flight experiences two forces, the air resistance (F_{drag}) and the weight (W).

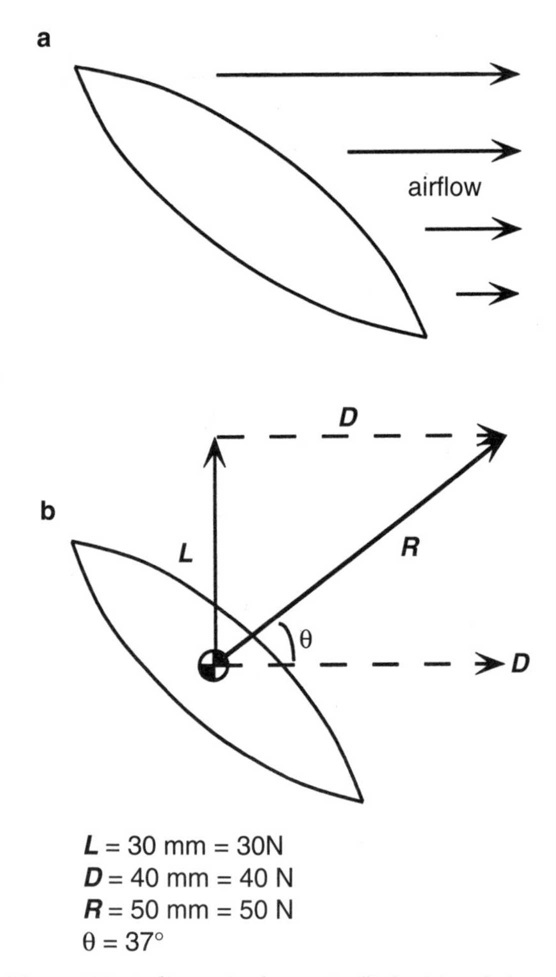

Figure 7.3 A discus is shown in flight (a) and the forces due to the flow of air around it are illustrated in (b). ***L*** is the lift force that resists the weight force whereas the drag force (***D***) slows the discus down. The sum of these two forces (***R***) is known as the air resistance.

drawn from the tail of the first one to the tip of the second one. If the two vectors are initially drawn to scale (for example 1 mm = 100 N) then the resultant force can be determined by measuring the length of ***R***. The direction of the resultant can also be obtained directly from such a scale diagram by using a protractor to determine the angle that ***R*** makes with the horizontal. Thus, in the example illustrated in Figure 7.3*b*:

> ***D*** = 40 N, ***L*** = 30 N and ***R*** = 50 N
> at an angle of 37° above the horizontal

This technique can be extended to include more than two forces. An example is shown in Figure 7.4 where four forces are applied and the resultant force is determined. In this example, the problem can be thought of as a series of steps by which two vectors ($\boldsymbol{F}_1$ and $\boldsymbol{F}_2$) are first added to form $\boldsymbol{F}_5$ (Figure 7.4*b*), which is in turn added to $\boldsymbol{F}_3$ to form $\boldsymbol{F}_6$ (Figure 7.4*c*). Finally, $\boldsymbol{F}_4$ is added to $\boldsymbol{F}_6$ to form the resultant force ***R*** (Figure 7.4*d*).

A faster graphical method than the one described above is to complete the polygon of forces by systematically joining all vectors tip to tail. When redrawing the vectors that are to be added, they must remain exactly the same size and be parallel to their original direction. The resultant force is the vector running from the tail of the first vector to the tip of the last vector in the chain, thus completing the polygon (Figure 7.4*e*). Once again, the magnitude and direction of the resultant force is determined by measuring its length and orientation relative to the horizontal.

To use the mathematical technique to add forces, it is necessary to introduce a concept known as resolution of a vector. To illustrate the process of resolving a vector into components, consider the force vector, ***F***, shown in Figure 7.5. If the two vectors, $\boldsymbol{F}_x$ and $\boldsymbol{F}_y$, which are

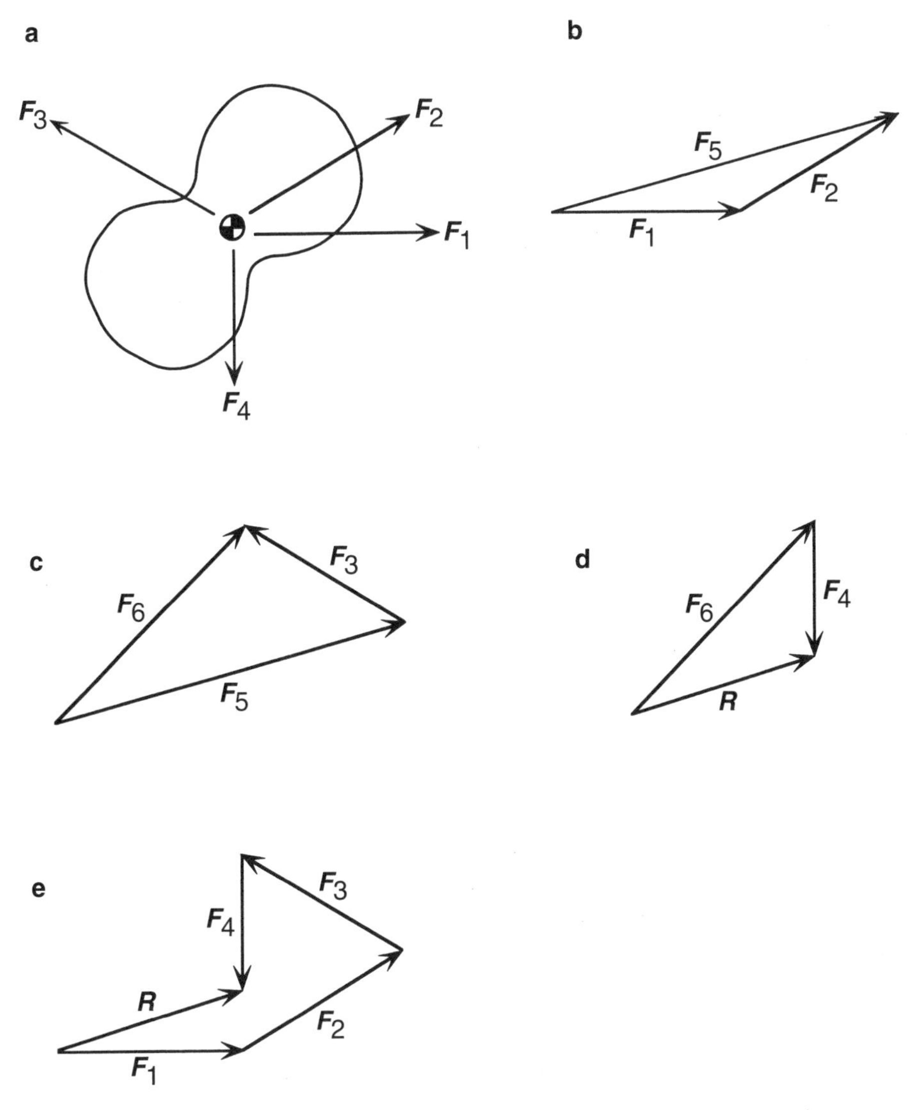

Figure 7.4 If four forces (F_1–F_4) are applied to a body (a) the net force can be determined by successively adding pairs of forces (b)–(d). Alternatively, the polygon of forces can be used to obtain the same result (e).

at right-angles to each other, are added, the resultant vector is $\boldsymbol{F}$. Thus $\boldsymbol{F}$ is said to have components $\boldsymbol{F}_x$ and $\boldsymbol{F}_y$. By using simple trigonometry, it is easy to demonstrate that these components have the following magnitude:

$$\cos\theta = \frac{\boldsymbol{F}_x}{\boldsymbol{F}};\ \boldsymbol{F}_x = \boldsymbol{F}\cos\theta$$
$$\sin\theta = \frac{\boldsymbol{F}_y}{\boldsymbol{F}};\ \boldsymbol{F}_y = \boldsymbol{F}\sin\theta \quad (1)$$

In these expressions the magnitudes of the vectors are used in the algebraic expressions. The directions of the components are along the axes of an XY coordinate system. By calculating the magnitude of, in this example, $\boldsymbol{F}_x$ and $\boldsymbol{F}_y$ the initial resultant vector has been resolved into its x and y components.

Adding vectors that are collinear or parallel is as simple as adding numbers. That is, the resultant vector of two collinear or parallel

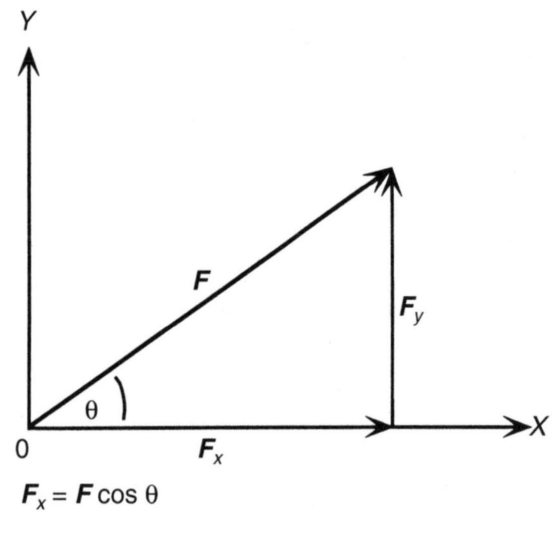

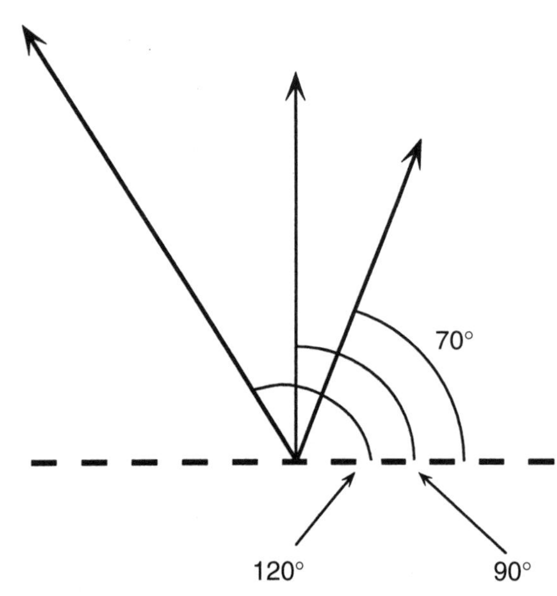

Figure 7.5 A force (F) is resolved into its horizontal (F_x) and vertical (F_y) components.

vectors is merely their algebraic sum. Thus to find the magnitude of the resultant for five (or more) forces acting on a particle, each of them can be resolved into its components and the parallel components simply added. Once the total forces in both directions (x and y) are known, the resultant force can be determined by calculating the vector sum of the two components. On a point of convention, vectors that point to the right or upwards are considered to have a positive magnitude whereas those that point to the left or downwards are said to be negative. Thus, if a vector was pointing down and to the left (toward the third quadrant), it would have negative vertical and horizontal components whereas a vector pointing to the second quadrant (up and to the left) would have a positive vertical and a negative horizontal component.

In summary, to add two vectors by adding their components you follow these steps:

1. Resolve the vectors into orthogonal components (usually along the axes of an XY coordinate system).
2. Algebraically add the components in each of the directions.
3. Use the theorem of Pythagoras to determine the resultant vector, that is, use the relationship:

$$c = \sqrt{(a^2 + b^2)} \qquad (2)$$

4. Calculate the inverse tangent of the ratio of the components to find the angle of the resultant relative to the horizontal.

These four steps can be summarised by the following two equations:

$$R = \sqrt{\left(\sum F_x\right)^2 + \left(\sum F_y\right)^2}$$
$$\theta = \tan^{-1}\left(\frac{\sum F_y}{\sum F_x}\right) \qquad (3)$$

The summation sign (Σ) in equation (3) is simply a shorthand way of designating addition. For example, suppose there are five forces (F_1 ... F_5) all of which have components in the X direction (F_{1x} ... F_{5x}). To determine the magnitude of the resultant force in this direction, we merely add all the components together. That is:

$$\sum F_x = F_{1x} + F_{2x} + F_{3x} + F_{4x} + F_{5x} \qquad (4)$$

Another example may help to illustrate these concepts. In Figure 7.6 a gymnast is pictured doing an iron cross. In order to perform this activity successfully, the gymnast needs very strong deltoid (shoulder) muscles since they hold the body in position. The forces produced by the three heads of the deltoid muscle (Figure 7.6*b*) are represented by three vectors, F_1, F_2 and F_3, acting at the angles indicated relative to the horizontal. The task is to find the magnitude and direction of the resultant force, R, in the deltoid muscle. In the diagram, each vector has a magnitude and direction indicated. For instance, F_1 = 120 N and is directed 120° above the horizontal. To solve this problem mathematically, it is necessary firstly to resolve each of the forces into *x* and *y* components, and then to use equation (3) to find the net force. That is:

$$\begin{aligned}\sum F_x &= F_{1x} + F_{2x} + F_{3x} \\ &= 120 \cos 70 + 150 \cos 90 + 210 \cos 120 \\ &= -64\ N\end{aligned} \quad (5)$$

$$\begin{aligned}\sum F_y &= F_{1y} + F_{2y} + F_{3y} \\ &= 120 \sin 70 + 150 \sin 90 + 210 \sin 120 \\ &= -445\ N\end{aligned} \quad (6)$$

The resultant force (R) and its direction relative to the horizontal is therefore:

$$\begin{aligned}R &= \sqrt{\left(\sum F_x\right)^2 + \left(\sum F_y\right)^2} \\ &= \sqrt{-64^2 + 445^2} \\ &= 450\ N\end{aligned}$$

and (7)

$$\begin{aligned}\theta &= \tan^{-1}\left(\frac{\sum F_y}{\sum F_x}\right) \\ &= \tan^{-1}\left(\frac{-64}{445}\right) \\ &= 98^\circ \textit{ above the horizontal}\end{aligned}$$

Thus the net force in the deltoid acts mostly superiorly (or upwardly) and slightly posteriorly (or backwards).

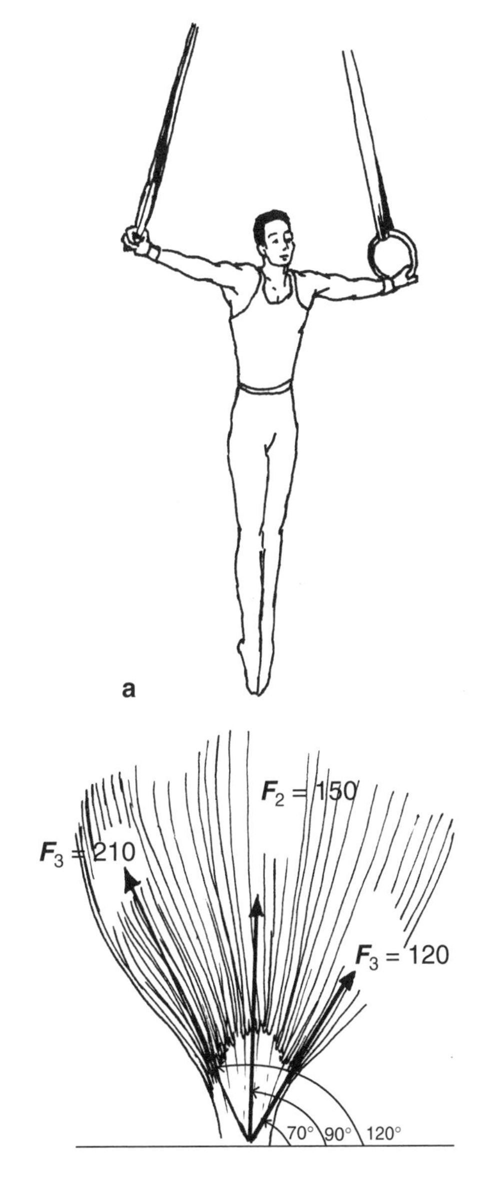

Figure 7.6 To perform an iron cross (a), the gymnast needs strong deltoid muscles. The force generated by each head of the deltoid is represented by a force vector oriented to the horizontal (b).

Types of load

It is often useful to classify forces according to the effects they have on a body. One such scheme organises loads into two categories, *impulsive* and *static*. Impulsive loads are forces that are applied for only a very brief period of time. Examples of impulsive loads include a tennis ball being struck with a racquet, a billiard cue striking a billiard ball, karate players or boxers experiencing punches, a basketball bouncing off the floor. In many sporting applications, these forces are applied for extremely short periods of time. For example, the impact time for a golf club with a golf ball is of the order of 0.4 ms (4/10 000th of 1 s) and during this short time the force changes in magnitude rapidly. In running, the force experienced during the initial contact of the heel with the ground (heel strike) has been described as an impact peak and has the same characteristics as an impulsive force being applied and removed in an extremely short time period. This impact peak is only of duration 5–40 ms (less than 1/20th of a second).

When a force is applied slowly (the forces that are consciously produced by skeletal muscles) or continuously without causing motion (such as the reaction forces applied by the body when maintaining a hand-stand position), the force is often assumed to be static in nature. When skeletal muscles contract and motion results the load is often referred to as active. On the continuum from impulsive to static, active loads are closer to the static pole than the impulsive end.

Finally, when describing the forces created by skeletal muscles, three types of contraction or force are observable: *concentric*, *eccentric* and *isometric* muscle contractions. These classifications are based on the tension developed in the muscle and the length changes it experiences. In concentric contractions, the muscle is contracting and shortening. That is, the muscle is being given electrical (nervous) signals to shorten in length and that is exactly what happens. In isometric (*iso* = one, *metric* = length) contractions, the force produced in the muscle does not cause the muscle length to shorten. The muscle does not shorten in length because other external loads balance out the tension in the muscle. In eccentric contractions, neural signals are transmitted to make the muscle fibres shorten but because the effect of the external load(s) is greater than the effort produced by the force developed in the muscle, the muscle lengthens.

This classification scheme should not be confused with one that was developed in classical mechanics. In this scheme, forces are categorised on the basis of their line of action. If the line of action passes through the body's centre of gravity (balance point) the load is known as a centric one. If, however, the line of action of the load does not pass through the centre of mass, the force is denoted as an eccentric force. Note that while the word eccentric is used in both systems, one refers to a type of muscle contraction whereas the other refers to a type of force.

Force development in muscles

There are many factors that influence the force that a muscle can produce including the number and ratios between the different types of muscle fibres, the state of training, fatigue, and the sources of available energy. However, from the biomechanist's perspective, two relationships are most important. These are the length–tension relationship and the force–velocity relationship.

The characteristic length–tension curve for skeletal muscle is shown in Figure 7.7. This diagram shows that the overall length–tension relationship is the sum of two relationships: the

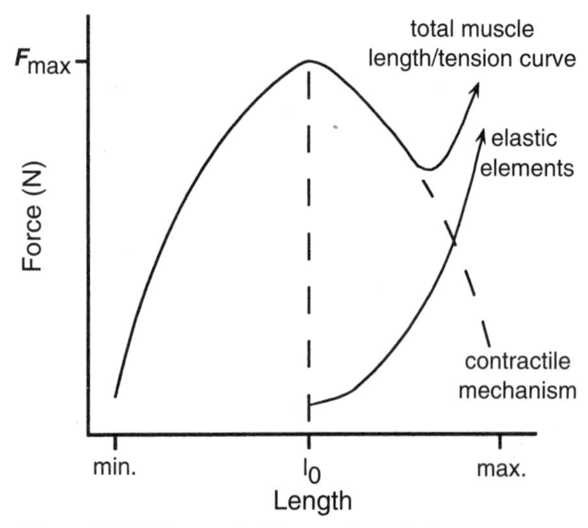

Figure 7.7 Schematic illustration of the force–length relationship of muscle

length–tension curves for both the contractile mechanism and the elastic, connective tissues. The contractile element is capable of producing greatest force at the mid range of motion (l_0) and has an inverted U-shape. At lengths shorter than resting length, it is believed that the cross-bridges between the actin and myosin filaments interfere with each other, while at lengths longer than l_0, there are insufficient bonding sites and therefore the muscle cannot produce tension. In contrast, the elastic elements, which can be modelled as a stiff rubber band, play no role in the development of force until the muscle's length is longer than l_0. However, at lengths greater than l_0, the elastic fibres stretch and produce tension in the muscle. The total muscle length–tension curve is the sum of the two functions.

The second major relationship in a muscle's force-producing potential, the force–velocity relationship, is illustrated in Figure 7.8. The midline of the graph, when $\boldsymbol{v}$, the velocity of shortening or lengthening, is zero corresponds to an isometric contraction. Values to the right of this vertical line relate to concentric contractions in which the muscle length decreases during contraction. Values to the left of the midline correspond to eccentric contractions, the situation in which the muscle is being lengthened by an external load as the muscle tries to shorten. From this graph it is clear that the greatest tension that can be produced by a muscle occurs when it is acting eccentrically and lengthening quickly. Isometric contractions produce the next highest force output and concentric contractions result in the least amount of tension available.

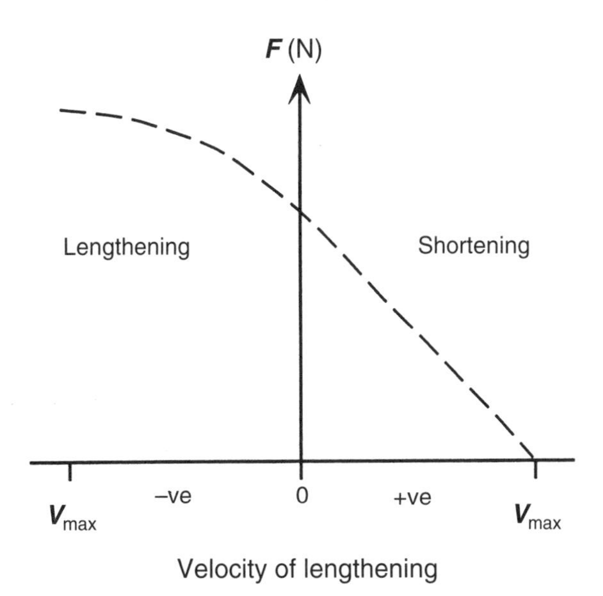

Figure 7.8 The characteristic force–velocity relationship of skeletal muscle. In the lengthening region the muscle is contracting eccentrically whereas in the shortening area the muscle is concentrically contracting. At a velocity of zero, there are no length changes and thus the muscle is acting isometrically.

Bodies

Biomechanists are concerned with the action of forces on different types of systems or bodies. The simplest of these systems, the *particle*, is a body of negligible dimensions, such that the total mass of the body is contained at this one point. In many instances in biomechanics, the whole human body is modelled as a single point located at the centre of mass of the body. This type of representation is useful when trying to simplify the study of motion of the body and can be satisfactorily used in applications such as vertical jumping.

A very popular current practice involves modelling of the human body as a system of interconnected *rigid* links. These links can be thought of as rods that have the same mass as the various body parts. For example, Figure 7.9 shows how various body segments such as the arm, forearm, leg, thigh, foot and so forth can be represented by a rigid body of appropriate dimensions and inertia. A body is considered to be rigid when the displacements of its elements, the particles that make up the body, are small compared with the motion of the body itself. For instance, when forces are applied to the thigh, the displacements of the bone elements and to a lesser extent the muscle relative to the bone are negligible in comparison with the overall motion of the thigh. In running, the knee may move through a distance of nearly 1 m under the action of various muscle forces but the length changes in the femur are so small that they can be ignored.

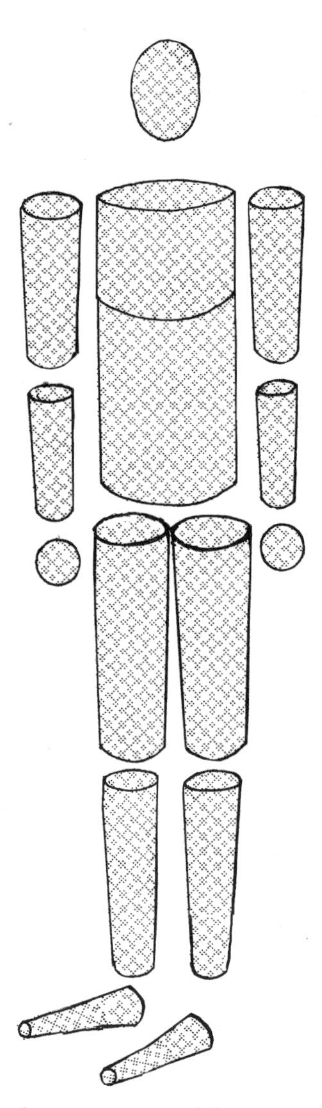

Figure 7.9 The Hanavan model has 15 segments of mathematically defined shapes such as spheres, cylinders and frustra of cones.
Source: Based on Hanavan, E.P. (1984), *A mathematical model of the human body*, MS Thesis, Air University, USAF.

Inertia

In the early days of physics, Kepler believed that the inertia of a body was the force necessary to bring it to rest. The term inertia is derived from the Latin word meaning 'inactive' or 'sluggish'. These days, however, it is known that the inertia of the body is its resistance to a change in state of motion. Newton's first law, the law of inertia, in modern-day language states that:

> *A body will move with constant velocity unless compelled to change by a net force*

That is, if a body is at rest it will tend to stay at rest. For example, everyday experience shows that an inanimate object will not change position unless a force is applied to it. Similarly, once an object is moving it continues to move with constant velocity unless it is forced to change its direction or speed. Such change can only come about if a force is applied to the object. In linear motion, the inertia of a body is the amount of matter it possesses—the mass of the object. The larger the mass, the greater the inertia and vice-versa. Many resistance (strength) training devices work on this principle—to increase the resistance to motion (inertia) a greater mass is added to the device. For example, additional plates are added to a barbell in order to increase the resistance provided by the barbell.

In angular motion, a body's resistance to rotation is dependent on two factors, the mass of the object and how this mass is distributed relative to the axis about which rotation is occurring. A good example of this phenomenon is illustrated in Figure 7.10. When a baseball bat is held well down the handle (it is 'choked up') (7.10*a*) it is easier to swing (rotate) than when it is held near the top of the handle (7.10*b*). In this example, the mass of the bat is unchanged yet its resistance to rotation is changed! Careful examination of this example indicates that in Figure 7.10*b* there are more particles of mass located at a greater distance from the axis of rotation (the hands) than in (Figure 7.10*a*). Thus, the object is difficult to rotate when its mass is a large distance from the axis of rotation. Formally, a body's resistance to angular motion is known as its *moment of inertia* (***I***) and is defined as:

$$\boldsymbol{I} = \sum_{i=1}^{N} m_i d_i^2 \qquad (8)$$

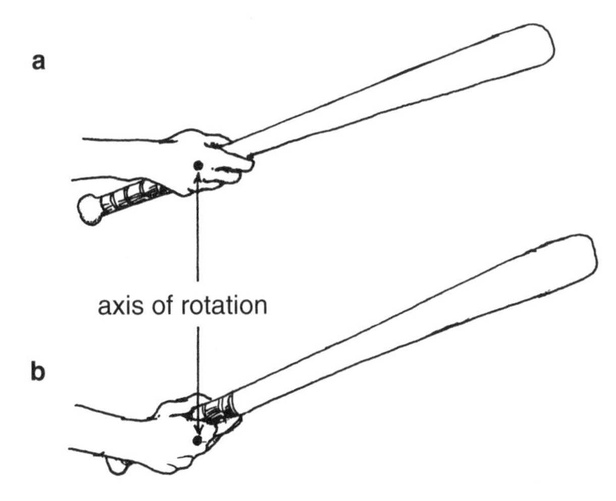

Figure 7.10 When a baseball bat is 'choked up' (a) the moment of inertia is smaller than when it is held near the butt (b).

where m_i is the mass of each particle in the body
d_i is the distance of the ith mass particle from the axis of rotation
N is the number of individual mass particles.

Whereas the inertia of a body is a fixed quantity defined as its mass, the moment of inertia of a body can vary depending on the location of the axis of rotation. Figure 7.11 shows the moment of inertia for a gymnast in various body configurations. Examination of the definition given for moment of inertia (equation (8)) indicates that the distance term (d) is critical because it is a squared term in the relationship. Thus the moment of inertia is highly dependent on how the mass is distributed, and a small change in d

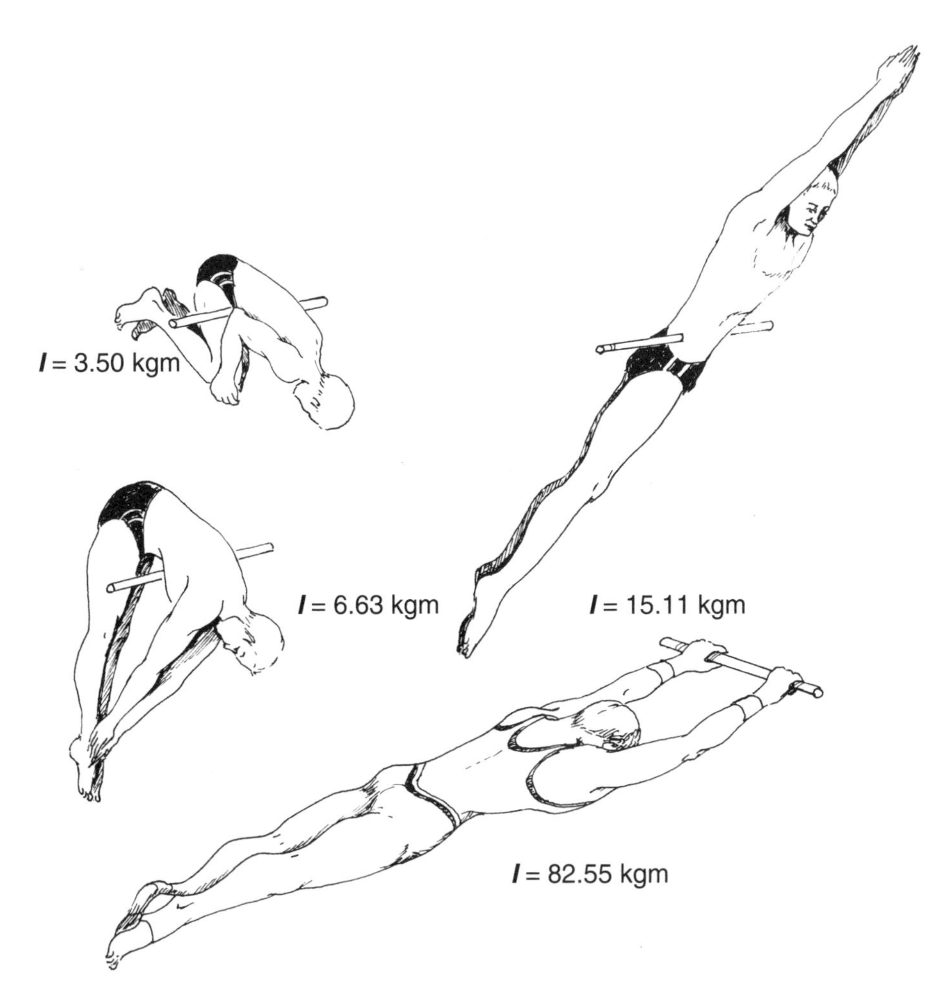

Figure 7.11 The moment of inertia of the body varies by a factor of over 20 depending on the axis of rotation and how the mass of the body is oriented. *Source*: Hay, J.G. (1978), *The Biomechanics of Sports Techniques*, Prentice–Hall, New York, p. 147. (Copyright 1978 by Prentice–Hall; redrawn with permission.)

will create a large change in I. Thus if athletes or patients wish to minimise their resistance to angular motion, they must move the mass close to the axis of rotation. This process is used during the swing phase of locomotion, that is, the time during which the non-support limb is brought forward ready to accept the weight of the body during the next step. At low speeds (walking) the hips and knees do not flex very much but at sprint speeds, both the hip and knee flex considerably to bring the mass of the entire lower limb as close as possible to the hip joint, and thereby reduce its moment of inertia. Reducing the moment of inertia in this way allows the recovery to occur rapidly and with minimum energy expenditure.

An excellent example of the effect that altering the moment of inertia of the body has on the motion of the body is provided by an ice skater performing an axial spin. When the skater starts spinning, the non-support leg and the arms are often extended away from the longitudinal axis of the body. In this configuration, the skater spins relatively slowly. However, as the arms are brought into the chest and the non-supporting limb is drawn as close as possible to the supporting limb, the speed of spin increases remarkably. This phenomenon comes about because the mass is drawn close to the axis of rotation and therefore the moment of inertia is reduced considerably.

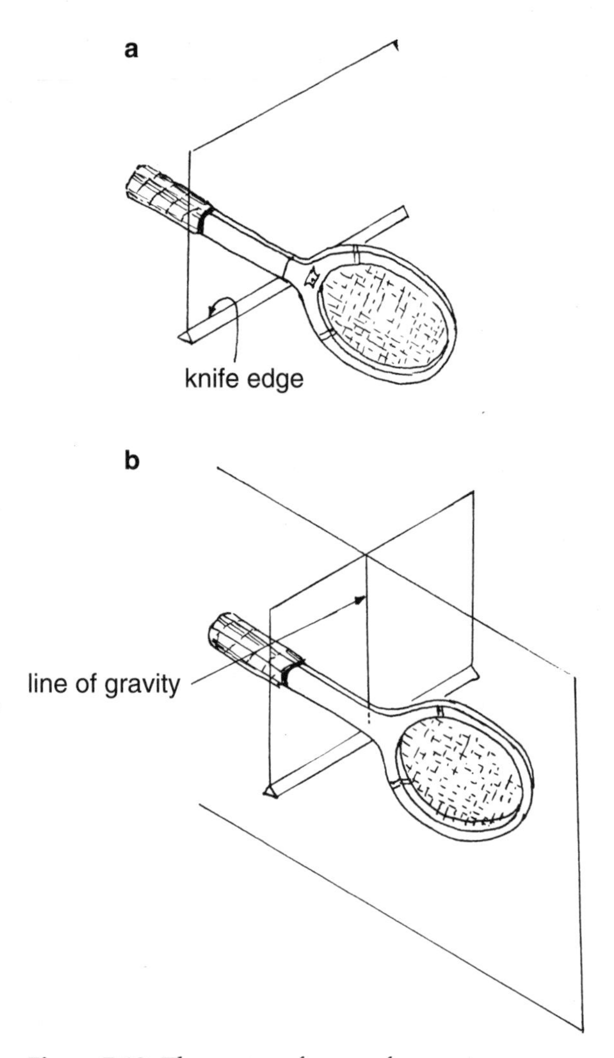

Figure 7.12 The centre of mass of a tennis racquet is found using the balance method. One plane is found first (a) and then the object's orientation is altered to find another plane (b). The intersection of these two planes defines the line of gravity.

Centre of mass

The *centre of mass* or the *centre of gravity* is a theoretical point at which the total mass of a body appears to be concentrated. In most diagrams used to represent bodies in motion, the letter G is used to designate the centre of gravity of the body. In many problems in biomechanics, the motion of the centre of mass of the system being studied is needed to fully define the motion of the body. Although we all know intuitively that the mass of an object is not located at a single point, experience has also shown us that at times it can appear to act at one point. Take, for example, the tennis racquet, pictured in Figure 7.12, balanced on a knife edge. To be able to balance the racquet, the upward force provided by the knife edge must be equal in magnitude but opposite in direction to the weight force (downward force due to gravity) and the two forces must act along the same line. Thus, since the racquet does not move, the centre of mass must lie in the plane containing the knife edge. If the racquet is balanced in a number of different orientations (a minimum of three is required to obtain a unique solution) the exact location of

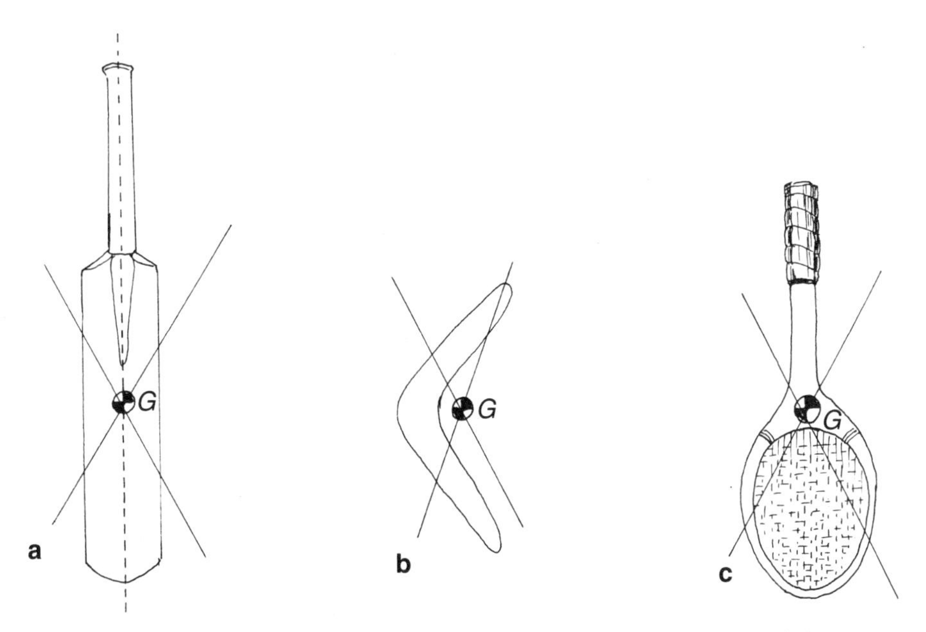

Figure 7.13 The suspension method can be used to find the location of the centre of gravity of a cricket bat (a), a boomerang (b) and a tennis racquet (c).
Source: Hay, J.G. (1978), *The Biomechanics of Sports Techniques*, Prentice–Hall, New York, p. 123. (Copyright 1978 by Prentice-Hall, re-drawn with permission.)

the centre of gravity can be determined by locating the intersection point of three planes. This process is known as the balance method for determining the location of the centre of mass.

The suspension method also can be used to locate the centre of mass. This technique is shown in Figure 7.13 where the centres of mass of a boomerang, a cricket bat, and a tennis racquet are being determined. By suspending the object from a number of points and determining the point of intersection of the lines of gravity passing through each of the suspension points, the exact location of the centre of gravity can be determined. From this diagram, it can be seen that the centre of mass may lie within or outside the object. High jumpers use this result to their advantage by arching their bodies over the bar. Using such a technique allows them to clear the bar with their body but not necessarily require that their centre of mass clear the bar. In fact, the very best jumpers' centres of mass often pass beneath or through the bar when they clear the height. An example of a jumper using a Fosbury flop style is illustrated in Figure 7.14, showing that the centre of gravity is below the height of the bar.

For rigid bodies such as human body segments, a cricket bat or a golf club, the location of the centre of mass is fixed. The position of the centre of mass of the entire human body is not fixed and is dependent on the positions of the arms and legs in relation to the trunk.

Figure 7.14 The high jumper pictured clears the bar even though her centre of gravity passes below the bar!

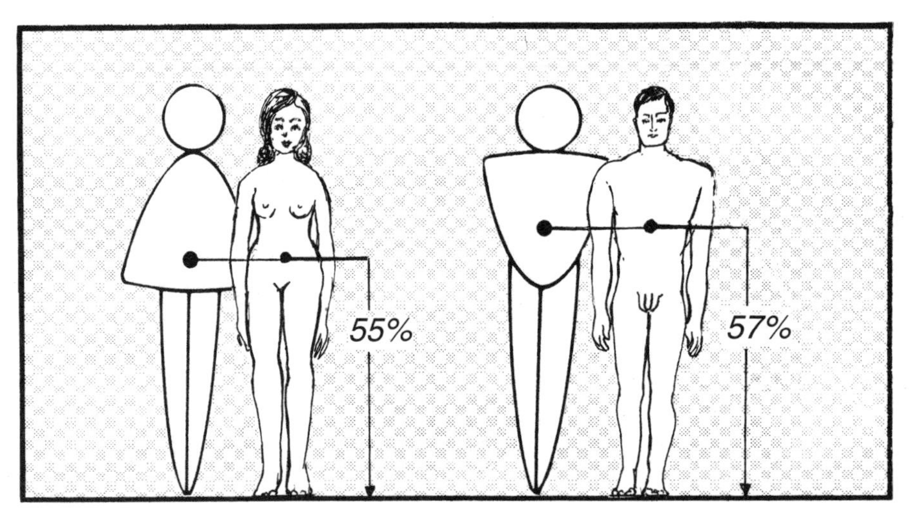

Figure 7.15 The centre of gravity location for males is approximately 57 per cent of standing height whereas for females it is only 55 per cent.

Lifting the arms up raises the height of the centre of gravity as does lifting up the legs. Moving one arm to the side displaces the centre of mass in the direction of motion of the arm. Thus, as the body posture changes, so too does the location of the centre of gravity. There are well established differences in the centre of gravity location of males and females because of the differing ways in which their masses are distributed (Figure 7.15). In adult males, the chest and shoulders are more massive than in women. Thus the position of the centre of mass in males is higher than females (approximately 57 per cent of standing height compared with 55 per cent for females).

The process of locating the centre of gravity of the human body is considerably more problematic than for rigid objects. A number of techniques, including the reaction board and segmentation methods, are used to locate the centre of mass of living human subjects. The reaction board is useful for finding the plane in which the mass centre lies (Figure 7.16). If a large reaction board and two scales, to measure two of the three reaction forces, are used (Figure 7.17), the line containing G can be determined for any particular posture. A problem with this technique is that the fluids in the abdomen, particularly, are displaced relative to their position during normal standing. For example, when standing up the fluids tend to 'drop' toward the pelvis whereas when lying supine they tend to move toward the spine. As we have discussed previously, moving the positions of the masses of the body causes the centre of gravity position to change. Thus, when abdominal fluid and viscera move, the position of the centre of gravity also moves.

The most commonly used method to determine G is the segmentation method. In this

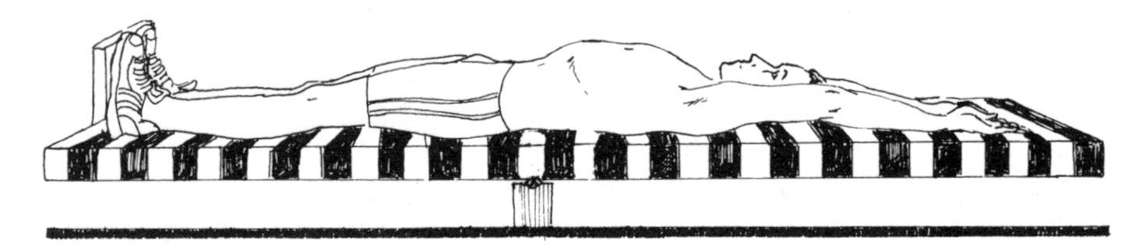

Figure 7.16 A reaction board can be used to find the plane in which the centre of gravity of a human body lies.

Figure 7.17 If two scales and only three points of support are used on a reaction board, the line of the centre of gravity can be calculated.

technique, the body is represented as a system of interconnected rigid links. The location of the centres of mass of each of the individual links is usually determined as a percentage of the segment length (based on numerous anthropometric studies) and the location of the centre of gravity of the whole body is estimated using the principles of static equilibrium (Figure 7.18). This method allows any body orientation or posture to be studied and also allows researchers to investigate body centre of mass movements during activities. Films of the action are taken and the centre of mass location is determined for every film frame. Plotting these locations on a graph visualises how the centre of gravity moves relative to the ground or any other reference point.

Modelling the human body

One of the explicit processes that all biomechanists use is that of modelling the human body. In its simplest form, the whole body is replaced by a single point mass, the centre of mass. Such a model is appropriate when simplistic analyses are required, such as determining leg power through a vertical jump test.

Viable and realistic models of the human body are required for detailed biomechanical analysis of movement. The most commonly used model is one in which the body is represented by rigid rods connected by simple, frictionless joints. The joints chosen to interlink the rods possess the same degrees of freedom as the real joints of the human body. For example, the wrist joint only allows flexion/extension and abduction/adduction and so has only two rotational degrees of freedom. A realistic model of this joint would only allow the same motions between the links. In a similar vein, the shoulder has three rotational degrees of freedom (flexion/extension, internal/external rotation, and ab–adduction). In most models of the body the shoulder joint is represented as a ball and socket joint.

Describing the motion of a linked system representing the human body is a relatively straightforward task and one that is familiar to many people these days. High-speed film or video cameras are often used in conjunction

Figure 7.18 The most common way of determining the centre of gravity location of the human body is the segmentation method.
Source: Hay, J.G. (1978), *The Biomechanics of Sports Techniques*. Prentice–Hall, New York, p. 137. (Copyright 1978 by Prentice-Hall; re-drawn with permission.)

with high-speed computers to generate stick figures of the body as it moves through space. The links of the stick figures correspond to the body segments. However, to estimate the forces that caused the observed motion is much more involved than merely describing the motion (Figure 7.19) and goes beyond the scope of this text book.

Torques

Figure 7.20 shows a simplified model of the elbow joint in which the elbow flexors (biceps brachii, brachioradialis and radialis) are replaced by a single muscle capable of producing a force $\boldsymbol{F}_m$. The insertion of this group of muscles is approximately 5 cm distal of the joint centre. That is, the tendon attaches to the bone at a point about 5 cm toward the hand from the elbow joint. Further, the axis of rotation runs into the page and is shown as a single dot, E, on the diagram. Motion at the elbow joint is described as flexion/extension, representing the rotation of the forearm relative to the arm.

When our forearm flexor muscles are stimulated to contract *all* that they are capable of doing is creating tension between their origin (the point where the muscle attaches at its end that lies closer to the trunk) and insertion points. Observation indicates that this tension is converted into angular motion but how does this come about? To answer this question careful examination of the way in which forces can be used to cause rotation is required.

Figure 7.19 Modern high–speed video systems have fast computers, video processors and cameras capable of taking 200 pictures per second.

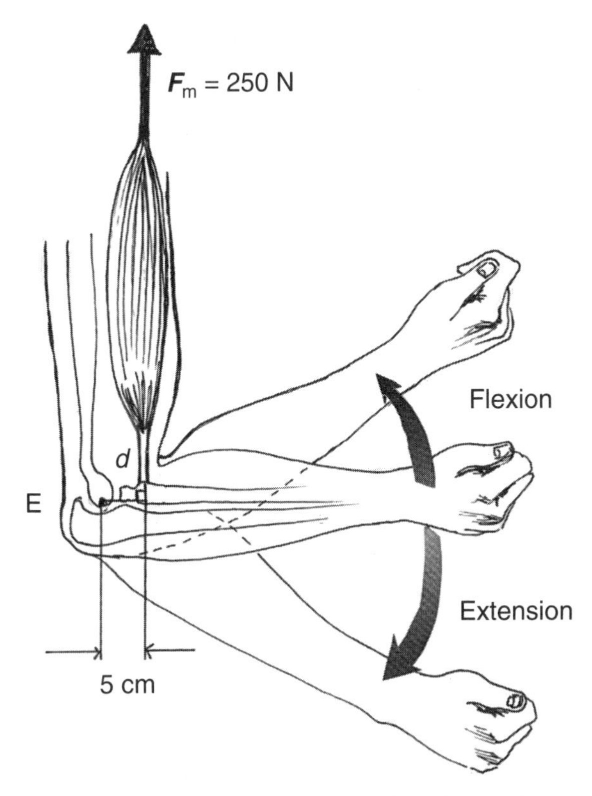

Figure 7.20 The elbow flexors (F_m) produce a force of 250 N, which is applied 5 cm distal to the elbow. A torque about the elbow joint results.

The term used by biomechanists to describe how effectively a force causes rotation is *torque*. Other terms used to refer to the same variable include moment of force or, in short-hand jargon, moment. Torque is dependent on two factors:

- the size of the force
- the moment arm length.

Moment arm length is the perpendicular distance between the line of action of the force vector and the axis of rotation. Mathematically, the torque (represented by the Greek letter tau, τ) is the product of the moment arm and the force. That is:

$$\tau = \boldsymbol{F}d \qquad (9)$$

Remember that when using this equation, d is the shortest (perpendicular) distance between the line of action of the force and the axis. Maximum torque occurs when the moment arm length and the force are maximised whereas minimum torque results when both the force and moment arm are smallest. The *right-hand thumb rule* is used to determine the sign of the torque, which is a vector quantity. To use this rule, curl the fingers of the right hand in the direction of the rotation that the force would tend to cause on the body. The thumb will then point in the direction of the torque vector. Several examples of using the right-hand thumb rule are illustrated in Figure 7.21. If the torque produces a counter-clockwise rotation it is considered to be a positive torque whereas if the rotation is clockwise, the torque is negative. Remember once again that the sign of the torque, because it is a vector, merely indicates the direction of the rotation that it produces.

Returning to Figure 7.20, the moment arm, d, is the distance from the joint centre, E, to the line of $\boldsymbol{F}_m$. With the elbow flexed at 90° as shown in the figure, this distance is 0.05 m. If the force produced in the muscle was 250 N, the torque generated by this muscle force would be:

$$\begin{aligned}\tau &= \boldsymbol{F}d\\ &= 250\ \text{N} \times 0.05\ \text{m}\\ &= 12.5\ \text{Nm}\end{aligned}$$

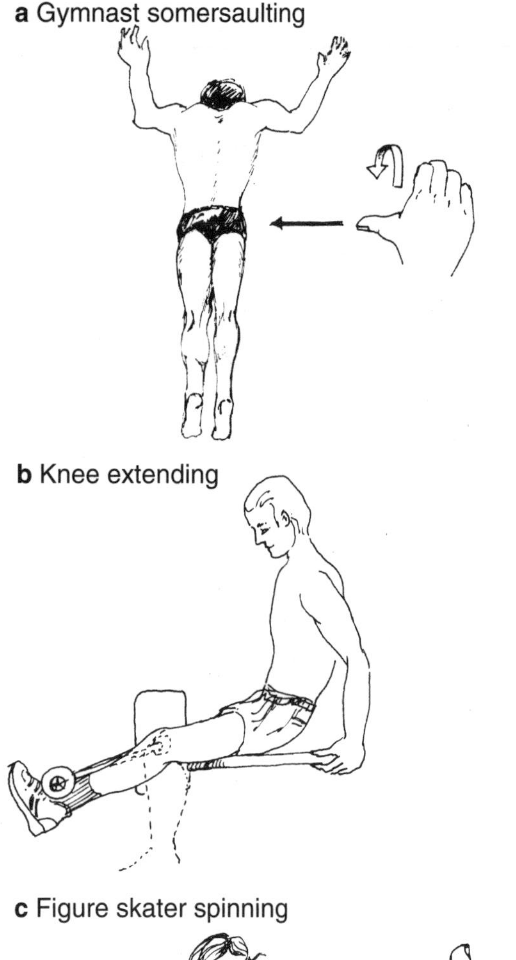

Figure 7.21 Illustrations of how the right-hand thumb rule may be used to define the angular motion vector

Another example is with the calculation of how much torque is produced on a bicycle wheel crank (Figure 7.22) at two different points during the cycle, 90° and 30° from top dead centre (TDC). TDC is the position during the pedal cycle at which the crank arm is vertical or the pedal is directly above the axle of the crank. In each of these cases the force produced by the rider is pointing straight down (vertically downward) and is 500 N. The length of the crank is 0.17 m. The solution to these problems is detailed below:

Case A, $\theta = 90°$: $\tau = Fd$
$= 500\ \text{N} \times 0.17\ \text{m}$
$= -85.0\ \text{Nm}$ (the sign is negative because the torque is in a clockwise direction)

Case B, $\theta = 30°$: $\tau = Fd$
$= 500\ \text{N} \times 0.17\ \text{m} \times \sin 30°$
$= -42.5\ \text{Nm}$ (see Figure 7.22*b* for detail on moment arm length)

Couples

The examples highlighted above are situations in which the applied force is eccentric, which, as noted earlier in this chapter, corresponds to a force that is not applied along a line passing through the centre of gravity of the body. When two forces of equal magnitude but opposite direction are applied to a body, pure rotation will result. When a person unscrews a tap or turns the steering wheel of a car (see Figure 7.23), two equal but opposite forces are applied but with different lines of action. The net result is that the tap and steering wheel undergo pure rotation. When pedalling a bicycle the cranks move through pure rotation as a result of the

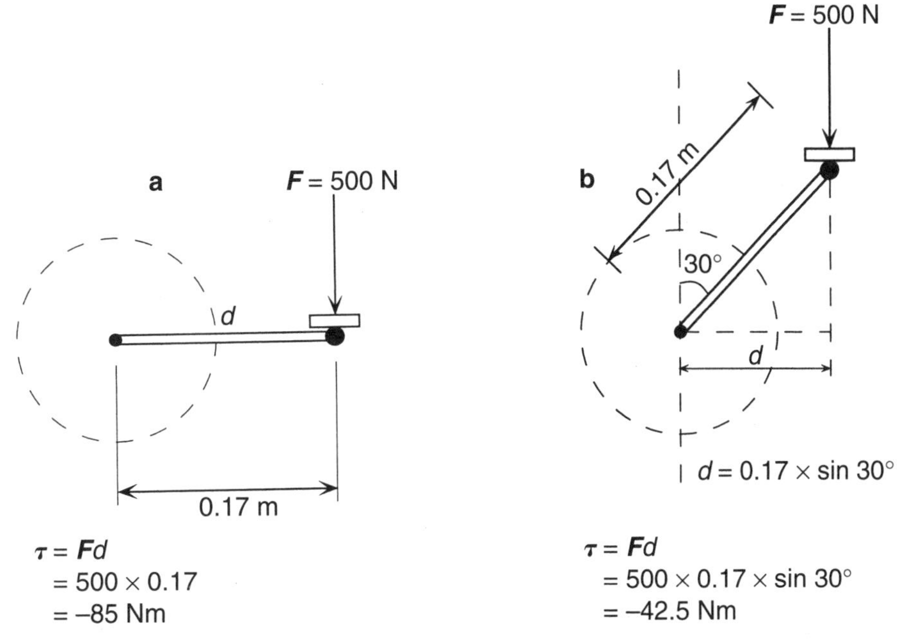

Figure 7.22 The torque produced by the foot on the bicycle crank is altered by the angle of the crank. When the crank is horizontal, the torque is –85.0 Nm (a) whereas it is only –42.5 Nm when the crank is 30° from the vertical (b).

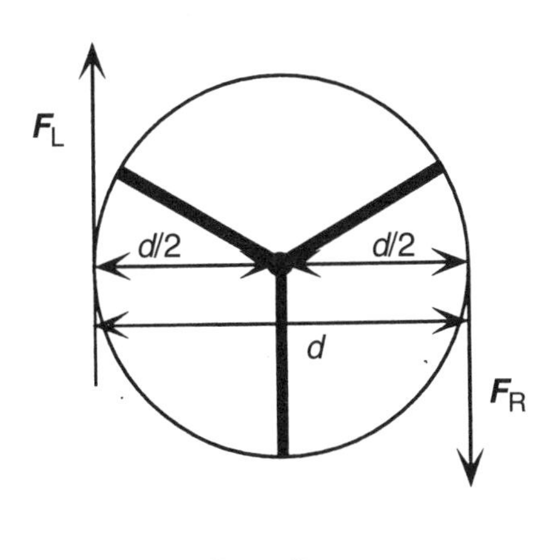

Figure 7.23 The left and right hands produce equal but opposite forces along different lines of action and create a couple.

force couple that is applied. In fact, a reaction force of equal magnitude but opposite in direction to the applied force generated by the foot is provided by the axle and so no translation of the crank results. The magnitude of the couple is the product of the perpendicular distance between the two forces and the size of the force.

Equilibrium

Equilibrium is that situation when a body experiences zero acceleration. In the linear sense, the requirement for equilibrium is that the resultant of all forces acting on the system equals zero. That is:

$$\sum_{i=1}^{N} F_i = 0 \tag{10}$$

This idea is often referred to as the first condition of equilibrium.

Although the net force on a body may be zero it can still be rotating and its rate of rotation can be changing. For a body to be in rotational equilibrium the sum of the torques acting on the body must also equal zero:

$$\sum_{i=1}^{N} \boldsymbol{\tau}_i = 0 \quad (11)$$

This notion is known as the second condition of equilibrium. For example, to maintain a stationary position such as an iron cross on the still rings or a handstand position, not only must the upward forces created by the upper limbs be equal to the weight force acting downward, the torques created by these forces must also sum to zero. If they did not sum to zero, the unbalanced force and torque would cause the body to begin to accelerate and it would no longer remain in static equilibrium.

Summary

A force is a push or pull that compels a body to change its state of motion whereas a body's resistance to a change in its state of motion is quantified by its inertia. Torques cause rotational motion and are dependent on the size of the applied force and the distance from the line of action of the force to the axis of rotation. The resistance to angular motion is known as the moment of inertia. When an eccentric force is applied to a body, the body undergoes both translational and rotational motion. The centre of mass of an object is the point at which the mass appears to be concentrated and is often referred to as the balance point or the centre of gravity. For a body to be in equilibrium, the sum of the forces and torques must equal zero.

CHAPTER 8

Basic concepts of kinematics

- Linear kinematics
- Angular kinematics
- Analogies between linear and angular kinematics

Kinematics is the branch of mechanics that deals with the description of motion. It can be further subdivided into linear and angular kinematics. As the names of these two divisions imply, linear kinematics is concerned with motion of a body when all parts of the object move through the same distance in the same time period, whereas angular kinematics describes that branch of kinematics in which objects or bodies undergo rotation. One of the most important principles underlying the study of motion of bodies is that their motion can be analysed as an independent combination of linear and angular motion. This statement means that any problem can be considered to be made up of two smaller problems, one involving only translation and the other involving only rotation. To describe the motion of rigid bodies such as the segments of the human body, the translation (linear motion) of the centre of mass of the segment and the rotation about the centre of mass must be known. In this chapter we introduce some basic concepts related to both linear and angular kinematics and draw analogies between these two branches of kinematics.

Linear kinematics

Displacement/distance; position

Displacement is defined as the change in position of a body and like all vector quantities has both direction and magnitude. To correctly specify someone's or something's displacement, both the size of the change and the direction in which the change took place must be considered. For example, to describe the displacement of a cyclist who rides from point A to point B, the straight-line distance (for example 40 km) and the direction (for instance North–West) must be supplied. Without both the magnitude and the direction enumerated, the displacement has not been correctly defined. In mechanics, the use of the Greek letter upper case *Delta* (Δ) is used in mathematical shorthand to represent change. Thus displacement (**s**) is written as:

$$\mathbf{s} = \Delta x = x_f - x_i \qquad (1a)$$

where x_f indicates the final position of the object, and
x_i indicates the initial position of the object

Using the variable x does not imply that displacement occurs only in the X direction of a coordinate system. Typically the vector quantity of displacement is considered to be the sum of three independent displacements along the axes of the coordinate system. That is:

$$\mathbf{s} = \mathbf{s}_x + \mathbf{s}_y + \mathbf{s}_z \qquad (1b)$$

If the motion occurs in a plane then the displacement along the Z axis is zero.

The displacement of a body is the straight-line distance between the starting and finishing positions and should not be confused with distance (d), a scalar quantity, that indicates the total distance covered along the path of travel. The metric unit of measurement for both displacement and distance is the metre (m). When solving problems in mechanics, it is conventional to use a coordinate system in which displacements to the right and upward are said to be positive and those to the left and downward are negative. This convention gives us a shorthand method of assigning the direction to the displacement vector.

Some examples may help to consolidate these ideas. Suppose that Penny, a 1st-year university student, decides that she is going to get into shape by doing some swimming training. During the first session she pushes herself and manages to swim 20 laps of the local 50 m pool and states (proudly!) that she swam 1 km (20 × 50 m = 1000 m = 1 km). Certainly the distance that she swam was 1 km but, what was her displacement? Her starting and finishing positions were the same, so her displacement was zero! That does not sound nearly as impressive as 1 km!

Here is another example. The Sydney to Hobart race is probably the most famous yacht race conducted each year in Australia. During this annual event, which starts on December 26, crews travel through distances of approximately 1300 km as they course their way down the coast of New South Wales then on to Hobart. The race record for the event is 2 days, 14 hours, 36 minutes and 56 seconds set by KiaLoa in 1975. The displacement of all the boats is exactly the same and substantially less than the actual distances that they travel. In fact, the displacement is only 1080 km SSW. Figure 8.1 illustrates the difference in the displacement and the distance that one of the yachts travelled.

Velocity and speed

Most people have a good concept of speed because it is very much a part of our everyday lives. Drivers must keep a close watch on the speeds at which their cars travel since most roads have restricted speeds. In vehicles, constant feedback on speed is given by the speedometer and this information reinforces an understanding of speed. Furthermore, most people have a good idea about average speed. For example, if someone travelled

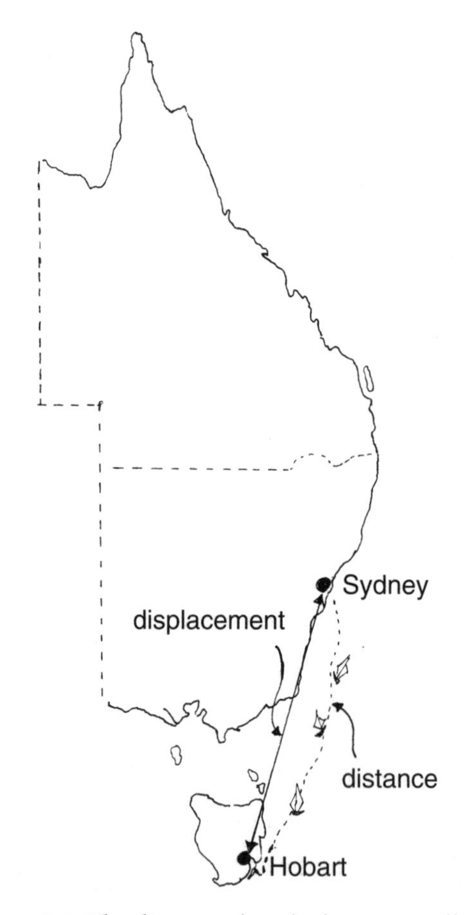

Figure 8.1 The distance that the boats actually travel in the Sydney to Hobart yacht race is much greater than the displacement.

120 km in 2 h, the average speed was 60 km/h. In other words, speed and velocity indicate how far a body travels in a fixed period of time and, therefore, the unit of measurement is m/s. But what is the difference between speed and velocity? Simply, speed is the rate of change of distance whereas velocity ($\boldsymbol{v}$) is the rate of change of displacement. Thus speed is a scalar quantity and velocity is a vector. In most situations involving the description of the technique a person may use in an activity, the speed is the magnitude of the velocity. As with displacement, the direction of the velocity is usually specified by the sign (+ or –). A positive velocity signifies motion upward and to the right while negative indicates a downward or left-moving object. Formally:

$$\text{speed} = \frac{\Delta d}{\Delta t} \qquad (2)$$

$$\bar{\boldsymbol{v}} = \frac{\boldsymbol{s}}{\Delta t} = \frac{\Delta x}{\Delta t} = \frac{(x_f - x_i)}{(t_f - t_i)} \qquad (3)$$

When a bar appears above a quantity, for example $\bar{\boldsymbol{F}}$, it indicates an average value of the quantity. Thus, $\bar{\boldsymbol{v}}$ means the average velocity. In the above equation, as Δt becomes smaller and smaller (or in mathematical jargon, approaches the limit of zero), the average velocity approaches the instantaneous velocity. In research applications, many high-speed film and video cameras have framing rates of 200 Hz, which means that 200 individual video fields or film images are taken every second. If velocity is calculated based on the change in position between successive video images of a tape that was recorded at 200 Hz, Δt would be equal to 0.005 s or 5 ms. Since Δt is very small, the velocity calculated would be considered to be instantaneous. For virtually all activities in human movement that are under voluntary muscular control a Δt of 5 ms is small enough to denote the calculated velocities as *instantaneous velocities*, and only in rare situations would framing rates above 200 Hz be necessary to capture the action. Two examples of situations that require extremely high sampling rates are the impact of the foot on the ground on landing from a jump and the impact of a tennis ball as it is struck by a racquet.

A graphical interpretation of the relationship between displacement and velocity will probably aid in understanding and consolidating these ideas. Figure 8.2*a* shows the displacement of a sprinter's centre of mass, plotted on the ordinate, starting from rest in the blocks ($t = 0$) to the 10 m mark. Time, of course, is plotted on the abscissa. In Figure 8.2*b*, the first section of the graph is expanded and the slope of the tangent to the position–time curve is calculated at two

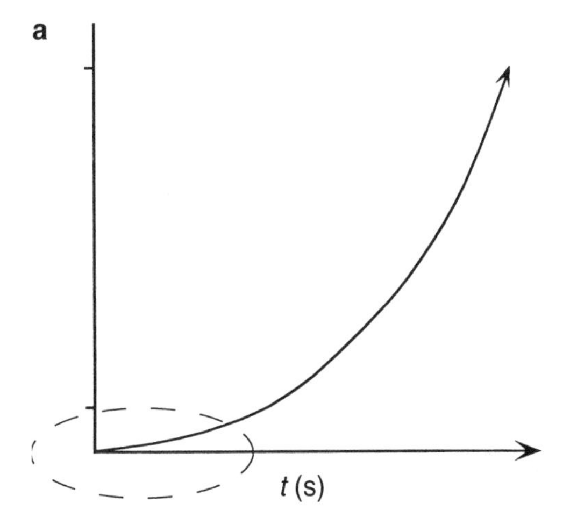

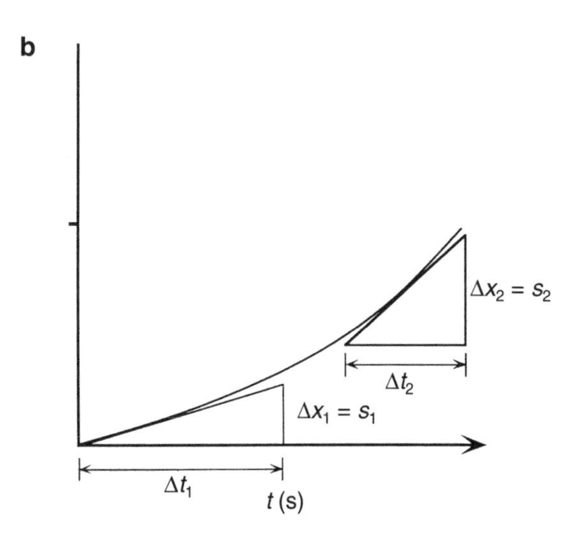

$$\text{slope} = v = \frac{\Delta x_1}{\Delta t_1} = \frac{0.25}{0.5}$$
$$= 0.5 \text{ metres per second}$$

Figure 8.2 The position of the sprinter's centre of mass plotted as a function of time over the first 10 m (a) and then for the first metre only (b)

points. Recall from high school mathematics that the slope, m, of a line is given by:

$$m = \frac{\text{rise}}{\text{run}} = \frac{\Delta y}{\Delta x} \quad (4a)$$

Substituting the quantities shown in Figure 8.2*b* into equation 4*a* yields:

$$m = \frac{\text{rise}}{\text{run}} = \frac{\Delta s}{\Delta t} \quad (4b)$$

But, from equation 3, $\mathbf{v} = \Delta x/\Delta t$, thus velocity can be considered as the slope of the displacement–time graph. Substituting the actual values from Figure 8.2*b* into this formula allows estimation of the velocity at the times shown:

$$\mathbf{v}_1 = \frac{\Delta s_1}{\Delta t_1} = \frac{0.25}{0.5} = 0.5 \text{ m/s}$$

$$\mathbf{v}_2 = \frac{\Delta s_2}{\Delta t_2} = \frac{0.4}{0.4} = 1.0 \text{ m/s}$$

If this process is repeated many times over for very small Δts, the graph illustrated in Figure 8.3 can be derived. This graph depicts the sprinter's velocity at every point in time. Thus an object's velocity can be derived if its displacement as a function of time is known.

The reverse process can also be used to determine a body's displacement if its velocity is known. Rearranging equation 3 indicates that $\mathbf{s} = \mathbf{v}\Delta t$, which means that displacement is the area under the velocity–time function. This notion is illustrated by two cases in Figure 8.4, where velocity is plotted against time and the area is calculated to give a measure of the

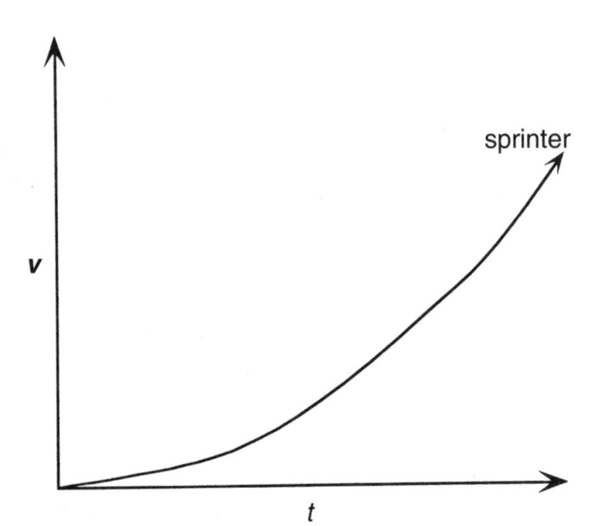

Figure 8.3 The sprinter's velocity over the first 10 m obtained by differentiating the displacement–time function

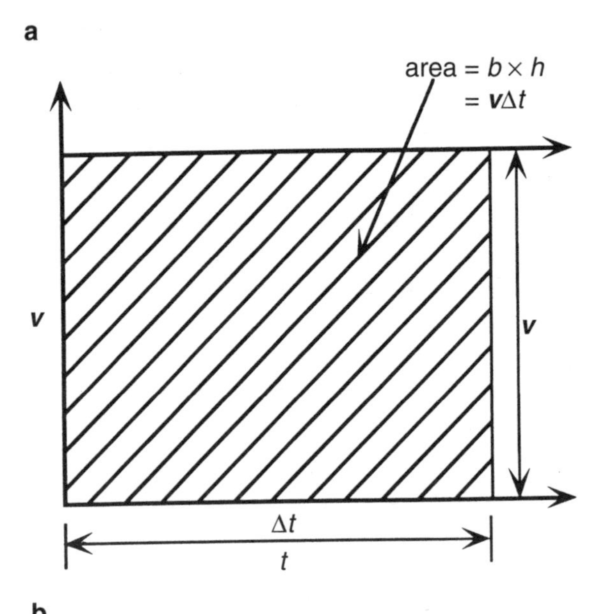

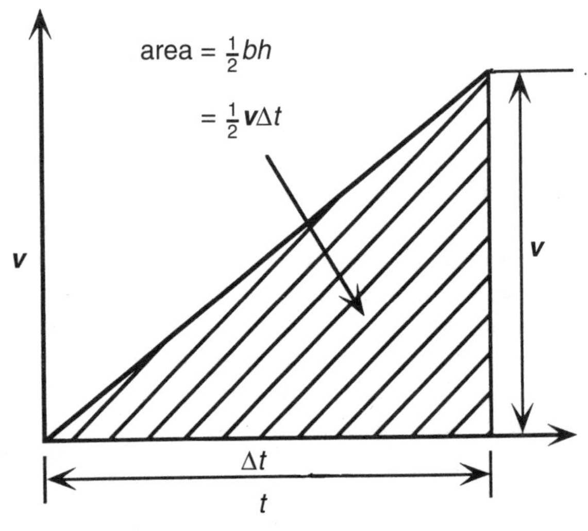

Figure 8.4 In the situation where velocity is constant (a), the displacement from the start to the finish of the interval (hashed area) is given by the area of the rectangle. When the velocity is linearly increasing (b), the displacement is the area of the triangle.

displacement. Although these two examples are very simple, namely the case of constant velocity (*a*) and linearly increasing velocity (*b*), the idea is completely general and can be extended to situations in which velocity changes in a complex way. To examine this process further is beyond the scope of this book and involves a branch of mathematics known as calculus, developed by Newton, to solve this exact problem.

Acceleration

There are not many situations in human movement when the velocity remains constant as a body moves through space; its velocity constantly changes both in direction and magnitude. The quantity used by biomechanists and physicists to describe how a body's velocity is changing is acceleration ($\boldsymbol{a}$) and it is defined as the change in velocity ($\Delta\boldsymbol{v}$) that takes place in a period of time (Δt). Formally,

$$a = \frac{\Delta \boldsymbol{v}}{\Delta t} \qquad (5)$$

The standard unit of measurement for acceleration is m/s^2 and a positive sign attached to the acceleration symbolises an increasing velocity upward or to the right whereas a negative sign indicates an increasing velocity to the left or downwards.

A relevant question to ask at this point is 'Why are we interested in finding out how fast displacement and velocity are changing?'. The answer to this question dates back to the days of Newton, who proved that there was a direct relationship between acceleration and force. This relationship, which has become known as Newton's second law of motion, states that:

$$\boldsymbol{F} = m\boldsymbol{a} \qquad (6)$$

This relationship is the fundamental law used in a quantitative study of moving bodies. Since many people have difficulty conceptualising acceleration (and because our sensory systems are not tuned to picking up accelerations) it is worth examining some examples using graphical techniques. Firstly, Figure 8.5*a* shows the

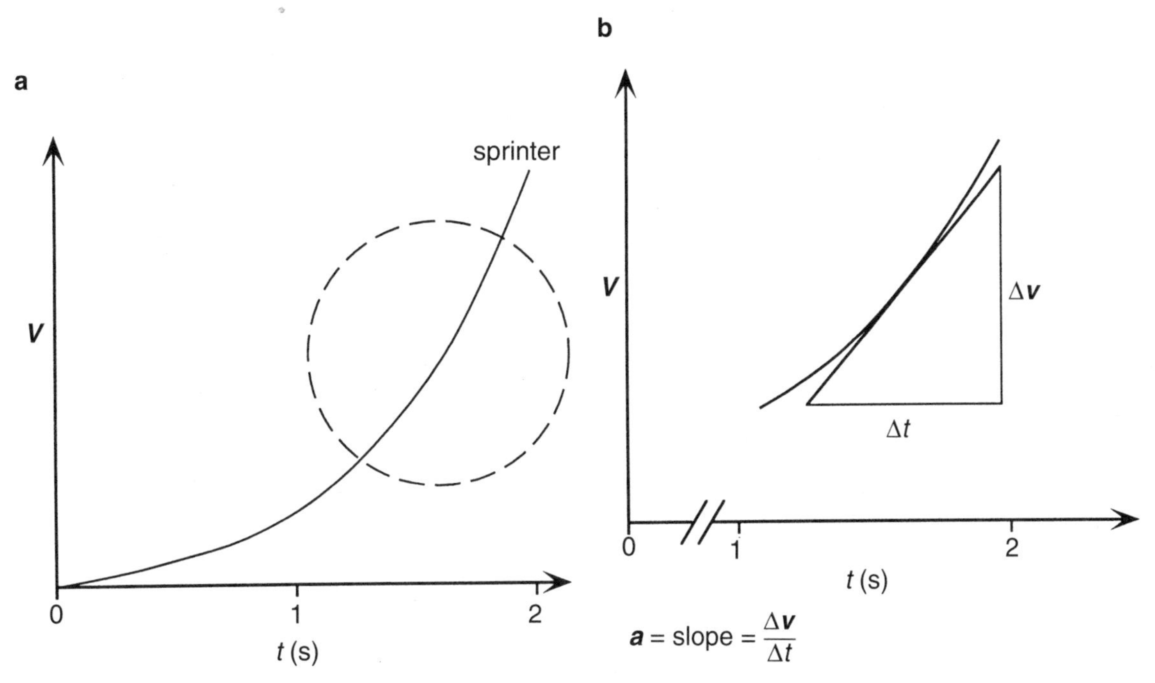

Figure 8.5 The velocity of the centre of mass of a sprinter shown over the first 2 seconds of a race (a) and for the time interval 1–2 s (b). The slope of the velocity–time function over this small interval is the acceleration experienced by the sprinter.

velocity–time graph of a sprint start over the first 2.0 s and Figure 8.5*b* illustrates an expanded view of the section between $t = 1.0$ and $t = 2.0$ s. The tangent to the curve is also shown in Figure 8.5*b* and the slope of this line is given by:

$$m = \frac{\text{rise}}{\text{run}} = \frac{\Delta \mathbf{v}}{\Delta t} \qquad (7)$$

Thus, acceleration is the slope of the velocity–time function. If the slope is calculated many times during the race, then the acceleration, too, can be plotted as a function of time (Figure 8.6). The acceleration–time graph signifies how the velocity changed during the event. In the sprint start, the velocity changed rapidly at first and although the sprinter was still increasing speed after 3.0 s, the rate of increase had decreased.

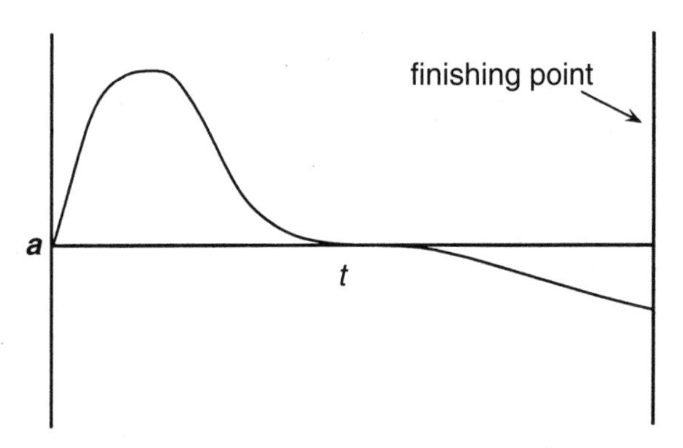

Figure 8.6 A hypothetical acceleration–time graph of a sprinter running for 10 s

A second example, showing the relationships among displacement, velocity and acceleration, is illustrated in Figure 8.7. In this example, a basketball player started from rest on the baseline of a court, sprinted to the half-way line and then ran backwards to the baseline, where she stopped. Figure 8.7*a* shows that the displacement increased at first and reached maximum when the player

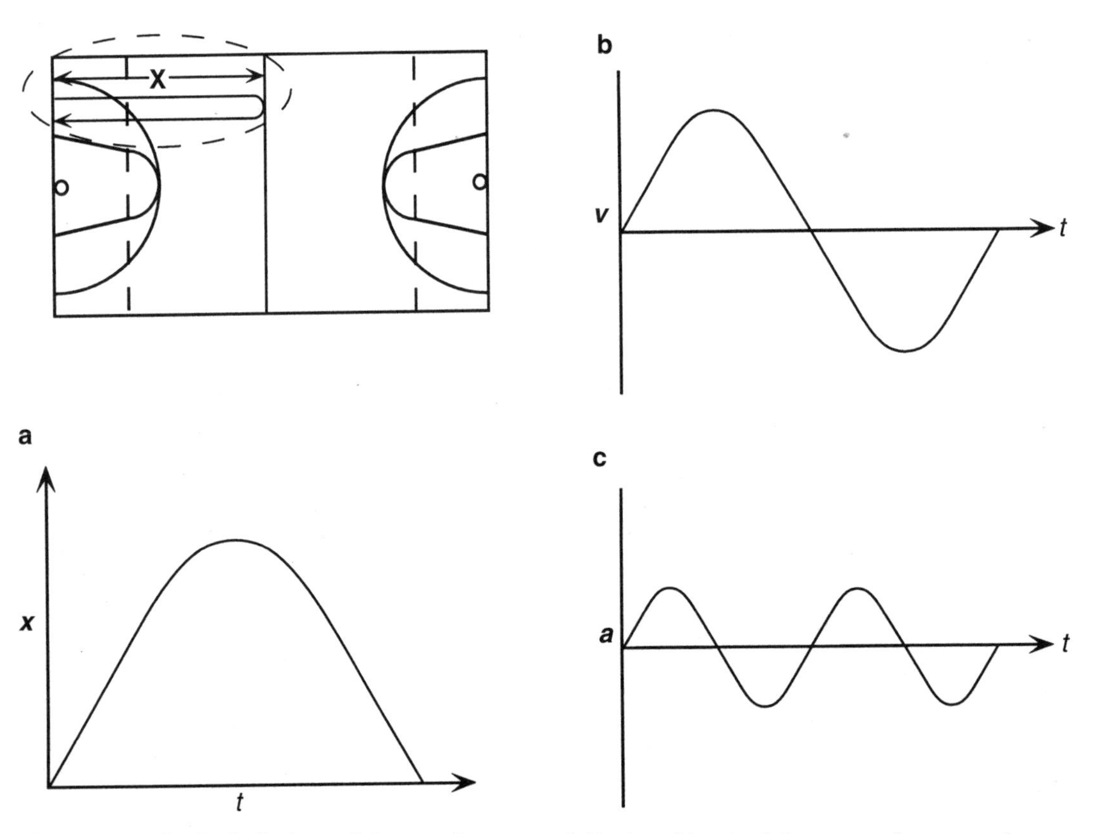

Figure 8.7 A basketball player did a wind-sprint and displaced her body's centre of mass as shown in (a). From these data, her velocity (b) and acceleration (c) were calculated.

touched the half-way line and returned to zero when the player arrived back at the baseline. The velocity–time graph (Figure 8.7*b*) depicts how the velocity increased at first and then became zero at the turn-around point, the half-way line. When the player was running back toward the baseline, the velocity increased but this time in the opposite (negative) direction. When the player arrived at the baseline and stopped, the velocity reduced to zero again. The acceleration–time graph shows how the velocity changed. When the player sped up the acceleration was positive until peak forward velocity was reached. As the player approached the half-way line and began to slow down, the acceleration was negative, that is, the velocity was decreasing. It remained negative as the basketballer then increased speed in the negative direction (the direction from half-way line to baseline) but as the player arrived at the baseline and stopped, the acceleration was once again positive, that is, the speed was decreasing in a negative direction (which is the same as increasing in the positive direction!).

In the above examples the model used for the human body was a particle and the motion of the various body segments was not analysed. Clearly this model is the simplest one that biomechanists use. Once problems involving particle motion can be routinely solved, complex models in which the body segments, such as the forearms, hands, arms and thighs, are modelled as rigid bodies can be used. Although it is beyond the scope of this text to discuss the mechanics of rigid bodies, some of the basic concepts that are relevant to understanding human movement will be discussed.

Momentum

Two of the fundamental laws governing the motion of bodies are the conservation of energy and momentum. The principle of momentum is one of the most important ones in physics generally and mechanics in particular. Most people have a good conceptual understanding of momentum. When given a choice (if one had to!) between being struck by a motor car or a bicyclist moving at 10 m/s we would intuitively choose the latter. Furthermore, if one had a choice between stopping a slap shot hit by a professional ice hockey player or one hit by an elementary school child, again, most of us would choose the latter. From these two examples it should be clear that a body's inertia (mass) and its velocity are implicated in momentum. In fact, the momentum ($\boldsymbol{p}$) is the product of the two. That is:

$$\boldsymbol{p} = m\boldsymbol{v} \qquad (8)$$

Momentum is a vector quantity with its direction specified by the direction of the velocity vector. Its units of measurement are kgm/s. Table 8.1 lists the momentum possessed by various sporting balls and objects as they move with velocities typical of real-life situations. Controlling one's momentum in many activities is paramount if success is to be achieved. Take, for example, down-hill skiing. If a skier builds up too much momentum by coming straight down the hill, he or she may not be able to turn sharply enough to pass through gates on the course or, worse still, may career into fences and spectators on the side of the course. For the weekend skier, the trees on the edge of the slope

Table 8.1 Typical momenta for various sporting implements and bodies

Object	Speed (m/s)	Mass (kg)	Linear momentum (kgm/s)
Football (punt)	29.1	0.415	12.1
Sprinter	12.0	80.0	960.0
Shot put	13.7	16.0	219.2
Volleyball serve	22.0	0.265	5.8
Tennis ball—forehand	40.0	0.055	2.2
Tennis ball—serve	52.8	0.055	2.9
Badminton	34.1	0.006	0.2
Soccer—kick	25.9	0.415	10.7
Soccer—head	12.8	0.415	5.3
Golf ball	76.2	0.046	3.5
Cyclist	13.9	75.0	1042.5
Cricket ball (bowled)	44.4	0.155	6.9

or fellow skiers might be the objects in danger of being struck!

There is a direct relationship between a body's momentum and the impulse of force applied to it. In fact, this relationship is a form of Newton's second law of motion, which states that the rate of change of momentum is proportional to the applied force ($\boldsymbol{F}$) and takes place in that direction. In other words:

$$\boldsymbol{F} = \frac{\Delta \boldsymbol{p}}{\Delta t} \qquad (9)$$

If we rearrange this equation and note that the impulse ($\boldsymbol{J}$) is the product of force and time, the impulse–momentum relationship evolves. That is:

$$\boldsymbol{J} = \boldsymbol{F}\Delta t = \Delta \boldsymbol{p} \qquad (10)$$

Thus any change in momentum experienced by a body is in direct proportion to the impulse that was applied.

Here is an example of applying this relationship. Suppose a golf ball (m = 0.046 kg) was struck by Greg Norman's driver. If the impact took 400 μs (400×10^{-6} s) and the average force was 9000 N, with what velocity did the ball leave the club? Substituting the values into equation 10, and noting that $\Delta \boldsymbol{p} = m\Delta \boldsymbol{v}$, yields:

$$\boldsymbol{J} = \boldsymbol{F}\Delta t = \Delta \boldsymbol{p} = m\Delta \boldsymbol{v}$$
$$9000 \times 400 \times 10^{-6} = 4.6 \times 10^{-2} \times (v_f - 0)$$
$$\boldsymbol{v}_f = 78.3 \text{ m/s}$$

Collisions, of which the above is an example, are an integral part of many sports and physical activities and a player's ability to control the speed and direction of a ball, a hockey puck or a shuttlecock, for example, can be good indicators of the skill level of the person. In snooker and billiards, the skill in the game includes the ability to make the cue (white) ball strike the object ball (the reds or colours) at exactly the right point and with precisely the correct speed so that it falls into one of the pockets of the table. At the elite level, the ability to control and predict where the cue ball will finish after it has struck the object ball and the cushions is a prerequisite of success. To do these tasks successfully (strike the object ball correctly and control the cue ball), a basic understanding of the physics of collisions is required. In other sports, such as golf, tennis, cricket and tenpin bowling where the speed of the ball is critical to successful performance, an understanding of the mechanics of collisions will help the performer to choose equipment that best suits him or her or may, if permissible, guide the modification of existing equipment so that it better meets the demands of the task.

The physical laws governing impacts were derived by Newton in the 1600s. His law of conservation of momentum maintains that in a closed system momentum is conserved. Formally, his law of conservation of linear momentum states that:

... when the net external force on a system is zero, the total linear momentum will remain constant

In a collision between a ball and a bat, where there are no external forces acting, the law can be re-stated as follows. Let m_1 and m_2 represent the mass of the bat and ball respectively and $\boldsymbol{v}_1$ and $\boldsymbol{v}_2$ their velocities. The total momentum of the system of bat and ball will be:

$$\boldsymbol{p}_t = \boldsymbol{p}_1 + \boldsymbol{p}_2 = m_1\boldsymbol{v}_1 + m_2\boldsymbol{v}_2 \qquad (11a)$$

If, after the collision, their respective velocities are $\boldsymbol{v}'_1$ and $\boldsymbol{v}'_2$, then:

$$\boldsymbol{p}_t = \boldsymbol{p}'_1 + \boldsymbol{p}'_2 = m_1\boldsymbol{v}'_1 + m_2\boldsymbol{v}'_2 \qquad (11b)$$

Since the momentum is conserved, the expression can be rewritten in the following form:

$$m_1\boldsymbol{v}_1 + m_2\boldsymbol{v}_2 = m_1\boldsymbol{v}'_1 + m_2\boldsymbol{v}'_2 \qquad (11c)$$

That is, the momentum prior to the collision equals the momentum after the collision. This relationship contains valuable information for people wishing to optimise performance in activities that involve striking or collisions. For example, if maximum speed of the ball after impact is the most important determinant of success in a particular sport, then there are a number of possible ways to increase post-impact velocity of the ball. These include:

- increasing the mass of the bat
- increasing the velocity of the bat prior to impact
- decreasing the velocity of the bat immediately after impact, in order to transfer as much momentum as possible to the ball
- decreasing the mass of the ball.

In most sports the player cannot control the size or the mass of the ball; typically these parameters are stipulated by the rules governing the sport. Of the remaining three methods of increasing post-impact ball speed, increasing the mass of the bat and the velocity of the bat are the ones over which the performer has most control. Often, however, these two are competing factors because increasing the inertia of the bat will make it difficult to accelerate and may result in a concomitant decrease in speed. Thus, increasing the mass of the bat must not be offset by an equivalent loss of speed. Similarly, increases in speed are most easily obtained by decreasing the inertia of the bat. If the proportional increase in speed is equal to the relative decrease in mass, the momentum will remain the same and no benefit will be gained. The best situation might be to increase the strength and power of the athlete enabling him/her to swing a heavy bat at a high speed, thus getting the benefit of increased mass and increased velocity!

Figure 8.8 shows three different high-speed impacts. It is interesting to note the large deformations of both the balls and the racquet. During the first part of the collision, energy is stored as strain energy in the deforming bodies, which is then returned or used later to restore the shape of the objects and force the masses apart. Some of the mechanical energy at impact is converted to heat and sound; thus impacts in real life are never perfectly elastic. In a perfectly elastic impact, no mechanical energy is converted to other forms. That is, if a ball were dropped from a height of 1.0 m and struck the ground in a perfectly elastic collision, it would rebound precisely to the height from which it was dropped.

A useful notion, the *coefficient of restitution*, has been developed to describe the relative elasticity of impacts. This coefficient ranges from zero, in which the objects do not separate after impact, through to one, indicating a perfectly elastic collision. Table 8.2 shows the coefficient of restitution of a number of balls. In defining these values, the characteristics of both objects, the ball and the surface, must be specified. Thus, a basketball landing on a concrete surface will bounce higher than if it is dropped from the same height onto a foam mat. Coefficients of restitution therefore indicate the amount of

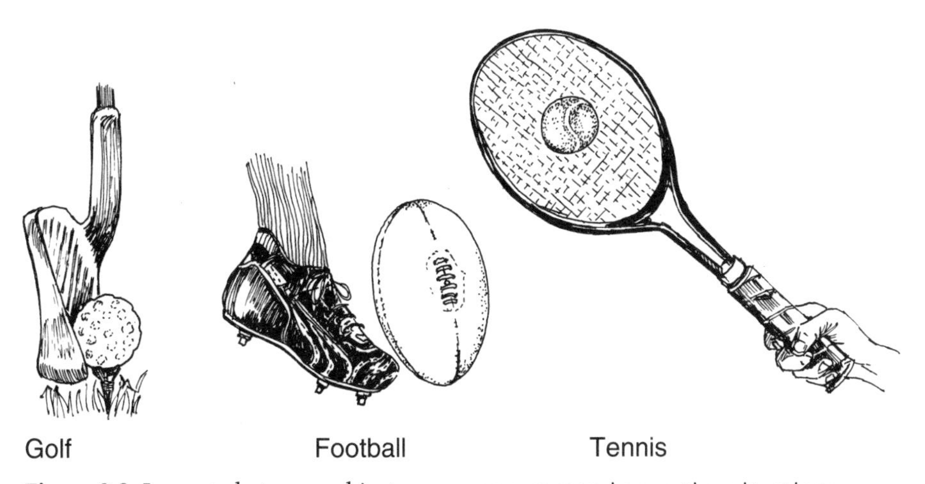

Figure 8.8 Impacts between objects are very common in sporting situations.

Table 8.2 Coefficients of restitution of balls dropped onto a wooden floor

Type of ball	Coefficient of restitution
Super ball	0.89
Basketball	0.76
Soccer	0.76
Volleyball	0.74
Tennis—worn	0.71
Tennis—new	0.67
Lacrosse	0.62
Field hockey	0.50
Softball	0.32
Cricket	0.31

Source: Data taken from Hay, J.G. (1978), *The Biomechanics of Sports Techniques*, Prentice-Hall, Sydney, p.80.

mechanical energy lost in any impact. Two factors influence the coefficient of restitution:

- Temperature—the coefficient of restitution is directly related to temperature and is evidenced most noticeably in squash, where the ball has to be *warmed up* before it can be used for play.
- Velocity of impact—the coefficient of restitution decreases as the relative velocity between the two striking surfaces increases.

Angular kinematics

As noted at the beginning of this chapter, problems in mechanics can be tackled as two independent, smaller problems involving linear and angular motion. To describe the linear motion of an object, a set of variables or characteristics that can be used to tell how far (displacement), how fast (velocity) and how quickly motion is changing (acceleration) were defined. What follows is the derivation of similar quantities used to describe the rotational motion of a body. Whereas linear motion variables are symbolised by English letters, angular motion variables have traditionally been assigned Greek letters such as θ, ω, and α.

Angular displacement/distance

Just as the linear displacement of an object is defined as its change in position, regardless of the path it takes to get to the final position, the angular displacement of an object is defined in a similar way. Figure 8.9 shows a golfer holding a golf club near the top of the backswing. The angular position of the club is described by the angle (β), which is the angle measured from the horizontal. This angle is positive if it is above the horizontal and negative if it is below the horizontal. Thus, the angular position can take on values between –180 and 180° ($-\pi$ to π when the angle is measured in radians; 1 rad ≅ 57.3°).

The angular displacement (θ), which is not a vector, is the change in angular position:

$$\theta = \Delta\beta = \beta_f - \beta_i \qquad (12)$$

and is the minimum of the two possible angles. In other words, the angular displacement does not take into account the direction of rotation and must always have a magnitude of less than 180°. In the example depicted in Figure 8.9*b*, the initial position of the club is approximately 45° and its final position is about 70°. Its angular displacement is the difference between the final and the initial positions, θ = 25°. The angular displacement is reported as the difference between the final and initial positions, regardless of the path taken whereas the angular distance accounts for the total angular distance through which the body rotated. For example, in a two-and-a-half forward somersault dive, the angular displacement is half a revolution (180°) whereas the angular distance (ϕ) is two-and-a-half revolutions or 900°.

Figure 8.10 shows the knee angle as a function of time during a single running stride. The angle plotted is the smaller of the two angles formed between the thigh and the leg. It can be seen from this diagram that the knee joint angle has periods when it increases (when the knee is extending) and when it decreases (the joint is flexing). Extension occurs at the end of the swing phase, when the limb is not in contact with the ground and is being moved into position for its next support phase, and during midstance, as the weight of the body is supported by this limb. The knee angular displacement pattern is very repeatable from stride to stride and exemplifies how consistent limb motion is when a skill has been well learned and practised.

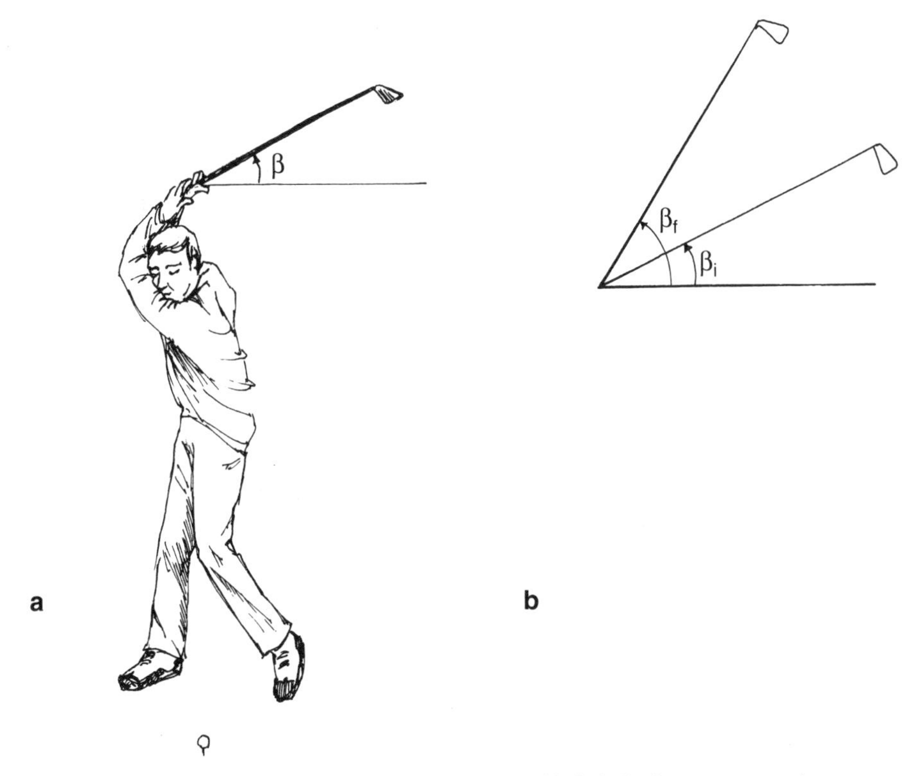

Figure 8.9 A golfer (a) angularly displaces the golf club (b) from its initial orientation (β_i) to its final position (β_f).

Angular velocity/speed

The average angular velocity of a body is defined as the rate of change of angular displacement in a given period of time. Amazingly, the instantaneous angular velocity (ω) is a vector quantity even though it is calculated from a scalar variable. The angular speed (σ) is similarly defined as the rate of change of angular distance with respect to time. Formally:

$$\omega = \frac{\Delta\theta}{\Delta t} \quad (13)$$

$$\sigma = \frac{\Delta\phi}{\Delta t} \quad (14)$$

Just as the linear velocity was calculated as the slope of the displacement–time graph, angular velocity is the slope of the angular displacement function. By definition then, the unit of measurement is rad/s or deg/s. The knee joint angular displacements plotted in Figure 8.10 were differentiated; that is, the slope was calculated many, many times for very small Δts, and then plotted as a function of time. The results of this process are shown in Figure 8.11 in which the knee joint angular velocity is plotted for one gait cycle. From this graph it can be seen that an increasing joint angle (extension) is associated with a positive angular velocity whereas when the joint flexes, the velocity is negative.

The right-hand-thumb rule is used to define the direction of the angular velocity vector. If the fingers of the right hand are curled in the direction of rotation, the thumb will point to the direction of the angular motion vector. It is helpful to think of this vector as running along the axis about which rotation is taking place.

Angular acceleration

How quickly does a rotating body change its velocity and why do we really care? To answer

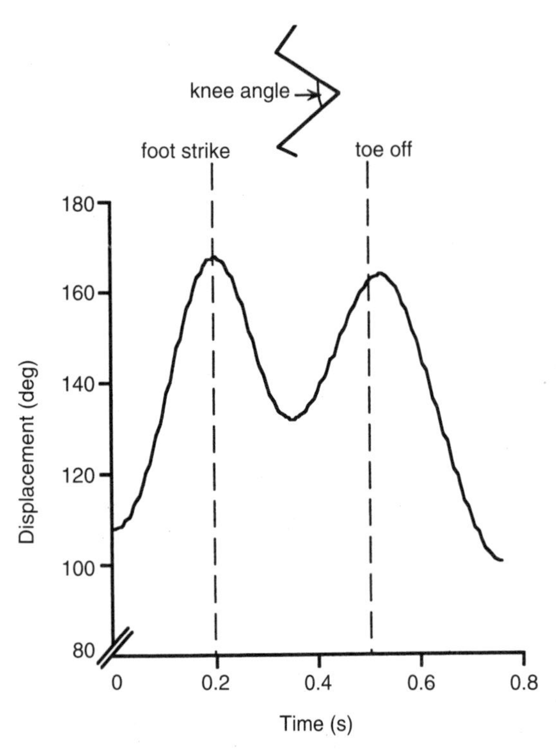

Figure 8.10 During a single running stride the knee angle changes considerably.

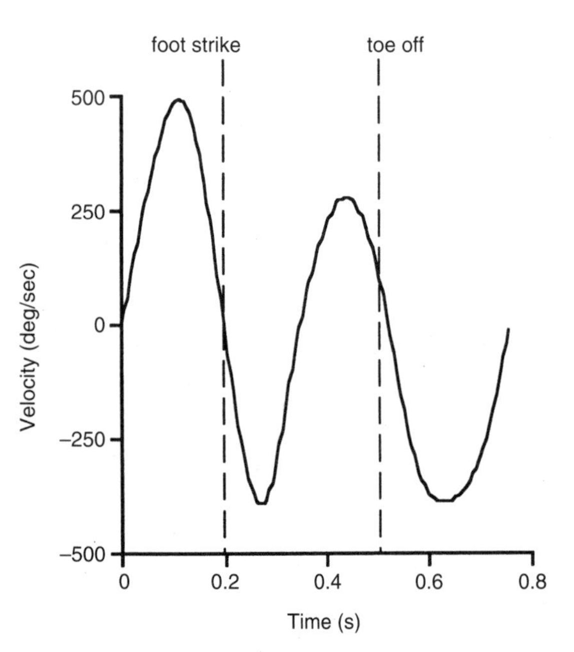

Figure 8.11 By differentiating the data in Figure 8.10, the angular velocity of the knee joint as a function of time can be calculated and plotted.

the second question first, just as Newton discovered the very simple but direct relationship between force and the rate of change of linear acceleration, he also demonstrated a very similar relationship between torque and the rate of change of angular momentum. In motion confined to a plane, where the moment of inertia remains constant, this relationship simplifies to a direct one between torque and angular acceleration. Thus, to fully describe and then understand the causes of angular motion, calculation and comprehension of angular acceleration are required. The angular acceleration (α) of a rotating body is the change in angular velocity that occurs in a finite time period. That is:

$$\alpha = \frac{\Delta\omega}{\Delta t} \tag{15}$$

Figure 8.12 illustrates the angular acceleration of the knee joint during a single running stride.

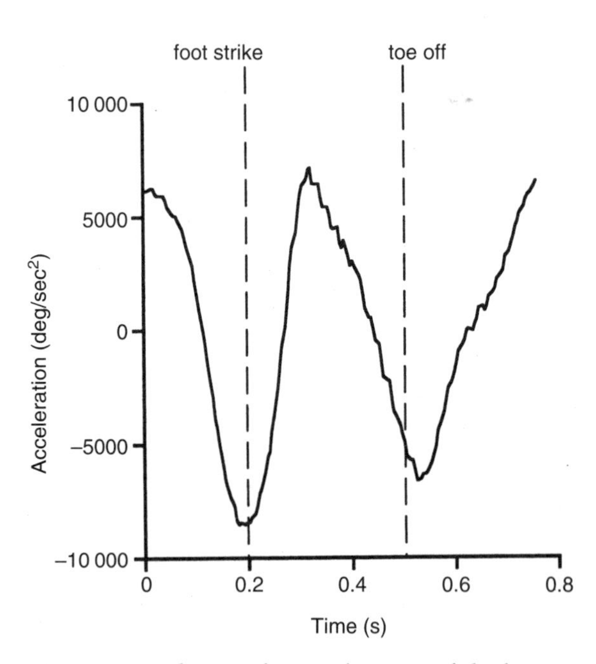

Figure 8.12 The angular acceleration of the knee joint as a function of time during one running stride. These data were obtained by differentiating the angular velocity–time function shown in Figure 8.11.

As noted above, in planar motion, Newton's second analogue states:

$$\boldsymbol{\tau} = \boldsymbol{I\alpha} \tag{16}$$

Where $\boldsymbol{I}$ and $\boldsymbol{\tau}$ are as defined previously.

This relationship is the fundamental one that mathematically describes the way in which a torque will cause a body to rotate.

The relationships among the angular displacement, velocity and acceleration of a rotating body are illustrated in Figure 8.13, where the knee joint motion of a cyclist is plotted on successive graphs. TDC stands for top dead centre and corresponds to that point at which the pedal reaches its maximum vertical position. These data were gathered using a high-speed video camera and processor that allowed the position of markers attached to the hip, knee and ankle to be determined every 5 ms. To compute the angular velocity from the angular displacement graph the differentiation process is used. That is, the instantaneous slope of the angular displacement function is determined for every point in time. The same process is then used to move from the angular velocity to the angular displacement functions.

The inverse process (integration) can be used to proceed in the opposite direction. In other words, to calculate the angular velocity from the angular acceleration graph, the area under the acceleration graph is determined. Similarly, to compute the angular displacement function, the area under the angular velocity–time graph must be calculated.

Angular momentum

A bicycle wheel can be made to spin by applying a force. In the absence of any retarding torques, such as air resistance or bearing friction, the wheel would continue to spin indefinitely at constant angular velocity. This situation is analogous to a body in linear motion. For a translating body a property called momentum was defined as the product of the body's inertia (mass) and its linear velocity (equation (8)). The angular analogue to this quantity is angular momentum ($\boldsymbol{L}$) and it too is the product of two

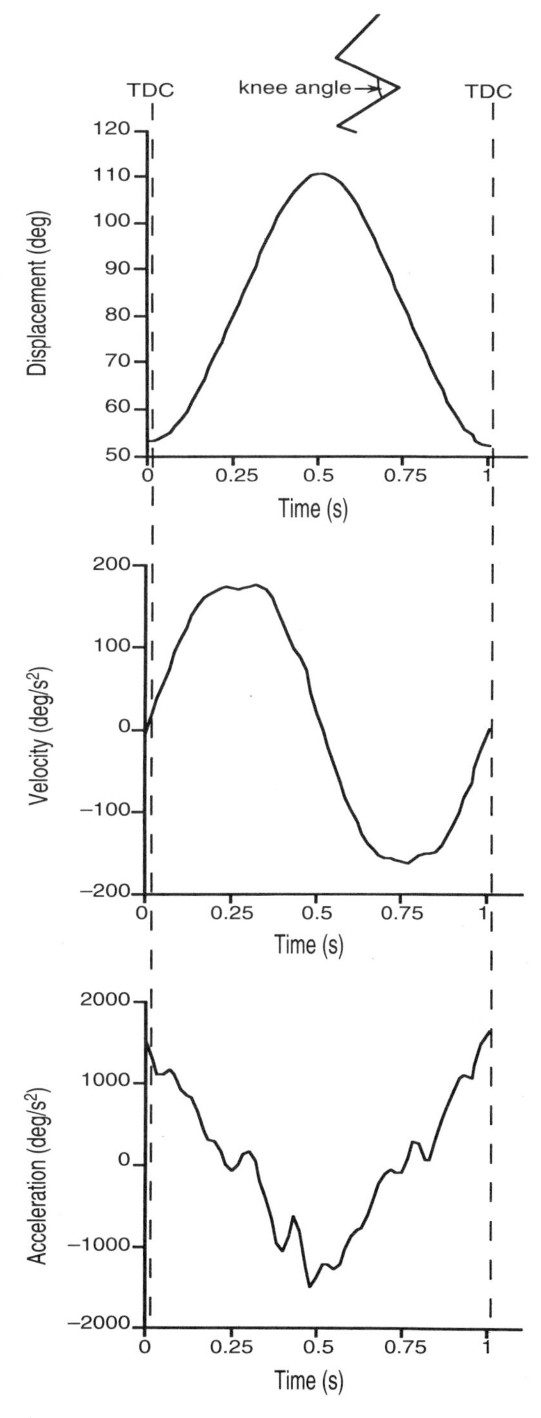

Figure 8.13 The relationships among angular displacement, velocity and acceleration are illustrated for the knee joint of a cyclist during one pedal revolution from top dead centre (TDC) to the next TDC.

terms, the body's moment of inertia (its resistance to a change in angular motion) and its angular velocity. That is:

$$\boldsymbol{L} = \boldsymbol{I\omega} \qquad (17)$$

When no net torque is applied to an object, the angular momentum remains constant. When gymnasts and divers perform airborne manoeuvres, no external torques are being applied and so their angular momentum remains constant. If they change their moment of inertia while in flight, there must be a concomitant change in their angular velocity so that their angular momentum remains constant. Thus, when they tuck tightly, their angular velocity increases (they spin faster!) and when they extend their bodies for landing their angular velocity decreases. Skilful gymnasts and divers carefully control their rates of spin so they can successfully execute their landings or 'entries' by manipulating their moment of inertia. Slowing their rotations at the appropriate time allows them to enter the water vertically (diving) or stay motionless on landing from a dismount

Figure 8.14 A diver manipulates his moment of inertia and thus controls his rate of rotation in preparation for entry to the pool.

(gymnastics). Figure 8.14 illustrates an example of this phenomenon in diving.

Analogies between linear and angular kinematics

Throughout this chapter there has been an emphasis on noting the similarities between linear and angular measures of motion. If one recognises that for every linear term there is an angular analogue it assists in remembering all the kinematic and kinetic quantities. Table 8.3 summarises the most important similarities.

Table 8.3 Similarities between the quantities used to describe linear and angular motion (Vector quantities are shown in bold)

Variable	Linear term	Units	Angular term	Units
Inertia	m	kg	$\boldsymbol{I}$	kgm^2
Displacement	$\mathbf{s}$	m	θ	rad
Distance	d	m	ϕ	rad
Velocity	$\mathbf{v}$	m/s	$\boldsymbol{\omega}$	rad/s
Speed	v	m/s	σ	rad/s
Acceleration	$\boldsymbol{a}$	m/s^2	$\boldsymbol{\alpha}$	rad/s^2
Momentum	$\boldsymbol{p}$	kgm/s	$\boldsymbol{L}$	kgm^2/s
Force	$\boldsymbol{F}$	N	τ	Nm

CHAPTER 9

Basic concepts of energetics

- Work
- Power
- Efficiency

Whether it is exhibited in the form of motion (kinetic energy) or as the potential to do work (gravitational potential energy, strain potential energy), conservation of energy, together with conservation of momentum, is one of the pillars of mechanics. Energy manifests itself in many forms including mechanical, electromagnetic, nuclear, thermal, and chemical energy. The work–energy theorem provides a convenient and useful means to relate the work done by a force in moving an object and the energy it possesses. By using this theorem, some problems can be easily solved without resorting to Newton's second law. It is worth noting, however, that no new underlying principles are introduced and that the work–energy relationship is based on Newton's laws. In this chapter we will examine the concepts of work, power and efficiency, which are central to the subdiscipline of biomechanics.

Work

Work done by a constant force

For most people, work is that arduous chore we do each day that gives us the money to then go and do other things in our free time. In classical mechanics, however, the term 'work' takes on a much narrower and more precise meaning. If a constant force, $\boldsymbol{F}$, is applied to an object and that object moves through a distance (its displacement), $\boldsymbol{s}$, parallel to $\boldsymbol{F}$, the work (W) of that force is defined as the product of the component of the force in the direction of the displacement and the magnitude of the displacement. That is, W is the scalar (dot) product of $\boldsymbol{F}$ and $\boldsymbol{s}$:

$$W = \boldsymbol{F} \cdot \boldsymbol{s} \qquad (1)$$

Although both force and displacement are vector quantities, work is a scalar and its unit of measurement is Nm or J. Consider the following example that is well known to bicycle riders. The retarding force due to rolling friction between the tyres and the road may be considered to be constant, independent of the speed of the cyclist. If rolling friction of the wheels is equal to −3 N (the rolling friction is in the opposite direction to the positive displacement of the bicycle, hence its sign is negative) and is applied constantly over a 40 km time trial, how much work does it do? Application of equation (1) gives the following result:

$$\begin{aligned} W &= \boldsymbol{F} \cdot \boldsymbol{s} \\ &= -3 \times 4.0 \times 10^4 \\ &= -120 \text{ kJ} \end{aligned}$$

Thus, the retarding force of friction does −120 kJ of work on the bicycle–rider system.

Equation (1) implies that for work to be done by a force, the displacement that the object experiences must be non-zero and the force must have a component in the direction of the displacement. For example, a sprint athlete who pushes against the starting blocks applies a force to the blocks but does no work because the starting blocks do not move. The reaction force produced by the blocks, however, does work on the sprinter because his or her centre of mass moves.

Further, if a force is acting perpendicularly to the motion of an object it is doing no work on the object because it does not have a component in the direction of the displacement. To illustrate the second situation consider a person riding on a push bike. In this example the vertical reaction forces at the wheels and the weight of the bicycle plus rider all act in a vertical direction. Since they have no component in the direction of motion (assuming that the bicycle is not moving up and down) they do no work! Thus the frictional forces that are applied at the points of contact between the road and the tyre are the forces responsible for doing the work on the bicycle–rider system. In this example, sliding friction opposes the backward motion of the rear wheel (at its point of contact with the ground) and does positive work on the system.

It is important, at this stage, to distinguish between mechanical work and physiological work. When a person performs an isometric contraction by, for example, hanging by the arms from a bar or holding a suitcase, no mechanical work is done because there is no displacement of the body ($\boldsymbol{s} = 0$). However, the active muscles fatigue because *internal* energy is being used to maintain the object in position. The muscles are 'working' converting chemical energy into tension. In other words, they are performing physiological work, but no motion occurs. Thus, despite the fact that one fatigues in these situations (the arm muscles eventually reach a stage that they can no longer resist the load) no mechanical work is being done because no motion in the direction of the applied force results.

The weight lifter depicted in Figure 9.1 is performing a bench press movement. As he lowers the bar to his chest (Figure 9.1*a*), the force that he applies through his hands ($\boldsymbol{F}_a$) is upwards (positive) and the displacement of the barbell is downwards (negative). The work done on the barbell by forces applied through the hands of the weight lifter is negative since the signs of the two quantities in the product are opposite. Similarly, the weight force of the bar itself acts downwards (negative) and its displacement, too, is downwards (negative). Thus the work done by gravity (W_g) is positive (the product of two terms with the same sign). The net work (W_{net}) done on the bar is the sum of these two works, the negative work done by the forces applied by the person and the positive work done by gravity. Generally, we can state that:

$$W_{net} = \sum_{i=1}^{n} W_i \tag{2}$$

where W_{net} is the net work done on the body by the external forces
W_i is the work done by the ith applied force
n is the number of external forces

Returning to the weight lifting example, if the barbell had a mass of 100 kg and the displacement was 0.7 m downwards, the work done by gravity would be:

$$\begin{aligned} W_g &= -(100 \times 9.8) \times -0.7 \\ &= 686 \text{ J} \end{aligned}$$

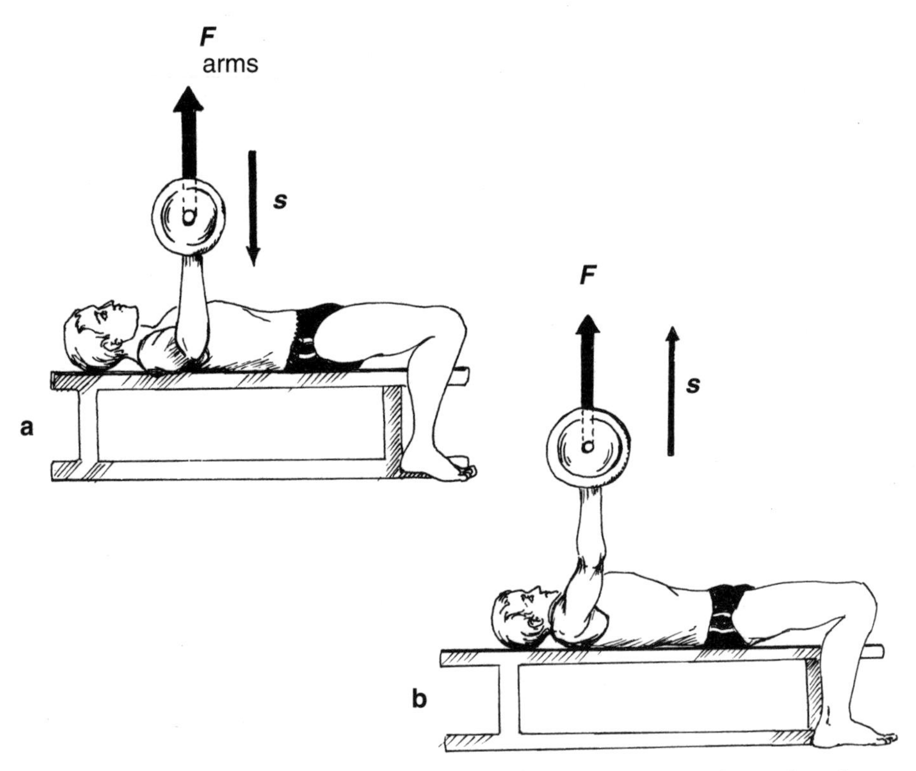

Figure 9.1 The weight lifter does negative work as he lowers the bar to his chest (a). When he raises the weight (b), and the displacement of the bar and the forces applied by the lifter are in the same direction, positive work is done.

If the weight lifter applied a force of 950 N upwards, his work would be:

$$W_a = 950 \times -0.7$$
$$= -665 \text{ J}$$

The net work done, W_{net}, is the sum of these two:

$$W_{net} = 686 - 665 \text{ J}$$
$$= 21 \text{ J}$$

This concept makes perfect sense since the resultant force (–980 + 950 N = –30 N) is acting downwards and the displacement of the bar occurs in that direction also. Thus, the work done by the resultant force is positive (the displacement occurs in the same direction as the force). In fact, this result can be generalised by expanding equation (2):

$$W_{net} = \sum_{i=1}^{n} W_i$$

but, $W_i = \boldsymbol{F}_i \cdot \boldsymbol{s}$ (3)

thus,

$$W_{net} = \sum_{i=1}^{n} \boldsymbol{F}_i \cdot \boldsymbol{s}$$

In terms of muscle contractions, when a muscle produces a force but the muscle lengthens (in response to a greater load than the one it is applying), the work done by the muscle force is negative (the signs of the displacement and the force are opposite) and the contraction is eccentric. Similarly, when the muscle shortens in length as it contracts (a concentric contraction), positive work is done by the muscle. If the muscle contracts (applies a force) but no displacement occurs, as is the case with isometric contractions, no mechanical work is done.

Work done by a varying force

In all the cases discussed above, the force has been assumed to be constant. The notion can, however, be extended to a situation in which the force varies. Figure 9.2 contains a graph in which the force, $\boldsymbol{F}_{ext}$, required to stretch a spring is plotted against the displacement ($\boldsymbol{x}$) of the spring. There are tissues in the human body, such as tendons and ligaments, that behave like springs, albeit non-linear ones. From the graph, the relationship:

$$\boldsymbol{F}_{ext} = k\boldsymbol{x} \tag{4}$$

is evident. The k in this expression is known as the spring constant and is the slope of the line in the graph. If k is small, the spring is not very stiff and does not provide much resistance to stretch. Conversely, if k is large, the resistance is great and the spring is considered to be stiff. If small displacements, $\Delta\boldsymbol{x}$, are marked off and vertical lines drawn up to the force–displacement curve, the work (ΔW) done over each small displacement can be calculated by assuming that the force was constant over that small interval. That is:

$$\Delta W = \boldsymbol{F}\Delta\boldsymbol{x} \tag{5}$$

which is the area of the rectangle bounded by $\boldsymbol{F}$ and $\Delta\boldsymbol{x}$. If all the small ΔWs are added together, the total work (W) done by the force is obtained. In fact, as $\Delta\boldsymbol{x}$ is made smaller and smaller, the total area under the force–displacement function approaches the area of a triangle with base length $\boldsymbol{x}$ and height $\boldsymbol{F}_{ext}$. In mathematical terms the relationship between the extension force applied to a spring, its displacement (stretch) and the work done is:

$$\begin{aligned} W &= \tfrac{1}{2}\boldsymbol{x}\boldsymbol{F}_{ext} \\ &= \tfrac{1}{2}\boldsymbol{x}(k\boldsymbol{x}) \\ &= \tfrac{1}{2}k\boldsymbol{x}^2 \end{aligned} \tag{6}$$

To evaluate the work done by a constantly changing force requires use of the integration concept from calculus as introduced in the previous chapter. The key elements of integration have also been introduced in the spring example described above in that the area of very small rectangles can be added to give the total area between the force–displacement curve and the displacement axis. Thus the concept of adding up the products of the force and the displacement over many, many small intervals is used to evaluate the work done by a constantly changing force. In situations where the force is continually changing, there will not be a simple relationship, such as the one in equation (6), between the work, the force and the displacement.

Figure 9.3 illustrates this idea with data collected using a force plate that measures the force produced by the ground in reaction to the force applied by the feet, as a person executes a vertical jump. In this example the subject started in a crouched position and was required to jump as high as possible without first lowering

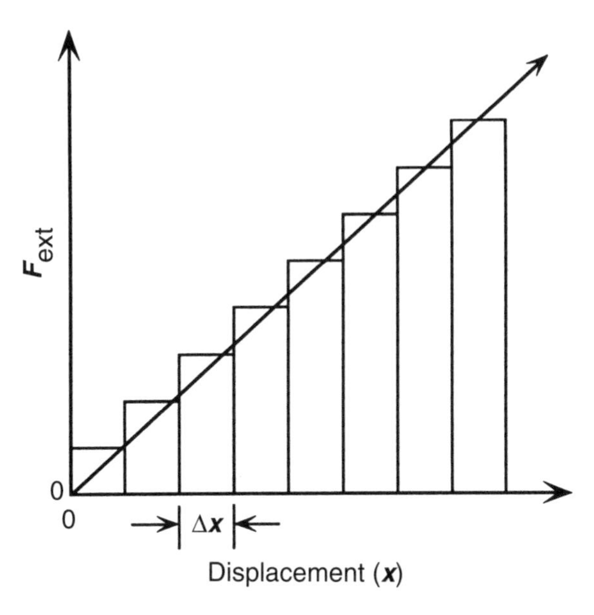

Figure 9.2 The force ($\boldsymbol{F}_{ext}$) required to stretch a linear spring is plotted as a function of the amount of stretch (displacement) of the spring.

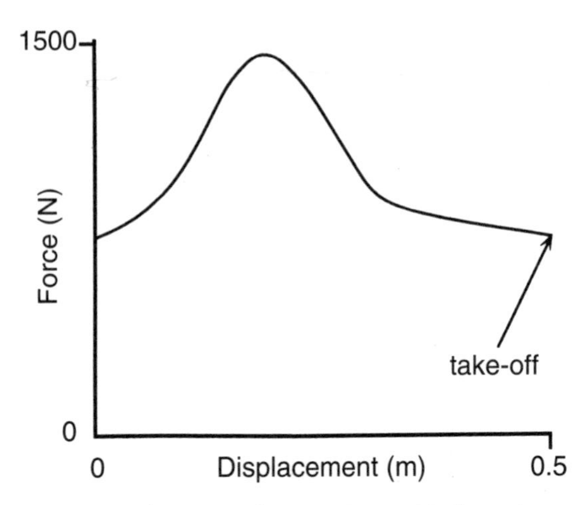

Figure 9.3 The vertical ground reaction force is plotted as a function of the displacement of the jumper's centre of mass during a squat jump.

the body. No counter-movement was permitted. The vertical component of the ground reaction force (the force applied to the body by the ground) is plotted on the ordinate (y-axis) and displacement of the centre of mass of the body is shown on the abscissa (x-axis). Zero on the displacement axis represents the start of the jump when the body's centre of gravity is at its lowest point. The area between the abscissa and the line showing the value of the applied force represents the work done on the body by the ground reaction force and is obtained by numerically integrating this function. This process is identical to the one described above in the spring example in which the areas of a large number of rectangles are summed to determine the total work. In this example, 560 J of work was done by the vertical component of the ground reaction force.

Work and kinetic energy

It was shown above that the resultant force applied to a body does work on it. In Chapter 7 Newton's second law, which relates the net force and the acceleration experienced by a body, was introduced. Consider now the situation in which a constant force, $\boldsymbol{F}_x$, acts on a particle of mass, m, moving in the X direction. Under the influence of this force it will experience an acceleration, $\boldsymbol{a}_x$ ($\boldsymbol{F} = m\boldsymbol{a}$). If we denote its initial displacement, $\mathbf{x}_i = 0$ and its final displacement $\mathbf{x}_f = \mathbf{s}$, then the work done by the force $\boldsymbol{F}_x$ is:

$$\begin{aligned} W &= \boldsymbol{F}_x \cdot \boldsymbol{s} \\ &= (m\boldsymbol{a}_x)\boldsymbol{s} \end{aligned} \qquad (7)$$

By using two of the equations of uniform motion ($\boldsymbol{s} = \frac{1}{2}(\boldsymbol{v}_i + \boldsymbol{v}_f)t$; $\boldsymbol{a} = (\boldsymbol{v}_f - \boldsymbol{v}_i)/t$) and doing a little algebra the following expression can be proved:

$$W = \tfrac{1}{2}m\boldsymbol{v}_f^2 - \tfrac{1}{2}m\boldsymbol{v}_i^2 \qquad (8)$$

The kinetic energy, the energy of motion, of a particle is defined as:

$$E_K = \tfrac{1}{2}m\boldsymbol{v}^2 \qquad (9)$$

Thus, the work done by a force is equal to the change in kinetic energy:

$$W = \Delta E_K = \tfrac{1}{2}m\boldsymbol{v}_f^2 - \tfrac{1}{2}m\boldsymbol{v}_i^2 \qquad (10)$$

This expression is known as the *work–energy theorem*. Kinetic energy, like work, is a scalar quantity and has the same unit of measurement, the joule. Kinetic energy is the energy that an object possesses as a result of its motion and is a measure of its ability to do work. Thus, the greater the kinetic energy of an object the greater its capacity to do work. Equation 10 illustrates that the work done by a constant force on a particle is equal to the change in its kinetic energy. Although the example above is shown for a constant force, this work–energy theorem is equally true for a varying force. If the work done by the resultant force is positive then the speed of the particle will increase (its kinetic energy will increase) whereas if the work of the force is negative, the particle's speed will decrease.

The work–energy theorem can be extended to include rigid bodies as well as particles. The kinetic energy of a rigid body must include not only the energy associated with the motion of the centre of mass of the object but also the energy coupled with its rotational motion. Thus, the kinetic energy of a rigid body is the

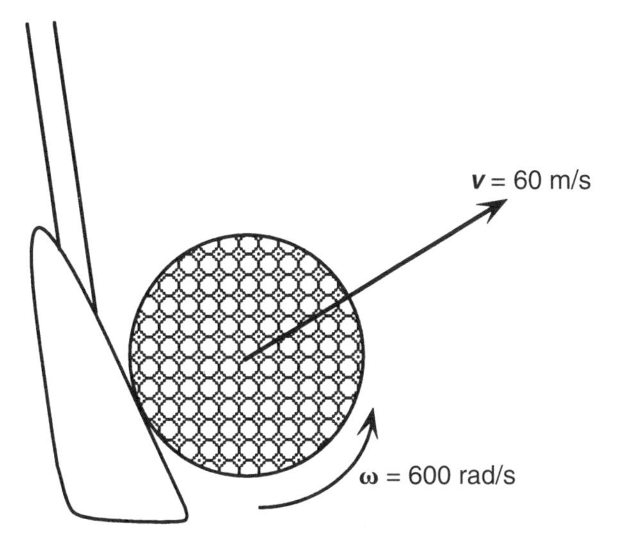

Figure 9.4 A golf ball struck by a club spins and translates in response to the forces applied. Its energy is the sum of the translational and rotational kinetic energy.

sum of its translational and rotational kinetic energies. Formally:

$$E_K = \tfrac{1}{2}m\boldsymbol{v}^2 + \tfrac{1}{2}\boldsymbol{I}\boldsymbol{\omega}^2 \quad (11)$$

Figure 9.4 shows a golf ball that is translating and spinning as it leaves the golf club. If the velocity of the centre of mass, $\boldsymbol{v}$, is 60 m/s and its angular velocity, $\boldsymbol{\omega}$ = 600 rad/s, how much kinetic energy does it possess at this instant (m = 46.0 g; $\boldsymbol{I}$ = 8.4 x 10^{-6} kgm^2)?

$$E_K = \tfrac{1}{2}m\boldsymbol{v}^2 + \tfrac{1}{2}\boldsymbol{I}\boldsymbol{\omega}^2$$
$$= \tfrac{1}{2} \times 0.046 \times 60^2 + \tfrac{1}{2} \times 8.4 \times 10^{-6} \times 600^2$$
$$= 82.8 + 1.5$$
$$E_K = 84.3 \text{ J}$$

From this example, most of the kinetic energy of the golf ball is associated with its translational motion and only a small percentage (2%) is rotational kinetic energy. In other situations where the moment of inertia of the rigid body is reasonably large and the velocity of the centre of mass is relatively small, the rotational kinetic energy contributes a much greater percentage to the total kinetic energy.

Potential energy

Before discussing potential energy it is necessary to introduce the concept of conservative and non-conservative forces. (This notion has nothing to do with political preference!) A force is conservative if the work done by that force as it moves a particle between its initial and final positions is independent of the path followed. For example, gravitational force is conservative because the work that it does depends only on the vertical displacement of the object between its initial and final positions. If a person were asked to move a 10 kg box from a 1.0 m high table to a point 3 m away on the floor (Figure 9.5), the work done by gravity (–10 × 9.8 × –1.0 m = 98 J) is the same regardless of the trajectory followed by the box.

Figure 9.5 A worker moves a 10 kg box from a 1 m high table to the floor.

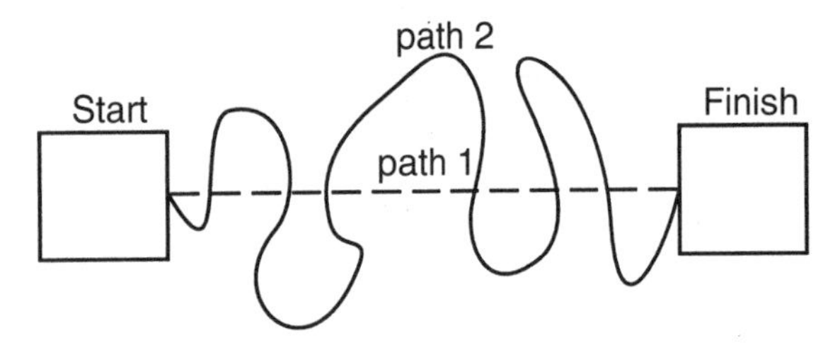

Figure 9.6 The friction experienced by the object as it moves along these two paths is different because friction always opposes the motion. Thus, the friction experienced will be greater along path 2 than path 1.

Friction, the force that opposes the motion or the impending motion, is non-conservative. If an object is dragged over the floor through two different paths (Figure 9.6), the work done by friction on the object will be greater for path 2, the longer of the two. That is, since the work done by sliding friction is the product of the force and the distance through which the object moves, the greater the path length, the greater the work done by friction.

For conservative forces a potential energy function, U, can be defined in which the work done by this force is equal to the decrease in potential energy. Stated mathematically, the work done by a conservative force W_c equals the negative of the change in potential energy:

$$W_c = \int_{\mathbf{x}_i}^{\mathbf{x}_f} \mathbf{F} d\mathbf{s} = -\Delta U = U_i - U_f \qquad (12)$$

where x_i and x_f are the initial and final positions of the object respectively
$\mathbf{F}$ is the applied force
$\mathbf{ds}$ is the displacement
ΔU is the change in potential energy
U_i and U_f are the initial and final potential energies respectively
$\int$ represents the integration operand (the area of a function)

But it was shown previously that the work done by a (conservative) force is equal to the change in kinetic energy of a particle (the work–energy theorem). Thus the following expression can be derived:

$$W = -\Delta U = \Delta E_K$$

thus,

$$U_i - U_f = E_{K_f} - E_{K_i} \qquad (13)$$

or

$$E_{K_i} + U_i = E_{K_f} + U_f$$

This last equation is the *law of conservation of mechanical energy*, which indicates that in a system that is acted on by conservative forces the total mechanical energy remains constant. Thus, any increase in kinetic energy is offset by a decrease in potential energy. The example of a ball being dropped illustrates this concept well (Figure 9.7). At the instant of release the ball has no kinetic energy and the total mechanical energy is in gravitational potential energy ($U_g = mgh$, where h is the distance above the ground). At the moment before the ball hits the

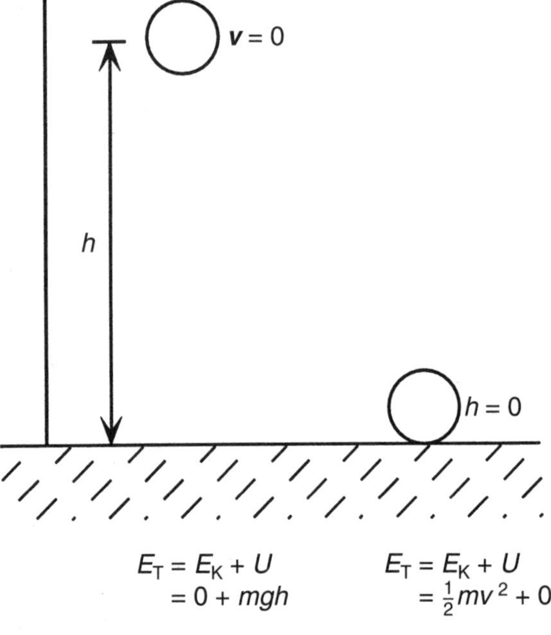

Figure 9.7 A ball is released from a height (h). At the instant of release (time 1) its total energy is in the form of gravitational potential energy whereas at the moment prior to impact with the ground (time 2), it only possesses kinetic energy.

ground, the gravitational potential energy is zero ($h = 0$) and the entire mechanical energy of the system is now expressed as kinetic energy ($\frac{1}{2}m\boldsymbol{v}^2$).

The work–energy theorem takes on a slightly more complicated form for non-conservative forces than for conservative ones. In this situation, the work due to the non-conservative forces, W_{nc}, equals the change in total mechanical energy of the system. Mathematically:

$$\begin{aligned} W_{nc} &= E_f - E_i \\ &= (E_{K_f} + U_f) - (E_{K_i} + U_i) \end{aligned} \tag{14}$$

If this equation is expanded fully for rigid body motion, recalling that the kinetic energy has both a translational and a rotational element (equation (11)), the following expression is obtained:

$$\begin{aligned} W_{nc} &= (E_{K_f} + U_f) - (E_{K_i} + U_i) \\ &= (\tfrac{1}{2}m\boldsymbol{v}_f^2 + \tfrac{1}{2}\boldsymbol{I}\omega_f^2 + U_f) - (\tfrac{1}{2}m\boldsymbol{v}_i^2 + \tfrac{1}{2}\boldsymbol{I}\omega_i^2 + U_i) \end{aligned} \tag{15}$$

This equation provides one of the axioms used to study the motion of rigid bodies or systems of linked rigid bodies such as the human. The application of this principle is used in many situations to evaluate the work done by various muscular forces and the flow of energy within body segments and between segments. Flow of energy within the segments refers to the change in the form of the energy, for example from kinetic to potential and vice-versa. Flow between segments represents the mechanism by which energy is passed from one body segment to adjacent ones.

Power

The ability to produce and utilise power has been described as the limiting factor in human performance. Sprint cyclists have been known to produce power outputs of up to 1800 W for very short periods of time, and it is reported that, for example, Miguel Indurain and Eddy Mercx, both champion endurance cyclists who have won the Tour de France 10 times between them, were able to maintain power outputs of approximately 500 W for over 1 h! But what is power? Very simply, it is the rate at which work is done. That is:

$$P = \frac{\Delta W}{\Delta t} \tag{16}$$

As indicated above, the unit of measurement for power is the watt, which is defined as 1 J/s. Recalling that $W = \boldsymbol{Fs}$, equation (16) can be rewritten as follows:

$$\begin{aligned} P &= \frac{\Delta W}{\Delta t} = \frac{\Delta \boldsymbol{Fs}}{\Delta t} \\ &= \boldsymbol{F}\frac{\Delta \boldsymbol{s}}{\Delta t} \\ &= \boldsymbol{Fv} \end{aligned} \tag{17}$$

Thus, power relates to strength (the force produced) and speed (the velocity). Power is a scalar quantity and it indicates how quickly energy is being consumed or produced.

One of the most commonly used tests of leg power is the Margaria stair climb test in which the time it takes a person to run up six steps (in two bounds) is determined. By knowing the displacement of the centre of mass and assuming that the kinetic energy remains constant, work being done by the forces applied by the legs can be calculated. If the time over which this work is done is known, the power can be determined. For example, if the time taken by a 55 kg person to run up the six steps (total vertical displacement of 1.03 m) was 0.22 s, how much power was used?

$$\begin{aligned} P &= \frac{\Delta W}{\Delta t} \\ &= \frac{mgh}{\Delta t} \\ &= \frac{55 \times 9.8 \times 1.03}{0.22} \\ &= 206\ W \end{aligned}$$

Efficiency

Vehicle manufacturers are concerned with efficiency, particularly for their small cars since this factor is one of the selling features. The most commonly used measure of efficiency in the automobile industry is the fuel economy, which is given in one of two forms: L/100 km or km/L. The fewer litres of petrol burned per 100 kilometres, the greater the efficiency of the car. One of the challenges in this industry is to maximise economy but still maintain reasonable power outputs. With the human body, in sprint activities and *power* sports, such as Olympic weight lifting, maximising power output is the goal, whereas in endurance activities such as long distance running, road cycling and cross-country skiing, the aim is to maximise the economy of performance. Thus endurance athletes attempt to maximise the amount of mechanical work done for the minimum physiological cost.

The efficiency with which humans can perform work is one of the most important determinants of one's physical work capacity. Efficiency is formally recognised as the physiological cost of doing mechanical work and economy is the physiological cost of doing a fixed amount of work. Of the energy produced by the chemical breakdown of adenosine triphosphate (ATP), only 25 per cent of it is available to do mechanical work; the rest is dissipated as heat. Considering the difference in temperature between the source, the human body and the heat sink (the environment) this figure is very high. Unfortunately, there is no simple way of measuring the total energy produced chemically by the body and so determining efficiency is particularly problematic. However, the amount of oxygen consumed and carbon dioxide produced, which are directly related to the chemical energy released, are relatively easily measured. Thus, by knowing information about the expired and inspired gases, the physiological cost of performing certain amounts of mechanical work can be estimated. These concepts are discussed further in Chapter 12.

Typically, in procedures used to measure the physiological cost of an activity, subjects run on a treadmill or ride on a bicycle ergometer. Resistance provided by friction on the flywheel, in the case of a bicycle ergometer, is the torque and if the angular distance through which the flywheel is moved is known, the work done by this force can be determined (provided the frictional force is measured!). It is assumed that the body is doing an equal amount of work to maintain the speed of the flywheel as friction is doing in resisting the motion. If the metabolic cost of doing this mechanical work is measured, too, estimates of economy and efficiency can be obtained. Sometimes efficiency scores are given by taking the quotient of the mechanical work done and the physiological cost associated with doing that work.

Such an approach is flawed, however, for a number of reasons including the fact that the body behaves as an articulated set of links and movement of the links themselves requires work to be done. The above approach of determining efficiency ignores the work done in changing the kinetic and potential energies of the body segments. Although the estimates of the external work done on the flywheel are good, the cost of moving the limbs through their ranges of motion is ignored. Only some of this work to move the body segments may be effectively used to propel the body. In cycling, the rotary movement of the legs results in translation of the whole body and the bicycle. There is a physiological cost associated with moving the links but only a portion of that physiological work actually moves the body in the desired direction. The task therefore in trying to measure efficiency is either to partition out the work done to move the body forward or to estimate the physiological cost of only the positive work done on moving the body's centre of mass. Both of these options require assumptions to be made and the validity of such assumptions is questionable.

Another difficulty in examining the efficiency of the human body is best illustrated by example. Consider a person who is doing dumb-bell presses (Figure 9.8) in which the movement of the arms is anti-symmetrical. That is, as one dumb-bell is raised the other in the contralateral limb is lowered. By the definitions of classical mechanics no work is being done on the system (the human body) because its total mechanical energy remains constant. As the potential energy of one dumb-bell increases there is a concomitant decrease in potential energy of the other one. The same situation is true for the

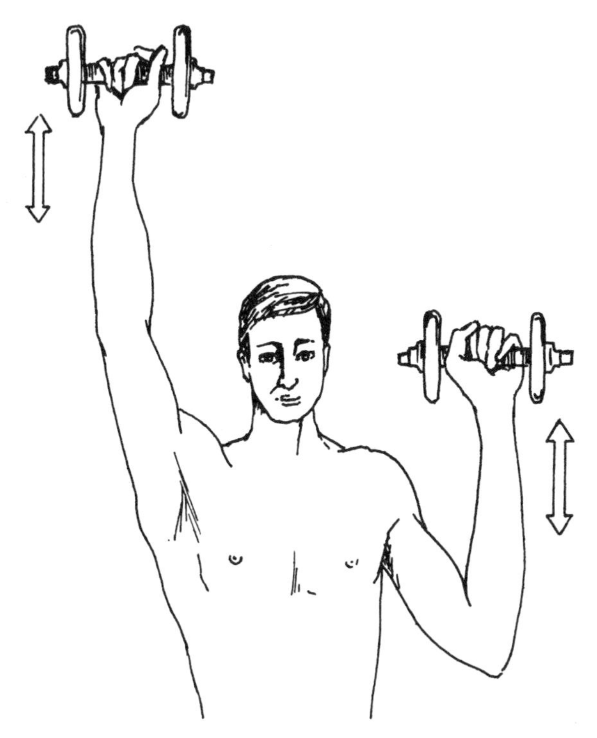

Figure 9.8 Illustration of a person doing dumbbell presses in which the limbs move in an anti-symmetrical manner

kinetic energy of the masses—increases in one are offset by decreases in the other. Everyone realises, however, that there is a considerable metabolic cost of doing this exercise. Is the efficiency of the system then zero? Clearly some modified definitions of the mechanical work of the system need to be introduced to handle this paradox.

What makes this problem even more difficult is that energy may be transferred within and between adjacent body segments and, in the case of muscles that span two joints, the work of the forces that they produce is exceptionally difficult to estimate. The notions of efficiency, energetics and transfer of mechanical energy are truly difficult problems and current research in biomechanics is trying to elucidate methods of accurately calculating work and defining concepts of power and work that make physical sense with respect to movement in and of the human body.

In summary, to increase efficiency we must:

- minimise the internal work, the work done in moving the body segments or links, in doing a fixed amount of work on the whole body
- improve the physiology of the cells and the cardiovascular system so that they can efficiently convert chemical energy to mechanical work
- use techniques of movement that allow storage of energy, for example pre-stretching a muscle before making it contract concentrically thus allowing energy to be stored in the elastic elements of the muscle and tendon, minimising the energetic cost of the movement
- perform activities using state of the art equipment, such as bicycles with three or four spokes, aerodynamically shaped helmets and running shoes that have an energy return system, thus minimising the external resistances to movement that must be overcome or introducing mechanisms of first storing and then returning mechanical energy to the system
- utilise techniques or movement patterns that optimise the transfer of energy within and among body segments, thus dissipating as little energy as possible in other forms such as heat and sound.

CHAPTER 10

Biomechanical changes throughout the lifespan

- Inertia changes
- Efficiency
- Scaling
- Adolescent awkwardness
- Sex differences

As the body grows and develops, changes occur that influence the mechanics of how it moves. For example, as a child grows, its mass increases and its limbs lengthen and these changes alter the loads that have to be borne by the various joints, muscles, ligaments and tendons. Understanding how growth changes alter the biomechanics of movement is an important element in gaining a full appreciation of human movement across the lifespan. In this chapter we bring some of the fundamental concepts of kinetics, kinematics and energetics introduced in chapters 7, 8 and 9 to focus on lifespan changes in inertia and efficiency. We also consider in this chapter the topic of biological scaling, which involves investigation of the size of different animals, including humans, and the relationships among the physical dimensions of animals and fundamental activities such as standing, running and maintaining homeostasis. Of particular importance to the study of human movement across the lifespan are changes in body dimensions in relation to the load-bearing limbs and changes in surface area and body volume with respect to heat loads on the body and the delivery of oxygen to the working cells.

Inertia changes

The inertia of a body (its resistance to a change in state of motion) is measured in the linear and angular sense by its mass and moment of inertia respectively. Bob Jensen, a Canadian researcher, has over the past 10 years provided the most up-to-date and comprehensive set of data on the way in which growth influences the inertia of the human. His work has used both longitudinal and cross-sectional methods to examine how the moment of inertia and the centre of mass and its location change between childhood and adulthood.

As a child grows, its mass increases and so too does its limb lengths. Thus, the inertia of each body part increases through the growing years. Furthermore, the proportional lengths of various body parts change, with the head becoming smaller, relative to stature, with age. Interestingly, the relative masses of various parts show considerably different proportions in children compared with adults. Jensen's work has demonstrated, for example, that the head mass at 4 years accounts for approximately 20 per cent of body mass; at 12 years it has reduced to 10 per cent and, by 20 years, it accounts for less than 7 per cent. On the other hand, the arm and forearm increase their proportions through the growing years whereas the mass of the hands, relative to total body mass, remains essentially unaltered. The feet follow a similar pattern to the hands, whereas the leg and thigh show initial increases until the end of puberty followed by a levelling out or even a decrease in proportional mass. The alterations that occur in the limbs (thighs, legs, arms and forearms) are quite small (< 2 per cent), particularly in comparison with the changes that occur in proportional head mass.

It is well documented that in falling or toppling accidents children are at greater risk of sustaining a head injury than are adults. Although it is arguable that adults, because of their superior movement control and greater experience (Chapter 18), are better able to cushion their falls than children, it is also likely that the large head dimensions of young children, relative to their overall body size, make the head a more probable site of injury for them. Further, because the head is relatively large (20 per cent of body weight) compared with that in an adult (7 per cent), the small child is *top heavy*, making it much less stable than an adult. This decreased stability means a predisposition to toppling accidents.

An analogy may help here. Consider two columns, one (A) that has uniform density and the other (B) that is top heavy but of equal total mass. These columns might respectively represent an adult (A) and a child (B) with their differing body proportions. Assume that the centre of mass of column A is at 50 per cent of its height whereas for column B, the centre of mass is higher, for example at 60 per cent of its full height. If these two columns are subjected to an external force (a perturbation) such that they begin to topple, the weight force provides a torque that angularly accelerates the column toward the ground. The size of this torque is dependent on the weight of the column and its distance from the base. At any angle between the upright position and the ground, the moment arm of the weight force of column B is greater that of column A. Thus, the torque causing the column to fall is greater for column B since the weight forces for both columns are equal. Thus, an appropriate conclusion from this analogy would be that once a child loses balance, it will accelerate toward the ground more rapidly than an adult because of its relatively greater head size and mass.

One of the most interesting findings of Jensen's studies relates to the position of the centre of mass and the length of the radius of gyration of the body segments. The radius of gyration is related to the moment of inertia of a rigid body and is the distance from the axis about which rotation is taking place to the point at which an equivalent point mass would have the same resistance to rotation as the distributed mass. When these two distances are expressed as a fraction of the length of the body segment, the ratios remain essentially unaltered during growth right through to old age. The radius of gyration stays constant at approximately 30 per cent of segment length despite absolute changes in mass and length. In practical terms, this finding indicates that the shape and density of the limbs remain similar throughout the lifespan. In this sense, the limbs of children do appear to be scaled down versions of adult ones.

Despite the fact that the relative location of the centres of mass of body segments and their radii of gyration remain in proportion throughout life, there are distinct implications of the absolute changes that occur in these parameters. The masses, lengths, and moments of inertia of the limbs and trunk of an adult are much greater than those of a child. What are the implications for movement? A long limb or segment is more difficult to accelerate than a short one, particularly if it is heavier than the short one, thus the frequency with which it can be oscillated is lower. In the running cycle, the limbs rhythmically oscillate and, at maximum speed, the major factor restricting further increases in speed is the ability to rapidly swing the non-support limb through for the next stance phase. That is, the frequency of movement controls the maximum speed of a person and frequency of movement is, in turn, constrained by the moment of inertia of the lower limb about the hip joint. The observation that adults have lower stride frequencies than children and that tall adults run with lower stride frequencies than their shorter counterparts for a given running speed has underpinnings that are related directly to the inertial properties of the body. Tall people (adults) cannot move their limbs as rapidly as short people (children)! However, children, because of their relatively short limbs, are required to take more strides to cover the same distance than adults. Thus, the reduced moment of inertia and increased ability to move limbs rapidly, which come with short limbs is offset, to some extent, by the fact that an increased number of strides are required to cover the same distance because of the reduced stride length.

Efficiency

From the previous chapter we can recall that efficiency is the ratio of the external mechanical work done (usually measured on an ergometer) to the physiological cost (estimated by oxygen consumption). A critical question in the context of this chapter is whether efficiency changes as a function of age. Although this question remains largely unanswered at this time there are some strong indications that children may be less efficient than adults in a range of activities, including running and walking.

Physiologists have demonstrated that the oxygen cost of doing a fixed amount of work, the economy of the body, changes as a function of age. For example, children require greater amounts of oxygen per kilogram of body mass to run at a fixed speed, than do adults. Furthermore, in growing children, the maximum oxygen uptake may remain constant, while economy improves. In practice, then, to run at a uniform speed, an 8 year old child is working at a greater percentage of his or her maximum than a 12 year old. (This concept will be discussed further in Chapter 14.) The reasons for this change in economy are many and may include factors such as:

- vertical displacement of the centre of mass; that is, as the child grows, there is less vertical oscillation
- increased stride length
- decreased lateral deviations
- greater ability to transfer energy among and within body segments
- altered body composition
- improved patterns of fuel utilisation and the ability to pace oneself (e.g. the older child or adult is better able to control running speed so that fatigue does not occur rapidly)
- changed muscle strength, elasticity, fibre type and vascularisation
- changed surface area to mass ratio.

When running, the objective is to move the body horizontally and therefore those who are most efficient do this task with as little energy cost as possible. If the body is moved up and down, work is done in raising the centre of gravity. Unfortunately, when this energy is returned as the body falls back to the ground, there are no good mechanisms available to store this energy and it is mostly dissipated as heat and sound. In other words, the muscles do work in raising the body but they cannot store this energy as, say potential strain energy, when the body comes back to earth. Thus, this work can be thought of as wasted since it does not propel the body forward. Such energy loss is clearly inefficient.

Increased stride length must be accompanied by a decrease in stride frequency if running speed is to remain constant (since running speed is the product of these two terms). As a child grows, natural or preferred stride length increases in response to increased limb lengths. Thus, stride frequency must decrease for uniform speed running. There are a number of possible reasons for economy improving as a function of increased stride length. One possible reason is that the amount of internal work done decreases. That is, if the stride frequency is low, the work done in moving the body segments up and down and forward and backward, as the centre of mass traverses forward, is minimal. The greater the number of strides necessary for a fixed amount of external work, the greater the internal work done. Since internal work has a physiological cost, efficiency is reduced because oxygen consumption is increased without any additional external work being completed.

Children have greater vertical oscillations of their centres of mass than do adults. Consequently they expend more energy moving their centres of mass up and down than adults and all of this vertical motion is wasted effort. As a child grows, running technique is altered and the vertical motion of the body is minimised, leading to an increase in running economy (see Figure 18.5).

In adults the side-to-side or lateral deviations of the body are very small whereas in children these excursions are large. An argument similar to the one presented above regarding the vertical motion of the body also holds true in this situation. Lateral movement of the body and its parts is counterproductive to efficiency since such deviations are orthogonal to the desired movement of the body and energy used in moving the body from side-to-side is wasted and inefficient. Lateral deviations are even more inefficient than upward and downward movements; however, because not only does work have to be done by the muscles to move the body in one direction, muscles also have to do an equivalent amount of work to move the body back in the opposite direction. At least, with vertical motion, gravity does the work when the body moves downward!

Patterns of movement of body segments of highly skilled performers are very consistent and are designed to utilise the elastic potential in structures such as tendons and muscles. Such usage increases the transfer of energy among and within body segments. Whenever energy can be stored in elastic tissues, particularly during eccentric activity, the body can then utilise that strain energy to generate kinetic energy at a later time. The trick in making this transfer most efficient is to minimise the time between the eccentric phase of the movement and the concentric phase. As skill increases, people learn how to move the body segments efficiently by timing the involvement of body segments in the action.

This notion of timing is often discussed among sports commentators when describing how a skilled batter or golfer seemed to play a shot effortlessly. Highly skilled sports people have learned how to involve their body segments at appropriate times during the action and to store elastic energy that can then be utilised to generate kinetic energy at a different time. In general, such efficient use of elastic energy is less apparent in the growing child than in the adult.

Altered body composition, improved patterns of fuel utilisation and the ability to pace oneself are discussed in detail in Chapter 14 as part of the consideration of physiological changes across the lifespan. It seems logical that as children lose adipose tissue, the forces generated by their muscles have to do less work moving tissues that have, at best, little physiological benefit. For example, if a young child has 20 per cent body fat and an adult has only 10 per cent, then the child has additional mass to be moved around and this mass does not contribute to the machinery for movement. Clearly, physiological cost is associated with the additional work required in moving this mass, and efficiency or economy suffers accordingly.

Changed muscle strength, elasticity, fibre type and vascularisation can lead to changes in efficiency. For example, if a muscle is working at its maximum it performs less efficiently than if it is working sub-maximally. The analogy of a motor vehicle may assist in conceptualising this idea. If a motor is working near its maximum power output, its efficiency will decrease due to factors such as increased heat production, the inability to completely remove the waste products of combustion and increased friction. A similar scenario is true for biological machines, as they become inefficient due to heat accumulation and the like when operating near maximum. Thus, if a child is working near maximum in order to complete a fixed amount of external work, that child will be less efficient than an adult, who need work only sub-maximally in order to complete the same amount of work.

As you will see in Chapter 14 many physiological factors change as a child grows. In most instances it is extremely difficult to tease out whether changes that occur in physiological work capacity across the lifespan are a result of biomechanical factors, such as improved technique or the capacity to store and utilise elastic energy, or whether they are due to physiological functions, such as improved ability to transport and utilise oxygen. This topic, like many others in human movement studies, crosses the boundaries of interest of two or more of the subdisciplines and thus requires well constructed, collaborative research to ascertain those changes that are due to technique adaptations and those that are due to altered physiological capacities. (Part 3 of this book provides a number of examples of topics, such as this one, that favour multidisciplinary and crossdisciplinary approaches to knowledge gathering.)

Scaling

Fundamentals

The childhood story of *Gulliver's Travels* is often used as a familiar example to illustrate a case in which the human form is scaled up and down. In Swift's story, the Lilliputians were exactly the same shape, that is, they had dimensions proportional to adult humans but they were 12 times smaller and the Brobdingnagians were also in exactly the same proportions as the adult human but they were 12 times larger. Animators also use scaling to create super-heroes or animals in cartoons that are scaled up versions of humans or other animals. Although a notion of simple linear proportioning is conceptually easy to comprehend and apparently straightforward to create, the whole notion is flawed when applied to both living and non-living matter as seen in nature. Unfortunately, it is not just a simple matter of merely linearly scaling all dimensions by some factor, to make the enlarged or reduced form. More complex scaling is apparent in biology.

Take the example of building a block of units. The engineer, when deciding on the foundations, considers how much weight or load will be applied to the footings, and then makes them of appropriate depth, length and width. If, however, a developer decided to erect another two levels above the existing ones, most people know, intuitively, that this would simply not be possible. In order to have additional floors the foundations would have to be made stronger to withstand the extra load of the weight of the materials in the second and third floors.

This same principle applies to animals as well as non-living structures. When comparison of the differences in the form and function of the smallest animals through to the largest ones is made, not only is a dramatic increase in size (of the order of 10 000 fold) apparent, but there is also a change in the shape of the load-bearing structures or tissues in the body. For example, when comparing the strengths and weights of the wing and leg bones of flying birds remarkable adaptations are apparent. The wing bones are hollow and extremely light whereas the leg bones are much thicker, heavier and stronger than the wings. These alterations represent adaptations to the stresses applied to the structures. During flight, the bird needs to minimise energy expenditure, so it makes sense to have very lightweight bones; however, when landing, the whole body weight must be supported and therefore the columns (legs) that provide the support, are 'over' designed, providing a large safety factor. This type of adaptation is typical of both the plant and animal kingdoms.

As we noted earlier in Chapter 3, the design of most bones in our bodies is a hollow form in cross-section because this type of structure gives greater strength than a solid structure made up of the same quantity of material. That is, a hollow bone is stronger in bending than a solid one of equal weight. Engineers have recognised this fact for a long time and modern-day bicycle manufacturers also use this principle when constructing the frames of racing bikes. The design brief for such machines is that they must be strong enough to withstand the stresses of riding but be of minimal weight. The best design to meet these requirements incorporates thin-walled, large-diameter hollow sections that minimise weight but maximise strength.

Appropriate scaling rules should also be used in modifying adult games, activities and sports for children. A number of important questions in children's sport relate to scaling. For example, in athletics how high should children's hurdles be and over what distances should running races be scheduled? In ball sports is it appropriate to give a child a scaled- down version of an adult tennis racquet or a cut-off golf club? A strong case can be made that the use of more appropriate designs than are currently in use, which take into account the size and strength of children, may increase the enjoyment, satisfaction and success that children derive from these games.

The study of the structural and functional changes in the shape and size of bodies is known as *scaling*. The field of scaling examines questions relating specifically to the ways in which dimensions such as mass, height, and strength change as a body alters in size. We will now look at dimensions and see how they are related. In doing so, we will discover that many dimensional changes can be ultimately related to growth or change in a single dimension of a body.

Geometric scaling is somewhat familiar to most of us. In Euclidean geometry, similar triangles are those in which the angles from one are exactly the same as the angles in the other, or the lengths of the three sides of one triangle are in the same proportion as the three lengths of the sides of a second triangle (Figure 10.1). The same similarity can be shown with a cube (Figure 10.2) in which the lengths of the sides remain in constant proportion. We can also see that the surface area of any one face of a cube does not vary with a face of another cube in the same linear fashion as their sides. In other words, the linear dimension of side length is scaled as 1:2, but the ratio for face surface area scales as 1:4 (a quadratic function). Furthermore, the volume varies in the ratio of 1:8 (a cubic relationship).

A simple mathematical way of expressing these relationships is to use an equation of the form:

$$y = ax^b \tag{1}$$

where y is the dependent variable
a is a constant
x is the independent (predictor) variable
b is a constant (for a particular relationship)

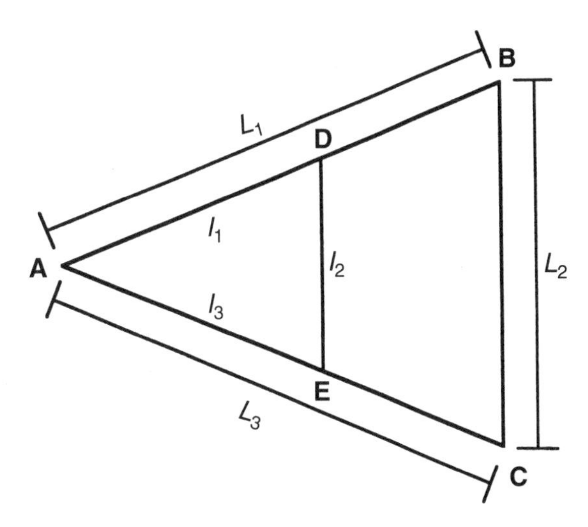

Figure 10.1 Similar triangles have identical angles and side lengths, which are in proportion.

Using this notation, the surface area and volume of cubes can be related as follows:

$$A = x^2$$
$$V = x^3 \tag{2}$$

where A is the surface area of the cube
x is the length of the side of the cube
V is the volume of the cube.

This notion can also be extended to plot the relationship between surface area and volume. Figure 10.3*a* shows this relationship graphically. When cube surface area is plotted against volume, the equation is:

$$A = aV^{\frac{2}{3}} \tag{3}$$

If, however, surface area per unit volume is plotted (Figure 10.3*b*) against volume, the relationship takes on the form:

$$\frac{A}{V} = aV^{-\frac{1}{3}} \tag{4}$$

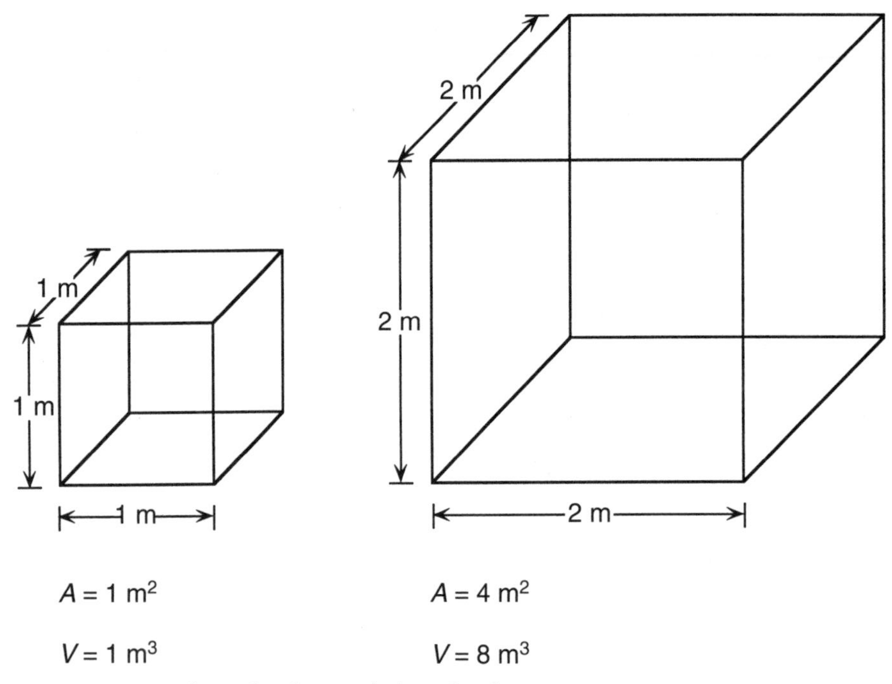

Figure 10.2 Similar cubes have side lengths that remain in constant proportion.

which indicates that surface area per unit volume decreases as the total volume increases. Again, this notion is probably familiar to you: the same volume of sugar dissolves more readily in a drink if it is in fine granules than in large chunks. With the fine granules, the surface area per unit volume is large, thus allowing the sugar to dissolve easily. Another familiar example is that children lose heat more quickly than adults because they have a greater surface area per unit of mass than adults.

Although geometric or isometric scaling is useful for geometric objects such as cubes, polygons, triangles and so forth, most biological specimens do not scale geometrically. Rather, they tend to scale allometrically (non-symmetrically). Despite this, relatively simple allometric relationships have been observed across a large range of animals of different sizes. One of the most useful tools for describing these relationships is the log–log graph, in which the dimensions that are being investigated are plotted on logarithmic scales. The beauty of this approach is that when the natural logarithms of exponential relationships, such as equation 1, are evaluated, the resulting relationship turns out to be a very familiar linear regression. Taking logs of equation 1 yields:

$$\begin{aligned} y &= ax^b \\ \log y &= \log a + b \log x \end{aligned} \qquad (5)$$

In this equation, $\log a$ is the y-intercept of the linear function and b (the exponent of equation (1)) is the slope of the line. Such equations are relatively easy to visualise, in that an increased slope (large b) means that the predictor variable increases rapidly in response to an increase in the independent variable. Similarly, a negative slope in a regression line is easily pictured. A negative slope indicates that the predictor variable decreases as the independent variable increases.

This method of taking logs of exponential relationships is used regularly to describe correlations among biological systems that vary

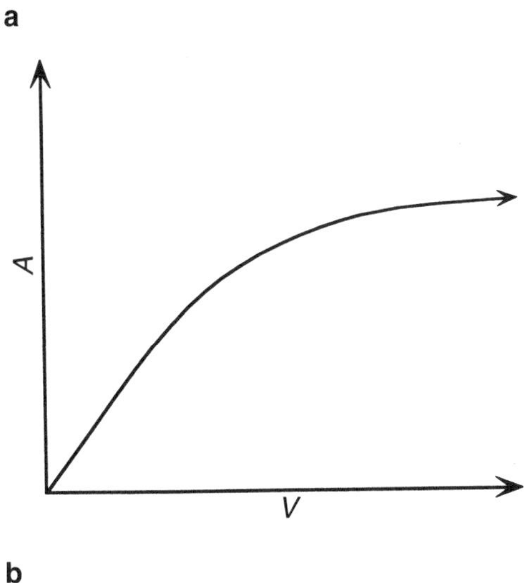

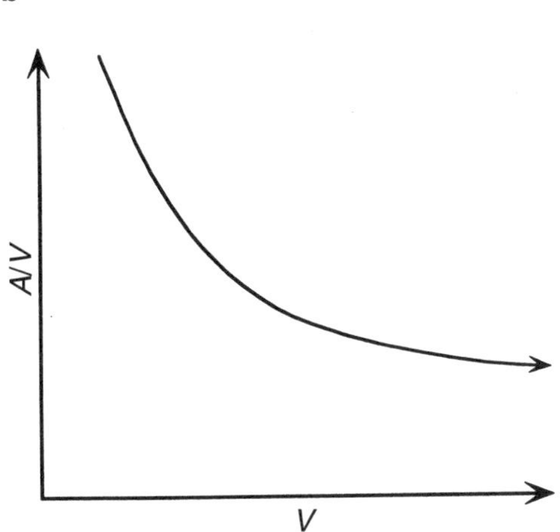

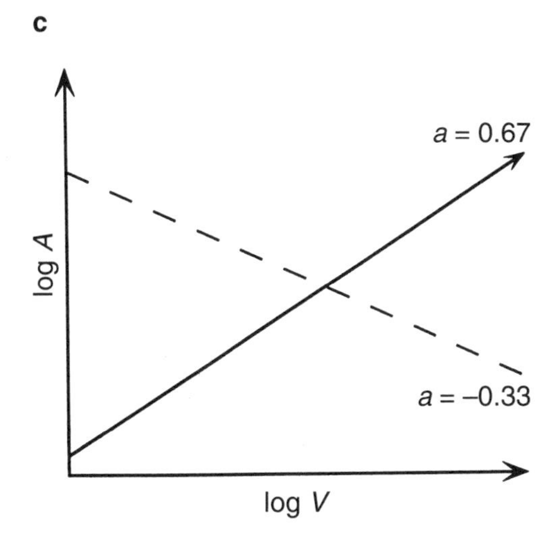

Figure 10.3 Illustration of the function obtained when the surface area of a cube is plotted as a function of its volume (a). The relationship takes on a different form if the surface area per unit volume is graphed as a function of volume (b). If the logarithms of these two relationships are taken and plotted (c), a direct linear relationship is obtained for area (solid line) and an indirect one is found for area per unit volume (dashed line).

substantially in size. Where there are differences of many thousands in mass, length, and volume of animals, using logarithmic scales transforms the data into a form in which relationships become simplified. Figure 10.3*c* shows how the volume–area relationships, discussed earlier for the cube, become simple linear regressions when plotted on logarithmic scales. The examples, described below, should help in understanding the importance of scaling in the biological world. However, before examining these examples, four basic relationships should be explicitly stated. These relate the mass of an object to its linear dimensions:

$$\begin{aligned} m &\propto l^3 = V \\ l &\propto m^{\frac{1}{3}} \\ A &\propto l^2 \\ A &\propto m^{\frac{2}{3}} \end{aligned} \qquad (6)$$

These relationships have major implications for the strength and size of bones relative to the mass they have to support, the amount of heat a body can lose and therefore its metabolic limits, temperature regulation, and the means of supplying oxygen for metabolism. Other derived relationships can also help to explain why children do not have the same functional capacities as adults.

Observation of the animal kingdom provides evidence of evolution within structures that enables animals to function with low risks of structural failure. For example, large animals such as elephants and rhinoceroses have short, squat legs to support their large body masses whereas other animals (antelope and giraffe) have very long, slender limbs. These two species have short, light bodies relative to their heights.

In many physical activities, structures in the human body are stressed to their limit and failure is a common problem. For example, Patrick Cash, a former Wimbledon tennis champion, completely ruptured his Achilles tendon (the largest tendon in the body!) during a normal game of tennis. Perhaps the human body has not been designed to withstand the types of load that are generated during high-level, competitive sports. Further, perhaps its tissues are not able to adapt quickly enough to the loads that must be produced to compete at the very highest levels in modern-day sports.

How does the strength of bones vary?

The skeleton is the support structure that both prevents the organism from collapsing and yet facilitates movement. It is, as described in Chapter 3, a system of articulated levers that allows rotations and translations. It also has many other biological functions such as formation of red blood cells and protection of vital organs. An interesting question to ask, therefore, is how does the skeleton scale in relation to the total mass of the body?

To answer this question, consider that the body of an animal, say the human, is supported on columns (the lower limbs). The total load that has to be supported by the columns is the mass of the body (M_b). If the compressive strength of the bone is assumed to be constant, increased strength can only be obtained by increasing the dimensions of the bone. Galileo realised this idea in the early 1600s! To increase the strength of the supporting columns, their cross-sectional area must increase in direct proportion to the load that they have to bear. That is, the slope of the logarithmic regression line must be 1. Alternatively, cross-sectional area must increase in proportion to M_b^1. As the body increases in mass, all of its linear dimensions will increase, therefore the length of the column increases in proportion to $M_b^{\frac{1}{3}}$ (see equation (6)). The volume (or mass — see equation (6) again) will be the product of cross-sectional area (M_b^1) and length ($M_b^{\frac{1}{3}}$), and then must be proportioned to $M_b^{1\frac{1}{3}}$. To summarise, the skeletal mass should vary in proportion to the mass of the body raised to the power of 1.33. In other words, to support the increased mass without risk of structural failure, the rate of increase of the mass of the skeleton must be greater than the rate of increase of the mass of the entire body.

The relationship between body mass and skeletal mass for a range of animals is shown in Figure 10.4. The regression line for these data is:

$$M_s = 0.061\, M_b^{1.09} \tag{7}$$

where M_s is the skeletal mass (kg)
M_b is the body mass (kg)

If the scaling was based on the requirement to withstand the gravitational load, it would appear, from these data, that the skeletons of the very small animals were over-dimensioned or alternatively that the large animals were under-dimensioned. (The predicted exponent of 1.33 is greater than the observed exponent of 1.09.) In other words, the rate of increase of skeletal mass as total body mass increases is

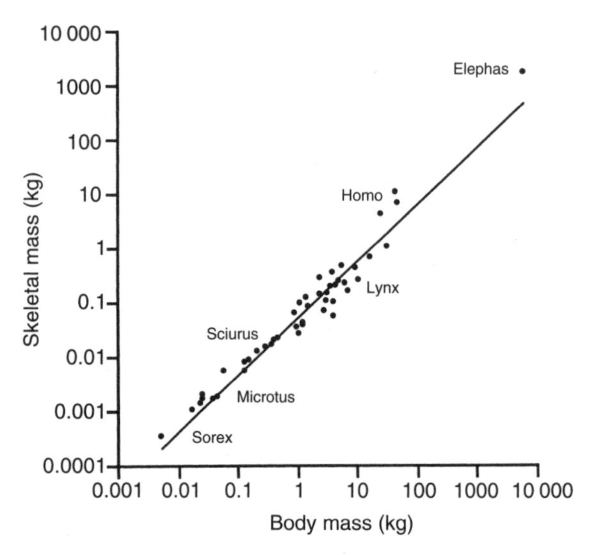

Figure 10.4 Skeletal mass is plotted as a function of body mass for a range of species from the animal kingdom.
Source: Prange, H.D., Anderson, J.F. & Rahn, H. (1979), Scaling of skeletal mass to body mass in birds and mammals, *The American Naturalist*, 113, 103–122. (Copyright 1979 by The University of Chicago Press; reproduced with permission.)

not as high as predicted on the basis of merely having to withstand gravitational load.

The eminent scientist, McNeil-Alexander, has provided an elegant and substantive solution to the apparent discrepancy between the predicted result and the empirical (observed) data of Prange, shown in Figure 10.4. McNeil-Alexander has shown that maximum stress during locomotion is the most likely criterion for scaling by demonstrating that stresses during hopping and running across a large range of species are fairly constant, between 50 and 150 MN/m^2. This range is close to the ultimate strength of bone and indicates that this is the likely basis for scaling, rather than merely the mass that has to be supported against gravity.

From a theoretical perspective based on observed data, the volume of a cylinder is proportional to its diameter squared and length. Using this assumption and the data of McNeil-Alexander, who found that leg lengths were related to $M_b^{0.35}$ and that bone diameter related to $M_b^{0.36}$, we find that skeletal mass should scale in proportion to $M_b^{1.07} = (M_b^{0.35} \times (M_b^{0.36})^2)$. This model agrees well with the data provided in Figure 10.4.

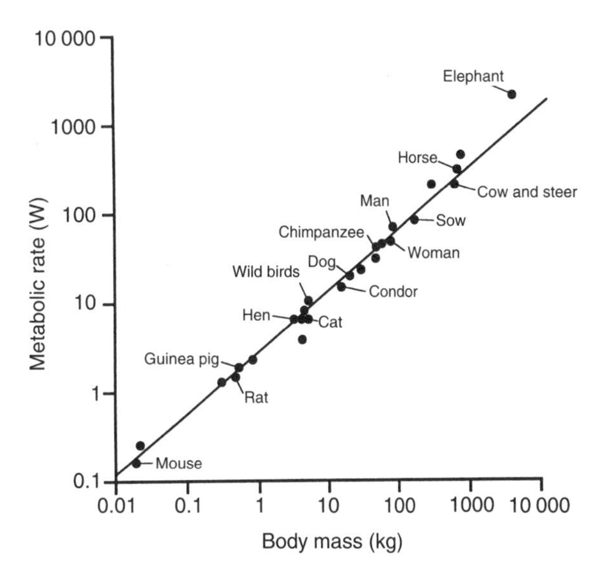

Figure 10.5 The mouse-to-elephant curve made famous by Benedict
Source: Schmidt-Nielson, K. (1991), *Scaling: Why is Animal Size so Important?* Cambridge University Press, Cambridge, p. 57.

How is metabolism influenced by size?

Considerable data have been collected and published relating metabolic rate and body mass. Although it is obvious that organisms use both anaerobic and aerobic metabolic pathways, the generally accepted method of assessing metabolic rate is to measure the rate of oxygen consumption (see Chapter 12). The data shown in Figure 10.5 (the now famous mouse-to-elephant curve adapted from Benedict, 1938) relate the body mass to the metabolic rate for a range of mammals differing in mass by over 200 000 times (the mass of an elephant is more than 200 000 times the mass of a mouse). The slope of the regression line is 0.75, which means that metabolic rate is proportioned to $M_b^{0.75}$. A simpler way of interpreting this relationship is to note that the metabolic rate does not increase as rapidly as body mass.

If a primary consideration for a mammal is to maintain constant body temperature, then we would expect that the oxygen consumption would be related to body surface area. Since surface area is proportional to l^2 or $M_b^{\frac{2}{3}}$ (equation 6) where l is a linear dimension of the body, then we would deduce that metabolic rate should be proportional to $M_b^{\frac{2}{3}}$. That is, smaller animals that have large surface areas relative to their masses should have higher resting metabolic rates than large animal that have small surface areas relative to their body masses, given that the primary mechanism of heat loss is through the surface of the body. This observation is correct, at one level, in that small animals such as mice do have higher metabolic rates (relative to their body mass) than large animals; however, the empirical data do not support perfectly the geometric prediction. The slope predicted on the basis of a surface area criterion would be 0.66 but the observed data have a slope of 0.75. What biological reasons account for the difference between the observed slope and the predicted slope?

Without going into the mathematical detail, it is sufficient to be aware that an elegant and logical solution does exist. If elastic similarity rather than geometric considerations is used as the scaling variable, the fit of the theoretical model to the empirical data is excellent. Elastic similarity is similarity between structures in the animals that are similarly threatened by elastic failure under their own body weight. Use of this criterion (rather than geometric or allometric considerations) indicates that length (l) scales in proportion to $M_b^{\frac{3}{8}}$. Since metabolic rate is proportional to the surface area through which heat is lost, or is proportional to l^2, it follows that skeletal mass is proportional to $(M_b^{\frac{3}{8}})^2 = M_b^{\frac{3}{4}}$. The empirical data fit nearly perfectly and lend support to this theory.

In summary, the criterion of elastic similarity appears to provide a plausible, biological basis for scaling metabolic rate to body mass. The research findings from a host of other physiological measures such as heart rate and frequency of breathing all support this model, with empirical data showing excellent fit with the theoretical model. In fact, the least squares fit of the log–log data shows linear regression slopes precisely equal to the predicted 0.75. Biological events therefore appear to fundamentally follow physical/mechanical laws.

Implications during growth

Use of geometric scaling is a useful means of gaining insight into the extent to which strength changes during growth are merely a consequence of the altered dimensions of the body. An example based on two people of different size is useful as a means of illustrating some of the key ideas here. Consider one child who is 100 cm tall and another who is 150 cm; their heights are in the simple ratio of 1:1.5. Using the fundamental relationships described above, any cross-sections, such as in bones, muscles, blood vessels and airways, will scale as 1:2.25 whereas volumes will scale as 1:3.375. That is, the taller person's cross-sectional areas will be 2.25 times larger and their volumes 3.375 times larger than those of the smaller person. Thus if the small person's mass was 20 kg the larger person's would be predicted to be 67.5 kg (3.375 × 20 kg).

Using these ratios, we would expect that the larger person would be able to lift 2.25 times as much weight as the smaller person since muscle force is proportional to cross-sectional area. Further, since work is the product of force and distance the larger person should be capable of doing 3.375 times as much work as the smaller person. However, if the task involves movement of one's own body mass as, for example, in doing chip-ups, the situation reverses and the larger person is handicapped by the additional mass. In fact, in such activities the scaling is proportional to the reciprocal of height. Thus if the height of the smaller person is considered to be 1, the strength per unit mass is 1. For the taller person who was 1.5 times taller, the strength per unit mass reduces to two-thirds (1/1.5).

Frequency of movement is inversely proportional to length, which is consistent with everyday observations: tall people (adults) use longer strides at a lower rate than shorter people (children). There are other good physical reasons why frequency is inversely proportional to length. Think about the following situation. A rod that is short and relatively thick, if swung back and forth rapidly, will withstand the internal stresses and will not fail. However, if a long, thin rod, similar to one of the long bones in the body, is moved at the same frequency its own inertia will cause it to fail. For this reason, humans cannot move their limbs as quickly as small animals such as birds and mice, and even if the muscles of one's body could somehow be altered to allow a moderate, say 25 per cent, increase in speed, the bones and probably the soft tissues also would be likely to fail!

Another interesting prediction made on the basis of the scaling of body dimensions relates to jumping performance. The ability to raise the body's centre of mass is directly proportional to the force that the muscles can produce and the distance through which the muscles can contract and inversely proportional to body mass. Using the relationships described earlier, jumping performance should be independent of size because $\boldsymbol{F}$ is proportional to l^2 and m is proportional to l^3. Thus, force times distance divided by mass leaves a ratio of 1 ($l^2 \cdot l \cdot l^{-3}$). Across a wide variety of animal species this relationship seems to be true although the taller animals have an advantage because their centres of mass start at a higher level than those in smaller animals. There is also an advantage for those people who have long lower limbs and a body mass that does not scale with l^3, but does so with a lower proportion, effectively making them tall for their weight. For these reasons adults are at an advantage over children with regard to high jumping performance.

Adolescent awkwardness

A common belief held by many people is that during the growth spurt all adolescents, particularly males, go through a stage at which they appear to become extremely clumsy and are described as having 'outgrown their strength'. Researchers have been interested in this phenomenon for a number of years dating back to the early 1920s. Systematic research of a large sample of Belgian boys in the 1970s and 1980s has provided quantitative data on adolescent awkwardness, some of which is summarised below.

During puberty there are rapid changes in the size and strength of bones and muscles and it has been proposed that, because of this rapid change in the dimensions of the body, performance on motor tasks requiring strength, power and co-ordination declines. The foundations for such a proposal include the contentions that: (i) because increased length and mass cause greater moments of inertia, this will in turn cause clumsiness, and (ii) because increases in muscle size and strength are often delayed relative to skeletal growth, performance in activities requiring strength and power is difficult.

The evidence is rather equivocal regarding these contentions. The studies that have been conducted on this topic indicate that a significant portion of boys show a decline in performance, during the growth spurt, on motor tasks requiring movements in which they have to work against or resist their own body weight. Interestingly, however, there is a very low percentage of boys who decline in their performance of explosive tasks such as vertical jumping. As has been pointed out in earlier chapters, during the growth spurt there is a direct, positive relationship between strength and age, indicating that absolute strength increases. However, because skeletal growth precedes strength increases, activities that require work against the body's own weight are affected negatively. Any decrements in performance that occur during the growth spurt are temporary and do not influence performance on such tasks as a young adult.

It has been proposed often that motor co-ordination and stability decline during the year in which peak height velocity occurs yet only a very small percentage of boys (1.6 per cent) in the Belgian study actually showed decline in motor co-ordination tasks such as plate tapping and balance during this period. This finding highlights that while there are a few boys who suffer temporary periods of clumsiness, this behaviour is not typical of adolescence. In fact, most boys (98.4 per cent) improve their motor co-ordination during the growth spurt, and this improvement continues through into early adulthood.

In summary, the concept of adolescent awkwardness is complex because of the ways in which it is defined and because some people will show a decline in performance on some tasks but not others. However, it is important to remember that the majority of adolescents improve on all aspects of co-ordination and strength during their major growth spurt. There is no compelling evidence to suggest that the notion of adolescent awkwardness is a particularly useful one, especially since there are even fewer indications for any loss of motor control during growth for girls.

Sex differences

Surprisingly little information is available on differences between males and females in terms of both inertial characteristics and how various biomechanical parameters change across the lifespan. Clearly males are more massive and have longer limbs and torsos than females but the few reliable research results available have shown no differences in the proportions of mass of various body segments, the distance to the centre of mass or the proportional radii of gyration.

One key biomechanical characteristic that obviously is different between the sexes is the location of the whole body centre of gravity. When the height of the location of the centre of gravity is expressed as a percentage of stature (height), females and males have percentages of 55 and 57 respectively (see Figure 7.15). These differences reflect the differing mass distributions in males and females, males tending to be larger in the shoulders and chest but, in absolute terms, equally wide across the hips as females. Whether these differences have any consequences in terms of motor performance is debatable. Theoretically, females with their lower centre of gravity should be more stable in static situations than males but whether a 2 per cent difference actually provides a functional advantage has yet to be determined.

CHAPTER 11

Biomechanical adaptations to training

- Optimisation
- Energetics/economy

In the previous chapter it was demonstrated how changes in growth throughout the lifespan substantially alter a number of key biomechanical determinants of human movement. In this chapter we examine how training and practice bring about adaptations that, in the main, enhance movement mechanics.

Optimisation

Fundamentals

A key foundation on which training and practice is based is that the body, as a biological system, functions more optimally as a consequence of the adaptations that occur in response to training. The notion of optimisation implies that the various sub-systems of the body responsible for movement become better able to contribute to movement as a result of training and come to function in a co-ordinated manner approaching their theoretical optimum. With training, the sum of the changes in the system components is usually greater than any of the improvements in the component processes considered alone. The effects of training may be pronounced with novice performers, with improved performance being readily apparent after a short period of time. For more skilled performers the optimising effects of training or practice may be less apparent but nevertheless significant. Small changes can make a big difference to the outcome of movements performed in the context of competition. If a swimmer improves his or her efficiency by 0.5 per cent the cumulative effect of 0.15 s per lap in a 1500 m race (an approximate 0.5 per cent improvement) can make a difference of the order of 4.5 s to the overall swim time, and this could certainly be the difference between a gold medal performance and finishing well behind the winner.

Another example may also help to demonstrate why the optimisation of movement is an interesting and important area of study. When learning to swing a golf club, precisely timed activation of muscles crossing many joints in the body is necessary to produce the smooth movement needed to generate maximum clubhead speed, orientation and direction at impact. In order to produce this co-ordinated movement, the muscles must be activated at exactly the right time, be at precisely the correct length, be lengthening/shortening at the appropriate rate and have the necessary strength and endurance qualities. Furthermore, for the co-ordinated movement to be translated to superior performance, the equipment being used must be matched with the performer. With the right combination of these events, an optimal performance can result. Performance optimisation, therefore, requires simultaneous control over many factors and as such has attracted the collaborative research of scientists from a number of different subdisciplines (see also Part 3).

One of the essential, and often difficult, aspects of optimisation is the ability to measure the specific changes that take place in response to training. When an athlete does strength training, for instance, it is relatively easy to determine the change in strength although the type of strength gain, as we shall see in Chapter 13, may be very much dependent on the nature of the training undertaken. However, measuring any accompanying changes in the strength of connective tissue, such as tendons, is much more difficult. Even using modern, non-invasive techniques such as magnetic resonance spectroscopy, the changes that occur in the tendon may be at a cellular or even molecular level and be imperceptible because of the tightness with which the atoms or molecules are bound. The only way to evaluate these changes would be to surgically remove the tissue and then perform histochemical and electron microscopic analyses to isolate the changes in the tissue structure. Such invasive procedures could clearly not be used on an ongoing basis. Being able to accurately and easily measure and monitor adaptations in response to training at the level at which the adaptation occurs is therefore frequently a challenge to biomechanists and others interested in optimisation.

Another important consideration in any optimisation regime is the choice of the criterion variable on which optimisation is to be sought. In some situations, such as endurance running, the criterion may be economy whereas, in other situations, such as weight lifting, the optimisation criterion may be maximum rate of force development. Knowing how the biomechanics of performance alter in response to training is a valuable source of evidence to guide the optimisation process. In the sections that follow we consider some examples of known biomechanical responses to training and biomechanical differences between skilled and novice performers.

Timing of segment involvement

One of the most striking features of skilled performance is the effortlessness conveyed by the performer. The master cricket batsman and the elite-level golfer are able to hit the ball with tremendous power but appear to do so with virtually no effort. Similarly, the ease with which a carpenter can hammer nails or a brick layer prepares and lays bricks conceals their highly developed skills. One of the keys to the success of these people is the timing of the involvement of their various body segments that allows them to maximise speed of movement and/or control of their body parts.

In activities in which an object is struck with a bat or racquet or a missile is thrown or kicked, the intention is most often to maximise the speed of the implement at the impact point or release. In tennis, the speed of the racquet head is critical in terms of success in serving and ground stroke play. Similarly, in kicking, throwing and hammering skills, the speed of the foot, hand and hammer, respectively, is one of the crucial factors in determining success. These types of activities see a proximal-to-distal body segment involvement. That is, the most proximal segments begin the movement, followed by the next most proximal until the very end segment of the linked chain (usually the hand or foot) is used. In throwing, the segmental movement begins with the lower body and trunk and ends at the wrist and fingers with release of the ball. The speed of the smaller and lighter body segments, with lesser inertia, is added to the larger ones with the effect of producing the greatest possible speed and force at the final segment endpoint. The small, lighter body segments, such as the hands and feet, are located distally (and hence involved toward the end of most co-ordinated movements) whereas the larger, heavier body segments, such as the trunk and chest, are located proximally (and hence are involved at the start of most co-ordinated movements).

The notion of adding segment velocities has become known as the *summation of speed principle* and a model of this principle is presented in Figure 11.1. In this figure, which represents the ideal pattern for an overarm throw, the distal segments begin to make their contribution only after the proximal one to which they are joined has reached maximum speed. For example, once the upper arm has reached its peak velocity, the elbow begins extending rapidly, thereby increasing the velocity of the forearm. Similarly, once the forearm has reached its highest attainable speed, the wrist then begins to make its contribution. Wrist joint motion can be flexion/extension or supination/pronation, depending on the type of sport or activity. Such a pattern optimises the use of the body's segments allowing those parts that are controlled by the small muscles to build on the speed developed by the large muscles of the trunk and thigh.

As stated earlier, the co-ordinated, effortless 'timing' utilised by elite level performers typifies

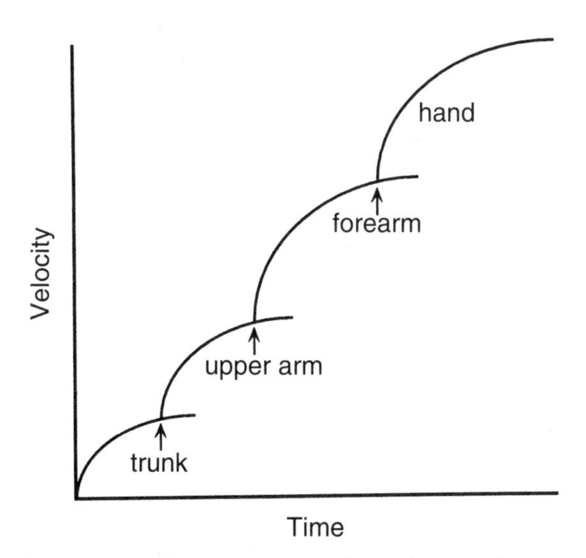

Figure 11.1 The summation of speed principle illustrated for a throwing example. Each successively distal segment begins accelerating when the contiguous, proximal one reaches its maximum.

this type of segment involvement and research scientists have shown that for a variety of sports such as kicking, tennis, squash, golf, javelin throwing, badminton and so forth, the summation of speed principle appears to hold. An interesting question to ask is what is the difference, in terms of segment interaction and contribution, between elite performers and novices or lesser skilled players?

Surprisingly, few studies have actually studied this phenomenon using 'real world' skills and settings. However, some recent work done with golfers has highlighted a number of interesting differences between elite performers and lesser-skilled players. With golfers, the rapid uncocking of the wrists immediately before impact differentiates those players who are able to hit the ball prodigious distances from those who can only hit the ball average lengths. The delayed wrist uncocking allows the generation of very high speeds of the clubhead by allowing the clubhead to continue to increase in speed until a few milliseconds before impact and by optimising the summation of forces originating from the movement of more proximal segments such as the trunk, back, legs and shoulders. If the wrists are uncocked early in the downswing, the upper limb speed is lost and this may actually begin to slow the clubhead down as it approaches the ball.

Another distinguishing feature between novice and expert golfers is that the novices have been shown to keep the range of movement of the shoulder and wrist joints to a minimum whereas the experts do not restrict motion at these joints. Novices do this, presumably, to limit the number of joints over which they need to exert control. Interestingly, in spite of the smaller amplitudes of movement and the number of joints actually contributing to the swing, novices are not as adept as experts in producing a consistent pattern of movement. Experts are much less variable in the angular displacement and velocity of the shoulder, wrist and elbow joints compared with novices.

Utilising stored elastic energy

The utilisation of stored elastic energy has been a topic of recent work by numerous research scientists as they try to elucidate how much energy can be stored in the elastic structures in the body and for how long this energy can be stored. As was discussed in Chapter 9, eccentric muscle contractions occur when the length of a muscle increases despite the contractile machinery being stimulated to shorten. In these types of contractions, the elastic tissues in the muscle and tendon are stretched. If the external load is then reduced so that a concentric contraction ensues, the potential strain energy stored in these elastic structures is available to do work. To what extent this potential energy is transformed into kinetic energy, and increases in efficiency consequently gained, depends on the type of activity and the skill of the performer.

In running it has been reported that as high as 50 per cent of the total energy requirement of running is available through elastic energy sources. This figure may be even higher for animals, such as horses and kangaroos, that have long tendons attached to small-bellied muscles, lying close to the body of the animal. In other activities, like the bench press in weight lifting, it is believed that elastic energy sources allow the performer to increase the total load by approximately 20 per cent of the maximum that can be lifted concentrically. The storage of elastic energy is short lasting. If a substantial delay occurs between the concentric and eccentric phases of a movement, the energy stored in the elastic structures is dissipated as heat.

Given this knowledge it is perfectly understandable why most activities that require high velocity at either release (such as softball pitching) or impact (such as chopping wood) utilise a long wind-up or back swing. By moving the body segments first in the opposite direction to that required for release or impact, the muscles that control the forward movement or swing are made to contract eccentrically, resulting in the storage of elastic energy. This energy can then contribute to the forward swing when the muscles controlling the swing eventually contract concentrically to accelerate the body segments in the desired direction. In throwing or tennis serving, the shoulder is horizontally extended and externally rotated during the wind-up phase, placing the internal rotators and the horizontal flexors on stretch. The energy stored in the elastic tendons and muscle fibres crossing the shoulder joint then becomes available during the acceleration phase to aid in forcing the throwing-side arm rapidly in the direction of the target. Similar practices are used in virtually all hitting and throwing tasks to enhance the speed of the forward movement or to increase the efficiency of the action.

The coaching adage of 'keep your knees bent and stay on the balls of your feet' used in many sports tasks takes on new meaning when examined from a biomechanical perspective. By adopting such a posture and rocking the weight of the body from side to side or front to back (as tennis players do in preparing to receive a serve), the muscles of the legs are involved in stretch–shortening cycles (eccentric followed by concentric muscle actions) that aid in increasing the force that can be generated when an explosive action is required. That is, when the tennis player has to move rapidly in response to the serve or the basketball player has to jump explosively to make a defensive play, the muscles will have stored as much elastic energy as possible during the stretching phase, which can then be released when needed.

One important aspect in relation to the use of stored elastic energy is that this energy is only available for brief periods of time, probably less than 250 ms. This fact has implications for novice performers and children, who do not have the specific muscle strength to overcome their own inertia or the inertia of a racquet or bat. If the equipment being used is too heavy or has a high moment of inertia, it may prevent the skill from being executed properly because the time between the start of the forward swing and impact may be delayed so much by the high inertia of the system that the energy is dissipated as heat. Once transformed into heat, the mechanical energy cannot reappear as kinetic energy. Thus the use of equipment that is matched to the performer is critical. More will be said about this later in this chapter.

Another important aspect in the utilisation of elastic energy is that the more the muscular and tendinous tissues are stretched, the greater is the energy storage. Thus, it is imperative to increase the flexibility of the involved joints so that the maximum amount of energy can be stored. A typical complaint of ageing golfers and tennis players is that they cannot hit the ball as far or as hard as they used to. Although loss of muscular strength, co-ordination and so forth may be part of the reason for this decrement in performance, another aspect is that people tend to lose their flexibility as they age, thereby limiting their energy storage potential. Thus, specific flexibility exercises are an important part of any training program, and they take on even greater importance if the participant wishes to maintain high performance levels over a life time. Skilled performers characteristically have good flexibility across those joints involved in force production for their particular activity.

Figure 11.2 Illustration of an extreme shoulder external rotation position adopted by a baseball pitcher in order to maximise the storage of elastic energy

Altering the movement pattern

With practice skilled performers establish movement patterns, often unknowingly, that are advantageous biomechanically. Three examples are used in this section to illustrate how the movement patterns of expert performers may provide mechanical advantages for them over the patterns used by less practised and less skilled performers.

One of the many features of highly skilled performers in striking and throwing activities is that they use much longer backswing or wind-up actions than those used by novices or less skilled players. The use of such movement patterns means that the distance over which the implement or the body can be accelerated is greater for the skilled performer and this creates the potential to increase the speed of the implement or object at the release or contact point. For example, in baseball pitching, elite pitchers use an exaggerated wind-up movement in which the leg is brought up to the chest and then the arm and forearm are rotated into extreme positions (Figure 11.2). This wind-up is followed by a very long forward step toward the plate. The consequence of these movements is that the hand goes through a large distance before the ball is released. If the time taken for the forward swing or acceleration phase to release is the same as for a player who uses an abbreviated action, then the velocity of the hand at the release point will naturally be higher for the pitcher whose hand has traversed the longer path. This will then translate into a higher release velocity for the ball and a faster pitch.

As a second example, consider the two different types of place kicking techniques that were once common in rugby league football—a soccer style round-the-corner technique and a front-on toe kicking technique. The soccer style, in which a looped back swing of the leg is used and the ball is struck with the instep is now used almost exclusively by the best kickers and there appears to be a number of good reasons for this. Firstly, the ball is contacted on its gently curved surfaces (its belly) by the relatively large instep of the foot. In comparison, the toe kicking style involves a planar motion of the limb in a short, staccato-like action, and the sharp end of the ball is contacted by the toe of the foot. The use of the instep striking the large surface of the ball increases the margins for error, thus increasing the likelihood of a successful result.

By comparison the toe contacting the small pointed end of the football has very little tolerance for positional errors of the foot. Furthermore, the use of body rotation and a looped or curved back swing of the leg means that the distance that the foot travels on its downward path to the ball is greater for the soccer style. As in the previous example, if the downswing time is constant then the foot's velocity will be greater in the soccer style compared with the toe kicking style as the foot moves through a larger distance and hence generates a larger velocity and force at contact. An additional advantage of long curved or loop backswings or wind-up movements is that the muscles and tendons are stretched to a greater length than if a short, abbreviated pattern of movement is used. As we noted previously, the greater the stretch placed on the muscles and tendons the greater the elastic energy that can be stored and the greater the potential for developing power during the acceleration and downswing phase.

A final famous example of altering technique to improve performance is that of Dick Fosbury, an American high jumper who developed the jumping style that now bears his name. Until the arrival of Fosbury, most high jumpers used a Western roll technique in which the front of the body was adjacent to the bar as the jumper cleared the bar. Fosbury's technique was radically different. His back was close to the bar, allowing the body to arch as he crossed over the bar. As noted in Chapter 7, this style allows the centre of gravity to pass through or underneath the bar whereas the Western roll method requires that the centre of mass pass above the bar. Thus, by altering technique an athlete may gain a mechanical advantage that allows him or her to reach a new standard of achievement. Changes in movement technique that occur with training/practice typically provide improvements in the fundamental biomechanics underlying the movement.

Muscle activation

Observable movement pattern changes with practice only come about through changes and reorganisation of the patterns of activation of the muscles that produce the observable movements. For this reason it is not surprising that muscle activation patterns have come under the scrutiny of a number of researchers in recent years. These scientists have tried to elucidate the differences in the ways in which experts and novices sequence and organise their muscle contractions. Although much of the research has been from a motor control perspective (see Chapter 19), there are also findings that bear interpretation from a biomechanical viewpoint.

A general finding is that expert performers in activities that do not require maximal force production tend to use as little muscle activation as possible, limiting activation to only those muscle groups essential for the performance of the skill. Such activation conserves energy, hence minimising fatigue. For example, in highly skilled golfers the elbow flexors of the left arm of right-handed golfers are silent during the downswing, whereas in novice golfers both the extensors and the flexors are activated at the same time. The novices' co-contraction is not efficient since one muscle group is trying to extend the joint while the other is trying to flex it. The net effect is that the joint becomes stiff; a likely reason why novices look clumsy and awkward compared with expert performers.

As we have noted above, trained performers have also learned how to turn the muscles off and on at appropriate times so that the actions they produce are smooth and continuous. In high-velocity or striking sports, the muscle activation patterns, consistent with the kinematics and the summation of speed principle, show a proximal-to-distal sequencing. There are numerous benefits from this type of organisation. Firstly, it is energetically more efficient to minimise muscle contraction since any muscle action has an associated metabolic cost. Secondly, the metabolic cost of eccentric muscle contraction is less than for isometric and concentric actions. Thus, the use of stretch–shortening cycles in which the muscles are first pre-stretched before concentric contraction means that not only does energy get stored for release later, but also that the cost of the eccentric work is smaller than for concentric work alone. Thus, there is a double benefit of using stretch–shortening techniques.

The specificity of adaptation—muscle strength versus muscle power

One of the most important features of training is that the body adapts specifically to the type of loading to which it is exposed. This specificity principle is discussed in great detail in Chapter 13 with respect to the subdiscipline of exercise physiology but it has important application in biomechanics as well, as the following example from resistance training demonstrates.

Resistance or weight training is now a feature of most athletes' training programs and highlights increased understanding by coaches of the benefits of such training and by scientists of the actual mechanical properties of the human body that such training modulates. Weight training used to be thought of merely as a means of increasing muscle size and strength. However, weight training is now used not only for these purposes, but also to increase muscle power and endurance, to provide variety in training, and to bring about adaptations in the body that would otherwise not be possible. The key to these changes is the realisation of the paramount importance of specificity of training. For example, if an athlete wants to increase leg power it is important to emphasise both strength and speed and not just one or the other.

Research has shown strong relationships between strength and/or power and performance levels in a number of sports including swimming, running, football, rowing and volleyball. Despite these strong relationships there are also studies that have shown increases in strength following a resistance training program but no concomitant increase in sporting performance. A deduction from such a result is that the training program was able to cause a strength adaptation in some muscle groups but that strength in these muscles was not the limiting factor in performance. For example, if the ability to generate power is the factor that discriminates between good and superior performance, the development of strength is unlikely to alter performance. Or, if the strength gains were not in the appropriate muscle groups, then it would be expected that performance would not be altered. For example, it would be unlikely that strengthening the muscles that dorsi-flex the ankle would lead to augmented kayaking performance.

One of the key issues, for which a grounding in basic biomechanics is invaluable, is the realisation that strength and power are different characteristics. Strength is the maximum force or torque that can be applied. Power, as defined formally in Chapter 9, is the rate of doing work and can be thought of as the product of force and velocity or torque and angular velocity. Thus, it is related to strength but also has an element that is related to speed. Consider the hypothetical relationships pictured in Figure 11.3 in which force is plotted against time for three people. The graph shows that person A has high strength but takes a long time to reach that level. Person B has moderate strength but is able to reach that peak quickly. Person C has low strength levels and moderate ability to apply that force quickly. This graph clearly illustrates very different performance attributes for the three people and the possibility to classify each individual on the basis of ability to generate force. Person A might

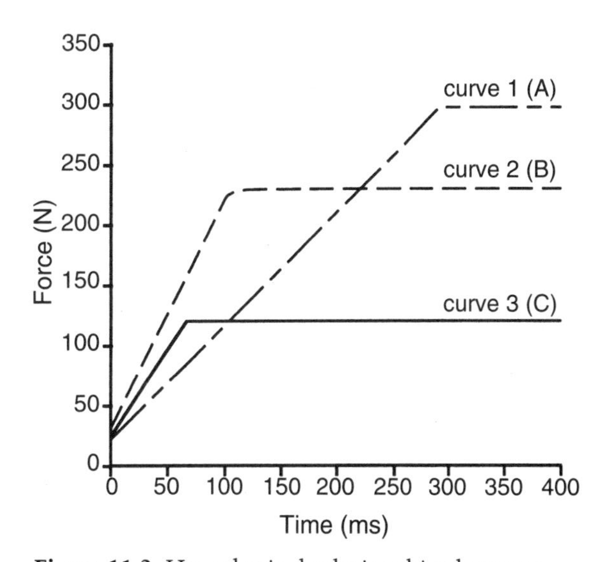

Figure 11.3 Hypothetical relationships between strength (force generating capacity) and time for three different people
Source: Bloomfield, J., Ackland, T.R. & Elliott, B.C. (1994), *Applied Anatomy and Biomechanics in Sport*, Blackwell, Melbourne, p. 112. (Copyright 1994 by Blackwell Scientific; adapted with permission.)

be a power lifter (an athlete who performs three weight lifting movements under rigidly controlled conditions) because extremely high levels of strength are required and speed of movement is not critical. Thus, the sport of power lifting is something of a misnomer as power is not actually an important attribute for excellent performance. Person B has moderate levels of strength but is able to generate force rapidly. Thus this person is probably best described as a power athlete because of the ability to rapidly apply large forces. Activities involving throwing (javelin, discuss and baseball pitching), sprinting and lifting would probably be suited to a person possessing this type of force-generating capacity. Person C, on the other hand, shows an ability to generate force quickly, which is a quality that is necessary for a power sport, but performance is going to be limited by low strength. Thus, to improve performance, this person probably needs to initially increase strength levels followed by specific training to enhance power.

Recall the force–velocity relationship presented in Chapter 7 for an isolated muscle (Figure 7.8). This relationship highlighted that muscle can produce greatest force when it is contracting eccentrically, followed by intermediate levels under isometric contraction conditions and least force at high shortening speeds (concentric actions). The power developed by a muscle can be readily determined from this type of graph by taking the product of the speed of contraction and the force. An example may be useful to illustrate this idea.

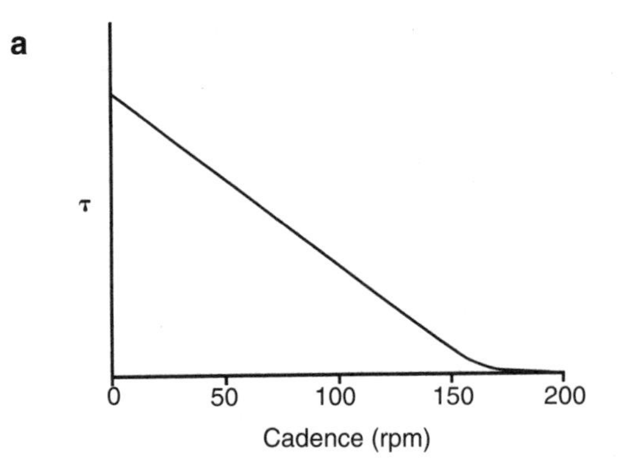

Figure 11.4 The torque produced by a cyclist as a function of cadence is illustrated in (a). By taking the product of the torque and angular velocity, the power–cadence function shown in (b) can be derived.

Consider a cyclist who is pedalling on a stationary bicycle. If only isometric and concentric muscle actions are considered, according to the force–velocity relationship, the average force or torque applied by the cyclist will be greatest when the pedals are not rotating (the muscles are contracting isometrically). As the angular velocity of the pedals or cadence increases, the average torque applied will begin to decrease until, at a very high cadence, the cyclist will be unable to apply a torque to the pedals. This relationship is depicted in Figure 11.4*a*. The power available at any cadence can be obtained by taking the product of the torque and the cadence. Despite the fact that the torque is highest under isometric conditions (the left-hand end of the graph), the power applied is zero! Similarly, when the cadence is at its maximum value, no torque can be applied and so the power is once again zero. If the points are plotted between these extremes, the relationship depicted in Figure 11.4*b* is observed and one can see that peak power is available when the torque is approximately 50 per cent of its maximum value and the cadence is approximately 80 rev/min. Unlike the linear relationship between torque and angular velocity depicted in Figure 11.4*a*, this relationship in biological systems is not linear. Thus peak power typically occurs at the point where the force is approximately 30–45 per cent of its maximum, isometric value.

Training can alter the torque–angular velocity (force–velocity) relationship at a joint by changing the torque that can be produced at any particular joint velocity. In fact, one of the training procedures normally adopted for those people who wish to increase their power output is first to increase strength (increase the torque-producing capacity of the muscles that cross the joint) and then enhance speed, the second component of power. During this training, the torque–angular velocity profile will alter, although the general shape of the function will persist.

The length–tension relationship of muscle also has implications for training. By altering the posture in which an exercise is done, the maximal resistance to movement can be changed. Take, for example, a biceps curl exercise. Normally these exercises are done standing and the elbow is flexed toward the chest. At the start of this action, the muscles are at their maximum length and they shorten to their minimum length. The maximum resistive torque occurs when the elbow is flexed to 90° and the weight has its greatest moment arm. The muscle is at its optimal length at this position and has maximum contractile capability. If, however, the exercise is done while the person lies face down on a bench, the length of the biceps muscle at the start and throughout the entire movement is shorter than in the standing position. The torque produced by the resistance is unchanged and, thus, the muscle is not at its optimal length when the resistive torque of the weight in the hand is maximum. The net result is that more effort is required to move the weight in this altered posture compared with the standing one, and a different adaptation to this type of training results in the muscle.

Modifying equipment and/or clothing

To date we have examined movement optimisation in the context of biomechanical adaptations resulting from training. An additional set of possible optimisations, but unrelated to training, come from the modification of equipment and/or clothing used during the performance of some movements. The examples we will examine here come from sport but comparable examples also exist in other fields of application of the biomechanical principles of human movement.

The popularisation and commercialisation of sport in recent times has been accompanied by increased efforts by scientists and engineers to improve the available equipment. There are many reasons for the increased efforts in this area ranging from the advertising, marketing and sales benefits afforded to those who base their product on scientific principles, to altruistic reasons that include the potential for increased performance and enjoyment for the participants. Irrespective of the reasons for the efforts, the design of modern equipment is mostly founded on sound engineering and mechanical principles, and this has generally improved both performance and safety standards in sports.

The past decade has seen tremendous changes in the design of racing bicycles, started initially in 1984 when disc wheels were first used at the Olympic Games. Engineers and scientists have worked together to minimise drag on bicycles by changing the shape of the frame, modifying the size of the front and rear wheels and altering the materials used in construction. Other design changes that have been beneficial for the rider include the development of indexed gears, three-spoked and four-spoked low drag wheels, clip-in pedals, shock absorbing devices on mountain bikes, and combined gear and brake levers. These changes have led to more efficient and safer riding.

One of the sports that has benefited most from improved application of mechanical principles to equipment design is golf. Fortunately, in this sport, it has been the average player who has benefited most. Modern golf clubs are designed with perimeter weighted clubheads that increase the area of the clubhead over which contact with the ball produces an acceptable result plus minimal jarring to the hands. Mechanical principles such as increased resistance to the rotation that occurs with an off-centre hit (increased moment of inertia) underpin the perimeter weighted clubs. Steel shafts are now becoming less and less common in golf clubs as new, light-weight but stiff materials such as boron, titanium and graphite take their place. The benefits of using light-weight materials in the shaft are that they reduce the inertia of the system, allow weight to be placed in the head of the club where it is most beneficial in terms of transferring energy to the ball, and allow shafts to be long and stiff but not heavy. Finally, and particularly with composite shafts made out of graphite and other synthetic fibre compounds such as kevlar, much greater variation is possible in the mechanical properties of the shaft. For example, the flex point can be altered substantially depending on how the matting of the fibres are laid or the stiffness can be made to vary over the length of the shaft. These alterations allow manufacturers to better match the playing characteristics of the clubs with the swings of the people for whom they are designed.

Tennis and squash racquets have also changed remarkably in design in the past decade or so as players have moved from the conventional small headed, wooden racquet to the oversized heads of modern-day racquets. As with golf, current technologies in fibreglassing and composite material science have meant that stiff, light but strong frames can be manufactured. The benefits of such designs include increased areas over which the ball can be struck successfully, decreased inertia and longer racquet life. Clearly these features are in the interest of all tennis and squash players but it is the average player who likely benefits the most.

Energetics/economy

Efficiency and economy of movement improve in response to training and these improvements are especially pronounced in gait, one of the most practised of all skills. Changes in economy of movement with training and practice come about for a variety of reasons, the most relevant biomechanical reason being the reduced internal work done in moving body segments by eliminating actions or motions that do not have a positive benefit on the desired outcome to move forward. For example, lateral movement of the body does not assist in forward movement, thus flicking of the legs laterally during the swing phase can be reduced or eliminated with training. Excessive vertical movement of the centre of gravity reduces efficiency (see Chapter 10) and this motion can also be minimised during training. Alterations in stride length and stride frequency also occur with both practice and maturation, which optimises the efficiency of running. As gait and other more complex skills are learned, the movement patterns become characterised by decreased variability within the movement kinematics and kinetics. In fact, research has demonstrated coefficients of variation (a single number used to indicate variability of patterns of movement) as low as 10 per cent for some key kinetic variables in gait.

Another good example of how efficiency can change in response to training can be seen in swimming. When people begin swimming for the first time their patterns of arm and leg movements are very inefficient. The head may bob up and down more than necessary, their legs may drop low in the water and have large lateral deviations, and their arm pull through the water may not make best use of hydrodynamic lift and drag principles. With training, people can improve many aspects of their technique and greatly reduce the drag experienced as they move through the water. The reduced drag, coming about because of improved biomechanics, results in their swimming being more efficient. It is likely, too, that as they learn these new patterns of movement there will be less stroke-to-stroke variation in the movement kinetics and kinematics. Decreased kinematic and kinetic variability leads to increased efficiency because unwanted or unnecessary movements that do not contribute to forward progression are eliminated from the pattern. Adaptations in movement with training occur in ways that are sound in terms of basic biomechanics.

Further reading

Chapter 7: Basic concepts of kinetics

Chaffin, D.B. & Anderson, G.B.J. (1991), *Occupational Biomechanics*, John Wiley, New York.

Enoka, R. (1988), *The Neuromechanical Basis of Kinesiology*, Human Kinetics, Champaign, IL.

Kane, J.W. & Sternheim, M.M. (1984), *Physics* (2nd edn), John Wiley, New York.

Miller, D.I. & Nelson, R.C. (1973), *Biomechanics of Sport*, Lea & Febiger, Philadelphia.

Winter, D.A. (1979), *Biomechanics of Human Movement*, John Wiley, New York.

Chapter 8: Basic concepts of kinematics

Hay, J.G. (1978), *The Biomechanics of Sports Techniques*, Prentice-Hall, London.

Hay, J.G. & Reid, J.G. (1988). *Anatomy, Mechanics and Human Motion*, Prentice-Hall, London.

Chapter 9: Basic concepts of energetics

Bloomfield, J., Ackland, T.R. & Elliott, B.C. (1994), *Applied Anatomy and Biomechanics in Sport*, Blackwell, Melbourne.

Jones, N.L., McCartney, N. & McComas, A.J. (eds) (1986), *Human Muscle Power*, Human Kinetics, Champaign, IL.

Chapter 10: Biomechanical changes throughout the lifespan

Schmidt-Nelson, K. (1991), *Scaling: Why is Animal Size so Important?*, University Press, Cambridge.

Chapter 11: Biomechanical adaptations to training

McNeil Alexander, R. (1992), *The Human Machine*, Natural History Museum Publications, London.

Section 3
Physiological bases of human movement: The subdiscipline of exercise physiology

Introduction

What is exercise physiology?

Exercise physiology is a subdiscipline of human movement studies that focuses on the physiological responses to exercise. Physiology can be defined as the study of the body's functions; hence, exercise physiology is the study of the responses of body function to exercise. Exercise physiology focuses on both the acute (immediate) responses to exercise as well as the chronic (long-term) adaptations to physical activity.

Exercise physiology draws on knowledge and techniques from a variety of speciality areas including other subdisciplines in human movement studies, physiology, biochemistry, endocrinology (the study of hormones and their functions), histology (the study of microscopic structure of tissues and cells), and, recently, cell and molecular biology.

Applications of exercise physiology

As Table I3.1 reveals, exercise physiology knowledge and skills are used in varied situations including sport, health promotion and preventive medicine, exercise rehabilitation, the workplace, school health and physical education, and the research laboratory.

Exercise physiology principles are used in designing effective assessment and training programs to increase the performance of both competitive and recreational athletes. Talent identification involves physiological and anthropometric profiling to identify young people with potential to excel in specific sports. Exercise physiology forms the basis of exercise prescription on an individual basis for healthy adults, as well as for special populations such as older individuals, cardiac patients or those with other conditions such as diabetes, pregnancy or arthritis.

Regular exercise also has a role in prevention of and rehabilitation after injury, including occupational injuries. Exercise physiology principles are used to assess the physical demands of some occupations, such as firefighting, and to develop fitness programs to effectively train employees. Coaches and physical education teachers working with children use exercise physiology to design training programs specifically for the young athlete. Finally, research in exercise physiology focuses on understanding the mechanisms underlying the physiological responses and adaptations to exercise. Basic research might include understanding the causes of fatigue during various forms of exercise, whereas applied research might focus on the best way to improve exercise capacity and performance.

Table III.1 Some applications of exercise physiology

Sport:
- Fitness profiling of athletes
- Talent identification
- Designing sports-specific training programs
- Evaluating training program effectiveness

Health promotion and preventive medicine:
- Physical activity for general health
- Exercise prescription in specific conditions, such as obesity, diabetes & pregnancy
- Corporate and community fitness programming and health promotion

Exercise rehabilitation:
- Exercise programming in patient groups, such as cardiac patients and chronic pain patients
- Injury rehabilitation
- Fitness programming to prevent re-injury

Worksite:
- Assessment of physical fitness demands of specific occupations, such as firefighting, military and police
- Exercise programming to meet occupational physical fitness standards
- Prevention of occupational injuries, such as back injury

School health and physical education and sport:
- Specialist teaching in health and physical education
- Coaching in community-based sport for children
- Adapting sport for children

Research laboratory:
- Physiological, biochemical and hormonal responses to exercise
- Mechanisms of training adaptations
- Mechanisms responsible for health benefits of physical activity

Typical questions posed and levels of analysis

Exercise physiology can be studied at many different levels from the whole body to the molecular. At the whole-body level, the exercise physiologist might ask whether a lifetime of exercise can increase lifespan. At the systemic level, exercise physiology attempts to understand the effects of regular exercise on the cardiovascular system; for example, how exercise helps to prevent heart disease. Understanding the effects of exercise on the pattern of fat (adipose tissue) distribution and implications for prevention of obesity represents work at the tissue level. At the cellular and subcellular level, the exercise physiologist may study the effects of endurance training on energy production or the mechanisms responsible for muscular hypertrophy following strength training. Finally, work at the molecular level might focus on how much of elite sports performance is genetically determined.

Historical perspectives

Just as the applications and levels of analysis of exercise physiology are multifaceted, so too are its historical origins. Recognition of the effects of exercise originate at least as far back as ancient Greek and Roman physicians such as Hippocrates (5th–4th c BC), Herodicus (5th c BC) and Galen (2nd–3rd c AD). For example, Herodicus, also a trainer of boxers and wrestlers, recognised that strength training caused muscle hypertrophy (growth) in wrestlers. Galen recognised the relevance of hygiene to health, and wrote about lifestyle factors such as air, food and drink, sleep, and motion (exercise), as well as the role of exercise in rehabilitation from illness.

In 1553, Spanish physician Christobel Mendez was the first to publish a printed book on exercise, entitled *Book of Bodily Exercise.* Mendez referred extensively to the writings of Galen, and he anticipated many of the modern developments in exercise physiology and sports medicine. For example, Mendez suggested that walking is the best form of exercise, recommended regular physical activity for women, and discussed exercise for children, injuries due to exercise, and exercise for the disabled.

The European interest in exercise and health continued well into the 19th century, and was adopted in the United States where early concern focused on ill-health among city dwellers. With the rise of university sport, particularly rowing, in mid-19th century England, 'trainers' (equivalent to today's coaches) began to recognise basic training principles. For example, in 1863, Charles Westhall recognised what we now call the concepts of specificity, individualisation of training, and overtraining.

Modern exercise physiology has its origins in work by 18th and 19th century European scientists working in the fields of physiology, biochemistry and nutrition. The advent of precise measuring devices led to quantification of physiological responses, including the responses to exercise. Work during the early part of this century centred on understanding energy metabolism and efficiency of movement during prolonged steady-rate exercise.

In the early 20th century, A.V. Hill in England was among the first to study the body's responses to non-steady-rate or maximal exercise in athletes. Hill related his findings on the energetics of muscular contraction in isolated muscle to understanding metabolism during maximal exercise in humans. Hill's work directly influenced the Harvard Fatigue Laboratory in the United States, which generated much of the foundation of modern exercise physiology. After the Second World War, scientists from the Harvard Fatigue Laboratory dispersed to various institutions mainly in North America, where they greatly expanded the application of exercise physiology to sport, medicine and health, physical education and the workplace.

In the 1960s, exercise physiology research adopted a biochemical and subcellular focus with reintroduction of the needle biopsy technique by Bergstrom, which allows the exercise physiologist to explore the metabolic responses of muscle cells to single bouts of exercise as well as the mechanisms responsible for cellular adaptations to exercise training in humans. For example, the importance of muscle glycogen (glucose) stores and dietary carbohydrate to exercise performance could only be identified by use of the biopsy procedure.

Although exercise physiology originated within the fields of medicine and physiology, since the 1970s there has been an increasing trend for exercise physiology research to be firmly placed within human movement or exercise sciences. At the same time, with recent advances in cell and molecular biology, exercise physiology has incorporated many of the tools of modern biomedical research. Exercise physiology now encompasses the study of the physiological and metabolic adaptations to exercise, from basic research at the molecular and cellular level through to applied research aimed at improving sports performance and health.

Professional organisations and training

The major international organisations representing exercise physiology are the International Federation of Sports Medicine and the American College of Sports Medicine. The latter offers professional specialist registration in two tracks—health and fitness and exercise rehabilitation. Exercise physiologists may also seek membership in any number of other professional or scientific organisations, which may represent regional, national, or international groups; for example, in North America, the Canadian Society for Exercise Physiology and the American Physiological Society; in Europe, the European Sports Medicine Association and the European College of Sport Science; and internationally, the International Congress of Sport Science and Physical Education. There are also many organisations representing interests in the applications of exercise physiology—the American Association of Cardiovascular and Pulmonary Rehabilitation and the British Association of Sport and Exercise Sciences are examples.

CHAPTER 12

Basic concepts of exercise metabolism

- Muscle is a unique tissue
- How is energy for exercise produced?
- Oxygen must be continually supplied to the working muscles for sustained exercise
- How is exercise capacity measured?
- Human skeletal muscle cells are not all the same
- Different activities require different amounts of energy
- Diet is important to energy metabolism and exercise performance

One important question asked by exercise physiologists is 'what factors limit performance?'. That is, what prevents a runner from running faster, what causes the athlete to become fatigued, what limits how much weight a power lifter can lift?

Exercise capacity is determined by how much energy the muscle cell can produce and how quickly this energy can be made available to the contractile elements within skeletal muscle. For example, maximum sprinting speed can only be maintained for 100–200 m (or 10–20 s). After the first 10–20 s, running pace slows because the muscle cells cannot maintain the desired rate of energy supply. Fatigue occurs when the rate of energy demand exceeds the rate of production in skeletal muscle. To continue our car analogy, the amount of energy produced by the muscle cell can be seen as the petrol that fuels a car's engine. The faster the car speed (exercise pace), the more fuel (energy) needed and the quicker the car (muscle) runs out of fuel. One way to improve perfomance is to increase the fuel stores (larger petrol tank).

It is important to understand those factors that limit performance. If we know what limits exercise capacity in a particular activity, we can use this knowledge to improve performance. For example, it is well known that depletion of stored glucose (glycogen) from skeletal muscle causes fatigue in long-duration events such as marathon running. As a result, endurance athletes now consume a high carbohydrate diet to enhance muscle glycogen stores and delay the onset of fatigue during training and completion.

In this chapter we examine a range of fundamental concepts related to exercise metabolism including discussion of how energy for exercise is produced, how exercise capacity is measured, and how metabolism changes for different muscle fibre types, for different physical activities and with different dietary manipulations.

Muscle is a unique tissue

As the body goes from rest to maximal exercise, the metabolic rate in human skeletal muscle can increase by up to 50 times. This increase in metabolic rate requires a tremendous capacity to transfer chemical energy to perform muscular work. The muscle also must be able to co-ordinate the rates of energy supply with demand. Because it is metabolically inefficient for cells to produce more energy than needed at any given time, the muscle cell does not store vast quantities of energy. During exercise, chemical energy must be continually supplied to the working muscles.

How is energy for exercise produced?

Muscular work requires transfer of chemical energy to mechanical energy. The chemical energy is supplied in the form of adenosine triphosphate (ATP). ATP is termed a 'high energy phosphate' molecule because energy is released on cleavage of its terminal phosphate bond (Figure 12.1*a*). ATP is split into adenosine diphosphate (ADP) and inorganic phosphate (P_i) in a reversible reaction. The energy released on cleavage of this bond is required for cross-bridge interaction between the thin and thick filaments of skeletal muscle, resulting in production of force in muscle (see Figure 3.12). During exercise and recovery, ADP can be re-phosphorylated to ATP provided sufficient chemical energy is supplied via metabolic pathways. Muscle cells contain all the necessary equipment to continually resynthesise ATP from ADP and P_i. ATP production occurs in both the cytosol and mitochondria of skeletal muscle cells, although far more ATP can be produced in mitochondria than in the cytosol. Anaerobic ('without oxygen') production of ATP occurs in the cytosol and aerobic ('with oxygen') production of ATP occurs in the mitochondria.

Three systems produce ATP

There are three energy systems for ATP resynthesis: (i) the immediate energy system, sometimes called the stored energy or high energy phosphagen system, (ii) the anaerobic system, sometimes called the anaerobic glycolytic, glycogenolytic or lactic acid system; and (iii) the aerobic or oxidative system (Table 12.1). The use of different systems to produce ATP means that energy is available for all types of exercise from very short bursts such as a power lift or 10 m sprint to sustained activity such as a triathlon. As described below, all three systems

Table 12.1 The three energy systems

	Immediate	**Anaerobic glycolytic**	**Oxidative**
Substrate(s)	ATP, PC	glycogen or glucose	glycogen or glucose, fat, protein
Relative rate of ATP production	very fast	fast	slower
Duration at maximal pace	0–30 s	20–180 s	>3 min
Limiting factors	PC depletion	lactic acid accumulation	glycogen depletion
Examples of activities	power/weight lifting	longer sprints	endurance events
	short sprints	middle-distance	ball games
	jumping, throwing	ball games	

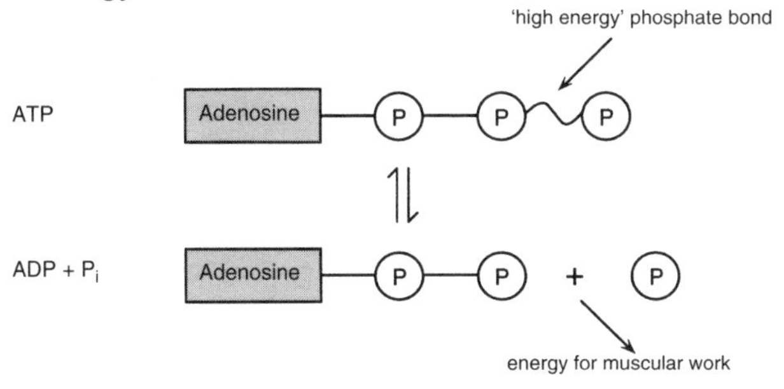

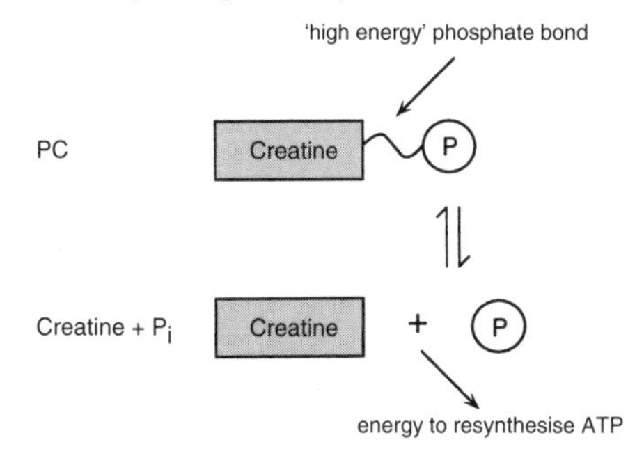

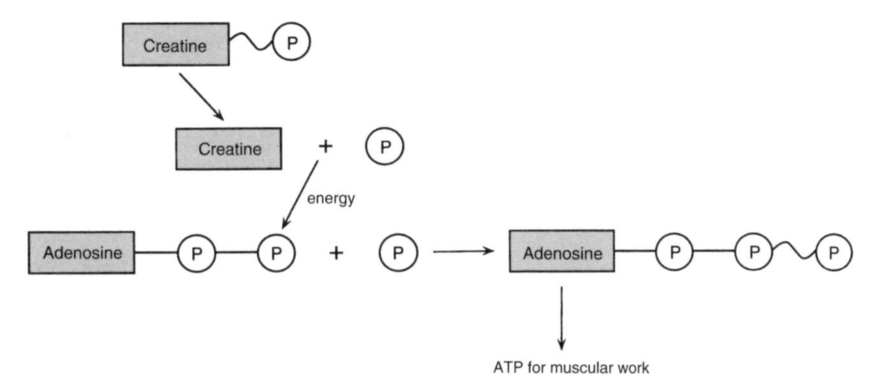

Figure 12.1 Schematic representation of 'high energy' phosphagens:

- When ATP is split into ADP + P_i, cleavage of the terminal ('high energy') phosphate bond yields energy needed for generation of tension within skeletal muscle (a).
- Cleavage of the high energy phosphate bond of PC yields energy that can be used to resynthesise ATP from ADP + P_i (b).
- At the onset of exercise and for very brief high-intensity exercise, PC provides the major means of regeneration of ATP from ADP + P_i (c)

operate simultaneously, and their relative contributions to ATP resynthesis depend on the intensity and duration of exercise.

The immediate energy system provides stored chemical energy in the form of phosphocreatine (PC), another high-energy phosphate molecule (Figure 12.1). As the name implies, stored PC provides an immediate source of energy for exercise. At the onset of exercise, the PC molecule is rapidly (within milliseconds) degraded, donating one of its phosphate groups to make ATP. PC provides the major fuel source at the onset of any exercise and for brief exercise lasting less than 30 s such as sprinting, throwing and jumping (Table 12.1).

PC is rapidly depleted during exercise—by 35 per cent after 6 s, by 65 per cent after 30 s, and nearly 100 per cent by 1 min of maximal exercise (Figure 12.2). PC is not replenished in the muscle until exercise stops; complete replenishment of PC requires up to 6 min after the end of exercise. As will be discussed in Chapter 13, effective sprint training must consider the time course of PC replenishment during recovery between exercise bouts.

The anaerobic glycolytic system provides the major source of ATP for maximal exercise, lasting

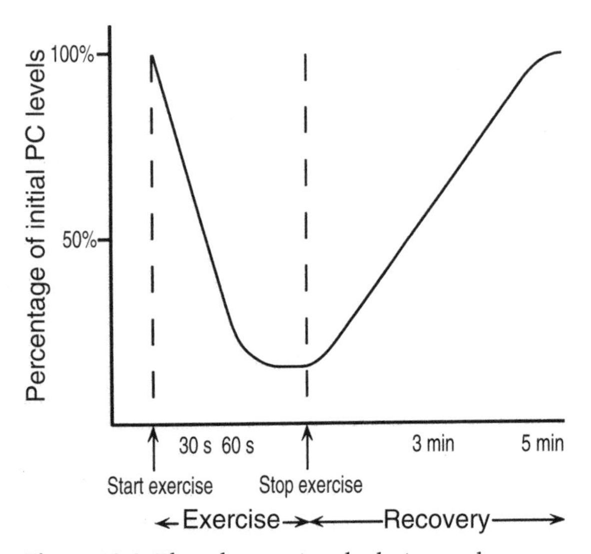

Figure 12.2 Phosphocreatine depletion and replenishment. PC is rapidly depleted during high-intensity exercise. Up to 5 minutes may be required to fully replenish PC stores after depletion

between 20 s and 3 min (Table 12.1). For example, during 30 s sprinting, anaerobic glycolysis provides 60 to 65 per cent of the needed ATP. As its name implies, energy is produced anaerobically (without oxygen) using glucose as the fuel (substrate). The glucose is derived from skeletal muscle glycogen stores and to a limited extent from blood glucose; glycogen is the intracellular storage form of glucose, consisting of a polymer of many glucose molecules. Lactic acid is a by-product of anaerobic glycolysis, and this system is sometimes called the lactic acid system.

One molecule of glucose, a six-carbon sugar, is degraded via a series of 13 steps to two molecules of pyruvic acid (Figure 12.3). Pyruvic acid can then either be converted to lactic acid or enter the tricarboxylic acid cycle (TCA cycle; sometimes called the Krebs cycle or the citric acid cycle). A net of two ATP molecules are produced anaerobically for each molecule of glucose converted to two pyruvic acid molecules.

Oxidative production of ATP occurs in the mitochondria, which are membrane-bound subcellular organelles present in most cells. Pyruvic acid, fatty acids and amino acids are further degraded via the TCA cycle, producing carbon dioxide, together with electrons and hydrogen ions (H^+) (Figure 12.3). The carbon dioxide diffuses out of muscle cells into the blood and is transported to be exhaled from the lungs. The electrons and hydrogen ions enter the electron transport chain, a series of proteins (enzymes) that eventually transfer the electrons and hydrogen ions to oxygen to produce water. This transfer of electrons and hydrogen ions along the electron transfer chain provides chemical energy to resynthesise ATP from ADP and P_i. Continued production of ATP via the electron transfer chain requires a constant supply of oxygen to the muscle cell.

The three energy systems operate as a continuum

The three energy systems operate as a continuum, not in an 'on–off' manner. Each system (immediate, anaerobic, oxidative) is always functioning, even at rest. What varies is the relative

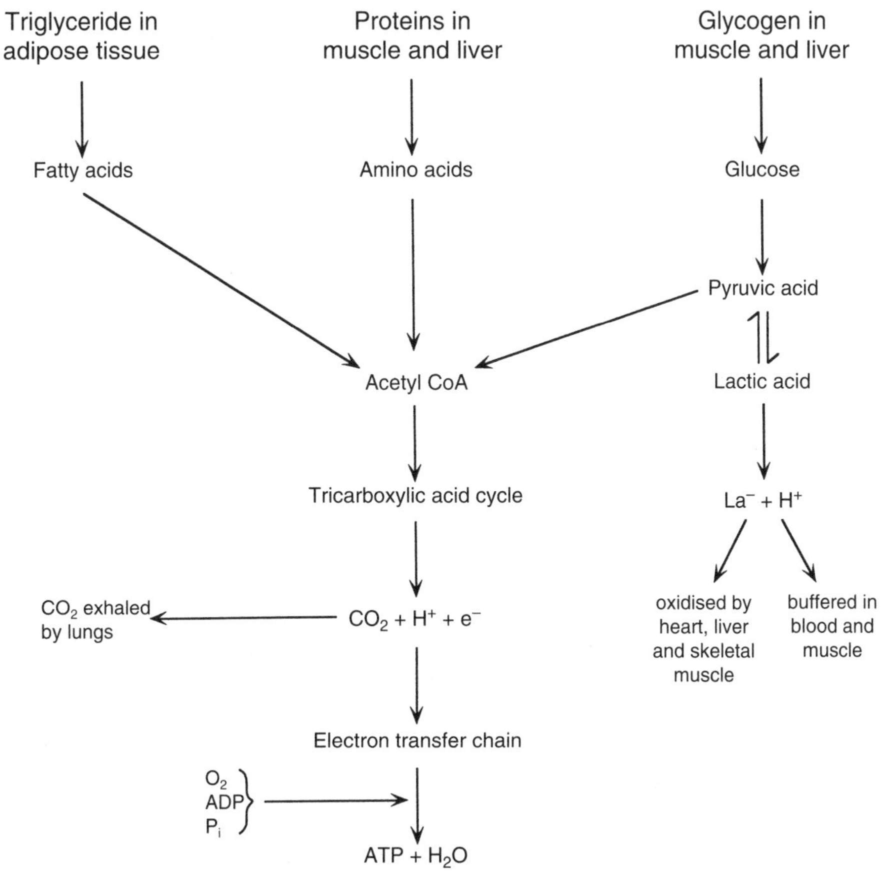

Figure 12.3 Simplified scheme of energy metabolism. The starting materials may be fats (fatty acids), proteins (amino acids), or carbohydrate (glycogen or glucose). Metabolism of fatty acids and amino acids requires oxygen, but carbohydrates may be metabolised either aerobically or anaerobically. The final end products are carbon dioxide, which is exhaled by the lungs; lactic acid, which may be metabolised by the heart, liver or skeletal muscle; ATP, which is used for tension generation within muscle; and water

contribution each system makes to total ATP production at any given time (Figure 12.4).

Even in the extremes of activity, such as marathon running or brief sprinting, all three systems are used. For example, in a marathon, the immediate system provides ATP at the start of the race; the anaerobic system provides much of the needed ATP for the first few minutes until the oxidative system reaches steady state. The anaerobic system also provides a significant amount of ATP during the race, for example for uphill running and for sprinting at the end. Over the course of the entire marathon, the oxidative system provides most of the ATP needed.

Fats, proteins and carbohydrates can be fuels for ATP production during exercise

As discussed above, ATP can be synthesised via metabolism of fats, proteins and carbohydrates. It is important to recognise that these nutrients are not transformed into ATP. Rather, degradation of these nutrients within the body releases

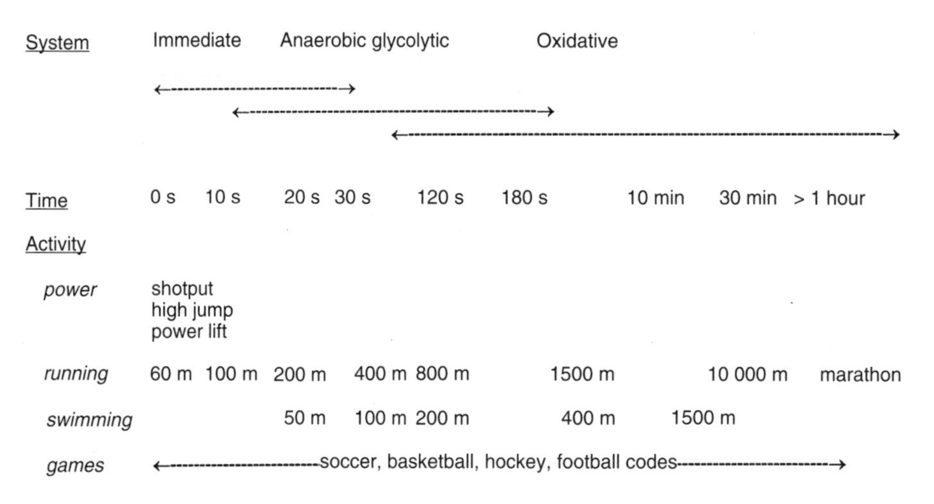

Figure 12.4 The energy system continuum. The relative contribution of each system to ATP resynthesis depends on exercise duration and intensity

energy contained in their chemical bonds; it is this energy that is used to synthesise ATP (Figure 12.3).

Carbohydrate, in the form of glucose, can be used to produce ATP either anaerobically or aerobically. In contrast, fats in the form of fatty acids and proteins in the form of amino acids can only be used to produce ATP aerobically. At any given time, the body will metabolise a mixture of these nutrients to produce ATP. However, the relative contribution of each nutrient to ATP production varies with exercise intensity and, thus, the metabolic rate.

At rest and during low-intensity exercise, fatty acids and glucose are used in approximately equal amounts as substrates for ATP production. As exercise intensity increases, ATP production progressively relies more on glucose and less on fatty acids (Figure 12.5). During maximal exercise, the muscles metabolise virtually only glucose, mainly derived from muscle glycogen. Amino acids generally contribute little to ATP resynthesis, usually less than 5 per cent, during moderate exercise. However, metabolism of amino acids may provide up to 20 per cent of energy production under some conditions, such as starvation or after several hours of prolonged exercise in which glucose supply to the muscle is severely limited.

Lactic acid—friend or foe?

Lactic acid is produced as a byproduct of anaerobic glycolysis. During exercise, lactic acid concentration may increase within muscle from 2 mmol/L at rest up to 30 mmol/L during maximal exercise. Excess lactic acid (in the form of the lactate ion—see below) is transported across the muscle cell membrane into the blood and circulated throughout the body. Blood lactate concentration may increase from 1–2 mmol/L at rest up to 15 mmol/L during maximal exercise.

Excess lactic acid produced during exercise is associated with muscular fatigue. Lactic acid produced during exercise rapidly dissociates into a lactate anion and free hydrogen ion (H^+) (Figure 12.3). An increase in H^+ concentration increases the acidity (lowers the pH) of muscle and blood. Although tissues and blood contain substances that partially buffer the increased acidity, the pH of muscle may decrease from 7.4 to as low as 6.7 during intense exercise—nearly a 10 fold increase in acidity. The anaerobic glycolytic system is sensitive to changes in acidity, and the decrease in pH inhibits or slows the anaerobic pathway. Thus, excess lactic acid accumulation resulting from anaerobic glycolysis inhibits further ATP production. Although the excess lactic acid causes fatigue during intense exercise, this inhibitory effect is a protective

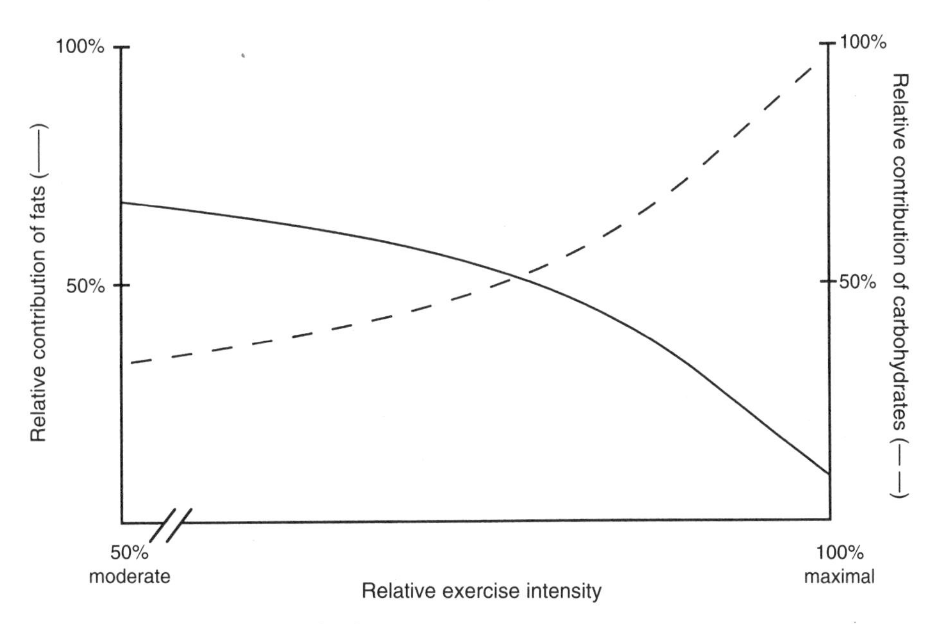

Figure 12.5 Fats and carbohydrates as substrates. The relative contribution of fats and carbohydrates to energy production varies with exercise intensity. Fats are the major substrate for low-intensity exercise; as exercise intensity increases, the contribution of carbohydrate increases, and of fatty acids decreases

response, since excess acidity can lead to cell death.

During and after exercise, excess lactic acid is removed from the working skeletal muscles and circulated to tissues such as the heart, liver, kidney and other skeletal muscles. Lactic acid is not inert, rather it can be converted back to pyruvic acid and degraded via oxidative metabolism to produce ATP in these tissues (Figure 12.3). Thus, excess lactic acid produced via anaerobic glycolysis can become a fuel for further ATP production in skeletal muscle.

After exercise ends, excess lactate is also reconverted back to glucose in the liver; this newly made glucose can then be used to resynthesise glycogen depleted during exercise. It takes approximately 20–60 min to fully remove lactic acid produced during maximal exercise. The rate of lactic acid removal is faster during active compared with passive recovery. Active recovery to optimise lactic acid removal consists of light activity, such as slow jogging at approximately 30–60 per cent maximum pace. During active recovery, the working muscles use the excess lactic acid as a fuel for ATP production. It is important that the pace of recovery be low enough to ensure that more lactic acid is not produced. As discussed in the next chapter, interval training programs should consider the rates of lactic acid accumulation and removal during exercise.

Oxygen must be continually supplied to the working muscles for sustained exercise

The aerobic energy system provides most of the ATP for sustained exercise lasting longer than approximately 3 min. Oxygen supply to the working muscles is an important limiting factor to endurance exercise capacity. If oxygen supply is insufficient to sustain ATP production, the muscles must increasingly rely on the anaerobic system, with consequent build-up of lactic acid and inhibition of further ATP production. When lactic acid production exceeds the maximal rate of removal, fatigue results and exercise pace must be reduced.

Oxygen consumption is an important measure of energy expenditure during exercise. A standard curve of oxygen consumption during exercise has several components (Figure 12.6). During the initial few minutes of exercise, oxygen uptake is not sufficient to provide all the energy needed, and the body is said to go into oxygen 'deficit'. During this time, ATP is supplied primarily by the two anaerobic systems—stored phosphagen (PC) and anaerobic glycolysis. The oxygen deficit occurs because of a lag in adjustment of the cardiorespiratory system to meet the increased energy demand at the onset of exercise.

During submaximal exercise of constant intensity, a steady state may be achieved in which oxygen uptake is sufficient to provide virtually all the ATP needed (Figure 12.6*a*). Theoretically, exercise could proceed at this rate indefinitely. During exercise of increasing intensity, maximum exercise capacity will eventually be reached (Figure 12.6*b*). Oxygen consumption will plateau at a value termed $\dot{V}O_{2max}$ or maximum oxygen consumption, described in detail below. (The dot over the V in $\dot{V}O_{2max}$ indicates a volume rate, that is, volume per unit time, e.g. L/min.)

At the end of exercise, oxygen uptake does not return to resting levels immediately, but takes some time to return to pre-exercise levels. This slow return of oxygen uptake after the end of exercise is called the 'elevated post-exercise oxygen consumption' or EPOC. EPOC was called the 'oxygen debt', a term that implied the oxygen consumed after exercise is used to 'repay' the deficit occurring at the onset of exercise. Although some of EPOC goes to resynthesising PC used at the start of exercise and to removing lactic acid produced during exercise, EPOC is a more complex phenomenon than simply repaying an anaerobic deficit. Consequently the term oxygen debt is no longer used. The excess oxygen goes toward removing lactate, and re-synthesising muscle stores of glycogen, PC and ATP. Extra oxygen is also needed because

a Submaximal, steady-rate exercise

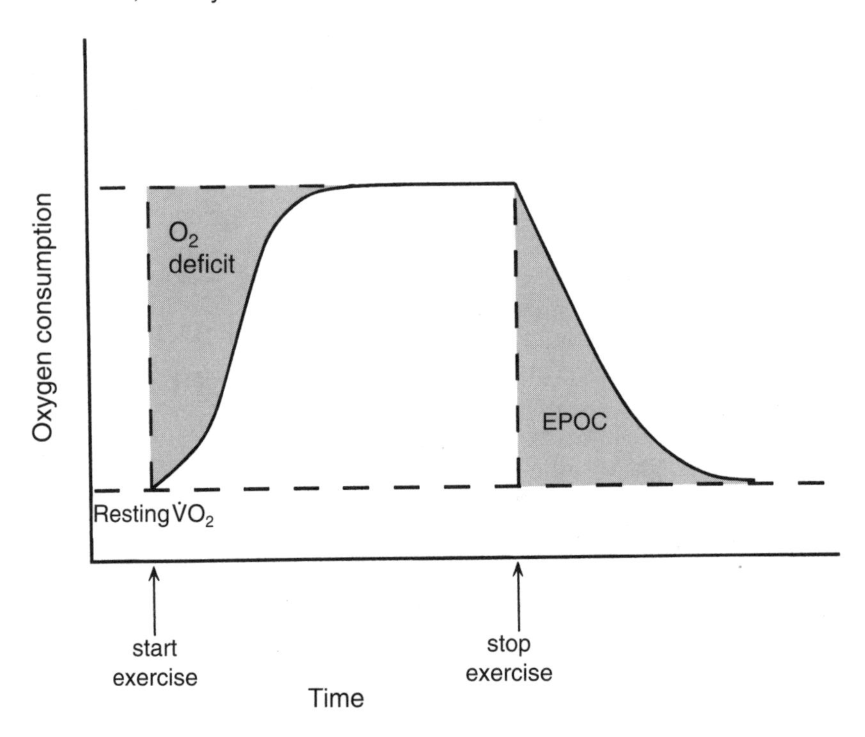

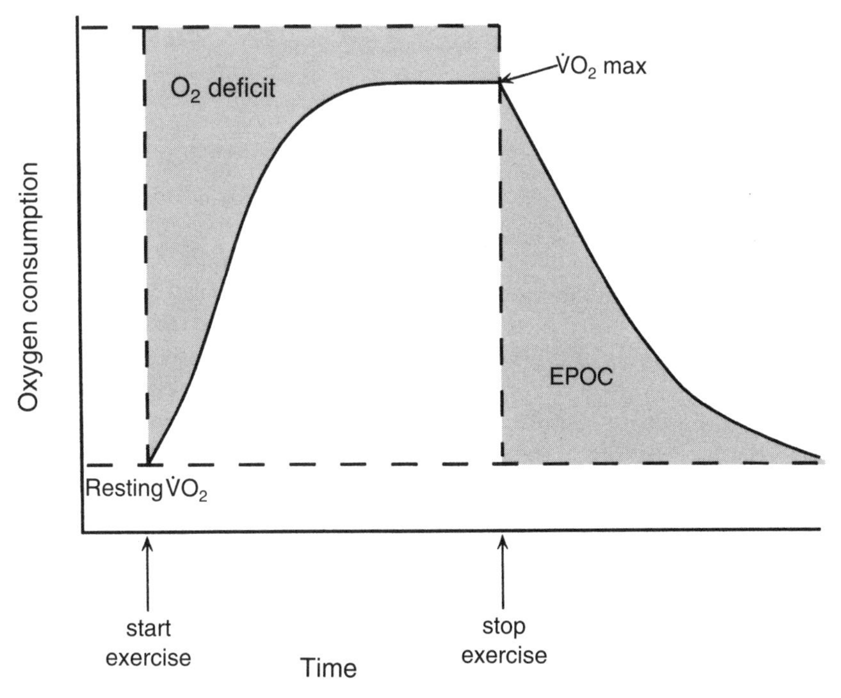

Figure 12.6 Oxygen consumption during exercise:

- Submaximal, steady-rate exercise (a). During steady-rate, submaximal exercise, $\dot{V}O_2$ reaches a plateau at which the ATP produced via oxidative metabolism equals the energy demands of exercise.
- Maximal or supramaximal exercise (b). During maximal or supramaximal exercise, oxygen consumption continues to increase with work rate until a maximal value ($\dot{V}O_{2max}$) is achieved. During supramaximal exercise, additional ATP above that produced by oxidative metabolism is generated via non-oxidative pathways, primarily anaerobic glycolysis

elevated body temperature raises metabolic rate throughout the body. In addition, the heart must work harder to maintain blood supply to the skeletal muscles and skin; since the heart relies almost exclusively on oxidative metabolism, excess work of the heart requires additional oxygen.

The amount of oxygen consumed during EPOC and the time needed for oxygen consumption to return to pre-exercise levels vary with exercise intensity and duration. Generally, intense prolonged exercise is associated with a large EPOC and extended time for oxygen consumption to return to baseline levels. Oxygen consumption may remain slightly elevated for several hours after intense prolonged exercise such as running a marathon.

$\dot{V}O_{2max}$ is a good indicator of endurance exercise capacity

$\dot{V}O_{2max}$ represents the maximum amount of oxygen an individual can consume per minute during exercise and is called aerobic power. The more oxygen consumed the higher the capacity for ATP production via aerobic metabolism. Thus, $\dot{V}O_{2max}$ gives an indication of endurance exercise capacity or the ability to continue exercising for a long time.

Due to a combination of heredity and training, elite endurance athletes generally have high $\dot{V}O_{2max}$ values (Figure 12.7). It has been estimated that up to 40 per cent of $\dot{V}O_{2max}$ is genetically determined. In addition, $\dot{V}O_{2max}$ may be increased by up to 40 per cent by aerobic exercise training. Elite endurance athletes are, to a certain extent, genetically predisposed toward excelling in endurance events (see Box 13.1). However, although a high $\dot{V}O_{2max}$ indicates broad potential for endurance exercise, $\dot{V}O_{2max}$ is not by itself a good predictor of exercise performance. That is, the individual with the highest $\dot{V}O_{2max}$ value will not necessarily be the top performer in endurance events. Other factors, such as training, motivation, skill, mechanical efficiency and the ability to maintain exercise for long periods at a high percentage of maximum capacity contribute to the quality of performance among elite athletes.

The effects of training on $\dot{V}O_{2max}$ will be discussed in the next chapter, and changes with age and sex differences in $\dot{V}O_{2max}$ will be discussed in Chapter 14.

How is exercise capacity measured?

Exercise capacity can be precisely measured in the laboratory using various work monitors or ergometers ('work meters') to measure the amount of work performed and energy expended during exercise; exercise capacity can also be estimated or measured directly in sport-specific field tests. Standardised testing procedures have been developed to measure different types of exercise capacity, for example endurance exercise capacity, muscular strength and power, and anaerobic work capacity.

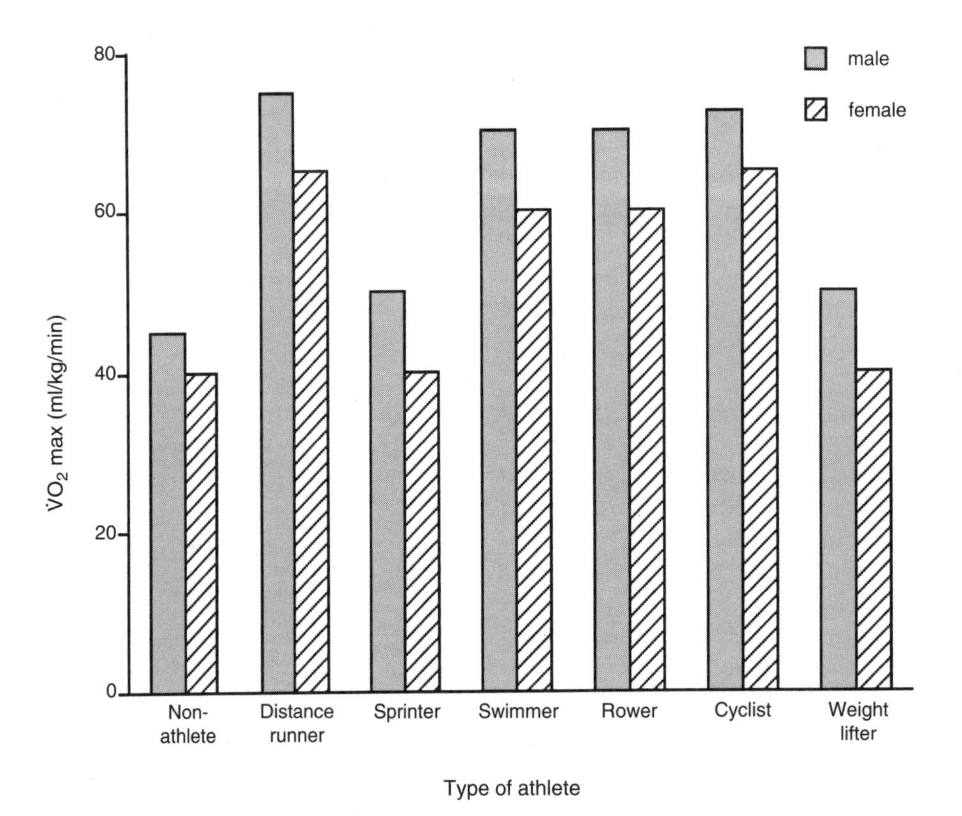

Figure 12. 7 $\dot{V}O_{2max}$ values for young adult athletes. Average $\dot{V}O_{2max}$ levels for high-performance male and female athletes in different sports

Aerobic or endurance exercise capacity

$\dot{V}O_{2max}$ is considered a measure of aerobic power, whereas endurance exercise capacity is a performance measure, such as the maximum time an individual can exercise at a given speed or the total amount of work that can be accomplished in a given time. $\dot{V}O_{2max}$ is assessed in the laboratory where precise equipment can sensitively monitor oxygen consumption during exercise. $\dot{V}O_{2max}$ can be measured during any mode of exercise, but is most frequently performed using stationary running on a motorised treadmill or cycling on a cycle ergometer (Figure 12.8*a*). Sports-specific ergometers have also been developed to simulate activities such as rowing, kayaking, or swimming.

The mode of exercise used to test $\dot{V}O_{2max}$ should be specific to the athlete's training. For example, a cycling test provides a more accurate measure of $\dot{V}O_{2max}$ for trained cyclists than a treadmill test. Similarly, although logistically difficult, measurement of $\dot{V}O_{2max}$ during swimming is more accurate than during running or cycling for assessing aerobic power in swimmers. Measuring $\dot{V}O_{2max}$ on the treadmill is most commonly used for the average individual who is generally comfortable with walking and jogging, and for athletes in sports involving running, such as distance running, triathlon, soccer, football codes or hockey.

In a $\dot{V}O_{2max}$ test, the subject starts to exercise at a comfortable pace, after which exercise intensity increases progressively, usually in 1–3 min increments, until the subject can no longer continue to exercise (volitional fatigue). Exercise intensity may be increased by increasing the speed and/or incline of the treadmill, or by increasing the resistance against which the subject pedals on the bicycle ergometer. For other types of ergometers, the pace of exercise may be increased until fatigue.

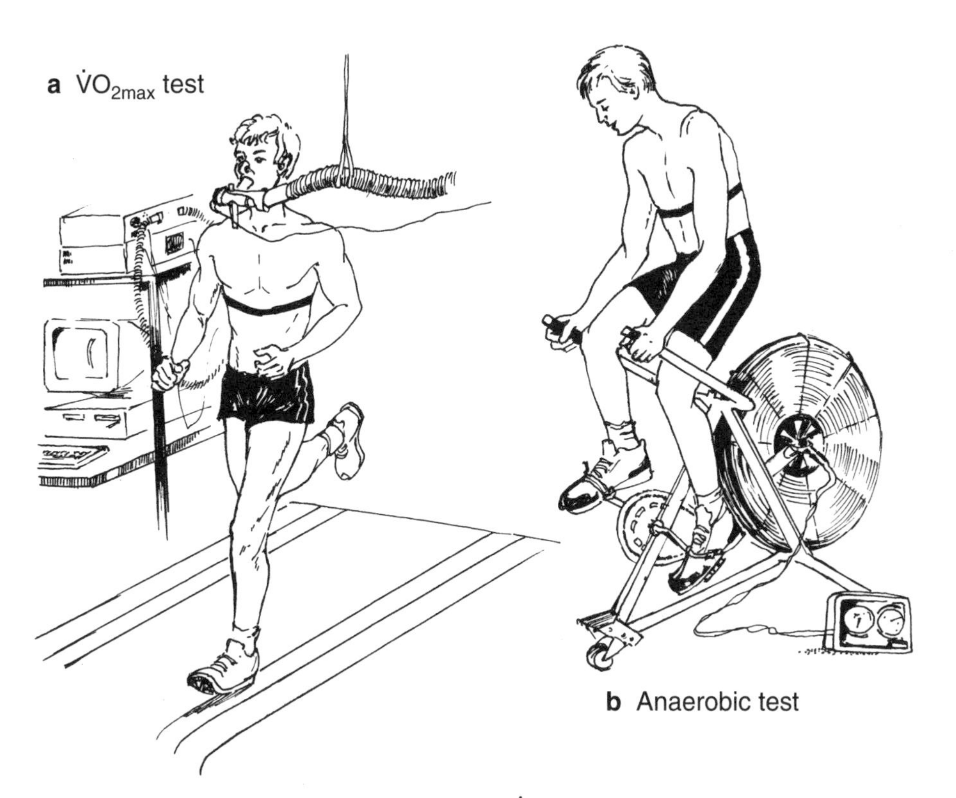

Figure 12.8 Exercise tests: treadmill test of $\dot{V}O_{2max}$ (a); cycle ergometer test for anaerobic power and capacity (b)

Throughout the $\dot{V}O_{2max}$ test, the subject breathes into a two-way, light-weight mouthpiece or mask connected via a long tube to gas analysers. The amount of oxygen, carbon dioxide, and total volume of air breathed are measured throughout the test, and then used to calculate oxygen consumption. $\dot{V}O_{2max}$ is defined as the highest value of oxygen consumed during exercise, and is usually achieved during the final minute of exercise, just before fatigue.

$\dot{V}O_{2max}$ is expressed as either absolute volume in litres of oxygen consumed per minute (L/min), or when adjusted for body mass, as millilitres of oxygen consumed per kilogram of body mass per minute (ml/kg/min). $\dot{V}O_{2max}$ in absolute terms (L/min) is closely related to body size, with larger individuals having higher absolute $\dot{V}O_{2max}$ values than smaller individuals. Adjusting for body mass takes into account differences in body size, and more accurately reflects endurance exercise capacity.

Anaerobic exercise capacity

As the names imply, anaerobic power and capacity refer to exercise capacities in activities requiring energy production primarily by the two anaerobic systems (immediate and anaerobic glycolytic), that is, brief high-intensity exercise. Anaerobic power is the maximum or peak power, expressed in watts (W), that can be achieved in an all-out exercise test. Peak power is usually achieved within the first 2–3 s of all-out exercise. Anaerobic capacity represents the total amount of work, expressed as kilojoules (kJ), that can be accomplished in a specified time, usually 10–90 s.

Anaerobic power and capacity are important factors in many sports and activities requiring rapid and powerful movement, such as sprinting, throwing, jumping and striking. There are several procedures to measure anaerobic power and capacity, some general and others sports-specific. Among the most commonly used are the 10 and 30 s cycle ergometer tests in which the subject pedals as fast as possible on a special cycle ergometer equipped with a work monitor to measure power and work (Figure 12.8*b*). Other general tests include vertical jumping, sprinting or stair climbing as measures of explosive power. Sports-specific tests have been developed to measure power and work on special ergometers, for example the rowing ergometer. Alternatively, the time required to perform a certain activity can be measured, for example in tennis, the time required to sprint between various designated places on a tennis court.

Why measure exercise capacity?

There are many reasons to measure exercise capacity. Precise measurement of aerobic and anaerobic power or anaerobic capacity in athletes enables a coach and athlete to evaluate the athlete's current state of fitness as well as the effectiveness of a training program. Because certain types of exercise capacity are very much related to genetics, testing can also be used in talent identification for some sports.

$\dot{V}O_{2max}$ is an essential measurement for research in the exercise physiology laboratory; it provides an accurate and reproducible means to standardise exercise intensity in the study of how the human body responds to exercise. For example, exercise intensity is set at a given percentage of maximum, say at a work rate eliciting 70 per cent of $\dot{V}O_{2max}$. The absolute work rate may vary between subjects, but each subject will be working at a similar rate relative to his or her own ability.

Setting exercise intensity relative to an individual's $\dot{V}O_{2max}$ is also widely used to prescribe exercise both for athletes during training and for developing 'health-related fitness' among the general population. For example, athletes often train at a high percentage of maximum aerobic power (80–100 per cent $\dot{V}O_{2max}$, see Chapter 13), while a lower exercise intensity (50–80 per cent $\dot{V}O_{2max}$) is recommended for developing health-related fitness in the average non-athlete (see Chapter 15).

$\dot{V}O_{2max}$ is often measured in the non-athlete to determine initial fitness level from which to develop an individual exercise prescription. The fitness test can be repeated at later times to determine the client's progress; clients are often fascinated and motivated by measures of their fitness level. $\dot{V}O_{2max}$ is frequently measured in the clinical setting to assess exercise tolerance. For example, cardiologists measure $\dot{V}O_{2max}$ and the heart's response to exercise in heart patients. These measures provide the cardiologist with a good indication of the extent of heart disease as well as the patient's progress during treatment.

Anaerobic power and capacity are important factors in many sports and activities, especially those requiring explosive movements such as sprinting and game-type sports. Accurate assessment of these factors is important in talent identification and in development and evaluation of effective training programs.

The cardiorespiratory system is essential to oxygen supply during exercise

The cardiorespiratory system consists of the lungs, the breathing tubes that carry air to the lungs (the trachea, bronchii and smaller branches), the heart and blood vessels (Figure 12.9). Oxygen, which comprises approximately 20.9 per cent of inspired air, diffuses from the smallest air sacs in the lungs (alveoli) to blood contained in capillaries surrounding the alveoli. Gas exchange between the alveoli and capillaries is so rapid that, even during maximal exercise, arterial blood is almost fully saturated with oxygen. Blood is carried via the pulmonary veins to the heart, then via the arterial system through progressively smaller arteries to capillaries, where gas exchange occurs within tissues. Oxygen delivery via the circulatory system to the working muscles is a major limiting factor for endurance exercise performance.

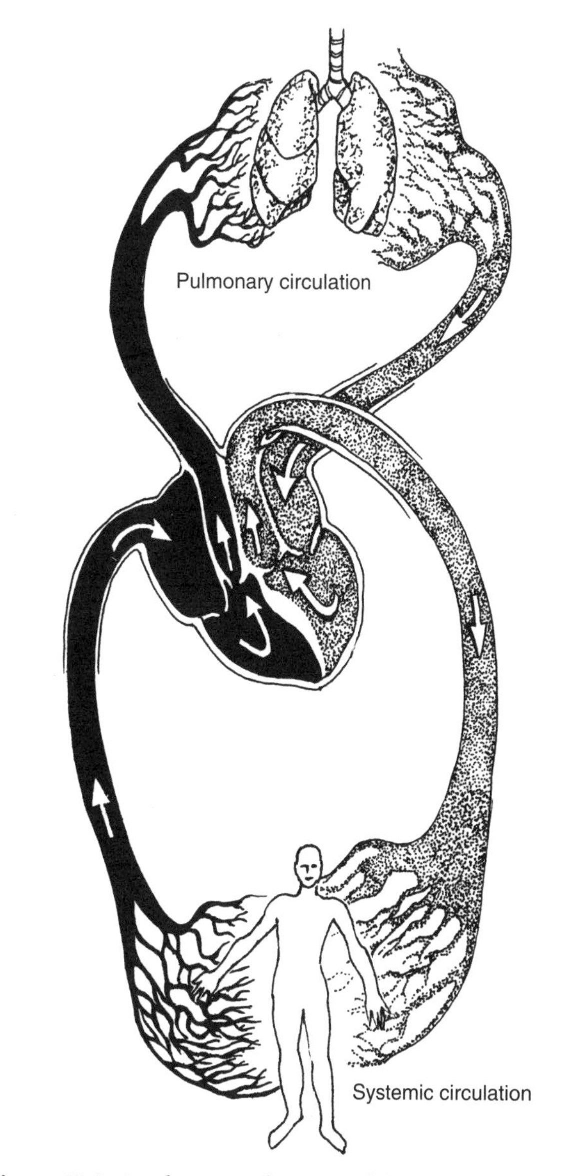

Figure 12.9 A schematic diagram of the circulatory system

As most athletes know, even before the onset of exercise, heart rate and respiration increase (Figure 12.10). This 'anticipatory' response is due to neural input into centres controlling the heart and respiration. In general, heart rate, blood flow to the heart and skeletal muscles and respiratory rate increase proportionately with increasing exercise intensity.

Heart rate increases linearly with increasing work rate up to a maximum value, which is age-dependent (Figure 12.10*a*). Maximum heart rate can be estimated by the equation 220 minus age, although there is large individual variation at any given age. Heart rate is controlled by input from the central nervous system, and responds to changes in acidity (pH), oxygen and carbon dioxide content, and temperature of the blood as well as body movement.

Stroke volume (*SV*) is the volume of blood pumped by the heart with each contraction. During exercise, stroke volume increases in response to neural input to the heart causing

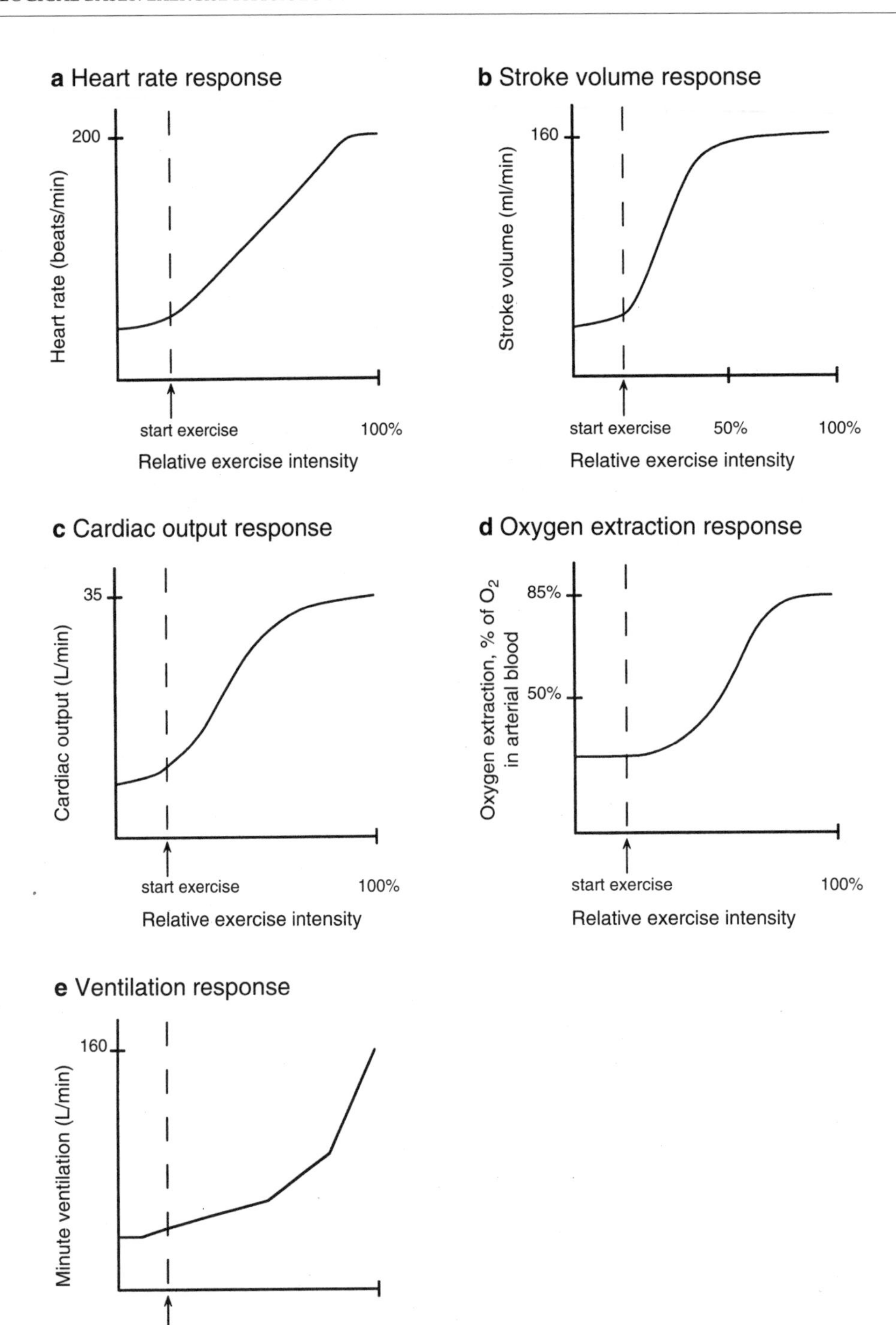

Figure 12.10 Cardiorespiratory system responses to exercise: heart rate (a); stroke volume (b); cardiac output (c); oxygen extraction by skeletal muscle (d); minute ventilation (e). Values are approximations for a young adult male endurance athlete. (See text for details)

the heart to contract more forcefully, which in turn increases return of blood to the heart and the volume of blood entering the ventricles. Stroke volume does not increase linearly with increasing work rate, but increases early in exercise and reaches peak levels at 40–50 per cent of $\dot{V}O_{2max}$ (Figure 12.10*b*). Thus, increased heart rate is primarily responsible for increases in cardiac output at exercise intensities above about 50 per cent $\dot{V}O_{2max}$.

Cardiac output (Q), the volume of blood pumped throughout the body per minute, is a function of both heart rate and stroke volume (Q in litres per minute = HR in beats per minute × SV in millilitres per beat). As with heart rate, cardiac output increases linearly with increasing work rate, reaching a plateau at maximum exercise capacity. During maximum exercise, cardiac output may reach values approximately four to eight times resting levels (Figure 12.10*c*). Cardiac output represents the ability of blood to circulate and to deliver oxygen to the working muscles, and is a major limiting factor for endurance exercise capacity. Cardiac output and oxygen consumption are linearly related.

Oxygen extraction by skeletal muscles also increases during exercise (Figure 12.10*d*). At rest, only about 25 per cent of the oxygen contained in blood is extracted by the tissues, but during maximal exercise skeletal muscles may extract up to 75–85 per cent of the oxygen in blood. Thus during exercise, increased blood flow to, and oxygen extraction by, the working muscles ensure adequate oxygen delivery.

Minute ventilation, the volume of air brought into the lungs per minute, is a function of both respiratory rate and the depth of each inspiration, both of which increase during exercise. Ventilation increases linearly with work rate up to about 55–75 per cent of $\dot{V}O_{2max}$, after which it rises disproportionately with increasing work rate (Figure 12.10*e*). During maximal exercise, ventilation may increase to 20–25 times above resting values. The large increase in ventilation ensures that blood flowing through the lungs is almost fully saturated with oxygen, even at maximum exercise. Thus, ventilation is not considered a limiting factor to exercise capacity, although fatigue of respiratory muscles may contribute to general fatigue during high-intensity exercise.

The combination of all the cardiorespiratory changes listed above ensures that, during exercise at intensities up to $\dot{V}O_{2max}$, oxygen supply to the working muscles is closely matched with oxygen use and energy demand in the skeletal muscles. At exercise intensities above $\dot{V}O_{2max}$, oxygen supply is insufficient to fully meet the demand for oxygen and ATP within skeletal muscle, and the excess energy is supplied via the non-oxidative pathways.

Where does the increased blood flow go?

During exercise the increased cardiac output is not uniformly distributed throughout the body. Rather, blood is redirected through the circulation so that, with increasing exercise, proportionally more blood flow goes to working muscles (Figure 12.11). At rest, only about 20 per cent of blood flow goes to skeletal muscles, with most going to the brain and visceral (internal) organs. At the onset of exercise, redistribution of blood flow occurs via vasoconstriction or narrowing of small arteries in the viscera, reducing blood flow to these organs. At the same time, vasodilation or opening of small arteries in the working muscles and skin increases blood flow to these areas. Circulation to the heart increases in proportion to heart rate and work rate. During maximal exercise, nearly 90 per cent of cardiac output may go to the working muscles. The redirection of blood flow is important in the maintenance of oxygen supply to the working muscles and heart; in providing glucose and fatty acids to the skeletal muscles; to removing carbon dioxide and lactic acid from the muscles; and to ridding the body of excess heat produced during exercise.

Human skeletal muscle cells are not all the same

Human skeletal muscle cells (called muscle fibres) are not homogeneous. Rather, in the

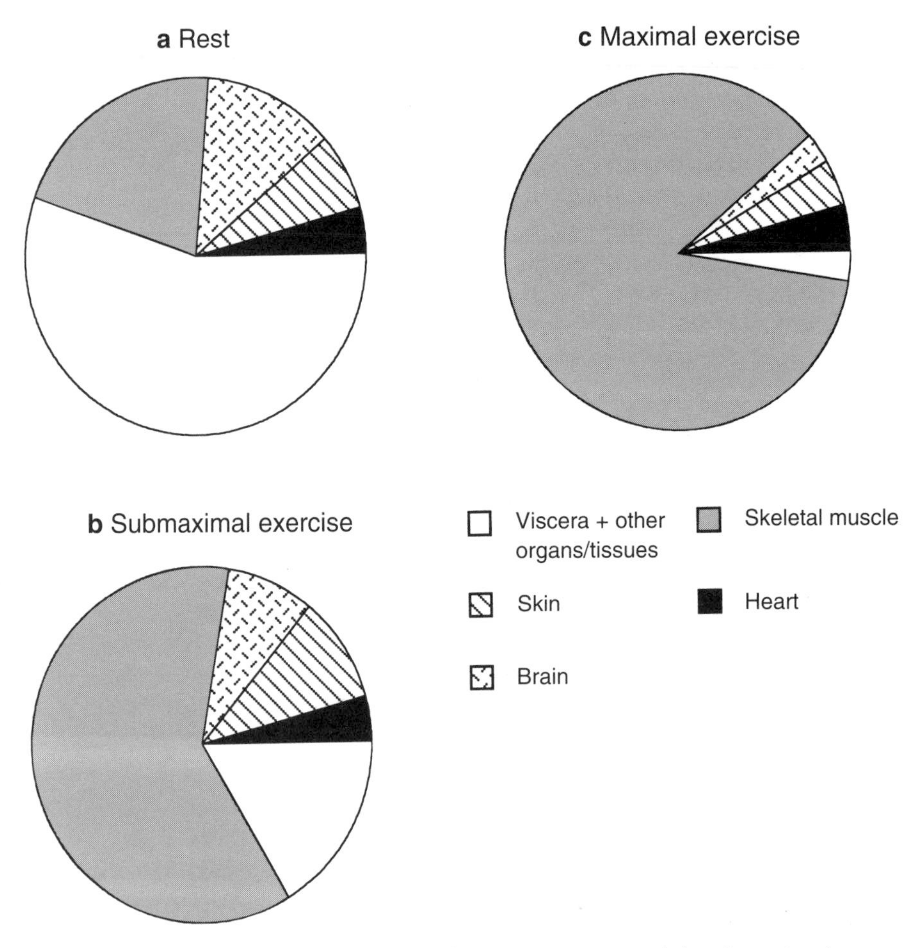

Figure 12.11 Blood flow redistribution during exercise: rest(a); submaximal exercise (b); maximal exercise (c). Relative proportions are expressed as a percentage of cardiac output. (See text for details)

human and many mammals there are three types of muscle fibres, each adapted for specialised function. Muscle fibre types are classified according to their physiological, biochemical and histological properties (see Table 12.2).

Muscle fibres can be classified according to their relative speed of contraction (physiological property): slow fibres (type I) contract and relax within 100 ms, whereas fast fibres (type II) contract within 50 ms. Muscle fibres can be also differentiated by the predominant energy system used to produce ATP (metabolic property): glycolytic fibres have a high capacity for anaerobic (glycolytic) metabolism whereas oxi-

Table 12.2 Human skeletal muscle fibre types

Characteristic	Type I	Type IIA	Type IIB
Fibre size	small	large	large
Contraction speed	slow	fast	fast
Force	low	high	high
Glycolytic capacity	low	high	high
Oxidative capacity	high	high	low
Capillary supply	high	moderate	low
Fatigue-resistance	high	moderate	low

Sources: Compiled from McArdle, W.D., Katch, F.I. & Katch, V.L. (1991), *Exercise Physiology: Energy, Nutrition and Human Performance* (3rd edn), Lea & Febiger, Philadelphia; Lamb, D.R. (1984), *Physiology of Exercise: Responses and Adaptations* (2nd edn), Macmillan, New York.

dative fibres have a greater capacity for aerobic metabolism. Fast (type II) fibres are glycolytic, and slow (type I) fibres are oxidative. By a combination of physiological (fast or slow) and metabolic (glycolytic or oxidative) properties, three designations describe human skeletal muscle fibre types:

- type I, slow oxidative (SO) fibres
- type IIA, fast oxidative glycolytic (FOG) fibres
- type IIB, fast glycolytic (FG) fibres

Each muscle fibre is innervated by only one motor neuron (nerve to the muscle cell), and it is the motor neuron that determines fibre type. Each motor neuron may innervate up to several hundred muscle fibres; one motor neuron and all the muscle fibres it innervates are called the motor unit (Figure 3.15). Because only one motor neuron innervates a motor unit, all muscle fibres within a motor unit are of the same fibre type.

In the average human, skeletal muscles are composed of approximately 50 per cent type I and 50 per cent type II fibres; about 25 per cent of muscle fibres are type IIA and 25 per cent type IIB. Most human skeletal muscles consist of a mixture of muscle fibre types, although particular muscles may contain a higher percentage of one fibre type. Athletes in particular sports may exhibit different fibre type distribution compared with the average individual.

Type I fibres are well suited to activities requiring low force generated over a long time. Muscles used primarily for posture and endurance activities (for example the soleus muscle, the deeper muscle in the calf) contain a high proportion of type I fibres. Muscles used for forceful contractions (for example the gastrocnemius, the superficial muscle in the calf) generally contain a higher percentage of type II fibres. The various fibre types are mixed within the muscle so that a microscopic view of a muscle in cross-section reveals a 'mosaic' of the different fibre types (Figure 12.12).

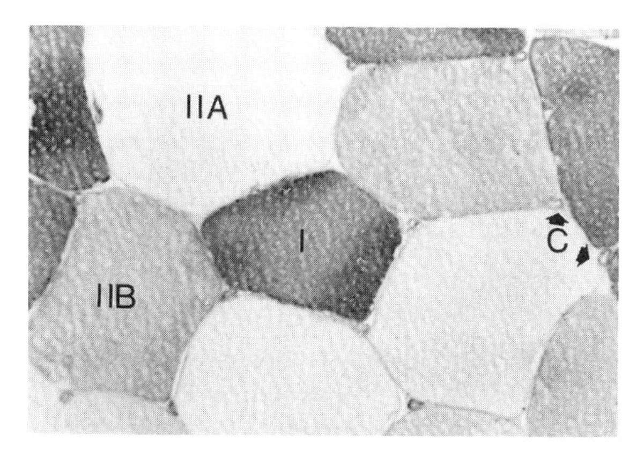

Figure 12.12 Muscle biopsy—photomicrograph (photograph through a light microscope) of a cross-section of human skeletal muscle from the vastus lateralis (thigh) muscle. The sample has been histochemically stained to delineate the three fibre types: type I (slow oxidative), type IIA (fast oxidative, glycolytic) and type IIB (fast glycolytic). Capillaries (C) appear as small round profiles between muscle fibres. (Photomicrograph courtesy of Dr PRJ Reaburn, The University of Queensland.)

As can be seen in Table 12.2, type I fibres are smaller, slower to contract, and are not capable of generating as much force as type II fibres. Type I fibres are well suited to sustained contraction because they are fatigue-resistant; that is, they can continue to contract repeatedly without fatigue. These fibres contain many mitochondria and are surrounded by several capillaries, ensuring a generous supply of oxygen. Thus, type I fibres have a high capacity for oxidative metabolism and are used primarily for endurance activities. It is also believed that, during exercise, type I fibres take up and metabolise lactic acid produced by type II fibres.

Type IIB fibres are the largest, fastest, and most forceful of the three fibre types. These fibres are well suited to activities requiring maximum force, such as weight lifting. Type IIB fibres have a low oxidative, but high anaerobic glycolytic capacity, and are capable of producing large amounts of lactic acid; these fibres fatigue easily, that is, they are not fatigue-resistant.

Type IIA fibres exhibit properties of both type I and type IIB fibres. They resemble type IIB fibres in that they are large, fast and capable of forceful contraction, with high glycolytic capacity; however, type IIA fibres are also similar to type I fibres in that they have more mitochondria, a better capillary supply, and higher oxidative capacity than type IIB fibres. Type IIA fibres are used in activities such as sprinting and lower intensity weight lifting.

What determines which fibre types are active?

When a skeletal muscle contracts, not all muscle fibres are activated or recruited. Muscle fibres are activated in proportion to the amount of force required. Muscle fibre recruitment follows a pattern, called the 'size principle', in which type I fibres (which are smaller than the other types) are activated first, followed by type IIA then type IIB fibres; type IIB fibres are activated only during very forceful contractions requiring maximum muscular force (Figure 12.13). Type I fibres are activated during all contractions, and although type II fibres may provide most of the force during maximal contractions, type I fibres are still activated and contribute to force production. This principle can be applied to designing exercise training programs. For example, low-intensity endurance exercise may only recruit type I fibres; higher intensity endurance exercise must be performed in order to recruit and thus induce adaptations in type IIA and B fibres. Similarly, only high-intensity weight training, near maximum muscular effort, will recruit type IIB fibres (as well as type I and IIA fibres). The application of muscle fibre recruitment patterns to training for sport will be discussed in the next chapter.

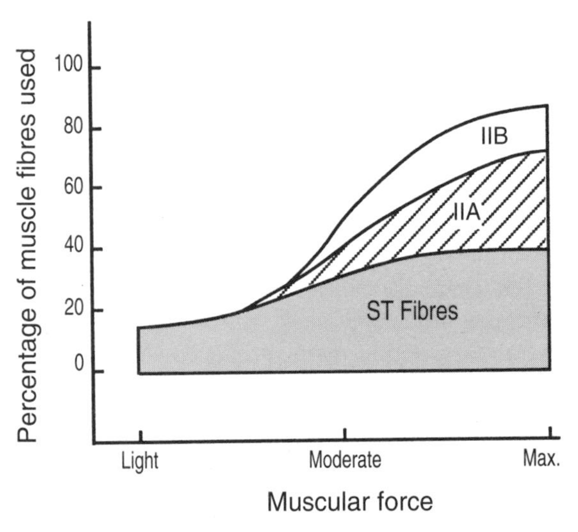

Figure 12.13 Skeletal muscle fibre recruitment during exercise. Smaller type I fibres are activated (recruited) first. Type II fibres are recruited during contractions requiring higher muscular force, with type IIB fibres recruited only at near maximum force
Source: Wilmore, Jack H. & Costill, David L. (1994), *Physiology of Sport and Exercise*, Human Kinetics, Champaign, IL, p. 37. (Copyright 1994 by Jack H. Wilmore & David L. Costill; reprinted by permission.)

How is skeletal muscle 'fibre typing' performed?

In humans, muscle 'fibre typing' is performed on a small (20 mg, about the size of a grain of rice) sample of muscle obtained by muscle biopsy technique. Under local anaesthesia, a small (1 cm) incision is made through the skin and underlying tissue down to the muscle layer. A thin muscle biopsy needle is inserted into this incision and the muscle sample is removed together with the needle. The incision is then closed with a stitch. The muscle biopsy procedure is painless and safe, since there are no pain nerve endings in skeletal muscle and the skin incision is performed under local anaesthesia; the entire procedure is performed under sterile conditions, and has become quite common in exercise physiology research. The athlete can usually resume training within a day of having a muscle biopsy. However, because obtaining a muscle biopsy is an invasive procedure, it is not generally used for routine testing of athletes. Muscle biopsies remain primarily a research tool to understand the physiological and metabolic responses and adaptations to exercise.

Once obtained, the muscle sample is frozen in a special medium that preserves its structure and allows it to be sectioned (cut) while frozen on a specialised instrument called a cryotome. The muscle sample is oriented so that the cylindrical muscle fibres are cut in cross-section, yielding circular profiles. Sections are then mounted on glass slides and stained with special dyes to visualise the biochemical properties of the muscle fibres, such as glycogen content or specific muscle proteins.

Why are there different types of muscle fibres?

The different fibre types provide for specialisation in muscle function, permitting muscle fibres to adapt to a wide range of physiological and metabolic demands. For example, recent work on concurrent or simultaneous strength and endurance training indicates that muscle cells cannot adapt completely to the full range of demands imposed by both types of training. The specialisation of fibre type allows certain fibres to adapt optimally to one type of demand, say increasing in size in response to strength training, while another fibre type can adapt optimally to another set of demands, say increasing metabolic capacity in response to endurance training.

How important is muscle fibre type to sports performance?

As shown in Figure 12.14, the relative proportions of fibre types differ between different types of athletes. For example, high-performance distance runners have a higher percentage of type I fibres, whereas elite sprinters and power athletes have a higher percentage of type IIA, and competitive strength athletes such as weight lifters exhibit a higher percentage of type IIB fibres. These differences in fibre types would imply that muscle fibre type is important to elite performance in specific sports. However, as for $\dot{V}O_{2max}$, top performance is related to many factors. Muscle fibre type gives only a broad indication of potential in those sports at the extremes of the energy system continuum, such as distance running, sprinting or power lifting. A mixture of fibre types would be advantageous in other sports requiring use of all energy systems, such as soccer or football codes, as well as skill-based activities, such as golf.

The question of whether skeletal muscle fibre type and number can be changed with exercise training will be discussed in Chapter 13.

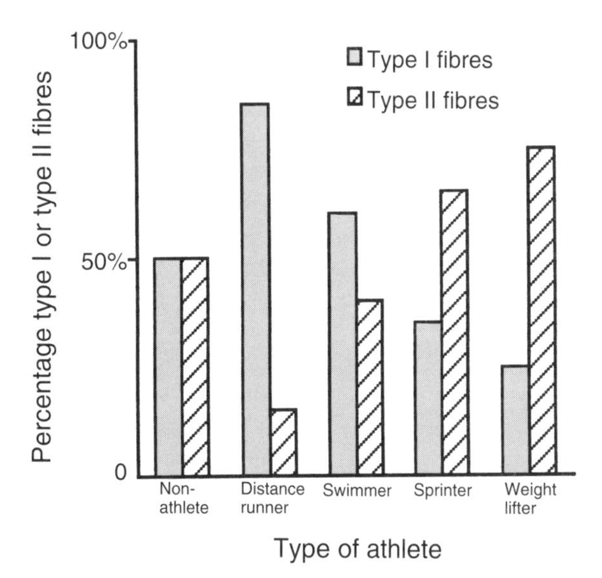

Figure 12.14 Skeletal muscle fibre type distribution in non-athletes and various types of athletes

Source: Compiled from data in: McArdle, W.D., Katch, F.I. & Katch, V.L. (1991), *Exercise Physiology: Energy, Nutrition and Human Performance*, (3rd edn), Lea & Febiger, Philadelphia; Wilmore, J.H. & Costill, D.L. (1988), *Training for Sport and Activity*, (3rd edn), Wm. C. Brown, Dubuque, IA; Wells, C.L. (1992), *Women, Sport and Performance: A Physiological Perspective*, (2nd edn), Human Kinetics, Champaign, IL.

Different activities require different amounts of energy

The energy cost of activity depends on the activity, intensity or pace of exercise, mechanical efficiency or technique and, for non-supported activities, the individual's body mass. Environmental factors such as temperature, wind or terrain may also influence the energy cost of activity.

As described in Chapter 9, mechanical efficiency is the energy cost of performing a specific amount of work; the human body is only about 10–20 per cent efficient. Mechanical efficiency may vary between individuals, and the energy cost of activity at a specified rate will be lower in the more mechanically efficient individual. In many activities, mechanical efficiency is an important determinant of performance. This effect is apparent in activities requiring a great deal of skill or technique, such as swimming.

Table 12.3 gives values for the energy cost of various activities. For activities in which body mass is supported such as cycling or swimming, energy cost is relatively independent of body mass. However, for activities in which body mass is not supported, such as walking, running or aerobic dance, the energy cost of a particular activity increases in proportion to body mass. The energy cost is highest in activities that use the entire body or large muscle groups, such as running, cycling or swimming.

Table 12.3 Energy cost of various activities

Activity	**Energy cost** (kcal/kg body mass/min)
Aerobic dance	0.10–0.16
Circuit training	0.09–0.13
Road cycling, 16–21 km/h	0.11–0.17
Football, soccer, hockey	0.10–0.20
Gardening	0.05–0.13
Golf	0.03–0.09
Jogging/running, 9–16 km/h	0.14–0.30
Squash	0.14–0.25
Swimming, freestyle	0.15–0.25
Walking, 3–8 km/h	0.04–0.12

Sources: Compiled from Inge, K. & Brukner, P. (1986), *Food for Sport*, William Heinemann, Australia, Richmond, NSW; McArdle, W.D., Katch, F.I. & Katch, V.L. (1991), *Exercise Physiology: Energy, Nutrition and Human Performance* (3rd edn), Lea & Febiger, Philadelphia; Pollock, M.L. & Wilmore, J.H. (1990), *Exercise in Health and Disease: Evaluation and Prescription for Prevention and Rehabilitation* (2nd edn), W.B. Saunders Co, Philadelphia.
Energy cost varies with pace of exercise, level of expertise, and position in the case of team sports such as football. The values represented in this table range from beginner to advanced levels, or according to pace as specified.

Why is it important to know the energy cost of activity?

Knowledge about the energy cost of activity can be used in several ways. For example, many people exercise to control body mass (weight); knowing how much energy is expended in certain activities is useful in prescribing the appropriate type, duration and intensity of exercise for fat loss. The energy cost of training must also be considered when planning diets for athletes. Athletes, especially endurance athletes, often have difficulty consuming sufficient energy and carbohydrate to meet the increased metabolic demands of training. Measurement of the energy cost of exercise is important in understanding the physiological responses and regulation of metabolism during exercise.

Diet is important to energy metabolism and exercise performance

Most athletes and coaches are now aware of the importance of diet to exercise performance and the special dietary needs of athletes. For example, athletes need to consume large amounts of energy and carbohydrate to fuel prolonged exercise and dietitians are often employed by sports teams to provide nutritional advice for athletes.

Most athletes need to consume a high carbohydrate diet

It is now well accepted that the availability of certain fuels is important to exercise capacity. Research in Scandinavia in the 1960s showed that diet has a significant impact on endurance exercise capacity. As shown in Figure 12.15, exercise capacity is directly related to both pre-exercise muscle glycogen stores and the amount of dietary carbohydrate consumed. Low carbohydrate diets are associated with low muscle

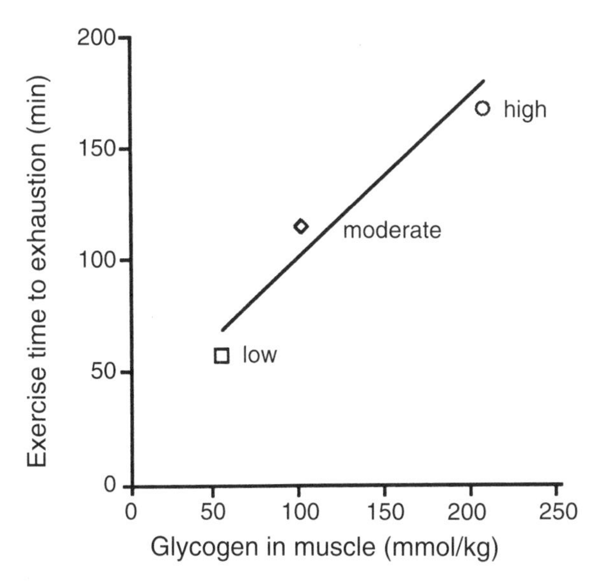

Figure 12.15 Dietary carbohydrate and exercise performance. Increasing dietary carbohydrate enhances muscle glycogen stores. Higher muscle glycogen content increases exercise time to exhaustion by delaying the point of glycogen depletion and, thus, the onset of fatigue
Source: Wilmore, Jack H. & Costill, David L. (1994), *Physiology of Sport and Exercise*, Human Kinetics, Champaign, IL, p. 352. (Copyright 1994 by Jack H. Wilmore & David L. Costill; reprinted by permission.)

glycogen stores and poor exercise capacity whereas high carbohydrate diets are associated with higher muscle glycogen stores and better exercise capacity.

Muscle glycogen stores can be depleted by as little as 40 minutes of intense prolonged exercise, such as distance running; fatigue occurs at the point of depletion of muscle glycogen. Since glucose is no longer available to the muscle once glycogen stores are depleted, the body must subsequently rely almost entirely on fatty acids, and to a limited extent amino acids, for ATP production. Because of differences in the chemical composition of fats and glucose, less energy is produced per given amount of oxygen with fat as a substrate compared with glucose (Table 12.4). Once glycogen is depleted and fat becomes the predominant substrate, exercise

Table 12.4 Energy released per litre of oxygen consumed

Substrate	Energy per litre of oxygen (kcal (kJ))
Carbohydrate	5.05 (21.2)
Fat	4.70 (19.7)
Protein	4.82 (20.2)

pace slows because relatively less ATP is produced. Thus, glycogen depletion is associated with an inability to maintain the rate of exercise and the perception of fatigue; in marathon running, this point is referred to as 'hitting the wall'.

Recent work has shown that muscle glycogen stores can also be depleted by prolonged periods of interval exercise in which anaerobic glycolysis is the predominant ATP-producing system. For example, 30 s of all-out sprinting may deplete 25 per cent of muscle glycogen, and ten 1 min sprints may deplete 50 per cent of muscle glycogen. Sports in which glycogen depletion may occur include soccer, football codes, basketball or any activity requiring repeated high-intensity sprinting over a prolonged period. Weight lifters training over extended periods of time may also experience fatigue due to glycogen depletion. Thus, dietary carbohydrate is important to many athletes, not just those participating in continuous, prolonged events.

Once glycogen is depleted from muscle, 24–48 h is generally required to fully restore glycogen levels provided adequate dietary carbohydrate is consumed. Athletes who train intensely on a daily or more frequent basis may thus be chronically glycogen depleted and have difficulty maintaining training and competitive performance.

Fortunately, one adaptation to high-intensity exercise training is that skeletal muscle stores more glycogen; moreover, in athletes, muscle glycogen stores can be further increased by consumption of a high carbohydrate diet. After depletion, the rate of muscle glycogen repletion varies with the type of diet, specific type of carbohydrate consumed, and how soon a meal is consumed after exercise. In general, glycogen

Box 12.1 Replacing muscle glycogen after exercise—does the type of food matter?

It has long been known that intense endurance exercise can deplete muscle glycogen stores leading to fatigue. A common practice among endurance athletes is to consume a high carbohydrate diet, especially during the few days before and after competition or long training sessions, to help replace the muscle glycogen used during exercise. However, it is only recently that scientists have looked at whether the type of carbohydrate foods eaten influences the rate and amount of glycogen replacement after glycogen depletion.

In a recent paper by Australian nutritionists and exercise physiologists Louise Burke (Australian Institute of Sport), Greg Collier (Deakin University) and Mark Hargreaves (University of Melbourne), the effect of the glycemic index of carbohydrate foods on muscle glycogen stores was studied. Glycemic index (GI) is a measure of the postprandial (postmeal) impact of a food, measured by the blood glucose response in a fasted individual after ingesting a specific amount of carbohydrate food. Foods with a high GI, such as bread or mashed potatoes, elicit a higher blood glucose and insulin response after ingestion. In contrast, ingestion of foods with a low GI, such as rolled oats or legumes, elicits a lower blood glucose and insulin response. It was thought that the higher levels of blood insulin and glucose caused by high GI foods would enhance glucose uptake and thus glycogen synthesis by skeletal muscle.

To deplete muscle glycogen stores, well-trained male cyclists exercised by cycling for 2 h at 75% $\dot{V}O_{2max}$ followed by four 30 s sprints on two occasions. For 24 h after each exercise bout, the cyclists consumed a high carbohydrate diet, consisting of high GI foods after one exercise session and low GI foods after the other. Blood samples were taken before and after each meal to measure the

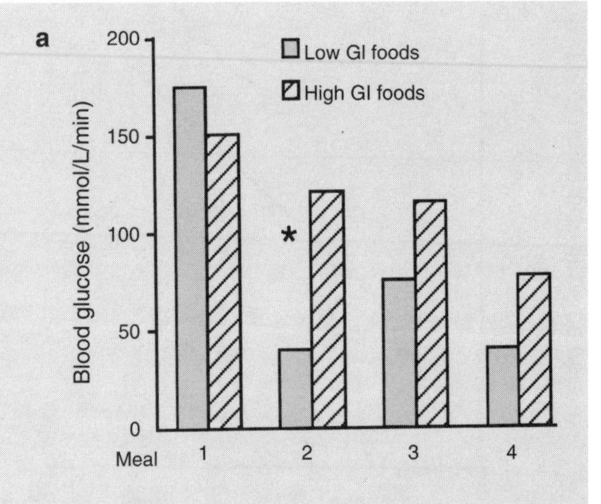

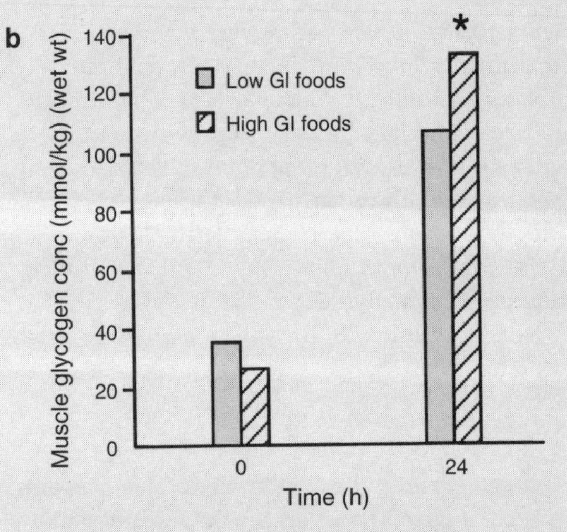

Figure 12.16 Glycaemic index and muscle glycogen replenishment. Blood glucose levels (a) were higher after exercise when a high glycaemic index (GI) diet was eaten; muscle glycogen content (b) was higher 24 hours after exercise when a high GI diet was consumed
Source: Burke, L.M., Collier, G.R. & Hargreaves, M. (1993), Muscle glycogen storage after prolonged exercise: effect of the glycemic index of carbohydrate feedings, *Journal of Applied Physiology*, 75(2): 1019–23, p. 1021. (Copyright 1993 by the American Physiological Society; adapted by permission.)

effect on blood glucose and insulin levels. Muscle biopsies were taken immediately after exercise and after 24 h of recovery to measure muscle glycogen levels.

Blood glucose and insulin were higher during the 24 h after exercise when the high GI foods were eaten compared with those for the low GI diet. There was a greater amount of muscle glycogen stored during the 24 h recovery period with the high GI diet (Figure 12.16). These results suggest that muscle glycogen stores are replenished faster when high GI foods are eaten in the 24 hours after intense endurance exercise. In practical terms, consumption of foods with a moderate to high GI during the day after intense endurance exercise may be preferable for athletes needing to replace muscle glycogen stores rapidly, for example during periods of daily high-intensity training or during competition extending for several days.

Reference: Burke, L.M., Collier, G.R. & Hargreaves, M. (1993), Muscle glycogen storage after prolonged exercise: effect of the glycemic index of carbohydrate feedings, *Journal of Applied Physiology*, 75, 1019–23.

repletion will be faster and more complete if a diet emphasising simple carbohydrates is begun as soon as possible after exercise (see Box 12.1).

It is now generally recommended that athletes who train intensely for several hours each day should regularly consume a high carbohydrate diet. A high carbohydrate diet consists of 60–80 per cent of daily energy intake as carbohydrate, equivalent to 6–8 grams of carbohydrate per kilogram of body weight per day or 420 to 560 grams of carbohydrate per day for a 70 kilogram person.

Do athletes need extra protein?

It is common for many athletes, especially strength and power athletes such as weight lifters and body builders, to consume a high protein diet and to supplement their diets with protein or amino acid powders, drinks and pills. Athletes do so in the belief that the extra protein is incorporated in newly synthesised muscle and will thus aid muscle growth or hypertrophy. Although it is true that athletes need more protein in the diet compared with the average non-athlete, a well-balanced diet is more than adequate to meet the protein needs of virtually all athletes.

For the average healthy non-athlete, the recommended daily intake of protein is 12–15 per cent of daily energy consumption, or 0.75 g protein/kg/day. For a 70 kg person that is about 52.5 g protein/day. (The protein content of certain foods are listed in Table 12.5). Recent research suggests that both power/strength and endurance athletes need about 50–100 per cent more protein per day than the non-athlete.

Table 12.5 Protein content of some foods

Food	Protein (g)	Serving size
Milk	9	250 ml
Cheese	5–7	30 g
Steak	26–34	100 g
Fish	18–26	100 g
Chicken	30–34	100 g
Wholemeal bread	3.5	1 slice
Cereal	2–6	30 g
Nuts	4	25 g
Egg	6	1 egg
Fruit	1	1 piece
Vegetables	1–2	30 g

Sources: Compiled from Inge, K. & Brukner, P. (1986), *Food for Sport*, William Heinemann Australia, Richmond, NSW; McArdle, W.D., Katch, F.I. & Katch, W.L. (1991), *Exercise Physiology: Energy, Nutrition and Human Performance* (3rd edn), Lea & Febiger, Philadelphia; Stanton, R. (1988), *Eating for Peak Performance: Gaining Your Full Potential at Work and Leisure*, Allen and Unwin, Sydney.

However, because athletes expend a large amount of energy during training, they also consume at least 50 per cent more energy per day than the non-athlete. The additional protein needs of the athlete can easily be met by a well-balanced diet (that is 12–15 per cent of daily energy as protein), simply by increasing the total amount of food consumed. These protein needs of athletes are depicted in Table 12.6. To provide the recommended 1.2–1.7 g protein/kg/day, the 70 kg athlete would require approximately 84–119 g protein/day. A daily diet of 3750 kcal (15 750 kJ) containing 12 per cent of total energy as protein would provide sufficient carbohydrate and protein to meet the energy and protein needs of the athlete.

Excess dietary protein is not incorporated into muscle, but is either excreted by the kidney or used to synthesise triglycerides (fat) deposited in adipose tissue. Neither is desirable, since excess protein excretion places an extra burden on the kidneys and excess fat impairs performance in athletes. Moreover, high protein diets and protein supplements are costly. In addition, a diet high in protein is generally inadequate in providing sufficient carbohydrate needed by both power/strength and endurance athletes to fuel extended training sessions. By emphasising protein at the expense of carbohydrate in the diet, the athlete may fatigue early or be unable to train at a high intensity for prolonged periods of time. Moreover, muscular hypertrophy may be compromised because there is inadequate energy available to fuel protein synthesis in the muscle cell.

It is important to replace water lost during exercise

Water is an essential part of the diet, although it is rarely thought of as a nutrient. Because the human body is relatively inefficient, much of the energy produced during activity is not used to perform work but rather appears as heat. The body can withstand only a relatively narrow range of core temperature. To prevent core temperature from rising dangerously during exercise, excess heat must be transferred from the working muscles to the body's external surfaces and then to the environment.

Heat produced during exercise is transferred from muscles to blood, then through the circulation to the blood vessels in the skin. Heat is then most effectively lost by evaporation of sweat from the skin; evaporation is the primary mechanism for heat loss during exercise. Depending on the individual, exercise intensity and duration, and environmental conditions, sweat rates may vary between 0.5 and 2.0 L/h. Even with evaporative cooling, an athlete's core temperature may increase from normal 37°C up to 40°C during exercise.

During prolonged exercise, redistribution of water within the body and loss of body water via sweating may significantly reduce blood volume. Loss of body water equivalent to 4–5 per cent of body mass may adversely affect thermoregulation and exercise capacity. The cardiovascular system adjusts to this loss of blood volume by increasing heart rate to offset the concomitant decline in stroke volume and

Table 12.6 Meeting protein needs through the diet

	Non-athlete	**Athlete**
Total energy consumed (kcal (kJ)/day)	2500 (10 500)	3750 (15 750)
Body mass (kg)	70	70
Percent energy intake as protein	12	12
Energy intake as protein (kcal (kJ)/day)	300 (1260)	450 (1890)
Grams of protein per day*	300/4.2 = 71	450/4.2 = 107
Grams of protein per kilogram body mass per day	71/70 = 1.0	107/70 = 1.5

*1 g protein yields 4.2 kcal when metabolised by the body.

Source: Compiled from Burke, L. (1992), Protein and amino acid needs of the athlete, *State of the Art Review* number 28, Australian Sports Commission, Canberra.

cardiac output. Thus, the same exercise will cause greater stress on the cardiovascular system when performed in a hot environment.

If prolonged exercise in the heat continues without replacement of body water, blood volume may drop significantly and the body may be unable to lose excess heat. Body temperature may increase dangerously, above 42°C. Heat illness or heat stroke may occur, in which the athlete's cardiovascular and thermoregulatory systems are severely impaired. Heat stroke is life-threatening if not treated promptly.

Athletes should consume water at regular intervals during prolonged exercise, especially when exercising in the heat. The general recommendation is to drink approximately 500 ml plain water during the 20 min before exercise, and 250 ml water every 15 min during exercise. Rehydration after exercise is especially important, and it may take several hours to completely replace water lost during exercise. Solutions containing glucose and electrolytes are often used by athletes in very prolonged exercise in which muscle glycogen may be depleted. A modest amount of glucose (less than 8 per cent) in solution will help to improve performance without compromising water replacement in very long events lasting longer than 2 h. A low concentration of electrolytes promotes faster absorption of water.

CHAPTER 13

Physiological adaptations to training

- What limits exercise performance?
- The energy systems respond to different types of training
- Muscular system changes after strength training
- Do muscle fibre number and fibre type change after training?
- Basic principles of training
- Training for cardiovascular endurance
- Does training need to be continuous?
- Exercise for health-related fitness
- Methods of strength training
- Combining muscular and aerobic fitness in circuit training
- What causes muscle soreness?

Effective training to enhance exercise capacity must take into account the different energy systems used to produce ATP in skeletal muscle. This is true regardless of whether the goal of training is to improve performance in an athlete, to enhance health in the average individual or to rehabilitate a patient with a particular disease. As we shall see in this chapter, the outcomes of any training program depend very much on the type of training undertaken. One of the challenges for exercise physiologists is to identify the relative contributions of the different energy systems to a particular activity and to use this information to develop training programs that maximise adaptations.

What limits exercise performance?

Factors that limit exercise capacity and performance are closely related to the predominant energy system(s) used during a particular activity. For example, factors limiting sprinting performance are different from those limiting performance in endurance events.

For power and speed activities in which maximal power is exerted for up to 20–30 s, such as power lifting, 100–200 m sprinting and throwing events, performance is related to the limited amount of ATP and PC stored in the muscles. As discussed in the previous chapter,

PC provides a rapid means of ATP synthesis but is soon depleted during maximal effort. Given that PC is resynthesised only after exercise, ATP must be supplied by the two other systems (anaerobic glycolysis, oxidative metabolism); consequently, the rate of ATP replenishment will not equal that of ATP breakdown and the pace of exercise must slow. Elite sprinters are characterised by an ability to use PC to resynthesise ATP at a faster rate than non-elite sprinters and non-athletes.

During maximal exercise lasting between 30 s and 2–3 min, such as 400–800 m running or 100–200 m swimming, anaerobic glycolysis is the major source of ATP synthesis. As discussed in Chapter 12, anaerobic glycolysis provides ATP at a relatively fast rate, but is limited by lactic acid accumulation. Excess lactic acid increases acidity within muscle cells, inhibiting anaerobic glycolysis and further accumulation of lactic acid. Performance in very high-intensity exercise is thus limited by the ability to buffer (neutralise) and remove excess lactic acid from the muscle. Recent evidence suggests that fatigue is also related to disturbance of the chemical and electrical gradients across the muscle cell membrane due to changes in the intracellular and extracellular distribution of electrolytes such as potassium.

Middle distance events lasting between 3 and 10 min, such as 1500–3000 m running, 400–800 m swimming and 4000 m cycling, are limited by a combination of lactic acid accumulation and consequent decrease in pH, moderate glycogen depletion, and disturbance of electrolyte distribution.

Longer duration events lasting between 10 and 40 min, such as 10 km running or 1500 m swimming, are limited by a combination of moderate lactic acid accumulation and partial glycogen depletion, as well as disturbance of the chemical and electrical gradient across the muscle cell membrane.

In very long duration events lasting more than 40 min, such as road cycling, marathon running and triathlon, performance is limited by a combination of nearly complete glycogen depletion, dehydration, and an increase in body temperature; the latter two factors progressively increase demand on the cardiovascular system to concurrently provide high rates of blood flow to both the skin and working muscles.

Activities that rely on all the energy systems over an extended time, such as basketball or football codes, are limited by a combination of factors similar to those described for longer and very long duration events. For example, glycogen depletion may occur after repeated high-intensity sprinting of the type required in many stop–start sports such as hockey, football codes and basketball. The extended duration of a match (60–90 min) may also increase body temperature and cause dehydration.

Exercise training delays the onset of fatigue

The purpose of exercise training is to induce metabolic and physiological adaptations in order to delay the onset of fatigue. Compared with the untrained individual, the trained athlete can perform more work, or exercise at a faster pace or for a longer time, before the onset of fatigue. Some examples of how training alters factors that limit performance are described below.

Muscle glycogen stores are increased by endurance training. As discussed in the preceding chapter, glycogen depletion from skeletal muscle is associated with fatigue in endurance events. Increasing pre-exercise glycogen stores can delay glycogen depletion and thus the onset of fatigue. In other words, the endurance-trained athlete will be able to exercise for longer before glycogen depletion causes fatigue.

Muscle PC stores may be increased by power and short-sprint training. As discussed in the previous chapter, PC provides the major means of ATP production during short, maximal exercise such as power lifting and short-sprint events. Since PC is rapidly depleted, increasing PC stores via training would enable more ATP to be synthesised before depletion during short explosive activities. However, not all research studies have found increased PC stores after power and sprint training.

An endurance-trained athlete begins to sweat earlier and sweats more during exercise than an untrained individual. These differences enable the endurance-trained athlete to avoid a precipitous rise in body temperature, which could compromise performance, during long-duration events.

The energy systems respond to different types of training

The three energy systems respond specifically to the metabolic demands of training. Improvements in performance and changes in the muscle cells themselves reflect the adaptations of the energy systems in response to the demands of training.

Immediate and anaerobic system changes after strength and sprint training

Metabolic changes in the anaerobic pathways occurring after sprint and strength training are summarised in Table 13.1. Sprint and strength training increase PC, ATP and glycogen stores within muscle fibres, especially type II fibres. These increases facilitate a higher power output in short-duration exercise through an increased capacity for ATP resynthesis via the breakdown of PC and through anaerobic glycolysis.

The activity of the enzymes controlling the rate of anaerobic glycolysis also increases with training, enhancing the amount of ATP that can be generated by anaerobic glycolysis. In addition, sprint training causes an increase in muscle glycogen storage, thus enhancing glycogen availability during repeated sprints.

Table 13.1 Metabolic and structural adaptations to sprint and strength training

Adaptation	Consequence
Increased muscle ATP ad PC	more ATP at onset of exercise
Increased muscle glycogen	delays onset of fatigue
Increased anaerobic enzymes	more ATP synthesised via glycolysis
Increased lactic acid buffering	higher capacity to tolerate high lactic acid levels
Increased muscle fibre size	increased muscular strength and power

The muscle's capacity to generate and to tolerate high levels of lactic acid during maximal exercise increases with sprint training. The increase in lactic acid production observed after sprint training is due to the enhanced rate of breakdown of glucose to lactic acid via anaerobic glycolysis. Exercise can continue despite high lactic acid levels because of an increased capacity to buffer (neutralise) the hydrogen ions (H^+) that dissociate from lactic acid.

Muscle fibre size increases, especially in type II fibres, with training. Because there are more active cross-bridges generating force, power output increases as a result of muscle fibre hypertrophy (growth).

Changes in aerobic metabolism after endurance training

Changes in the aerobic energy system resulting from endurance training are summarised in Table 13.2. Generally, 6 weeks of endurance training will elicit a 20–40 per cent increase in $\dot{V}O_{2max}$ due to changes in both the cardiovascular system and skeletal muscle cells.

The activity of mitochondrial enzymes controlling the TCA cycle and electron transfer chain greatly increases, often by more than 100 per cent. Oxygen uptake and oxidative production of ATP are enhanced, especially in type I fibres; oxidative capacity will also increase in type IIA fibres provided training is at a pace sufficient to recruit these fibres. Increased reliance on oxidative metabolism means less lactic acid is produced via anaerobic glycolysis during endurance exercise.

The capacity of skeletal muscle fibres to utilise fatty acids to produce ATP increases after endurance training. This means that, at any given submaximal exercise intensity, more fatty acids and less glycogen will be used to produce ATP in the trained compared with untrained muscle. Increased capacity to use fatty acids spares glycogen stores, delaying glycogen depletion and the onset of fatigue during prolonged exercise. In addition, muscular stores of glycogen increase after endurance training, providing more glycogen for prolonged higher intensity aerobic exercise. The extra glycogen

Table 13.2 Metabolic adaptations to endurance training

Adaptation	Consequence
Increased $\dot{V}O_{2max}$	greater endurance performance
Increased muscle glycogen	more work before onset of fatigue
Increased mitochondrial enzymes	increased oxidative capacity
Increased use of fats as substrate	less reliance on glycogen, less glycogen depletion
Enhanced lactic acid removal and oxidation	more work before onset of fatigue
Increased lactic acid threshold	more work before onset of fatigue
Increased capillary number	more blood, oxygen and substrates delivered to muscle; more lactate and carbon dioxide removed from muscle
Increased oxygen extraction by muscle	more oxygen available for ATP production
Increased muscle myoglobin content	more oxygen delivered to mitochondria

allows the muscles to work for longer before depletion of glycogen and the onset of fatigue.

The capacity of type I muscle fibres and other tissues to remove lactic acid from the blood increases following endurance training. The increased capacity to use lactic acid to produce ATP during exercise reflects the increased oxidative capacity of type I fibres after training. (Remember, lactic acid can be a used as a fuel or substrate by the TCA cycle and electron transfer chain). Enhanced removal of lactic acid keeps blood and muscle lactic acid levels low during exercise, preventing early fatigue.

Endurance training also increases blood capillary numbers within skeletal muscle, especially around type I and IIA muscle fibres. Increased blood supply enhances the delivery of oxygen to, and removal of carbon dioxide and lactic acid from, working skeletal muscle fibres.

Endurance-trained skeletal muscles contain more myoglobin than untrained muscles. Myoglobin is an iron-containing protein that transports oxygen through the skeletal muscle cell. Increased myoglobin content enhances the rate of oxygen delivery from the cell periphery to the site of oxidative metabolism in the mitochondria.

It is important to note that these metabolic adaptations are specific to the type of training and recruitment pattern of muscle fibre types. Only endurance training will increase the oxidative capacity of skeletal muscle. Similarly, only high-intensity speed or power training will increase intramuscular stores of PC.

Metabolic changes will only occur in those muscle fibres recruited during activity. For example, only type I fibres will show changes in oxidative capacity after low-intensity endurance training; training must be of higher intensity in order to recruit and train type IIA fibres. Similarly, resistance (weight) training must include work near maximum strength to recruit and cause adaptations in type IIB fibres. Sprinting recruits predominantly type I and IIA fibres, and these fibres exhibit the most profound changes after sprint training. Lower intensity endurance training will not change muscle fibre size, but higher intensity endurance training will induce hypertrophy of type I and possibly type IIA fibres.

Cardiorespiratory system changes enhance aerobic capacity after endurance training

In addition to the cellular metabolic changes noted above, the cardiorespiratory system also adapts to endurance training. These adaptations enhance oxygen delivery to skeletal muscle and the muscle's ability to extract and utilise oxygen to produce ATP during exercise. A summary of these changes is provided in Table 13.3.

Oxygen consumption

Resting oxygen consumption is related to body size, and in the absence of large changes in body size or muscle mass, remains essentially

Table 13.3 Cardiorespiratory system responses to endurance training

Adaptation	Consequence
Increased $\dot{V}O_{2max}$	increased endurance performance
Decreased resting and submaximal heart rate	less work done by the heart
Increased resting and exercise stroke volume	increased cardiac output during maximal exercise; lower heart rate for same cardiac output during submaximal exercise
Increased maximal cardiac output	increased blood and oxygen delivery to muscles
Increased blood volume, red blood cell number and haemoglobin content	increased oxygen delivery to muscle cells
Increased oxygen extraction from blood	increased oxygen delivery to mitochondria
Decreased blood viscosity	easier blood movement throughout body
Increased maximal minute ventilation	increased removal of carbon dioxide

unchanged after endurance training. Submaximal oxygen consumption at the same absolute work rate also remains unchanged after training, unless there is a change in mechanical efficiency; efficiency may change in sports such as swimming in which technique improves with training. In contrast to resting and submaximal oxygen consumption, $\dot{V}O_{2max}$ may increase by up to 20–40 per cent after endurance training.

The extent to which $\dot{V}O_{2max}$ improves after endurance training depends on initial fitness level and previous training, genetics and the type of training program. In general, previously unfit individuals show large relative gains in $\dot{V}O_{2max}$ with training because they are farther from their genetically determined upper limit of $\dot{V}O_{2max}$. Heredity is also an important factor, and there is individual variability in the extent of improvement in $\dot{V}O_{2max}$ after endurance training (see Box 13.1). The type of training is also important in that higher intensity, and frequent and longer duration training will induce larger and faster improvements in $\dot{V}O_{2max}$. An improved $\dot{V}O_{2max}$ after training is due to a combination of metabolic changes such as increased oxidative capacity and blood supply within muscle.

Heart rate

It is not unusual for an endurance athlete to have a resting heart rate of 40–50 beats per minute; the resting heart rate in a non-athlete is about 70 beats per minute. This decrease in resting heart rate is called *training bradycardia* (slowing of heart rate). During submaximal exercise at the same absolute work rate, heart rate will be lower in the trained than in the untrained individual. Because of changes in stroke volume (discussed below), the heart is more efficient; that is, the heart is able to pump the same amount of blood within fewer contractions. Thus, the same exercise is less stressful on the heart after training. The decreases in resting and submaximal heart rate result from changes in the heart's neural control. In contrast to resting and submaximal heart rate, maximal heart rate is more a function of age, and remains essentially unchanged by training.

Stroke volume

Stroke volume, the volume of blood pumped with each heart beat, is higher at rest and at all intensities of exercise following training. Endurance training over a period of years increases the size and strength of the ventricles, increasing the amount of blood that can be pumped with each contraction. At rest and during submaximal exercise, the higher stroke volume coincides with a lower heart rate, resulting in the same cardiac output.

Cardiac output

Cardiac output, or the volume of blood pumped each minute, is a product of stroke volume and heart rate. Resting cardiac output is related to body size, and does not change with endurance training. During submaximal exercise at the

Box 13.1 Heredity and athletic ability—how much of sports performance is genetically determined? (Or are great athletes born that way?)

It is obvious that some genetically determined factors are important for sports performance. For example, basketball, volleyball and netball players tend to be taller than average. But what about other important aspects of exercise performance such as $\dot{V}O_{2max}$, muscular strength and power, and anaerobic power and capacity?

Claude Bouchard and associates at Laval University in Quebec, Canada have for many years studied twins to determine the role of heredity in determining exercise capacity. Identical (monozygotic, MZ) twins share identical genetic makeup; non-identical (dizygotic, DZ) twins are no more similar than siblings in terms of their genetic make-up, but would share similar environmental influences, having been born and raised together. If exercise capacity is genetically determined, then identical twins should perform similarly in tests of exercise capacity, whereas performance of non-identical twins would not be as closely related.

Bouchard's group studied endurance exercise capacity in siblings (42 brothers), non-identical (33 pairs, both sexes) and identical twins (53 pairs, both sexes). Variables measured included $\dot{V}O_{2max}$, total work output over 90 min continuous exercise, maximum heart rate and minute ventilation, and body composition (fat free or lean body mass).

On all variables measured, MZ twins were more closely related than DZ twins or siblings; the genetic effect was estimated to range from 40 to 70 per cent depending on the variable measured. For example, the influence of genetics was calculated as 40 per cent for $\dot{V}O_{2max}$; 50 per cent for maximum heart rate; and 70 per cent for total work accomplished in 90 min of exercise. Thus, at least half of endurance exercise performance can be accounted for by genetic influence. Bouchard's group has also estimated that at least half of anaerobic exercise capacity is also related to heredity.

Not only are anaerobic and aerobic exercise capacity influenced by heredity, but the capacity for improvement with training also appears to be very much genetically determined. In a training study on twins and siblings, Bouchard's group found that about 60 per cent of the improvement in $\dot{V}O_{2max}$ after endurance training could be attributed to genetic influence (Figure 13.1). There are,

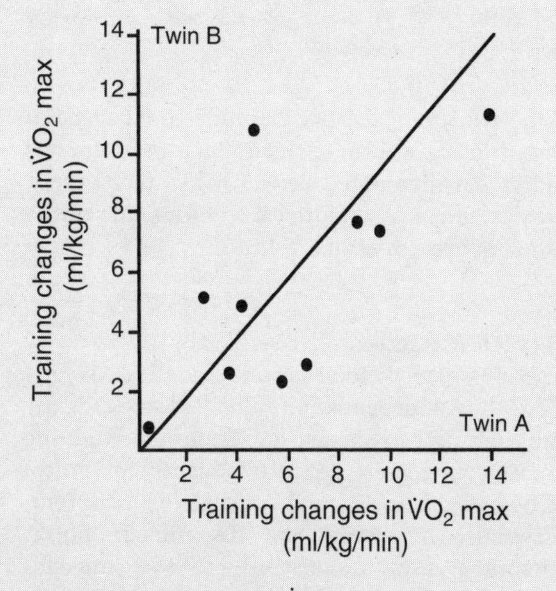

Figure 13.1 Changes in $\dot{V}O_{2max}$ in monozygotic twins following endurance training. Ten pairs of monozygotic (identical) twins trained by cycling for 40–45 min four times per week for 20 weeks; $\dot{V}O_{2max}$ was measured before and after training. Data are presented as the relative improvement in $\dot{V}O_{2max}$ for each pair of twins. Although the range of improvement varied from 0 to 41 per cent among all subjects, the relative improvement in $\dot{V}O_{2max}$ was similar for both twins in each pair (twins A and B).
Source: Bouchard, C., Boulay, M.R., Simoneau, J.-A., Lortie, G. & Perusse, L. (1988), Heredity and trainability of aerobic and anaerobic performances: An update, *Sports Medicine*, 5, 71. (Copyright ADIS Press Ltd; adapted by permission)

however, some physiological variables that are not so closely related to genetics. For example, skeletal muscle fibre type and activities of key metabolic enzymes in skeletal muscle appear to be only about 25–35 per cent genetically determined.

Although exercise performance is to a great extent genetically determined, this is only relevant for the elite athlete, and should not be viewed as discouraging wide participation in sport. There are only a few sports, such as distance running, in which performance is determined mainly by physiological factors. Most sports require a combination of physiological capacity, skill and technique, optimal training, and motivation. Moreover, regardless of genetic predisposition, initial fitness and skill level, all individuals are capable of improving exercise and sports performance with training.

Source: Bouchard, C., Boulay, M.R., Simoneau, J-A., Lortie, G. & Perusse, L. (1988), Heredity and trainability of aerobic and anaerobic performances: an update, *Sports Medicine*, 5, 69–73.

same work rate, cardiac output is unchanged or somewhat lower in trained than in untrained individuals. In contrast, maximal cardiac output may double after endurance training, due to the large increase in stroke volume.

Oxygen extraction

The ability of skeletal muscle to extract oxygen from blood depends on blood flow and the muscle's oxidative capacity. Endurance training greatly increases oxidative capacity, mitochondrial density and myoglobin content, especially in type I and IIA muscle fibres. Increased use of oxygen by the muscles facilitates diffusion of oxygen from blood to muscle. Endurance training increases the ability of skeletal muscles to extract oxygen during both submaximal and maximal exercise.

Other cardiovascular system adaptations include changes in the composition of blood. Endurance training causes increases in the total volume of blood in the body, number of erythrocytes (oxygen-carrying red blood cells) and content of haemoglobin (oxygen-carrying protein within erythrocytes). These changes provide for increased oxygen delivery to the working muscles. In addition, blood becomes less viscous due to a proportionally greater increase in the volume of solution (plasma) compared with the increase in erythrocyte number. Lower blood viscosity means less resistance to blood flow, which enhances blood supply to the working muscles during exercise and reduces the work of the heart. These adaptations also increase the reservoir of fluid available for thermoregulation during exercise in the heat.

These cardiovascular system adaptations greatly enhance oxygen delivery to the working muscle. The increase in oxygen delivery, coupled with the higher oxidative capacity of skeletal muscle, mean that the body relies more on oxidative, and less on lactic-acid-producing anaerobic metabolism during exercise. Also, these changes improve the capacity to use fat for ATP production, which will help to delay glycogen depletion and the onset of fatigue during prolonged exercise. Thus, the capacity for endurance exercise greatly increases in response to aerobic-type training.

Changes in the respiratory system

Due to adaptations in the respiratory muscles, maximum minute ventilation, or the maximum volume of air inspired each minute, increases with endurance training. Although ventilation is not believed to limit endurance capacity, increased ventilatory capacity means that, in the trained individual, the respiratory muscles are less likely to fatigue during endurance exercise. Moreover, the increased volume of air inspired and expired enhances the lung's ability to rid the body of carbon dioxide and to buffer lactic acid produced during exercise.

The lactic acid threshold is a good indicator of endurance exercise capacity

The amount of lactic acid produced during exercise is a function of exercise intensity and state of training. As can be seen from a curve of blood lactic acid level plotted against exercise work rate (Figure 13.2), blood lactic acid level does not increase linearly with increasing work rate. Rather, the level remains low until a certain exercise intensity, above which blood lactic acid concentration increases exponentially.

The point at which the blood lactic acid level begins to increase has been given many names, including the lactic acid threshold, anaerobic threshold, and onset of blood lactic acid accumulation (OBLA). The lactic acid threshold generally occurs at 50–65 per cent of $\dot{V}O_{2max}$ in untrained, and 70–80 per cent of $\dot{V}O_{2max}$ in endurance-trained individuals. Thus, compared with untrained individuals, endurance-trained athletes can exercise at a higher intensity before blood lactic acid begins to accumulate.

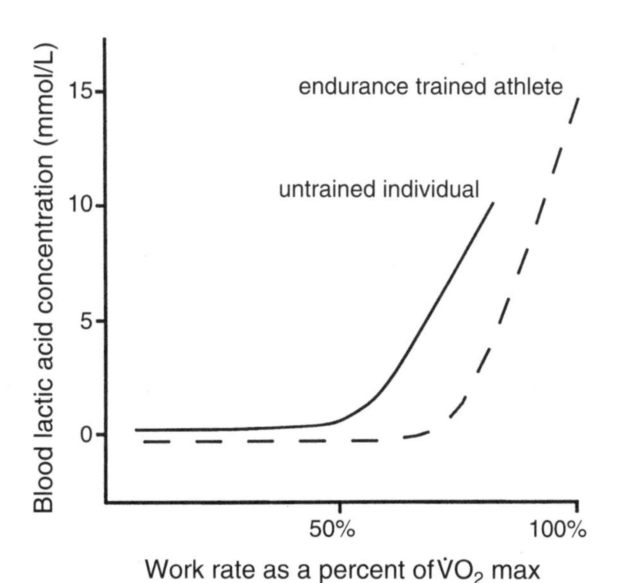

Figure 13.2 Lactate threshold—general trend of blood lactate concentration plotted against relative work rate, expressed as a percentage of $\dot{V}O_{2max}$. Endurance training moves the point of inflection of blood lactate toward a higher relative percentage of aerobic power. That is, endurance-trained athletes can exercise at a higher intensity before blood lactic acid begins to accumulate

The lactic acid threshold indicates the intensity of exercise that can be maintained without significantly taxing the anaerobic glycolytic system. This threshold represents the exercise intensity that can be maintained at the upper limit of aerobic metabolic capacity without a major contribution of the anaerobic system. Below this threshold, ATP production occurs almost entirely via aerobic pathways; above the threshold, ATP production relies on an increasing contribution from anaerobic metabolism. Recall from Chapter 12 that lactic acid produced during high-intensity exercise can cause fatigue by increasing muscle acidity and slowing the rate of ATP production via anaerobic glycolysis. Thus, the lactic acid threshold represents the intensity of exercise below which an individual can, theoretically, maintain exercise indefinitely without fatigue.

Although $\dot{V}O_{2max}$ is a good general indicator of endurance exercise capacity, exercise intensity that coincides with the lactic acid threshold is a better predictor of performance among elite endurance athletes than $\dot{V}O_{2max}$. The endurance athlete who can maintain a faster pace of exercise with lower levels of lactic acid will be the better performer. Although measurement of the lactic acid threshold is complex, regular measurement can provide an important means of evaluating an athlete's response to a training program.

Muscular system changes after strength training

Elite power lifters and body builders are excellent examples of the tremendous capacity of skeletal muscle to adapt to strength training. In the past, strength and power training were used only by athletes in sports that specifically involved weight lifting. It has recently become clear, however, that virtually all athletes benefit from some type of strength training.

Muscular strength may be defined as the maximum force that can be produced in a single movement or contraction. *Muscular power*

represents the product of force times speed. *Muscular endurance* is another important aspect of muscular fitness, and is defined as the ability of muscle to repeatedly develop and maintain submaximal force over time. Muscular endurance is important to performance in sports such as swimming, rowing, kayaking and cycling. Muscular strength, power and endurance are all inter-related, since each is a function of the maximum amount of force that can be applied by a particular muscle group or in a specific movement.

Muscular strength

As we have noted in previous chapters, muscular strength is closely related to the size or cross-sectional area of muscle (Figure 13.3), as well as to muscle fibre type distribution, neural factors and hormones, such as the male hormone testosterone, which stimulate muscle growth after periods of strength training. Muscular strength may increase by 20–100 per cent over several months of resistance training.

The increase in strength is due to a complex interaction of many factors, including neural, structural and metabolic adaptations in skeletal muscle. The relative contribution of each of these adaptations to strength gain varies by individual. For example, in older individuals, the capacity for muscular hypertrophy may be limited and strength gain may depend more on neural factors. (This issue will be discussed further in the next chapter.)

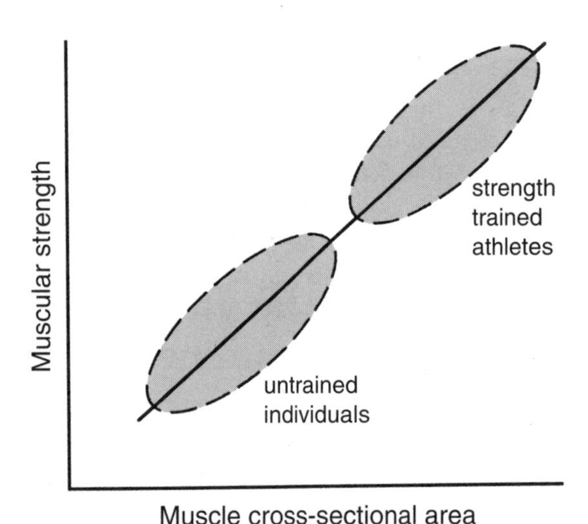

Figure 13.3 Relationship between muscular strength and size—general trend of the relationship between muscular strength and cross-sectional area. Strength training causes muscle hypertrophy, which increases both muscle cross-sectional area and strength

As with anaerobic and endurance training, muscular adaptations to strength training are specific to the intensity and volume of training. According to the size principle of motor unit recruitment, the largest and strongest type IIB fibres are only recruited for forceful contractions near maximum strength. Thus, strength training programs must include maximal or near maximal lifts to induce adaptations in type IIB fibres and to maximise strength gains.

Neural adaptations

Neural adaptations account for much of the early strength gain before muscle hypertrophy occurs. Strength training enhances synchronous recruitment and reduces inhibition of motor units. Motor units do not generally fire at the same time, but fire in sequence or asynchronously. Synchronous recruitment of motor units results in a summation of force because more muscle fibres are activated at the same time. In addition, strength training appears to relieve some of the natural inhibition of motor units. Under normal conditions, each motor neuron receives input from several other neurons, some of which are inhibitory. Reduced inhibition by strength training permits more motor units to become active and thus generate force. The first 4 weeks of a strength training program also improve skill and co-ordination in lifting heavy weights, both of which contribute to early improvement in strength.

Muscle hypertrophy

Muscle hypertrophy begins after approximately 6–8 weeks of strength training, and is the major contributor to continued strength gain after this time. As was noted in Chapter 6, muscle hypertrophy occurs via increases in the average diameter of muscle fibres, and the amount of connective tissue between muscle fibres. The increase in muscle fibre size is due to increases in the number of myofibrils and contractile filaments, so that there are more cross-bridges available for force generation. The rates of protein synthesis and degradation increase in skeletal muscle, although the increase in synthesis exceeds that for degradation, resulting in a net increase in skeletal muscle protein.

The extent of hypertrophy is specific to both the intensity and volume of strength training as well as to muscle fibre type. Low-intensity strength training, which recruits mainly type I and to a certain extent type IIA fibres, will induce hypertrophy only in these fibres. Type IIB fibres are the largest and strongest, with the highest capacity for hypertrophy (twice that of type I fibres). Maximum hypertrophy occurs only when type IIB fibres are recruited through training at nearly maximum strength.

Metabolic adaptations

Metabolic changes induced by intense strength training include increases in ATP, PC and glycogen content in type II fibres (Table 13.1). Increases have also been noted in the activity of several key enzymes involved in PC breakdown and production of ATP in type II fibres.

Muscular power and endurance

Power relates to the ability to generate a large amount of force rapidly; that is, power involves both strength and speed. The amount of muscular force that can be generated depends on several factors, including the speed of contraction. Essentially, force and speed of contraction are inversely related and, for a given muscle, increasing the speed of movement reduces the amount of force generated (see Figure 7.8). Thus, maximal power depends on a balance between speed and force.

Muscular endurance, or the ability to maintain force or repeated contractions over time, is very much related to strength. Increasing strength also improves muscular endurance because, after strength training, a lower proportion of maximal force is required to maintain a given submaximal force. In other words, the same submaximal force can be maintained over time with recruitment of fewer motor units, and thus less fatigue after training.

Do muscle fibre number and fibre type change after training?

It is generally accepted that human skeletal muscle fibre number and fibre type are genetically determined and cannot be changed appreciably through normal activity such as exercise. However, recent research suggests that both fibre type transformation as well as an increase in fibre number may occur, although the evidence is stronger in experimental animal models than in humans.

Hyperplasia, an increase in fibre number, has been induced by intense, high-volume strength training in experimental animals. Research on humans also provides indirect evidence for hyperplasia after intense, high-volume strength training. For example, some experienced weight lifters and body builders have very large muscles but with only average size muscle fibres. It has been proposed that high-volume, high-intensity strength training stimulates division of skeletal muscle fibres, which increases the number of fibres, each of which remains about average size. However, even if hyperplasia does occur, hypertrophy and changes in neural factors are the two predominant mechanisms responsible for increases in strength after resistance training.

Muscle fibre type transformation may occur as a result of training. The most likely transformation is between type IIA and IIB fibres. There is some evidence for transformation from type IIB to IIA fibres after high-intensity endurance training, and from type IIA to IIB fibres after high-intensity resistance training. It is unlikely, however, that exercise training can induce complete transformation between type I and type II fibres. Moreover, as with hyperplasia, if fibre type transformation does occur, its contribution to improvements in performance is relatively small compared with changes in key metabolic, structural and neural factors resulting from exercise training.

Basic principles of training

There are certain basic principles common to all types of training, from conditioning at the highest (elite) level of sports competition to programming for the average adult who exercises for the health benefits.

Specificity

Skeletal muscle responds specifically to the physiological and metabolic demands of exercise. Thus, training must reflect the specific energy demands of the activity. For example, only endurance-type training will increase the oxidative capacity of skeletal muscle and only strength training will induce hypertrophy in type II muscle fibres. The concept of specificity also extends to movement patterns and neural adaptations, and training should, as far as possible, simulate the speed, force and timing of the activity.

Overload

The body adapts to the physiological and metabolic demands of training over time. Training loads must progressively increase to induce continued improvement in exercise capacity. Athletes frequently experience a plateau in performance if training loads are not continually adjusted to accommodate recent training adaptations and increased work capacity. Overload is normally achieved by altering the combination of frequency, intensity, time (duration) and type of activity.

Individualisation

Training adaptations are best achieved with individualised programs specific to the needs of the athlete. Although general principles of training are applicable under most situations, it is important to recognise that the rate or extent of adaptations may vary between individuals. As discussed later in this chapter, individualisation is important to prevent the overtraining syndrome.

Reversibility

Training adaptations last only as long as the physiological and metabolic demands are continued. Detraining, a reversal of training adaptation, begins within days of cessation of training. Once a certain level of fitness is achieved, a minimum amount of regular exercise is necessary to maintain training adaptations. In general, maintenance of training adaptations requires less exercise than is needed to induce the initial changes. Provided that training intensity is maintained, training volume may be decreased by up to 50 per cent with little loss of fitness.

Periodisation

Athletes frequently train in weekly, monthly or seasonal cycles, called 'microcycles'. Periodisation ensures that the athlete reaches peak form at crucial times, such as major competition (Figure 13.4). Periodisation also allows for variety in training to prevent boredom and overtraining, and permits continued adjustment of the training program in response to the athlete's progress or other factors such as injury or illness.

Periodisation allows the athlete or team to focus on various aspects of the sport at different times in the season. For example, in team sports such as soccer or hockey, early season training may emphasise conditioning for aerobic fitness and strength. As the season progresses, training will focus less on endurance capacity and strength and more on speed, power, skill, strategy and team work.

Overtraining

Overtraining refers to excessive training leading to prolonged fatigue and poor performance. The athlete is often unable to maintain training loads or to perform at the standard expected. Overtraining often results from increasing training volume or intensity too rapidly and not allowing adequate recovery between training sessions. Prolonged rest and a reduction in training loads over several weeks or months are necessary to restore performance in an overtrained athlete.

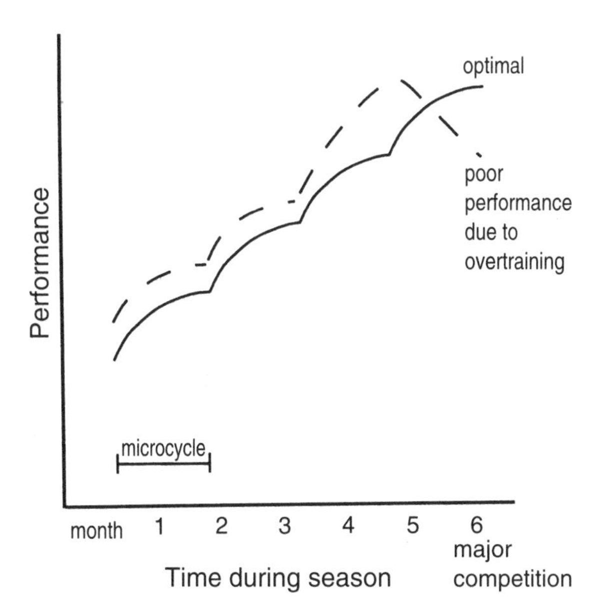

Figure 13.4 Training microcycles—theoretical model of training microcycles, each approximately 6 weeks long. Performance begins to plateau at the end of each microcycle, requiring increased training volume and/or intensity for further improvement. Ideally, performance should peak during major competition; however, excessive training early in the season may cause overtraining and a decline in performance late in the season and during major competition

Optimal training must balance all of the principles discussed above. Indeed, training of the top athlete is both an art and a science, based on scientific principles underlying training adaptations together with the coach's intuition and observation as to how an individual athlete adjusts to overload induced by training. Optimal performance depends on maximising the positive training adaptations while minimising the potential negative effects of excessive training loads.

Training for cardiovascular endurance

The type of cardiovascular endurance training program depends on the individual's objectives and desired outcomes. For example, the older individual seeking health benefits from exercise requires a program different from that of the high-performance distance runner. Thus, exercise programs should be individualised, although certain basic principles apply to all types of endurance training.

There appears to be a minimum threshold in terms of exercise intensity, frequency and duration for improving aerobic exercise capacity. To improve $\dot{V}O_{2max}$, the average healthy young adult needs to exercise for at least 15 min duration, at a minimum of 60 per cent $\dot{V}O_{2max}$, at least three times a week. Improvements may occur at a lower exercise intensity in older, less fit individuals or in those with disease (see Chapter 15). In general, improvements in $\dot{V}O_{2max}$ and endurance exercise capacity are related to the total amount of exercise performed; exercise intensity, frequency and duration may be manipulated in many ways to increase aerobic power and endurance exercise capacity. However, according to the principle of specificity, training intensity and duration for the endurance athlete should approximate those required in competition.

Does training need to be continuous?

Training may involve continuous or interval exercise. As their names imply, continuous training is performed without any rest breaks, while interval training involves alternating intervals of exercise and rest. Table 13.4 summarises the beneficial and detrimental aspects of each type of training.

Both continuous and interval training can be employed to develop certain aspects of fitness. For example, short interval training involves repeated high-intensity intervals at maximal pace and induces adaptations in the immediate energy system; in contrast, longer intervals at lower intensity primarily train the aerobic energy system.

Continuous training

Continuous training may be further defined by exercise intensity. Lower intensity continuous

Table 13.4 Types of training

Type	Advantages	Disadvantages
Continuous	• time efficient	
	• trains cardiovascular and muscular endurance	athletes may need higher intensity
	• easy to follow routine program	can be monotonous
	• can be sports-specific, e.g. distance running	may not be specific to some activities, e.g. team sports
	• less chance of injury because lower intensity	
Fartlek	• adds variety of pace	may not be sports-specific
	• train at higher intensity than in continuous training	higher intensity may increase risk of injury
Interval	• adds variety of pace and duration	more time is needed
	• can be very sports-specific, e.g. sprinting	increases risk of injury due to higher intensity
Circuit	• trains cardiovascular and muscular endurance as well as strength	not maximal improvements in endurance and strength
	• time efficient	need access to equipment
	• can be very sports-specific	
Aerobic circuit	• time efficient	need access to equipment
	• can train at high intensity	higher risk of injury
	• can be very sports-specific	

training is usually in the range of 70–80 per cent $\dot{V}O_{2max}$ for athletes, and 50–70 per cent $\dot{V}O_{2max}$ for those seeking health-related fitness. Lower intensity continuous training can be used in a variety of situations, including development of health-related fitness for the average adult or during early-season aerobic training in many sports. Higher intensity continuous training, above 80 per cent $\dot{V}O_{2max}$ for the athlete and 70 per cent $\dot{V}O_{2max}$ for the average adult, is generally recommended only for the very fit or for the competitive athlete. The athlete often exercises near race pace in this type of training.

Continuous training need not be at a constant pace. Fartlek (Swedish for 'speed play') training is a type of continuous training that includes short bursts of faster work interspersed with longer periods of slower paced exercise. The shorter bursts of higher intensity exercise train both the anaerobic glycolytic and aerobic energy systems, whereas the longer periods of slower exercise induce adaptations in the aerobic system, which enhances removal of lactic acid produced during higher intensity exercise. Fartlek training also adds variety to continuous training, which may become repetitive and monotonous.

Interval training

Interval training involves alternating periods (intervals) of exercise and rest. As shown in Table 13.5, the length of both the exercise and rest intervals is manipulated to induce specific training adaptations.

Interval training offers several advantages over continuous training. For competitive athletes, the duration and intensity of exercise intervals can be varied to train specific energy systems, and training may be performed at or near race pace. Because of frequent rest intervals, interval training permits the athlete to perform more exercise with less fatigue than in continuous training. Moderate intensity interval training is also an effective means of gradually introducing exercise to the untrained adult interested in health-related fitness. There is less risk of

Table 13.5 Interval training variables

Variable	Type of interval		
	Short interval	**Intermediate interval**	**Longer interval**
Work interval	5–30 sec	30–120 sec	2–5 min
Intensity/pace	> 95% race pace	90–95% race pace	60–90% $\dot{V}O_{2max}$
Rest interval	3–5 times work interval	2–3 times work interval	1–2 times work interval
Major energy system	immediate	anaerobic glycolysis	oxidative
Major effects on:	speed and power	speed and power, muscular endurance, lactic acid buffering and tolerance	cardiorespiratory endurance, muscular endurance, lactic acid removal

Source: Compiled from Rushall B.S. & Pyke, F.S. (1990), *Training for Fitness and Sport*, Macmillan, Sydney; Wilmore, J.H. & Costill, D.L. (1988) *Training for Sport and Activity* (3rd edn); Wm. C. Brown, Dubuque, Iowa.

overuse injury, muscle soreness and an abnormal response in moderate interval training compared with higher intensity continuous training. In addition, exercise is often perceived as less intense and therefore more comfortable. One disadvantage of interval training is that, because of the rest periods, interval sessions are typically longer than continuous training sessions.

There are three general categories of interval training, each corresponding to the predominant energy system used.

Short interval training, or anaerobic interval training, consists of work intervals lasting 5–30 s. Short interval training relies predominantly on the immediate energy system, and is used to develop muscular strength and power; exercise is performed at or near race pace. Since PC levels may be depleted by more than 50 per cent after 30 s maximal exercise, rest intervals must be long enough (3–5 min) to restore PC. Moderately high levels of lactic acid may accumulate in skeletal muscle and blood after 20–30 s exercise intervals.

Intermediate interval training, or anaerobic–aerobic interval training, consists of work intervals lasting 30 s to 2 min, and is performed at high intensity (>90 per cent of race pace). Intermediate interval training relies on PC breakdown and anaerobic glycolysis for energy production, and to a limited extent aerobic metabolism. Muscle and blood lactic acid levels will be very high at the end of each work interval, and longer rest periods (several minutes) are needed to restore PC levels and to remove lactic acid.

Long interval training, or aerobic interval training, consists of work intervals lasting 2–5 min; this type of training relies primarily on the aerobic system for ATP production. Longer intervals induce changes in the oxidative capacity of muscle, especially type I and IIA fibres. If performed at a higher intensity, at 70–90 per cent $\dot{V}O_{2max}$, the anaerobic glycolytic system will also be taxed, producing moderately high lactic acid levels. Higher intensity long interval training enhances lactic acid removal and oxidation.

Exercise for health-related fitness

As the term implies, health-related fitness describes physical activity performed primarily for health benefits rather than improvement in performance or physical work capacity. Health-related fitness encompasses an ability to perform the physical tasks of daily living as well as reduction in the risk of diseases associated with a sedentary lifestyle, for example heart disease or obesity as discussed in Chapter 15.

The American College of Sports Medicine (ACSM) has issued a Position Statement based on extensive review of many years of research

Table 13.6 Summary of American College of Sports Medicine recommendations for exercise for health

- Any activity using large muscle groups that can be maintained for a prolonged period of time, such as walking, jogging, cycling, swimming
- Intensity from 40 to 85 per cent of $\dot{V}O_{2max}$ or from 55 to 90 per cent of age-predicted maximal heart rate
- Duration from 15 to 60 minutes of continuous or interval training
- Frequency of from three to five times per week
- Include moderate intensity strength training of the major muscle groups twice per week

Source: Compiled from American College of Sports Medicine (1991), *Guidelines for Exercise Testing and Prescription* (4th edn), Lea & Febiger, Philadelphia; Pollock, M, L. & Wilmore, J.H. (1990), *Exercise in Health and Disease: Evaluation and Prescription for Prevention and Rehabilitation* (2nd edn) W.B. Saunders, Philadelphia.

on the health benefits of exercise. The ACSM recommendations, summarised in Table 13.6, focus on the appropriate type and amount of exercise suitable for maintaining health in the normal adult population. There are several reasons for these specific recommendations put forward by the ACSM.

Type of recommended exercise

Whole-body exercise or activities that use large muscle groups, such as walking, jogging, swimming and cycling, are recommended. This type of exercise promotes loss of body fat and favourable changes in cardiovascular disease risk factors. In addition, whole body or large muscle group exercise is less likely to cause abnormal cardiovascular responses such as excessively high blood pressure or irregular heart beats. Intense exercise using isolated body parts, such as power weight lifting, may induce large increases in blood pressure, which may be dangerous for individuals with high blood pressure (hypertension).

Intensity of exercise

During low to moderate intensity exercise in the range of 50–70 per cent $\dot{V}O_{2max}$, fatty acids provide the major fuel source for ATP production; regular exercise in this range, therefore, enhances fat loss by mobilising fatty acids from adipose tissue. Low to moderate intensity exercise can be maintained for prolonged periods of time (up to 60 min), increasing energy expenditure during exercise, which is important for loss of body fat and body mass. Low to moderate intensity exercise is also less likely to cause musculoskeletal injury. Moreover, activity in this range of intensity is more attractive to the average adult, increasing the probability of continued participation. Very fit individuals may choose to exercise in the higher intensity range of 70–80 per cent $\dot{V}O_{2max}$, particularly since exercise intensity is critical to improving and maintaining endurance exercise capacity.

Duration of exercise

At least 15 min of moderate exercise is needed to induce training adaptations. For the average adult, beneficial effects of training do not appreciably increase when exercise sessions extend beyond 60 min, but the risk of overuse injury increases with exercise duration. Training effects are a function of both exercise intensity and duration; lower intensity training requires longer duration and vice versa. Although a training effect may occur in response to only 10 min of high-intensity exercise, above 90 per cent $\dot{V}O_{2max}$, this intensity of exercise is not recommended for the non-athlete. Longer duration, low-intensity exercise is also recommended to increase total energy expenditure and thus enhance loss of body fat.

Frequency of exercise

A minimum weekly training frequency of 3 days is recommended to induce a training effect. For the average adult, only marginal gains in fitness and health benefits occur with more than five exercise sessions per week, but the risk of injury increases greatly with frequency. Three to five sessions per week provide a good balance between optimal beneficial training effects and potential risk of injury. However, risk of injury is not generally associated with low-intensity, low-impact activities, such as walking or swimming, and these activities may be undertaken daily.

Resistance exercise

Appropriate resistance exercise, such as low to moderate intensity weight training, will also induce health benefits, such as alteration in heart disease risk, more commonly associated with endurance-type training. Moreover, resistance exercise helps to maintain muscle mass, counteracting the normal declines in muscular strength, bone density and functional capacity that occur with ageing. (These effects will be discussed in more detail in the next two chapters.)

Methods of strength training

Strength training should be an important component of all fitness programs, from strength and power athletes through to individuals who exercise for the health benefits. Of course, athletes in sports requiring strength and power, such as weight lifting, body building and sprinting must emphasise resistance training. However, many other athletes also benefit from strength training, especially those in sports requiring a high level of muscular endurance, such as rowing, swimming and cycling. Strength training is also used in prevention and rehabilitation of sports-related injuries.

Muscular strength exercises can enhance health-related fitness. Moderate-intensity resistance training has been shown to confer health benefits such as favourable changes in body composition and blood lipids (fats) related to heart disease. Moderate resistance training may also help to prevent and treat some types of lower back pain and other conditions such as arthritis and osteoporosis (see Chapter 15); maintain lean body mass, muscular strength and mobility with ageing (see Chapter 14); and improve muscular 'tone' and body shape, an important aesthetic consideration for many adults who exercise.

Muscles can contract in different ways

As we have seen in earlier chapters, there are two general types of muscle contractions—static and dynamic (Figure 13.5). In a static, or isometric, contraction the muscle generates force but does not change length because resistance to that force is stationary. Isometric contractions help to stabilise joints during activity. Isometric strength training is primarily used when joint mobility is limited, for example after an injury or in a joint affected by arthritis.

In a dynamic contraction, the muscle's length changes while generating force against a movable resistance. Dynamic contractions, as we have noted previously, may be either concentric, in which the muscle shortens during force development, or eccentric, in which the muscle generates force while lengthening. Most force-producing movements occur via concentric contraction, for example the propulsive action in running, jumping or throwing. Eccentric contractions are used to stabilise or decelerate the body especially when counteracting gravity, for example in landing after a jump or in slowing arm movement at the end of throwing.

There are different types of strength training

Strength or resistance training programs may include static or dynamic contractions. Given that gains in strength are specific to the type of program, resistance-type activities should match the individual's objectives and desired outcomes. As indicated above, isometric training has limited applications, primarily because the gain in strength is limited to the particular joint angle at which training occurs.

Dynamic strength training may be further classified by the type of resistance against which force is applied. Isotonic training is performed against a constant resistance, for example a

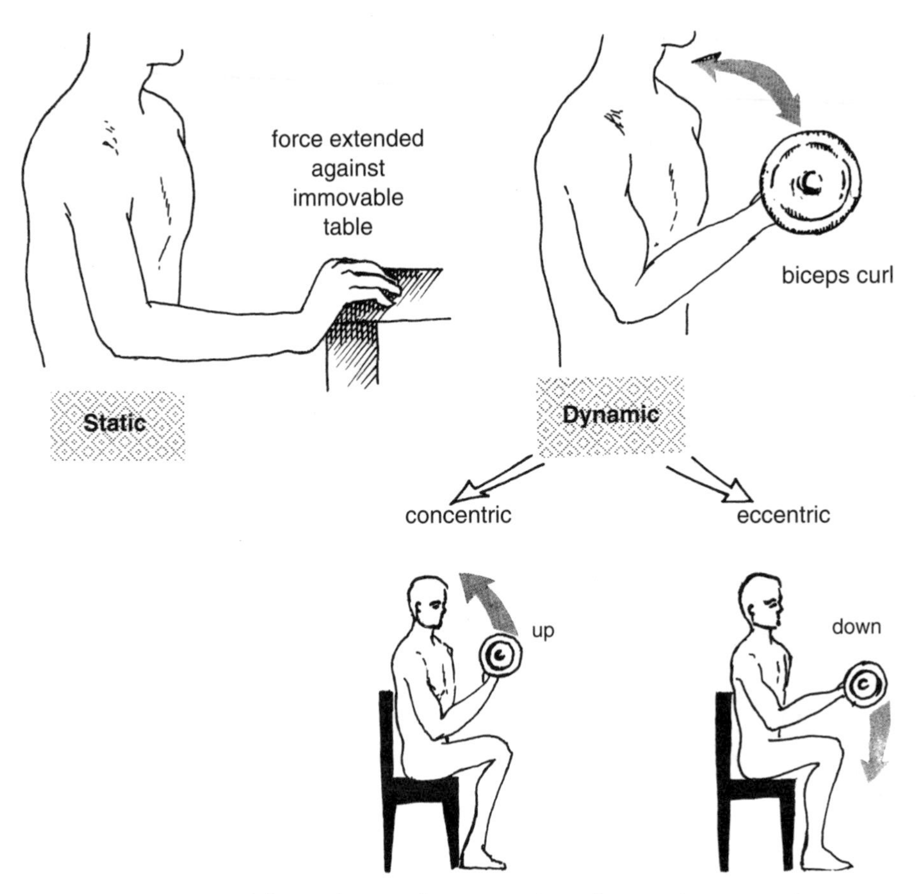

Figure 13.5 Static and dynamic muscle contractions. In a static contraction, tension is generated but muscle length does not change. In a dynamic contraction, the muscle (in this case the biceps brachii) may shorten or lengthen while developing tension; these are referred to as concentric and eccentric contractions respectively

barbell or fixed weight (pin-weight) machine. Isokinetic training is performed with a constant speed of movement against variable resistance. Specialised equipment, such as the Cybex or KinCom isokinetic dynamometers, which control the velocity of movement, is required for isokinetic training. The machine controls the speed of movement while offering resistance equal to the force being applied; maximal force is measured at each joint angle throughout the range of motion. The speed may be pre-set to simulate movement of a particular activity.

Table 13.7 compares the three different types of resistance training (isometric, isotonic, and isokinetic). Isometric training is low cost and low risk, but gives minimal gains in strength. Isotonic training can be made very sports-specific and yields excellent improvements in strength, but may increase the risk of injury and muscle soreness because of the eccentric component. Isokinetic training gives excellent strength gains with minimal risk of injury, but is less sports-specific since movement is generally limited to simple flexion and extension at certain joints only, and natural movements rarely occur at constant speed. Isokinetic training is often used in the clinical setting to assess muscular strength throughout the range of motion, to identify weakness within a certain muscle group or imbalance between muscle groups, and for

Table 13.7 Comparison of types of strength training programs

Variable	**Training program**		
	Isometric	**Isotonic**	**Isokinetic**
Strength gain	very little	moderate to high	high
Cost	very low	moderate	high
Equipment needed	none	easy to access	expensive, difficult to access
Sport specificity	low	high	moderate to high
Risk of soreness	low	high	low
Risk of injury	low	high	low to moderate

Source: Compiled from Wilmore, J.H. & Costill, D.L. (1988), *Training for Sport and Activity* (3rd edn), W.C. Brown, Dubuque, Iowa; McArdle, W.D., Katch F.I. & Katch, V.L. (1991), *Exercise Physiology: Energy, Nutrition and Human Performance* (3rd edn), Lea & Febiger, Philadelphia; Lamb, D.R. (1984), *Physiology of Exercise*: *Responses and Adaptations* (2nd edn), Macmillan, New York.

rehabilitation of injury. Isokinetic training is also used widely in the research laboratory.

Training specifically to improve muscular strength and endurance and to induce hypertrophy

As in other forms of training, adaptations to strength training are specific to training program variables such as the type of training, method of measurement, training intensity and duration. Improvement in strength also depends on the individual's age, fitness level and initial strength. In general, training via one type of contraction only improves strength when measured in a similar manner. That is, isotonic strength training improves strength only when measured isotonically and isokinetic strength training improves strength when measured isokinetically at a speed of movement similar to that used in training. Thus, strength training programs must be specific to the desired outcomes. Training program variables that can be manipulated include: (i) the number of repetitions or times a weight is lifted; (ii) sets, or groups, of repetitions; (iii) training volume or total amount of work performed, which is the product of the number of repetitions and the number of sets, and (iv) intensity of resistance.

Intensity can be expressed in two ways. Resistance can be prescribed as a percentage of maximal strength, or the weight that can be lifted once only, referred to as one repetition maximum (1 RM). For example, if an athlete's maximum strength in the bench press is 50 kg, a set might consist of five repetitions at 70 per cent of 1 RM; resistance would be 70 per cent of 50 kg, or 35 kg to be lifted five times. This method requires initial measurement of maximum strength (1 RM) for each muscle group to be trained, which may be inconvenient or contraindicated for some individuals. Resistance training intensity may also be prescribed in terms of the weight or resistance that can be lifted for a specified number of repetitions. For example, a set may consist of 10 RM, which is the maximum weight the athlete can lift only 10 times. Although this method does not require the individual to perform maximum lifts, it does require trial-and-error to estimate resistance for each muscle group.

The pattern of skeletal muscle fibre type recruitment and specific energy systems must be considered in prescribing strength training programs that are specific to the desired outcomes. Table 13.8 gives examples of different types of resistance training programs designed to improve muscular strength and endurance and to induce hypertrophy. Gains in maximum strength require recruitment of the larger type IIB fibres, which occurs only when near-maximum force is applied. Muscular hypertrophy appears to require high-intensity, high-volume training for months to years. Optimal development of muscular power first requires development of strength, followed by work emphasising speed and the explosive application of force that simulates the actual movement pattern.

Table 13.8 Prescription for strength training

Aim	Number of repetitions	Number of sets	Intensity	Rest between sets
Maximal strength	1–8	3–6	80–100% 1RM or 1–8RM	>3 min
Muscular endurance	15–25	1–4	>65% 1RM or 15–25 RM	<1 min
Muscular power*	3–6	1–6	75–90% 1RM or 3–6 RM with explosive movement	>3 min

*Best when performed directly after a set of maximal lifts

Source: Compiled from Wilson, G. (1992), Strength training for sport, *State of the Art Review*, Number 29, Australian Sports Commission, Canberra; Wilmore, J.H. Costill, D.L. (1988), *Training for Sport and Activity* (3rd edn), W.C. Brown, Dubuque, Iowa; Lamb, D.R. (1984), *Exercise Physiology: Responses and Adaptations* (2nd edn), Macmillan, New York.

Combining muscular and aerobic fitness in circuit training

As the name implies, in circuit training the athlete moves around a circuit of different exercise stations. Circuit training was designed for athletes as a way to develop all aspects of physical fitness, including muscular strength, power, endurance, agility and cardiovascular endurance. Because it is time-efficient and versatile, circuit training has also recently become a popular form of exercise in health and fitness centres.

Circuit training can be sports specific to train skills or movement patterns required of a sport. Weight lifters use circuit training to work on specific muscle groups. Circuit training can be modified for lower intensity work to develop health-related fitness, for example by alternating weight lifting with aerobic-type activity (Figure 13.6).

A major advantage of circuit training is its versatility. Stations can be changed regularly to provide variety and prevent overuse of a particular muscle group. For athletes, the stations may also be altered as training progresses through the season. One disadvantage of circuit training for athletes is the difficulty in designing a single circuit program to optimally train both muscular strength and cardiovascular endurance; more than one type of circuit may be needed for athletes in training.

What causes muscle soreness?

All athletes, and most people who exercise, have at some point experienced muscle soreness in the days following heavy, unaccustomed training sessions. This type of muscle soreness is called delayed onset muscle soreness (DOMS), because it begins the day after exercise and may last for up to several days.

DOMS is greatest after eccentric exercise, and is most painful in the muscles used during eccentric exercise. For example, prolonged downhill walking or running will cause DOMS in the quadriceps muscles (on the front of the thigh) because these muscles contract eccentrically to counteract the force of gravity. It is generally believed that the greater muscle soreness occurs because fewer muscle fibres are activated during eccentric compared with concentric exercise.

Over the years, several mechanisms have been put forward to explain DOMS. Contrary to popular belief, DOMS is not related to lactic acid accumulation. For example, lactic acid levels are not elevated in the types of exercise most likely to cause DOMS, such as downhill walking. Muscle spasms during and after exercise have also been proposed as a mechanism for DOMS, although there is little experimental evidence to support this concept. The most likely cause of DOMS is damage to both muscle fibres and connective tissue within muscle resulting from

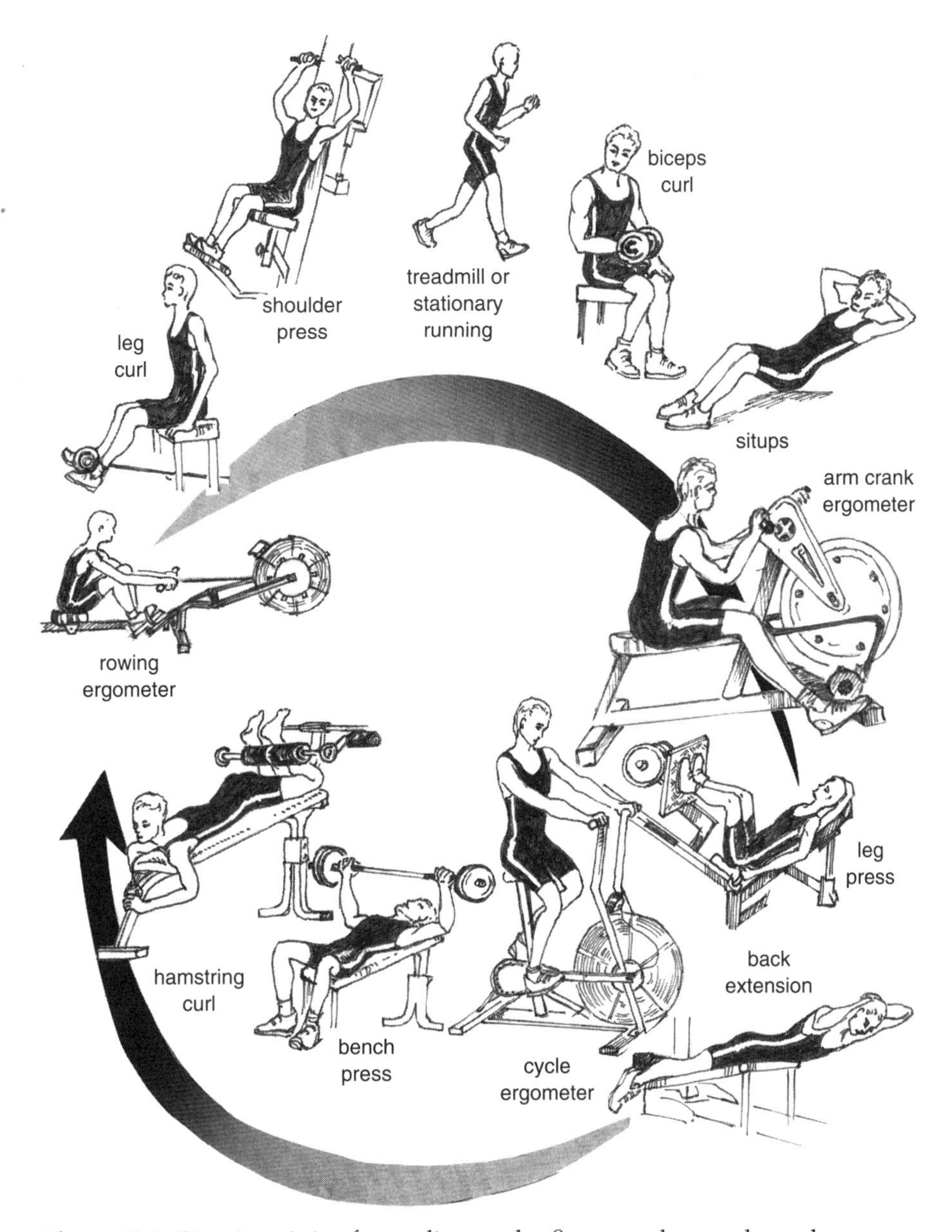

Figure 13.6 Circuit training for cardiovascular fitness and muscular endurance. Exercise includes alternating aerobic-type activities, such as running or cycling, and resistance work. The individual moves from station to station in a circuit, exercising for a given time, usually 30 to 90 seconds, with shorter rest intervals between stations

prolonged exercise. Damage to muscle fibres and connective tissue initiates an inflammatory response within muscle, causing influx of cells from the circulation, tissue swelling, and the sensation of pain.

The skeletal muscle damage associated with DOMS is temporary, and is usually repaired within a week. There is no evidence for permanent tissue damage associated with DOMS and it is believed that some level of breakdown and repair may be important in initiating training-induced adaptations in skeletal muscle. One interesting feature of DOMS is that a single session of eccentric exercise training provides protection for up to months against a subsequent occurrence of DOMS within the same muscle.

CHAPTER 14

Changes in physiological capacity and performance throughout the lifespan

- How does exercise performance change over the lifespan?
- Are children miniature adults?
- Growth versus training—are children trainable?
- Exercise prescription for children
- Exercise capacity during ageing
- Exercise prescription for the older individual
- Sex differences in physiological responses and adaptations to exercise

Much of our basic knowledge of exercise physiology comes from research on young adult males, who comprise a relatively accessible group for research studies, especially in the university environment. Over the past few decades, however, work has focused on understanding the responses to exercise in a wider range of individuals—from children to the elderly, in both males and females—and identifying the mechanisms underlying any differences in their responses and adaptations to exercise and training.

This chapter examines how exercise performance changes across the lifespan for both males and females and the implications of these changes for exercise prescription in children and in the elderly.

How does exercise performance change over the lifespan?

Figure 14.1 shows general trends in United States record running performance by age for males and females, from children to older adults. At all ages males perform better than females and, for both sexes, performance is best in the age range of approximately 18 to 30 years. Factors responsible for age and sex difference in exercise performance are discussed in this chapter.

Are children miniature adults?

Children differ from adults in their metabolic, cardiovascular, respiratory, thermoregulatory and perceptual responses to exercise. Thus, children cannot simply be considered as small adults. A

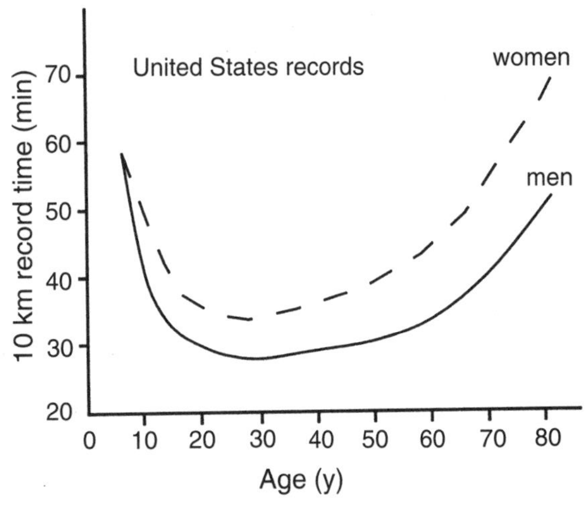

Figure 14.1 Trends for US records in the 10 000 m by age and sex. The fastest times are recorded in the age range between approximately 20 and 30 years, and for males compared with females
Source: Wilmore, J.H. & Costill, D.L. (1994), *Physiology of Sport and Exercise*, Human Kinetics, Champaign, IL, p. 425. (Copyright 1994 by Jack H. Wilmore & David L. Costill; reprinted by permission.)

different approach to exercise testing and training is needed when studying how children respond physiologically to exercise. For example, given that young children may lack the motivation or understanding to perform continuous exercise to exhaustion, the standard $\dot{V}O_{2max}$ test may be inappropriate for assessing aerobic exercise capacity in the young. In general, the acute responses to exercise and adaptations to training are qualitatively similar but quantitatively different in children and adults.

Aerobic exercise capacity in children

When expressed as an absolute rate (L/min), $\dot{V}O_{2max}$ is much lower in children than in adults, and increases proportionately with growth. In untrained males, $\dot{V}O_{2max}$ increases until age 16 to 18 years and remains relatively constant until the mid-20s. In untrained females, $\dot{V}O_{2max}$ increases until age 12–14 years, after which it may plateau or possibly decrease. In both boys and girls, much of the growth-related change in $\dot{V}O_{2max}$ is related to changes in body composition. On average, $\dot{V}O_{2max}$ tends to be higher in boys than in girls across all ages, although there is a great deal of overlap, especially up to age 10 years. Before puberty, absolute $\dot{V}O_{2max}$ in girls is 85–90 per cent of that in boys; after puberty, the gap widens, with $\dot{V}O_{2max}$ in girls approximately 70 per cent that of boys.

When adjusted for growth-related changes in body size (ml/kg/min), $\dot{V}O_{2max}$ is generally similar in young children compared with adolescents and young adults (Figure 14.2). The sex difference in $\dot{V}O_{2max}$ relative to body mass is smaller than for absolute $\dot{V}O_{2max}$ reflecting sex differences in body size and composition, especially after puberty. Although $\dot{V}O_{2max}$ relative to body mass remains constant after puberty in boys, it may actually decrease after puberty in girls. This decrease is probably due to both physiological and social factors, namely the increase in fat

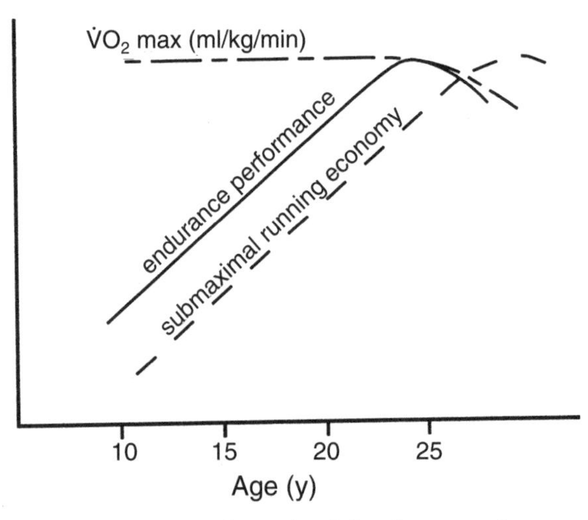

Figure 14.2 General trends of developmental changes in aerobic power, running economy and performance. Aerobic power expressed as L/min tends to increase with body size during growth. However, when expressed relative to body size (ml/kg/min), aerobic power is similar in children and adults. The increases in performance during maturation are thus related to improvements in running economy rather than aerobic power
Source: Rowland, T.W, 1990, *Exercise and Children's Health*, Human Kinetics, Champaign, IL (Copyright 1990 by Thomas W. Rowland. Reprinted by permission.)

mass in girls as well as declining participation in regular exercise after puberty.

Despite being similar in adults and children, $\dot{V}O_{2max}$ relative to body mass is not as good a predictor of endurance capacity in children as it is in adults. For example, some studies have shown that endurance training in children may improve 2 km run performance without markedly improving $\dot{V}O_{2max}$.

Although aerobic power may be similar in children and adults, there is a considerable difference in the metabolic cost of activity. For example, the oxygen cost of exercising at a set pace (such as running at a given speed) is much higher in children than in adolescents and young adults. As discussed in Chapter 10, economy and efficiency of movement are lower in children; moreover, the differences in economy between children and adults increase as a function of exercise intensity. The lower economy in children is most likely due to differences in technique and body dimensions; economy of movement increases as children grow.

Cardiorespiratory responses to exercise in children

As was noted in Chapter 12, maximum heart rate is a function of age, and children have very high maximal heart rates, frequently over 200 beats per minute. However, both submaximal and maximal cardiac output as well as stroke volume are lower in children compared with adults. The high maximal heart rate does not fully compensate for the lower maximal stroke volume, resulting in a lower cardiac output during exercise. Blood flow to working muscles is higher during exercise in children than in adults, and this may partially compensate for the lower cardiac output. However, compared with adults, the oxygen-carrying capacity of blood is lower in children due to lower levels of the oxygen-carrying protein haemoglobin. In contrast, during exercise, skeletal muscle oxygen extraction and muscle oxidative capacity appear to be similar in children and adults.

The respiratory system is less efficient in children than in adults. Children require a higher minute ventilation at any given oxygen consumption; that is, in children more air must be inspired by the lungs to provide the same amount of oxygen to the muscles as that available to muscles in adults. This inefficiency is due to the higher respiratory rate and shallower breathing (lower tidal volume) in children compared with adults. The respiratory muscles of children must work harder during exercise, and respiratory muscle fatigue contributes to the higher metabolic cost, feelings of discomfort and early fatigue experienced during intense exercise.

Anaerobic exercise capacity in children

Anaerobic capacity, as measured in 10–30 s all-out exercise, is much lower in children compared with adults. For example, anaerobic capacity at age 9 is less than half that at age 19 years (Figure 14.3). Anaerobic capacity increases throughout childhood and adolescence, with the greatest changes occurring between the ages of 9 and 15. As with aerobic capacity and strength, there are few sex differences before age 9, with sex differences emerging between the ages of 10 and 13. For example, by age 13, anaerobic capacity in girls is about 70–75 per cent that of boys. Anaerobic capacity peaks between ages 14 and 16 in girls, but continues to increase until at least age 20 years in boys.

The difference in anaerobic capacity between children and adults persists even when values are adjusted for differences in body mass, muscle mass, or muscle size. Thus, children have a much lower capacity to generate mechanical power over the short term, and this difference is not due simply to a smaller body size or less muscle mass. The increase in anaerobic capacity during adolescence exceeds the increase in muscle mass, indicating that muscle growth alone cannot fully account for changes in anaerobic exercise capacity.

The amount of ATP and PC stored in skeletal muscle, and the ability to use PC to regenerate ATP, are similar in children and adults, indicating that the phosphagen energy system is not deficient in children. However, the capacity of anaerobic glycolysis to produce energy for exercise is lower in children compared with adults. Skeletal muscle

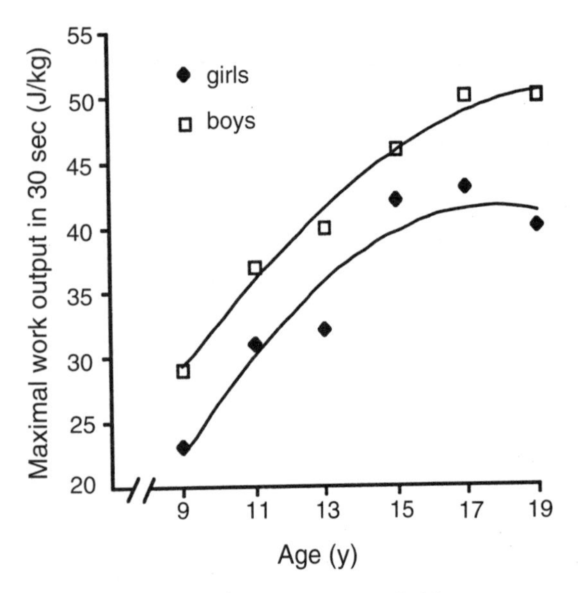

Figure 14.3 Anaerobic capacity in children measured as maximal work output, in joules per kilogram body mass, during a 30 second knee extension test. Work output increases with age and is significantly higher in males than females from age 11 years
Source: Saavedra, C., Lagasse, P., Bouchard, C. & Simoneau, J.-A. (1991), Maximal anaerobic performance of the knee extensor muscles during growth, *Medicine and Science in Sports and Exercise*, 23, 1085. (Copyright 1991 by the American College of Sports Medicine; adapted by permission.)

glycogen stores, glycolytic enzyme activity, and the ability to buffer acidity are lower in children than in adults. These differences are reflected in a 30 per cent lower production of lactic acid by children during brief maximal exercise. Although children have a lower capacity for anaerobic-type exercise, they appear to recover faster than adults after brief high-intensity exercise, possibly because of the lower amount of lactic acid produced. Thus, children may be better able to tolerate interval training and stop–start activities such as soccer or hockey.

Muscular strength in children

As described in Chapter 13, muscular strength is a function of muscle size or cross-sectional area. It is not surprising, then, that muscular strength increases with growth during childhood and adolescence. As with $\dot{V}O_{2max}$ and anaerobic exercise capacity, muscular strength is similar between boys and girls up to age 8–9 years. In boys, muscular strength increases linearly until age 13–14 years, followed by an accelerated increase during the adolescent growth spurt. In girls, strength increases linearly until age 14–16 years, after which it remains relatively unchanged.

Both the sex and maturational differences in muscular strength are related to differences in body composition, especially muscle mass. During childhood, body size and body composition are only modestly related to muscular strength. However, after puberty, strength becomes more closely related to body size, muscle mass and somatotype. For example, in post-pubertal boys, muscular strength is positively correlated with both endomorphy and mesomorphy and negatively correlated with ectomorphy. Body size, somatotype and muscular strength are more closely related to each other in boys than in girls.

The large increase in strength after puberty in boys is due to a higher production of the male hormone testosterone, which increases protein synthesis and muscle mass. In both boys and girls, gain in strength during and after puberty is also due to simultaneous maturation of the neural pathways that control movement. The rate of strength gain during maturation varies widely between individual children, and muscular strength measured in childhood is not a good indicator of strength in adulthood.

Thermoregulatory responses to exercise in children

Children produce more metabolic heat during exercise than adults, even when values are adjusted for differences in body mass. Thus, compared with adults, children must lose more body heat to avoid an increase in core temperature during exercise. During exercise and especially in a warm environment, the body loses heat mainly by evaporative cooling via sweating. Compared with adults, children begin to sweat at a higher relative work rate and they also sweat less during exercise. The lower rate of heat loss via sweating is partially compensated by a higher rate of heat loss via circulatory adjustments and conduction and convection from the skin's surface. This is because children have a higher surface area to mass ratio. However, the combination of higher metabolic heat production and poorer ability to remove heat via sweating means that children are less tolerant of prolonged exercise, especially in a warm environment. In addition, children require a longer time to acclimatise (adjust) to exercise in warm environments.

Growth versus training—are children trainable?

Aerobic endurance training

It has been debated whether pre-pubescent children exhibit adaptations to exercise training beyond those expected from growth alone. Some believe that a child's inherently high physical activity level acts as a training stimulus so that most children are already well trained and further increases in fitness in response to training are not possible. In other words, it has been reasoned that because young children are naturally active, they are near their biologically determined exercise capacity and cannot significantly improve further. It has also been argued that children are not able to endure prolonged exercise to a sufficient degree to benefit from a training effect. In support of this argument, several research studies have shown no improvement in $\dot{V}O_{2max}$ after endurance training programs in children. However, most of these studies used less vigorous exercise than that generally recommended to improve cardiorespiratory fitness. In contrast, many recent studies have shown improvements in $\dot{V}O_{2max}$ and endurance exercise performance in children when training was based on principles recommended to improve fitness in adults, as described in Chapter 13. These studies indicate that children show the same relative improvement as adults in aerobic power, endurance performance and cardiovascular system variables with appropriately prescribed exercise training. In children, $\dot{V}O_{2max}$ may increase by 5–25 per cent with endurance training, which is similar to the range of improvement expected in adults. Other cardiorespiratory adaptations include decreased resting heart rate and heart rate at a given submaximal work rate; increased maximal cardiac output and stroke volume; increased work rate (as a percentage of $\dot{V}O_{2max}$) at the lactic acid threshold; and increased maximum minute ventilation.

Biochemical changes also appear to be similar after aerobic training in children and adults, although these are more difficult to study because of obvious ethical and safety issues in obtaining blood and muscle samples from children. Biochemical adaptations, including increased muscle glycogen storage and oxidative capacity in skeletal muscle, occur to the same relative extent in children as has been observed in adults after endurance training. Aerobic performance often increases more than $\dot{V}O_{2max}$ suggesting that economy and efficiency of movement (technique, skill) may also be enhanced by endurance training in children (Figure 14.2).

Anaerobic training in children

Anaerobic training has been associated with improved performance in brief high-intensity exercise, such as the 30 s all-out cycling test. Increased anaerobic exercise capacity enhances performance in both aerobic and anaerobic exercise by increasing the relative amount of work supported by anaerobic energy production. Biochemical adaptations include higher blood and muscle lactic acid concentrations during and after maximal exercise; increased skeletal muscle stores of glycogen and ATP–PC; and enhanced glycolytic enzyme activity in skeletal muscle.

Strength training in children

Until about 10 years ago, it was generally considered inappropriate for children and adolescents to participate in strength or weight training. This view was based on the belief that children have a limited capacity to increase strength beyond gains due to normal growth, and that strength training may be harmful to the growing musculoskeletal system. However, scientific research over the past decade confirms

Box 14.1 Strength training in children—Is it safe and effective?

Until the mid-1980s, children were advised against strength training. It was generally believed that growing bones would be injured by strength training and that children's muscles were incapable of hypertrophy and increasing strength beyond that due to growth alone.

Work by Arthur Weltman and colleagues in the United States was one of the first controlled studies to address the question of strength training in children. Results showed that a well-designed moderate training program can increase muscular strength as well as flexibility and exercise capacity without causing injury.

Twenty-six pre-pubescent boys, average age of 8 years, participated in a 14 week program consisting of strength-training exercises on hydraulic weight machines. Training sessions were 45 minutes in length, three times per week, and consisted of high-repetition, low-resistance work, with each exercise performed for 30 s followed by 30 s rest. Only concentric exercises were performed, since these were less likely to cause muscle soreness or damage. A control group of boys similar in age to the strength-trained group did not participate in the training program, but performed the same tests of strength, exercise capacity and flexibility before and at the end of the study. The control group was included to determine if any changes after the 14 weeks were due to growth alone or to the training program.

Before the start and at the end of the 14 weeks, isokinetic strength, flexibility and exercise capacity were measured in both the control and exercise groups. Compared with the control group, isokinetic strength increased in the exercise group by 18.5–36.6 per cent and muscular flexibility increased by 8.4 per cent. Vertical jump scores (a measure of explosive leg power) increased by 10.4 per cent. In addition, $\dot{V}O_{2max}$ in L/min increased by 19.4 per cent and in ml/kg/min by 13.8 per cent. Although the circumferences of some upper body parts increased modestly, suggesting muscular hypertrophy, most of the strength gains were not accompanied by increases in muscle size, indicating increases in strength in the absence of hypertrophy.

Importantly, all of the increases in muscular strength and exercise capacity occurred without evidence of musculoskeletal injury to the boys. Scintigraphy to visualise bone, and physician review of each subject, showed no evidence of injury or damage to bone including epiphyses, or to skeletal muscle. The authors concluded that a well-controlled supervised program of concentric resistance exercises emphasising muscular endurance (high repetition, low resistance) is safe and effective in increasing muscular strength and other performance measures in pre-pubescent boys.

Table 14.1 Percentage changes in aerobic power, exercise capacity and muscular strength following strength training in pre-pubertal boys

Variable	Percentage improvement after training
$\dot{V}O_{2max}$ (ml/kg body mass/min)	13.8
Exercise time on treadmill	14.0
Knee flexion strength	23.6
Knee extension strength	24.5
Elbow flexion strength	36.6
Elbow extension strength	32.1

Source: Compiled from Weltman, A., Janney, C., Rians, C.B., Strand, K., Berg, B., Tippitt, S., Wise, J., Cahill, B.R. & Katch, F.I., (1986) The effects of hydraulic resistance strength training in pre-pubertal males, *Medicine and Science in Sports and Exercise*, 18, 631. (Copyright 1986 by Williams and Wilkins, Baltimore; adapted by permission.)

Subsequent research has confirmed these findings that resistance training, if performed properly and under supervision, enhances strength, reduces the risk of sports-related injuries and improves sports performance among children and adolescents.

Sources: American College of Sports Medicine (1993), Current comment: the prevention of sport injuries of children and adolescents, *Medicine and Science in Sports and Exercise*, 25(8, suppl), 1–7; Weltman, A., Janney, C., Rians, C.B., Strand, K., Berg, B., Tippitt, S., Wise, J., Cahill, B.R. & Katch, F.I. (1986), The effects of hydraulic resistance strength training in pre-pubertal males, *Medicine and Science in Sports and Exercise*, 18, 629–38.

that suitably prescribed programs can improve muscular strength and overall fitness without increasing the risk of injury or reducing muscular flexibility. Both boys and girls can participate in strength training (Box 14.1).

Musculoskeletal injuries arising from strength training in children and adolescents usually occur during maximal lifts above the head or body in unsupervised programs; poor technique is a frequent cause of injury. Recommendations for strength training in children and adolescents include the following:

- Programs should be supervised and include spotting for all lifts above the head or body using free weights.
- Programs should be non-competitive and emphasise proper form and technique rather than the amount lifted.
- Training should involve high-repetition, low-resistance work with at least seven to ten repetitions per set.
- Children under age 16 years should not perform maximal lifts.

Exercise prescription for children

There are no set guidelines for exercise prescription in children. Since many responses and adaptations to training are similar in children and adults, exercise prescription in children should roughly follow those general recommendations for enhancing fitness in adults as described in Chapter 13. However, it should be kept in mind that, compared with adults, children have very different reasons and motivation for exercising. For example, adults usually exercise for reasons of health and appearance such as prevention of heart disease and the enhancement of weight loss; children tend to exercise for fun, peer recognition and social interaction.

Children seem to prefer interval-type exercise often in the context of games and sport, possibly because of the limited capacity of both the aerobic and anaerobic energy systems and thermoregulation, as described above. Alternatively, it is likely that interval-type exercise better suits a child's shorter attention span, and that children find continuous exercise monotonous.

Exercise capacity during ageing

It is well known that physical work capacity and sports performance decrease during ageing. Figure 14.1 showed changes in world record sports performance with age. After about age 30, performance begins to decline in both male and female athletes.

Changes in physical work capacity with age influence not only sports performance, but the capacity to carry out many tasks of daily living such as walking, lifting and climbing stairs. Extreme loss of physical work capacity may limit the ability of older individuals to maintain an independent lifestyle, resulting in institutionalised care of the elderly.

Until recently, a decline in functional capacity was regarded as simply an inevitable consequence of growing older. However, it has recently been questioned whether the dramatic changes in work capacity during ageing are due solely to the ageing process itself or to the increasingly sedentary lifestyle that usually accompanies ageing; older individuals tend to be less active during both leisure and work. Many of the age-related changes in functional capacity resemble those seen with detraining or cessation of exercise training. It is important to distinguish the inevitable effects of ageing from those due to inactivity. If age-related changes are due primarily to lifestyle, then maintaining physical activity patterns throughout adulthood into old age may prevent or delay many of the adverse physical consequences of ageing.

What happens to exercise capacity during ageing in sedentary individuals?

Aerobic capacity during ageing

Both cross-sectional and longitudinal studies indicate that aerobic power, as measured by $\dot{V}O_{2max}$, declines after about age 30 (Figure 14.4). The rate of decline averages about 0.5 ml/kg/min/year, but may vary depending on the group studied. Thus, over a 10 year period, say from age 40–50 years, the average sedentary person may expect a decline of 5 ml/kg/minute (or up to 10 per cent) in $\dot{V}O_{2max}$.

The decrease in $\dot{V}O_{2max}$ during ageing in a sedentary individual is related to several factors, including loss of lean body or muscle mass and increase in fat mass; decrease in maximum heart rate and cardiac output; and a decline in the capacity of skeletal muscle to extract oxygen during exercise.

In sedentary individuals, the age-related decline in $\dot{V}O_{2max}$ is generally correlated with changes in body composition. Loss of muscle mass and an increase in fat mass collectively decrease the amount of metabolically active muscle tissue that consumes oxygen during energy production for physical work.

About 50 per cent of the decrease in $\dot{V}O_{2max}$ may be accounted for by a decrease in maximum heart rate (see Chapter 13). Maximum heart rate decreases as a result of changes in neural input to the heart and does not appear to be altered by training at any age. Since cardiac output is a function of heart rate and stroke volume, the decrease in maximum heart rate results in a decrease in maximum cardiac output. As discussed in Chapter 12, cardiac output is an important determinant of $\dot{V}O_{2max}$. In contrast to decreases in cardiac output and heart rate, maximum stroke volume may not change much during ageing.

The ability of skeletal muscle to extract and consume oxygen during exercise also decreases with age in a sedentary population. This decrease is due to a reduction in the number of capillaries within skeletal muscle as well as changes in the capacity to redirect blood flow to skeletal muscle during exercise. During maximal exercise, skeletal muscle in older individuals has a limited capacity to increase blood flow, and thus oxygen delivery to working muscles is compromised.

Anaerobic capacity during ageing

There has been far less research on anaerobic capacity in older sedentary individuals. As described in Chapter 12, tests of anaerobic capacity involve 30 s sprints, often on a cycle ergometer. It is difficult and possibly unsafe for older individuals unaccustomed to intense exercise to exert a maximal effort in such a test.

The few studies that have been conducted on older sedentary people indicate that anaerobic capacity declines by about 6 per cent per decade. This decline is very much related to the loss of muscle mass, especially of the thigh muscles, which are the main source of power in cycling tests of anaerobic capacity. The decreases in anaerobic capacity and muscle mass are due to selective atrophy of the larger, stronger type II fibres and possibly to loss of type II motor units.

Muscular strength

Muscular strength is important to sports performance as well as to basic functions such as posture and simple activities such as walking and climbing stairs. Dramatic changes in muscular strength accompany ageing, and may limit an older individual's ability to maintain an independent lifestyle.

Isometric, isotonic and isokinetic strength have all been shown to decrease by about 2–4 per cent per year in ageing sedentary individuals (Figure 14.5). The decline in strength is due to several factors. Lean body mass decreases

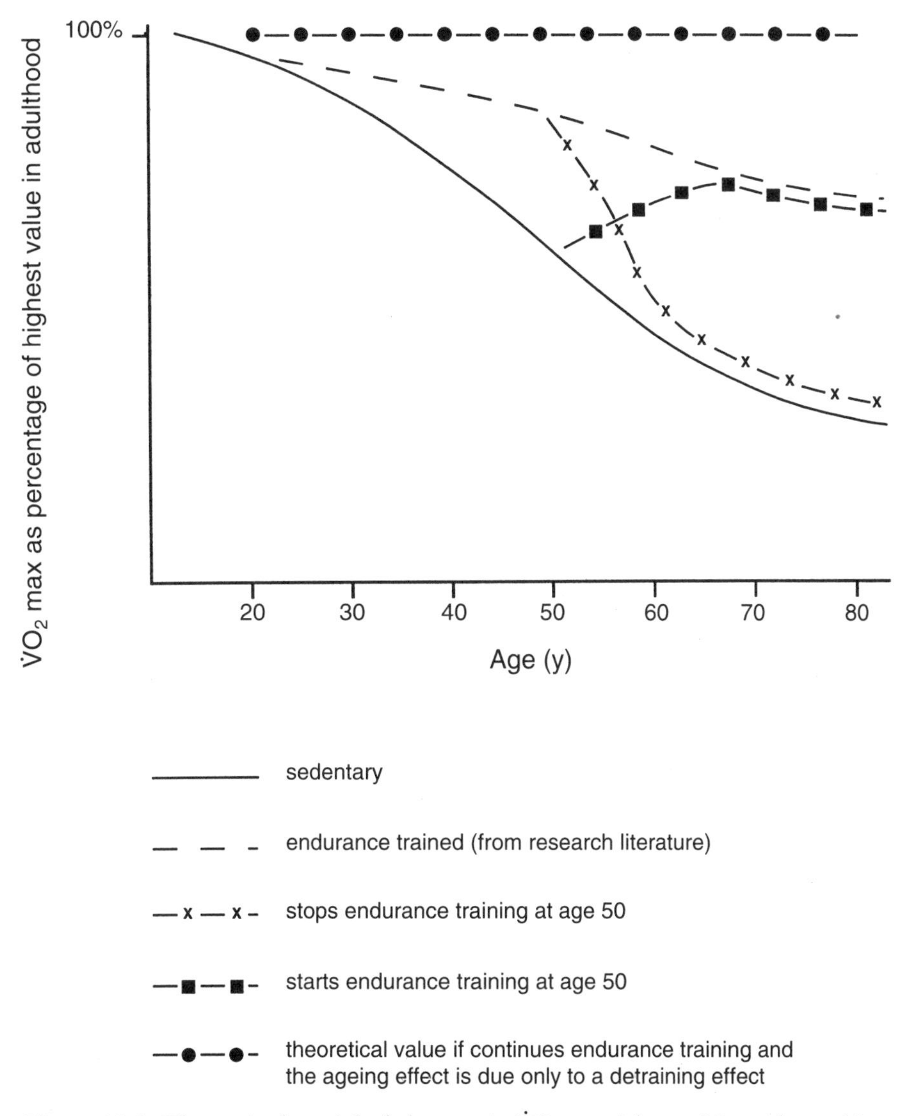

Figure 14.4 Theoretical model of changes in $\dot{V}O_{2max}$ with age. Normal trend is observed for the ageing sedentary individual and for veteran endurance athletes.The theoretical trend for the endurance-trained athlete who remains active, assuming that the decline in $\dot{V}O_{2max}$ normally observed during ageing is due only to a sedentary lifestyle is also shown. Changes in $\dot{V}O_{2max}$ for the endurance trained athlete who becomes sedentary at age 50 years [x--x], and for the sedentary individual who begins endurance training at the same age [•–•], show that aerobic power improves with training at any age, but that adaptations persist only as long as training is continued

gradually between ages 30 and 50, after which there is an accelerated loss of muscle mass. As explained in Chapter 13, muscular strength is generally related to muscle mass or cross-sectional area. During ageing, the decline in muscle mass is due primarily to atrophy of the larger stronger type II muscle fibres as well as possible loss of these fibres. In contrast, type I

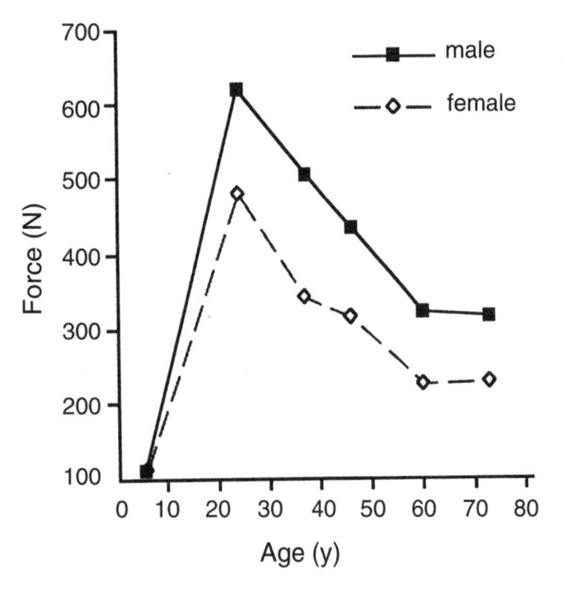

Figure 14.5 Changes in leg strength, or maximum force produced in N, during maximal voluntary vertical jumping in males and females age from 4 to 73 years

Source: Bosco, C. & Komi, P.V. (1980), Influence of age on the mechanical behavior of leg extensor muscles, *European Journal of Applied Physiology*, 45, 213. (Copyright Springer-Verlag Gmblt & Co; adapted by permission.)

fibres do not atrophy as much as type II fibres, most likely because type I fibres are recruited during normal activities such as standing and walking.

During ageing, the amount of connective tissue may increase within skeletal muscle; however, connective tissue does not contribute to active force generation. In addition, age-related changes in neural input, particularly loss of type II motor units to skeletal muscle, may influence recruitment patterns and reduce force generation. Finally, older untrained individuals may be reluctant to voluntarily exert maximal force, and this may result in lower strength readings during testing.

Can age-related changes in work capacity be prevented or reversed?

Since people tend to become more sedentary with age, it is unclear how much of the age-related changes in exercise capacity are due to a reduced level of physical activity and how much to the ageing process itself. In sedentary individuals, many of the age-related changes in exercise capacity resemble the effects of detraining, for example decreases in $\dot{V}O_{2max}$ and strength.

Comparison between masters (veteran) athletes and age-matched sedentary individuals indicates a much slower rate of decline with age of $\dot{V}O_{2max}$ and muscular strength in the athletes. Moreover, training studies generally indicate that older individuals exhibit relatively similar responses to exercise training as is seen in younger people, provided that similar training programs are used.

Aerobic capacity in masters (veteran) athletes

In the masters athletes, $\dot{V}O_{2max}$ decreases with age at half the rate observed in an ageing sedentary population. For example, although $\dot{V}O_{2max}$ may decline with age by 0.5 ml/kg/min/year in sedentary people, longitudinal studies in veteran endurance athletes who continued to train over a 25 year period show decreases of only 0.25 ml/kg/min/year (see Figure 14.4). Thus, about half of the decrease in aerobic power appears to be inevitable with ageing, and half seems to be preventable with continued training.

In the veteran athletes, the decrease in $\dot{V}O_{2max}$ is due mainly to the decrease in maximum heart rate; as described earlier, maximum heart rate declines with age regardless of fitness level. Relatively smaller decreases in maximum stroke volume and skeletal muscle oxygen extraction may also contribute to the slower rate of decline in $\dot{V}O_{2max}$ in the athlete. In contrast, other factors, such as skeletal muscle capillary number; muscle fibre size, especially type I fibres; and skeletal muscle oxidative activity appear to be maintained with continued training.

Anaerobic capacity and muscular strength in masters athletes

Few studies have examined anaerobic capacity and muscular strength in masters athletes. The limited research to date suggests that, as in aerobic capacity, continued training halves the rate of decline with age. Much of the age-related decline in muscular strength can be accounted for by loss of muscle mass even in masters athletes.

However, continued training partially prevents skeletal muscle atrophy, particularly of type II fibres, which normally occurs in sedentary individuals. In addition, strength-trained veteran athletes exhibit less connective tissue and fatty infiltration of skeletal muscle compared with sedentary people. Recent research suggests that, in veteran athletes, specific attention to strength training induces muscle hypertrophy and improves strength and anaerobic performance.

Despite inevitable changes with age, compared with the average untrained young individual, the veteran athlete exhibits higher levels in many physiological variables such as $\dot{V}O_{2max}$, muscular strength, and anaerobic capacity.

Aerobic training in previously sedentary individuals

Early studies on endurance training in previously sedentary older people found no changes in $\dot{V}O_{2max}$ despite other training responses such as reduced resting heart rate and increased physical work capacity. However, these studies were very short term (6–8 weeks), and perhaps a longer period is needed to show training effects in older individuals.

Later studies clearly show that $\dot{V}O_{2max}$ increases with endurance training in older individuals. As in younger individuals, changes in $\dot{V}O_{2max}$ are closely related to both the total amount of exercise as well as training intensity. Increases in $\dot{V}O_{2max}$ in the range of 20–40 per cent have been observed in older individuals following endurance training. Increases in maximum cardiac output (due to increases in stroke volume), and increases in skeletal muscle oxygen extraction and oxidative capacity accompany increases in $\dot{V}O_{2max}$. Relative improvements with training are similar for older and younger sedentary individuals.

Anaerobic training in previously sedentary individuals

Few studies have examined the effects of anaerobic training in older untrained people. The limited data suggest that high-intensity aerobic training up to 85 per cent $\dot{V}O_{2max}$ also improves anaerobic capacity, as measured in the 30 s sprint cycle ergometer test. The mechanisms are unclear at present, but may be related to increases in muscular strength and size, especially of type IIA fibres, as well as enhanced oxidative capacity in muscle.

Strength training in previously sedentary individuals

Increases in isometric and isokinetic strength after resistance training have been well-documented in older, previously sedentary people (Box 14.2). Early studies suggested that gains in muscular strength occurred in the absence of muscle hypertrophy, and must therefore have been due to neural adaptations. However, more recent work, using sensitive imaging devices to clearly measure muscle cross-sectional area, such as computerised tomography (CT scans) and magnetic resonance imaging (MRI), shows that muscle hypertrophy occurs with resistance training in previously sedentary older people. Muscle hypertrophy is due to hypertrophy of type II fibres, especially type IIA fibres. However, the increase in strength is greater than can be accounted for by hypertrophy alone, indicating that neural adaptation also contributes to strength gains in the elderly. It is generally believed that the capacity for hypertrophy is somewhat limited in the elderly and that neural factors contribute proportionally more to strength gain in older compared with younger individuals. The limited capacity for muscle hypertrophy in older compared with younger men may also be due to an age-related decrease in levels of the male hormone testosterone.

Exercise prescription for the older individual

Exercise prescription for the older individual follows the general guidelines for health-related fitness as discussed in Chapter 13, with some modification. A conservative approach to exercise prescription is needed, considering that older individuals often exhibit a low functional capacity and slower rate of adaptation to exercise training.

Box 14.2 Strength training in the elderly—how adaptable are ageing muscles?

During ageing, loss of muscle mass and strength can significantly impair functional capacity and possibly limit the ability to maintain an independent lifestyle. Poor muscular strength, loss of bone mineral density, and poor balance due to loss of muscle strength increase the risk of falling and sustaining a bone fracture in elderly women. A group of researchers from Tufts University, Harvard Medical School, and Penn State University in the U.S. have provided convincing evidence that resistance training can counteract some of these adverse changes.

In postmenopausal women (average age 61 years), one year of high-intensity strength training 2 days per week resulted in significant increases in bone mineral density (BMD) in the spine and hip as well as total body bone mineral content (BMC). Strength training included concentric and eccentric contractions on pneumatic resistance machines using muscle groups of the hips, thighs, back, abdomen, and upper body. Training intensity was set at 80% of 1 RM with 3 sets of 8 repetitions performed per session for a total of 45 min per session. A control group of women who did not exercise was included to compare normal changes occurring across the one-year period. In the exercise group, muscle strength and mass, BMD, and BMC all increased, whereas muscle mass, BMD, and BMC declined in the control group. The increases in muscle mass and strength were accompanied by enhanced dynamic balance and hormonal evidence of new bone formation in the trained group.

In a related study, it was shown that 2 sessions per week of high-intensity strength training resulted in continued improvement in muscular strength over one year. About 50% of the increase in strength occurred during the first 3 months, with smaller increases in strength throughout the remaining 9 months (Table 14.2). The continued improvements in strength suggest that, in older individuals, skeletal muscle is capable of continued adaptations in response to high-intensity resistance exercise. In another study, 10 weeks of high-intensity resistance exercise training in older (72–98 years) male and female nursing home residents resulted in increases in muscular strength and cross-sectional area, mobility, and spontaneous activity. Moreover, subjects weakest at the onset of the study improved most. Taken together, these studies suggest that, even in the elderly, significant improvements in muscle mass and strength are possible with appropriate resistance training and that these improvements help counteract some of the debilitating physical effects of ageing.

Sources: Nelson, M.E., Fiatarone, M.A., Morganti, C.M., Trice, I., Greenberg, R.A., & Evans, W.J. (1994), Effects of high-intensity strength training on multiple risk factors for osteoporotic fractures: A randomized controlled trial, *JAMA*, 272, 1909–1914; Morganti, C.M., Nelson, M.E., Fiatarone, M.A., Dallal, G.E., Economos, C.D., Crawford, B.M., & Evans, W.J. (1995), Strength improvements with 1 yr of progressive resistance training in older women, *Medicine and Science in Sports and Exercise*, 27, 906–912; Fiatarone, M.A., O'Neill, E.F., Ryan, N.D., Clements, K.M., Solares, G.R., Nelson, M.E., Roberts, S.B., Kehayias, J.J., Lipsitz, L.A., & Evans, W.J. (1994), Exercise training and nutritional supplementation for physical frailty in very elderly people, *New England Journal of Medicine*, 330, 1769–1775.

Table 14.2 Effects of high-intensity strength training over one year

	Percent of total year's improvement in strength			
Strength of:	**0–3 mo**	**3–6 mo**	**6–9 mo**	**9–12 mo**
Upper body	50%	20%	15%	15%
Lower body	45%	25%	20%	10%

Source: Morganti, C.M., Nelson, M.E., Fiatarone, M.A., Dallal, G.E., Economos, C.D., Crawford, B.M., & Evans, W.J. (1995), Strength improvements with 1 yr of progressive resistance training in older women, *Medicine and Science in Sports and Exercise*, 27, 906–912.

Table 14.3 Exercise prescription for the older adult

Variable	Consideration for exercise prescription
Low functional capacity	low-intensity exercise (40–50% $\dot{V}O_{2max}$), gradual progression, interval training to avoid early fatigue
Poor muscular strength and endurance	exercises to enhance muscular strength and endurance
Impaired co-ordination	simple movements, supported exercise, exercise machines
Increased risk of disease especially heart disease, obesity, hypertension, arthritis and osteoporosis	health-risk screening and medical examination, modification of exercise prescription according to condition
Prevalence of osteoarthritis	low-impact, low-intensity exercise, weight-supported or water-based activities
Low heat tolerance	replacement of fluids frequently, avoidance of exercise in the heat

Table 14.3 lists some of the special concerns to be considered when prescribing exercise for the elderly.

Sex differences in physiological responses and adaptations to exercise

Sex differences in aerobic exercise capacity

As mentioned above, $\dot{V}O_{2max}$ tends to be higher in males than in females at all ages. This difference is rather small in young children, becoming larger with age, especially after puberty, when aerobic power continues to increase in boys but plateaus in girls.

Boys also perform better than girls in endurance tests such as the 1 km run. The gap in endurance performance cannot be explained only by differences in $\dot{V}O_{2max}$, which suggests that other factors influence sex differences in endurance performance. For example, beginning about age 1, boys tend to be more active than girls, and the difference in activity levels increases with age. Thus, on average, boys may be considered better trained or closer to their biologically determined potential. The reasons for sex differences in activity patterns among children are complex, but probably relate more to social rather than biological factors, especially in young children.

After puberty and continuing throughout adulthood into old age, aerobic exercise capacity is higher, on average, in males than in females. The magnitude of the difference in aerobic power, however, is related to how $\dot{V}O_{2max}$ is expressed. For example, since absolute $\dot{V}O_{2max}$ (L/min) is related to body size, there is a large sex difference, approximately 40 per cent, between sexes. The sex difference narrows to about 20 per cent when $\dot{V}O_{2max}$ is adjusted for body size (ml/kg/min), and even further to less than 10 per cent when adjusted for differences in body composition (ml/kg lean body mass/min). This difference reflects the fact that oxygen uptake during exercise is largely a function of muscle metabolism; the more metabolically active the muscle tissue, the higher the oxygen use.

Even when $\dot{V}O_{2max}$ values are adjusted for body composition, there is still about a 10 per cent difference between men and women. This residual difference is most likely due to other cardiovascular factors influencing oxygen delivery to and extraction by skeletal muscle. As shown in Table 14.4, blood level of haemoglobin, the oxygen-carrying protein in blood, is lower in women than in men. In addition, heart size relative to body size is also somewhat lower, resulting in a lower stroke volume and cardiac output in women than in men. The combination of lower haemoglobin concentration and cardiac output means that, in women, proportionately

Table 14.4 Physiological sex differences and effects on performance

Female compared with male	**Effect on exercise performance**
Smaller body, shorter limbs, narrower shoulders, wider hips	lower muscular strength and power
Less muscle mass, proportionately more body fat	less metabolically active tissue, lower muscular strength and power, aerobic and anaerobic power
Smaller averge muscle fibre diameter	lower muscular strength and power
Smaller heart	lower stroke volume, cardiac output and $\dot{V}O_{2max}$
Lower blood haemoglobin level	lower oxygen delivery to muscle and lower $\dot{V}O_{2max}$
Lower $\dot{V}O_{2max}$	lower aerobic power and endurance performance
More reliant on circulatory adjustments than sweating to lose heat	lower sweat rate during exercise

Variables in which there are no sex differences:

- Distribution of muscle fibre types
- Lactic acid threshold as a percentage of $\dot{V}O_{2max}$
- Tolerance of exercise in the heat
- Metabolic capacity per gram of muscle
- Relative improvement in aerobic and anaerobic capacity and muscular strength after training
- Psychological factors influencing sports performance

Source: Compiled from Wells, C.L. (1992), *Women, Sport and Performance: A Physiological Perspective* (2nd edn), Human Kinetics, Champaign, Illinois; McArdle, W.D., Katch, F.I., Katch, V.L. (1991), *Exercise Physiology: Energy, Nutrition and Human Performance* (2nd edn), Lea & Febiger, Philadelphia.

less oxygen is delivered to working muscles during maximal exercise.

It should be noted that these sex differences in aerobic power and related cardiovascular system measures become very small, of the order of less than 10 per cent, when female and male well-trained endurance athletes are compared. In addition, males and females respond similarly to endurance-type training, and show similar improvements in $\dot{V}O_{2max}$ and endurance performance relative to initial levels.

Sex differences in anaerobic capacity

Compared with the large number of studies on sex differences in aerobic exercise capacity, there are fewer studies comparing anaerobic exercise capacity in males and females. When examining sex differences in anaerobic exercise capacity, it is important to include similarly trained individuals; many of the sex differences reported in the early literature arose because relatively fit males were compared with less fit females.

When comparing similarly trained males and females, total work output and peak power in a brief all-out exercise test are lower in females compared with males. These differences are due mainly to the smaller body size and amount of muscle in females. As in $\dot{V}O_{2max}$, sex differences in anaerobic exercise capacity become smaller when adjusted for body weight, but do not disappear completely, mainly due to differences in body composition and other factors such as relative limb length. Peak muscle and blood lactic acid levels may be lower after maximal exercise in females, again possibly due to the smaller muscle mass. However, during submaximal exercise at a given percentage of $\dot{V}O_{2max}$, blood lactic acid level and the lactic acid threshold do not appear to differ between males and females. Males and females generally show similar training responses provided that training programs are similar.

The sex differences in anaerobic power and capacity are reflected in performance differences in power events such as jumping, throwing and weight lifting. There are greater sex differences in performance in these power events, compared with other events such as sprinting or longer duration activities.

Sex differences in muscular strength

As discussed earlier in this chapter, strength does not differ between girls and boys until about age 11, after which muscular strength continues to increase in boys but plateaus in girls. In general, muscular strength in adult women is about two-thirds that in men, although there is large individual variability depending on body size and muscle mass as well as the muscle group measured and individual fitness level. As with other measures, sex differences in strength become smaller when adjusted for body size and limb size, indicating that much of the sex difference in strength is due to differences in muscle mass. When expressed relative to lean body mass or muscle cross-sectional area, strength is similar between males and females. Sex differences are also smaller when strength is compared in similarly trained male and female athletes. For example, strength differences between male and female power lifters are smaller than those for untrained males and females.

Sex differences in muscular strength also depend on the body segment and muscle groups compared. Muscular strength in females is closer to that of males when comparing the lower body rather than the upper body. For example, chest and arm strength relative to body weight in females is about 50–60 per cent of that in males; in contrast, strength of the hip flexors and extensors relative to body weight is similar in males and females. These differences by body segment are probably due to two factors—first, that men have proportionally more of their muscle mass in the upper body and, second, that men traditionally have participated in activities that train the upper body.

As with endurance training, males and females show similar improvement in strength relative to initial levels after resistance training, provided that similar training programs are used. It was once believed that, in females, skeletal muscle does not hypertrophy in response to strength training due to the much lower level of the male hormone testosterone. However, recent research using sensitive imaging techniques to visualise muscle cross-sectional area shows that muscle hypertrophy does occur in females, at least in short-term studies lasting up to several months. Muscle hypertrophy may not be as obvious due to the larger subcutaneous fat layer in the female. There have not been enough long-term studies to determine whether muscle size continues to increase with strength training in females.

Sex differences in sports performance—will female performance ever equal that of males?

Although there are sex differences in world record performance in most sports, the gap between males and females has been narrowing over the past few decades. For example, the difference between the men's and women's 100 m world records decreased from 11.40 to 7.74 per cent between the years 1956 and 1986. The reason for this change is that, over the past few decades, women's sports performance has improved at a much faster rate than men's, most likely due to social rather than biological factors. It has been suggested that, should this trend continue, there may in time be no sex differences in world records in some sports. It is likely that swimming will be the first sport in which this may occur. Sex differences in world record performance are smaller in swimming than for other sports because the higher body fat levels in females is advantageous because of the extra buoyancy. However, given the physiological and biomechanical differences between sexes as described previously, it is unlikely that sex differences will completely disappear in all sports, especially those involving primarily speed, power and cardiorespiratory endurance.

CHAPTER 15

Applications of exercise physiology to health

- What are the major causes of disease and death in developed countries?
- Physical inactivity costs millions of dollars each year
- Few adults exercise sufficiently to derive any health benefits
- Obesity
- Cardiovascular diseases
- Hypertension
- Diabetes
- Asthma
- Cancer
- Osteoporosis
- Arthritis
- Infectious illness in sport

Regular physical activity has many health benefits. Moderate exercise can prevent obesity, reduce the risk of heart disease and hypertension (high blood pressure), guard against or reverse adult-onset diabetes and possibly prevent osteoporosis (loss of bone mass). Exercise is considered part of a healthy lifestyle; indeed government, the medical profession, and other health professions recommend regular moderate physical activity for virtually all individuals, including children, the elderly and those with disease or disability. Regular exercise is encouraged among the general population for reasons other than merely enhancing lifestyle; reduced health care costs, increased industrial productivity and improvements in quality of life that accompany increased physical fitness benefit society as a whole. This chapter examines diseases linked directly to physical inactivity.

What are the major causes of disease and death in developed countries?

The causes of disease and death have changed dramatically in recent history. Whereas infectious diseases such as influenza ('flu') and tuberculosis were the major killers in the early part of this century, the major causes of illness and death in most developed countries are now heart disease, stroke and cancer, which collectively account for approximately two-thirds of all deaths each year.

There are many reasons for the change in causes of disease and death over the past century. The reduced incidence of infectious illness is due to advances in hygiene, such as water and waste treatment and food processing; in medicine, for example antibiotics and other drugs; and in public health, such as mass immunisation. The rise in the incidence of other diseases such as heart disease and cancer is partly attributable to an increasingly affluent and sedentary lifestyle.

Physical inactivity costs millions of dollars each year

Physical inactivity is more than just a personal lifestyle issue; physical inactivity is associated with many diseases and high health care costs. In many developed countries, health care costs have increased dramatically over the past 20 years and now consume approximately 10 per cent of national income. In most developed countries, health care costs are shared among the general population, through government-supported medical care, or through private and employer-supported health insurance funds. Thus, the high costs due to lifestyle factors such as physical inactivity are met by society as a whole. Heart disease alone costs billions of dollars each year in direct medical costs and in indirect costs such as loss of income and productivity.

Can disease be prevented?

It has been estimated that more than half of the incidence of the major killers such as heart disease, cancer, stroke and diabetes may be prevented, particularly by altering lifestyle. Increasing regular physical activity among the general population has been suggested as one means of reducing the incidence and costs of disease. For example, it has been estimated that as many as 250,000 deaths per year in the U.S. (about 12% of the total) are attributable to a lack of regular physical activity. North American studies suggest that over $1 billion per year could be saved in costs of geriatric care (care of the elderly) by increasing physical activity among older adults.

Few adults exercise sufficiently to derive any health benefits

Although most adults agree that regular exercise is important for good health, few exercise sufficiently to derive significant protective health benefits. For example, about 60% of Americans are not sufficiently active on a regular basis to derive health benefits; moreover, about 25% of Americans are considered inactive. Participation decreases with age and more than 35% of Americans over age 55 years are considered sedentary. Participation also declines dramatically during adolescence, especially in girls; nearly half of North American and European young people (12 to 21 years of age) are not vigorously active on a regular basis. At all ages, men are more likely to exercise than women.

In 1996, the U.S. Office of the Surgeon General published its first report addressing the issue of physical activity among Americans (*Physical Activity and Health: A Report of the Surgeon General*). The report's main message is that health and quality of life can be improved by including moderate amounts of daily physical activity in one's lifestyle. Moreover, the health benefits are proportional to the amount of physical activity so that any increase in activity level will help improve health. The report noted that regular physical activity reduces the risk of premature death in general, and of heart disease, high blood pressure, colon cancer, and diabetes mellitus in particular (see Box 15.1); physical activity also helps improve mental health and is important for continued health of bones, joints, and muscles. Thirty minutes of daily activity such as brisk walking or gardening is encouraged.

Why do some people choose to exercise and others do not?

For most people, simply recognising that exercise is important to good health is not enough to

Box 15.1 Quantifying the health risk of physical inactivity—How much can exercise help prevent disease?

It is now accepted that physical inactivity is associated with various health problems, such as heart disease. However, it is only recently that large research studies have specifically examined the relationship between physical activity and disease to more clearly understand and measure the degree to which physical activity may prevent disease. The Aerobics Center Longitudinal Study, originating from the Cooper Institute for Aerobics Research in Dallas, Texas, followed over several years more than 25,000 men and women who had undergone a preventive medical examination, including physical fitness testing, between 1970 and 1989. Data were analyzed to determine statistical associations between physical fitness and death due to cardiovascular disease and other causes.

Study participants were grouped into three levels of physical fitness based on their initial fitness test—low fitness (least fit 20%), moderate fitness (next 40%) and high fitness (highest 40%). Along with other risk factors such as cigarette smoking and elevated blood pressure and blood cholesterol, low fitness in men was associated with an increased risk of death due to all causes and cardiovascular disease; in women, low fitness and smoking were related to increased risk of death. Regardless of the existence of other risk factors, mortality declined with increasing fitness level. The greatest decline in death rates occurred between the low and moderately fit groups, suggesting that even moderate fitness confers significant health benefits. Moreover, the mortality rates in highly fit men and women with other risk factors were lower than in low fit individuals with no risk factors. Importantly, the study showed that low fitness is as much a risk factor for cardiovascular disease as the other major risk factors such as cigarette smoking, high blood pressure, and high cholesterol. These data indicate that, even in the presence of other risk factors, improving physical fitness through moderate activity is effective in lowering the risk of premature death due to cardiovascular disease and other causes. The authors conclude by stating their belief that physicians should counsel *all* sedentary individuals to improve their physical fitness by becoming more physically active.

Sources: Blair, S.N., Kampert, J.B., Kohl III, H.W., Barlow, C.E., Macera, C.A., Paffenbarger, R.S. Jr, & Gibbons, L.W. (1996), Influences of cardiorespiratory fitness and other precursors on cardiovascular disease and all-cause mortality in men and women, *JAMA*, 276, 205–210; Blair, S.N., Kohl III, H.W., Barlow, C.E., Paffenbarger, R.S. Jr., Gibbons, L.W., & Macera, C.A. (1995), Changes in physical fitness and all-cause mortality, *JAMA*, 273, 1093–1098.

maintain the habit of regular physical activity. Although many people enthusiastically begin exercise programs, fewer than 50 per cent will still be exercising regularly after 6 months.

As might be expected, people who exercise frequently tend to be young, non-smokers, fairly lean, relatively fit and have previously participated in physical activity or sport (Table 15.1). Regular exercisers tend to exhibit a higher degree of self-motivation especially towards health-related behaviour and to have family members who support their participation. Regular exercisers also tend to be more highly educated and of a higher socioeconomic level than those who do not exercise. Physically active people commonly exercise for the health benefits, for the effects on appearance and for the enjoyment derived from physical activity.

There are many reasons why people do not exercise regularly or drop out of exercise programs soon after beginning. Perceived lack of time and inconvenience of exercise facilities

Table 15.1 Factors influencing exercise participation and adherence

Positive effect on participation and adherence	Negative effect on participation and adherence
• Younger age	• older age
• Higher socioeconomic status	• illness, injury, disability
• Higher educational level	• perceived lack of time
• Spousal and peer support	• cigarette smoking
• Convenient facilities	• obesity
• Previous participation in exercise and sport	• inappropriate type of exercise
• Exercise in groups or with a partner	• boredom, lack of variety

are two commonly cited reasons. Many people have misconceptions regarding the level of exercise needed to derive the health benefits, believing far more exercise is needed than is actually necessary (see Chapter 13). People often drop out of an exercise routine because of injury or illness, boredom, lack of enjoyment or failure to meet unrealistic expectations due to inappropriate exercise programs. Lack of family support and feelings of selfconsciousness may also cause individuals to stop exercising. In addition, social and cultural norms may mitigate against physical activity in many groups. (A more detailed consideration of motives for exercising and adhering to exercise programs is made in Chapter 21.)

What can be done to improve exercise participation?

There has been much interest in identifying ways to increase exercise participation among the general population. Given that convenience and access to facilities greatly influence activity patterns, many companies now provide on-site employee exercise programs in an effort to enhance worker health and fitness. Work-based programs are not only convenient, but have added benefits to companies by reducing absenteeism and increasing productivity among workers. Convenient community-based health and fitness centres offering a variety of programs may also help to improve participation. More adults may participate in exercise with increasingly wide encouragement of exercise by the medical profession as well as government and health agencies. School physical education and sports programs offer an effective means to influence the activity patterns of entire generations; adults are more likely to exercise if they have participated in (and enjoyed) sport and physical activity as children.

The role of regular physical activity in preventing a number of specific diseases will now be considered.

Obesity

Obesity is often considered an aesthetic rather than a health issue. However, obesity is a major health concern because it contributes to hypertension and heart disease, certain forms of cancer (breast and bowel), osteoarthritis, and non-insulin-dependent diabetes mellitus. In the United States, an estimated 61 million people are overweight; this number represents between 31 and 50% of the population, depending on age, sex, and race or ethnic group. Moreover, the incidence of overweight and obesity has increased over the past 30 years. Twenty-one percent of young Americans between ages 12 and 19 years are considered overweight, a dramatic increase in the past 12 years.

Overweight and obesity can be defined in several ways (Table 15.2). When using the Quetelet Index (QI) or Body Mass Index (BMI) (see Chapter 4), overweight is defined as a QI over 25 and obesity as a QI over 30. If standard height–weight charts are used, overweight is usually defined as 10 per cent, and obesity as 20 per cent, over recommended weight for height. Because of the simplicity of measurement and calculation, these two definitions are appropriate for the general population. However, athletes

Table 15.2 Definitions of overweight and obesity

Measure	Overweight	Obese
Quetelet Index (QI) or Body Mass Index (BMI)	>25	>30
Height–weight tables	10–20% over recommended weight for height	>20% over recommended weight for height
Body composition	adult male 18–20% fat adult female 27–30% fat	adult male >20% fat adult female >30% fat
Waist-to-hip ratio	adult male >1.0 adult female >0.85	

often carry additional body mass as muscle, and many athletes, especially those in strength and power sports, are classified as overweight or obese using these simple definitions. Although more complex to obtain, assessing body composition is a better way to define obesity because it delineates fat from lean body mass. After all, it is the excess fat, not total body mass, that is associated with disease. For adults, overweight may be defined as between 18 and 20 per cent fat for men and between 27 and 30 per cent fat for women; obesity may be defined as a percentage of body fat above 20 for men and 30 for women.

Obesity is usually caused by an imbalance between energy intake and expenditure. In other words, more energy is consumed as food than is expended through normal activity. Obesity is rarely caused by hormonal factors, but it is now widely believed that it is strongly influenced by genetic factors.

The pattern of fat distribution across the body appears to be important in the health risk of obesity. The so-called male pattern of obesity of fat deposition primarily in the upper body (chest and waist) carries a higher risk of heart disease, stroke and diabetes than the female pattern of fat deposition primarily around the hips and thighs. This relationship holds true even when comparing fat deposition within one sex. For example, women with fat deposition primarily in the upper body are at higher risk of cardiovascular disease than women with fat deposition primarily in the lower body. The two patterns of fat deposition and obesity have been termed the 'apple' and 'pear' shapes for male and female patterns, respectively (Figure 15.1). The pattern of fat distribution may be measured using the waist-to-hip ratio (W:H ratio), calculated by dividing the circumference of the waist by that of the hips. The recommended W:H ratios are less than 0.85 for women, and less than 1.0 for men.

Exercise and obesity

Since obesity is caused by an imbalance between energy intake and expenditure, it follows that obesity can be prevented by decreasing energy intake, increasing energy expenditure or some combination of the two. Although dieting is often needed, weight reduction programs that rely solely on dieting are generally ineffective for permanent weight loss. For a variety of reasons, regular physical activity is important to long-lasting control of body weight (Table 15.3).

Research studies have shown that exercise alone or an exercise and diet combination are

Table 15.3 Benefits of exercise in weight loss

- Faster rates of weight loss when combined with dietary restriction
- Less dietary restriction required for moderate weight loss
- Greater fat loss compared with dietary restriction alone
- Lean body mass and basal metabolic rate maintained and possibly increased
- Favourable changes in body composition
- Higher probability of maintaining new (lower) body mass once achieved

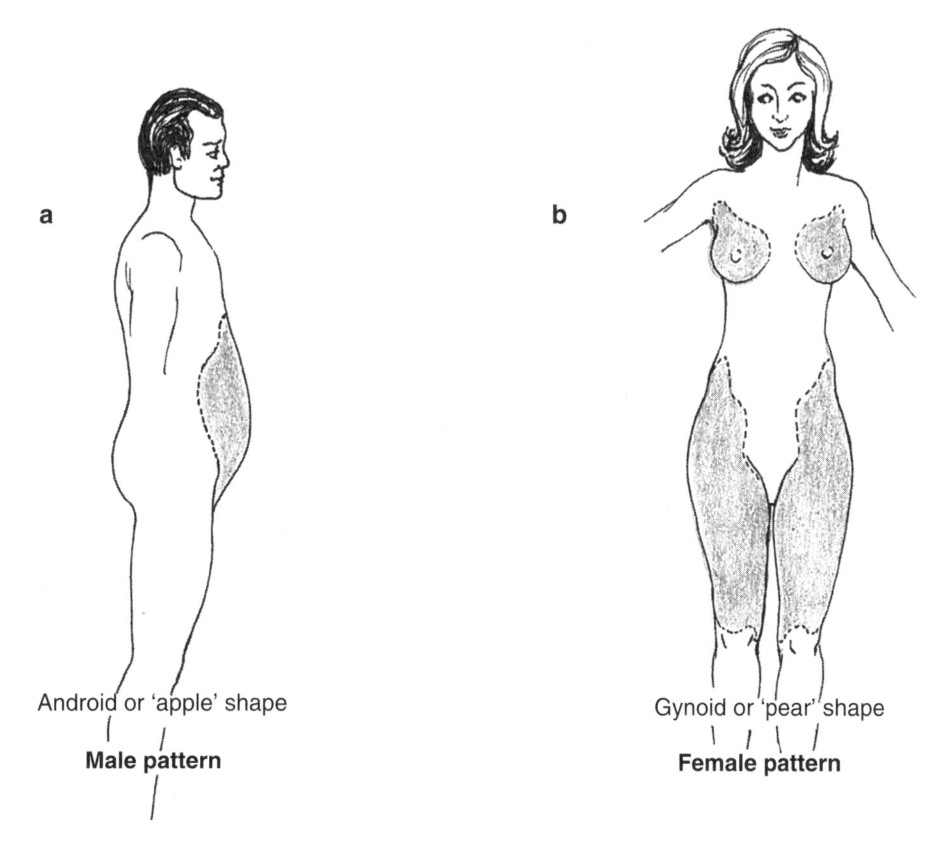

Figure 15.1 Pattern of fat distribution in the body—android or male pattern ('apple' shape) (a); gynoid or female ('pear' shape) (b)

more effective in reducing fat mass than dieting alone. Exercise to reduce body weight helps to maintain lean body mass, whereas muscle mass may be lost in weight reduction programs that involve dieting alone. It is important to maintain muscle mass because skeletal muscle is a major site of fat metabolism; the body's capacity to burn fat declines with decreasing muscle mass. Exercise is also important to maintaining the body's basal (or resting) metabolic rate (BMR or RMR). BMR, the minimum amount of energy needed to maintain bodily functions, accounts for a significant part (between 60 and 75 per cent) of total daily energy expenditure depending on the amount of physical activity. BMR is very much related to age, gender, body size and lean body mass. BMR may decrease by 10–20 per cent after significant loss of body mass through diet alone. This reduces the amount of energy needed by the body, and is one reason why maintaining weight loss is so difficult after dieting. In contrast to dieting alone, weight loss through exercise helps to prevent the decline in BMR, thus ensuring permanent loss of body mass and fat. Moreover, exercise will increase muscle mass, enhancing lean body mass and leading to a more favourable body composition.

Exercise prescription in obesity

Increasing total energy expenditure is vital to losing weight through exercise. However, exercise programs must also consider the limited exercise capacity of obese individuals. Exercise should be low impact to prevent injury to load-bearing joints. Examples include walking, low impact or water-based aerobic activities, swimming, cycling, and circuit-type resistance exercise using weight and stationary exercise machines. Each exercise session should be long enough to expend at least 300 kilocalories (1260 kJ) of energy, equivalent to at least 30 min exercise per session. Since energy expenditure increases with increasing duration of exercise, longer sessions approaching 60 min are recommended provided they do not lead to overuse injury or excessive fatigue. Exercise should be performed at least four times per week and daily if possible.

Since obese individuals may be in poor physical condition at the start of a program, low-intensity programs that progress gradually are recommended. Exercise intensities in the range of 40–60 per cent $\dot{V}O_{2max}$ or heart rate reserve, equivalent to approximately 50–65 per cent age-predicted maximum heart rate are appropriate at the onset of an exercise program.

Cardiovascular diseases

Cardiovascular diseases, including coronary heart disease, stroke and other circulatory system diseases, account for nearly one-half of deaths each year in most developed nations. Although the death rate due to cardiovascular diseases has been declining since the late 1960s in North America and much of Western Europe, deaths from cardiovascular disease still account for almost 50 percent of all deaths in both developed and developing countries. Moreover, the incidence of cardiovascular disease is still increasing in many regions, such as Eastern Europe.

Coronary heart disease is characterised by a narrowing of the inside (lumen) of the blood vessels supplying the heart. Cells lining the interior of these blood vessels become damaged, possibly in response to consistently elevated blood pressure and/or high levels of circulating cholesterol (a type of lipid or fat). Smooth muscle cells proliferate and move into the inside of the blood vessel, narrowing the vessel's inside diameter and restricting blood flow and delivery of oxygen and nutrients to various areas of heart muscle. This process is called atherosclerosis.

Although genetic predisposition plays an important role, much of the incidence of heart disease can be prevented by lifestyle modification. Risk factors for heart disease have been identified from decades of research. Primary or major risk factors are those factors that have been shown to be closely related to the incidence of disease; secondary risk factors are also related to the risk of disease, but the association is not as clear. Table 15.4 lists the major risk factors for cardiovascular disease. These factors are based on population studies, and denote the statistical probability of developing disease when risk factors are present; they cannot be used to predict with any certainty the risk of developing disease for a particular individual.

Elevated blood pressure and cholesterol level are associated with development of atherosclerosis and heart disease. Cholesterol is carried in the blood in several different molecules, usually combined with special proteins. Low-density lipoproteins (LDL) carry most of the cholesterol; high levels of LDL increase the risk of heart disease. High-density lipoproteins (HDL) are considered protective against heart disease, by removing cholesterol from the body. Thus, a favourable lipid profile for prevention of heart disease includes low blood levels of total cholesterol and LDL together with high levels of

Table 15.4 Cardiovascular disease risk factors

Risk factor	**Measurement**
Age	>45 years in men and >55 years in women
Family history of disease	before 55 years in close male relative and 65 years in female relative
Cigarette smoking	
Diabetes	IDDM >30 years of age; NIDDM >35 years of age
High blood pressure	> or = 140/90
High blood lipids	cholesterol >5.2 mmol/L or HDL-C <0.9 mmol/L
Physical inactivity	least active 25% of population

Source: Compiled from American College of Sports Medicine's (1995) *Guidelines for Exercise Testing and Prescription* (5th edn), Lea & Febiger, Philadelphia.

HDL. The ratio of total cholesterol to HDL cholesterol is an important measure.

Cigarette smoking also contributes to the process of atherosclerosis, although the exact mechanisms are unclear at present. Smoking may elevate blood pressure and certain chemicals contained in cigarette smoke may contribute to blood vessel damage.

There is also a genetic predisposition toward heart disease.

Exercise and cardiovascular disease

Regular moderate exercise is now widely recognised as an effective means to prevent cardiovascular disease. On average, the risk of cardiovascular disease is halved in those who participate in regular physical activity. Regular exercise appears to work both directly, by strengthening the heart and making it a more efficient pump, as well as indirectly by altering a number of other risk factors for cardiovascular disease. Table 15.5 summarises the effects of regular physical activity on several of these risk factors.

Table 15.5 Effects of regular physical activity on cardiovascular disease risk factors

Risk factor	Effect of exercise and implication for disease prevention
High blood pressure	lowers blood pressure in those with mild hypertension
Blood lipids	may lower blood cholesterol level; increases blood HDL levels; decreases ratio of cholesterol to HDL
Obesity	reduces body weight and body fat
Diabetes	prevents obesity; reduces fat levels; improves the body's sensitivity to insulin, thus glucose clearance
Stress	effective stress management strategy
Cigarette smoking	smokers who exercise may smoke less or stop smoking

Regular moderate exercise lowers blood pressure in those with mildly to moderately elevated blood pressure (see below). Regular moderate physical activity increases the protective high-density lipoprotein fraction and may lower the total cholesterol level in blood. As described above, exercise is an excellent means to prevent obesity by controlling body weight and fat levels, and maintain lean body mass. Physical activity also increases glucose tolerance, helping to prevent and/or treat non-insulin-dependent diabetes mellitus (see below). Finally, smokers who exercise regularly may be more likely to stop smoking once an exercise program is established.

Exercise prescription to prevent and treat cardiovascular disease

The general recommendations for exercise for good health, as described in Chapter 13, apply to prevention of cardiovascular disease. The risk of disease decreases proportionally with increasing amount of exercise performed. It is the total amount of exercise, rather than exercise intensity, that is most important in reducing risk. Moderate exercise such as brisk walking (in the range of 50–70 per cent $\dot{V}O_{2max}$) appears to be nearly as effective in reducing the risk of disease as more intense exercise such as jogging, running or organised sport; moderate exercise is certainly more attractive to, and safer for, most people. Recent research shows that brisk walking for 30–45 min, performed four or five times per week, markedly reduces the incidence of cardiovascular disease.

Regular moderate exercise is also advocated in the rehabilitation of patients with, or at high risk of, cardiovascular disease. Research indicates that exercise rehabilitation increases functional capacity and slows the progression of the disease in cardiac patients. Exercise rehabilitation programs must be individually prescribed based on the extent of disease and symptoms. Medically supervised programs staffed by exercise scientists are generally recommended at least during the early stages of an exercise program. Patients will generally have a low exercise capacity and, as a consequence, very moderate and gradual programs are needed.

Hypertension

Hypertension or high blood pressure affects 12–16 per cent of the Australian population; up to 60 per cent of older adults may exhibit high blood pressure. The World Health Organisation (WHO) defines hypertension as a chronically elevated blood pressure reading above 160/95, although readings over 140/90 are generally considered high. The first number is systolic blood pressure and represents the pressure in the arteries during the heart's contraction; the second number is diastolic blood pressure and represents the pressure in the arteries during relaxation of the heart. Hypertension is associated with increased risk of stroke, heart disease, renal failure and peripheral vascular (blood vessel) disease. In most cases of hypertension, there are no symptoms and no single identifiable cause, although there is a strong genetic predisposition; lifestyle factors such as diet, obesity, cigarette smoking, inactivity and life stresses are all associated with elevated blood pressure.

Exercise and hypertension

Recent research indicates that individuals with mildly to moderately elevated blood pressure in the range of 140–160/90–95 benefit from regular physical activity. Moderate exercise has been shown to reduce both systolic and diastolic blood pressure in these individuals, often to the range of normal blood pressure. It is not clear whether exercise has a direct effect on blood pressure, or whether the effect is indirect via other factors such as reducing body weight; weight loss by itself can help to lower blood pressure in individuals with mild hypertension. Regular exercise may directly influence blood pressure by reducing heart rate at rest and during activity and/or by damping the body's responses to stress hormones, which act to increase heart rate and blood pressure. Significantly elevated blood pressure (above 160/95) often requires medication in addition to lifestyle modifications such as diet and exercise.

Exercise prescription for hypertension

Because hypertension increases the risk of cardiovascular disease, individuals with hypertension should check with a medical practitioner before beginning an exercise program. A medically supervised exercise test may be needed to determine the patient's blood pressure response to various levels of exercise. Once approved, the general recommendations regarding exercise of low to moderate intensity apply to exercise for hypertensive individuals. Exercise should begin in the range of 40–60 per cent $\dot{V}O_{2max}$ and progress gradually. Whole-body exercise or exercise using large muscle groups, such as walking, swimming, and cycling are preferable, since blood pressure is less likely to increase dramatically during these types of exercise. Although high intensity resistance (weight) training is generally not recommended because it may cause blood pressure to rise dangerously, lower intensity resistance work that emphasises muscular endurance is now considered safe and effective for hypertensive individuals. For example, an appropriate program might include circuit training with intensity at each station in the 15–20 RM range (see Chapter 14). Since many individuals with high blood pressure are also obese, general guidelines for exercise to reduce body mass should be followed. It may also be prudent for the patient to monitor his or her own blood pressure response during exercise to ensure that levels stay within recommended limits.

Diabetes

Diabetes is a metabolic disease characterised by high blood glucose levels. Diabetes is the fifth leading cause of death among Australians, and is a contributing factor to other illnesses such as cardiovascular, peripheral vascular, and renal diseases.

There are two main types of diabetes: type 1 or insulin-dependent diabetes mellitus (IDDM) and type 2 or non-insulin-dependent diabetes mellitus (NIDDM). IDDM begins more often in childhood or adolescence, and is considered an autoimmune disease in which the body's immune system attacks its own cells, in the case of IDDM, the insulin-producing cells of the pancreas. In contrast, NIDDM does not appear until later in life, during middle-age or old-age, and is associated with lifestyle factors such as obesity and physical inactivity.

The incidence of NIDDM has increased over the past 30 years, due possibly to an increasing prevalence of obesity and physical inactivity. NIDDM more than doubles the risk of developing cardiovascular disease. It is believed that there is a genetic predisposition to both types of diabetes, although 60–95% of NIDDM may be prevented or reversed by changes in lifestyle.

High blood glucose levels may be caused either by insufficient production of insulin by the pancreas, as in IDDM, or by resistance to insulin in peripheral tissues, as in NIDDM. Insulin, a hormone released from the pancreas, is necessary for glucose uptake from the blood into most tissues. Removal of glucose from the blood is limited if there is insufficient insulin and/or resistance to insulin's action, thereby keeping blood glucose levels high. IDDM is treated with injected insulin to replace what is normally produced by the pancreas. NIDDM is treated with a combination of dietary and lifestyle modifications and sometimes medication to increase the pancreatic production of insulin and the body's response to the insulin that is produced.

Exercise and diabetes

Provided that blood glucose levels can be controlled to some extent, regular moderate exercise is an important part of treatment of both types of diabetes, and especially NIDDM. Given that exercise stimulates tissue uptake of glucose from the blood and that a single exercise session may enhance blood glucose uptake for up to 48 h, regular exercise may improve glucose tolerance in both IDDM and NIDDM diabetics. Moreover, because NIDDM is associated with obesity, regular moderate exercise is an important part of treatment. In obesity, tissues such as adipose (fat) tissue are resistant (do not respond) to the insulin produced by the pancreas. Exercise training not only stimulates glucose uptake by these tissues but, as described above, helps to control body mass.

Exercise prescription for diabetes

Regular moderate exercise is recommended for improved glucose tolerance in individuals with either form of diabetes. It is best if exercise intensity and duration, as well as time of day, are relatively consistent from day to day. Individuals with either form of diabetes may need to adjust carbohydrate intake before or after exercise. For the IDDM individual, exercise should be avoided during the peak time of insulin activity to avoid a precipitous drop in blood glucose levels. The athlete with IDDM should work closely with his or her medical practitioner to monitor possible changes in insulin and glucose requirements.

For the individual with NIDDM, exercise prescription should follow the same general recommendations as described previously for obese individuals: 30–60 min low to moderate intensity whole-body exercise, performed daily if possible, but at least four times per week. Activities such as walking, cycling, swimming, low-impact or water-based aerobics, and low-intensity circuit exercise are recommended. Since many individuals with NIDDM are also obese, high-impact activities such as running should be avoided, at least until a significant amount of body weight is lost.

Asthma

Asthma is characterised by an increased airway responsiveness to selected stimuli, which causes a narrowing of the airways (bronchoconstriction) and difficulty in breathing. Asthma is a frequently occurring respiratory disorder, especially among children, and is one of the most common reasons for consulting a medical practitioner. The incidence of asthma varies by country, but may be as high as 5–10% of the population. Asthma should not prevent participation in exercise; many countries report that 10% of their Olympic athletes are asthmatic.

Exercise and asthma

Asthma can be brought on by many stimuli including cold dry air, allergens, dust, smoke, and exercise. Most asthmatics will develop exercise-induced asthma (EIA) during physical activity. EIA is characterised by a reduction in the volume of air that can be forcefully expired (forced expiratory volume), resulting in breathlessness and inability to maintain high levels of exercise. The severity of EIA is generally related to exercise intensity, and the most asthmogenic (most likely to cause EIA) form of exercise is intense continuous activity (above 75 per cent $\dot{V}O_{2max}$) lasting 6–8 min.

Although exercise can bring on an episode of asthma, regular exercise is widely encouraged for the asthmatic individual and especially for children. Regular exercise enhances respiratory and cardiovascular function, helps to prevent obesity, and is important for motor development in children; participation in sport and physical activity is also an important aspect of social development especially at school. By following certain guidelines in combination with modern medications, most asthmatic children and adults can participate in the full range of physical activity and sport.

Exercise prescription in asthma

Each asthmatic individual may respond differently to exercise, and the response may also vary from time to time, depending on the type and intensity of exercise as well as environmental factors such as temperature and humidity. Appropriate use of preventive medication can significantly reduce the frequency of EIA. Each asthmatic individual, especially competitive athletes, should work closely with a medical practitioner to best determine the optimal combination of medication and exercise (Box 15.2).

Swimming, walking and other moderate activities are considered less asthmogenic than more intense activities such as jogging and running (see Table 15.6). In general, asthmogenicity increases as the demands on the respiratory system increase. For the non-athlete with asthma, the general guidelines for exercise training for health, as described in Chapter 14, are recommended. That is, exercise should include moderate whole-body or large muscle group activity in the range of 50–85 per cent $\dot{V}O_{2max}$ or 60–85 per cent of age-predicted maximum heart rate. Competitive athletes will need to train according to the demands of the sport. Asthmatic athletes have been successful in virtually every sport, providing evidence that few modifications need be made for training the athlete with asthma.

The warm-up prior to exercise is very important, since it allows the cardiorespiratory system to adjust gradually to the demands of exercise. Warm-up should include light aerobic-type activity and muscular flexibility exercises. The aerobic part of a work-out should begin at low intensity and increase gradually. Interval training and games involving intermittent

Table 15.6 Exercise for the asthmatic individual

Recommendation	Reason
Maintain regular exercise	onset of EIA is more predictable and controllable
Exercise intensity $<80\%$ $\dot{V}O_{2max}$ (in non-competitive athletes only)	lower minute ventilation is less likely to cause EIA
Proper warm-up before all activity	warm-up may prevent EIA or lessen its severity
Avoid cold, dry air	EIA is more likely in cold dry environments
Swimming and/or interval exercise	higher humidity around swimming pools; interval exercise is less asthmogenic
Avoid exercise during or immediately after a respiratory infection	EIA more likely to occur after respiratory infection
Use prescribed medication before exercising	may prevent EIA

Source: Compiled from Morton, A.R. & Fitch, K.D. (1993), Asthma, pp. 211–28 in J.S. Skinner, (ed.), *Exercise Testing and Exercise Prescription for Special Cases* (2nd edn), Lea & Febiger, Philadelphia.

Box 15.2 Management of asthma in high-performance athletes

Over the past 20 years, Alan Morton and Ken Fitch of the University of Western Australia have studied the responses to exercise in asthmatic athletes, including children as well as adults. Up to 10 per cent of Australia's Olympic athletes have some form of asthma, and since intense exercise can trigger an attack, management is vital to the elite asthmatic athlete.

In the 1970s and 1980s Morton and Fitch were among the first to show that exercise-induced asthma is less likely to occur in intermittent compared with continuous exercise and in swimming compared with running. Their work provides an important scientific basis for practical guidelines for exercise in asthmatics as discussed in this chapter. Recent advances in medication to treat and prevent asthma have improved management of the disease, especially for athletes. However, some drugs also influence exercise performance, and it is important to determine whether any new medication is an ergogenic aid (performance enhancing); if so, then international sports organisations must decide whether to permit or ban the use of that drug among athletes. Salbutamol, commonly prescribed for asthmatics, has recently been suggested to have ergogenic properties. However, as Morton and Fitch point out, banning use of salbutamol would eliminate many asthmatic athletes from competing at the elite level.

Morton, Fitch and associates investigated whether the use of salbutamol by non-asthmatic athletes has any beneficial effect in standard exercise tests. Seventeen distance runners and 17 power athletes performed the same tests on three separate occasions—using no medication, using salbutamol, and using a placebo that appeared similar to salbutamol but contained no active drug. There were no significant effects of salbutamol on any measure including $\dot{V}O_{2max}$, time to exhaustion in an endurance running test, post-exercise blood lactic acid levels, maximum heart rate, isokinetic strength, or anaerobic power and capacity as measured in the 10 and 30 s sprint cycle ergometry tests (see Table 15.7). These results indicate that salbutamol exerts no ergogenic effects in high-performance athletes, and the authors concluded that salbutamol should remain a legitimate drug for treatment of asthmatic athletes.

Selected references: Morton, A.R. & Fitch, K.D. (1993), Asthma., pp. 211–27, in J.S. Skinner (ed.), *Exercise Testing and Exercise Prescription for Special Cases*, Lea & Febiger, Philadelphia; Morton, A.R., Papalia, M.S. & Fitch, K.D. (1992), Is salbutamol ergogenic?: The effects of salbutamol on physical performance in high-performance non-asthmatic athletes, *Clinical Journal of Sports Medicine*, 2, 93–97; Morton, A.R., Papalia, M.S. & Fitch, K.D. (1993), Changes in anaerobic power and strength performance after inhalation of salbultamol in non-asthmatic athletes, *Clinical Journal of Sports Medicine*, 3, 14–19.

Table 15.7 Exercise test results after inhalation of salbutamol and placebo in non-asthmatic athletes*

Variable	Placebo	Salbutamol
$\dot{V}O_{2max}$ (ml/kg/min)	75.7	74.6
Exercise time to exhaustion (min)	10.3	10.3
Maximum minute ventilation (L/min)	136.0	129.2
Anaerobic power (W/kg)	18.5	18.6
Anaerobic capacity in 10 s test (J/kg)	150.0	152.0
Peak isokinetic leg extension torque (kg.m)	23.4	23.3

*There were no differences between placebo and salbutamol on any variable tested

Source: Compiled from Morton, A.R., Papalia, S.M. & Fitch, K.D. (1992), Is salbutamol ergogenic?: The effects of salbutamol on physical performance in high-performance nonasthmatic athletes, *Clinical Journal of Sports Medicine*, 2, 94 (copyright 1992 by Lippincott-Raven Publishers); Morton, A.R., Papalia, S.M. & Fitch, K.D. (1993), Changes in anaerobic power and strength performance after inhalation of salbutamol in nonasthmatic athletes, *Clinical Journal of Sports Medicine*, 3, 16. (Copyright 1993 by Lippincott-Raven Publishers, adapted by permission.)

activity, such as soccer or hockey, are often recommended, since EIA is less likely to occur during shorter work intervals. Avoiding exercise during the coldest and driest part of the day, or exercising indoors in a more controlled environment with fewer allergens or pollutants than outdoors, may also reduce the risk of EIA. Swimming is highly recommended because the more humid environment of a pool reduces the likelihood of EIA. Respiratory infections, such as the common cold, increase the susceptibility to asthma, and intense exercise should be avoided during and for several weeks after a respiratory infection.

Cancer

After heart disease, cancer is the second leading cause of death in many developed countries. Although there is a strong genetic predisposition toward cancer, lifestyle factors are important in determining cancer risk. It has been estimated that half of all cancer deaths in developed countries may be prevented by changes in lifestyle, such as reducing cigarette smoking and alcohol consumption, altering diet, and reducing exposure to environmental and occupational carcinogens. Cancer is a complex and multi-faceted disease with potential to occur in many sites in the body. Regardless of site, cancer is characterised by uncontrolled cell growth (malignancy), which eventually overwhelms the body's normal functioning.

Exercise and cancer

Compared with our understanding of how exercise influences other major diseases, such as heart disease and diabetes, relatively little is known about the relationship between exercise and cancer. The strongest evidence for a protective effect of exercise comes from large population studies on colon cancer. These studies show that people who are physically active over many years, either at work or during their leisure time, are 25–50 per cent less likely to develop colon cancer compared with inactive individuals. Even less is known about exercise and prevention of other forms of cancer. There are some studies suggesting that regular physical activity may also protect against breast and reproductive system cancers in women and prostate cancer in men.

The mechanisms by which exercise may protect against cancer are not known at present. It is likely that protection may occur via different mechanisms, each specific to the site of cancer. For example, it has been suggested that exercise reduces the risk of colon cancer by enhancing movement through the bowel, thus limiting exposure of the bowel to potential carcinogens in faeces. Regular physical activity may reduce circulating levels of some sex hormones, which may be causal factors in reproductive system cancer in both men and women. In addition, exercise reduces body fat levels, and excess body fat is linked with increased risk of breast and colon cancer. Physically active people are likely to adopt healthier lifestyles, for example choosing to be non-smokers and to maintain an appropriate body weight. Regular physical activity may also stimulate the body's immune system's response to cancer cell growth.

Exercise prescription in cancer

Exercise has aroused recent interest as a possible treatment for cancer patients in addition to standard medical treatment. Regular moderate exercise may counteract some of the debilitating effects of cancer and its treatment, such as excessive loss of body mass, especially lean tissue, and decreased functional capacity. Moderate exercise is also associated with enhanced psychological state, which may help some cancer patients to cope with side effects of treatment such as nausea and fatigue.

Exercise prescription for cancer patients follows general recommendations for individuals with low functional capacity. Very low-intensity and low-impact exercise is recommended. Exercise should be avoided when the patient experiences excessive fatigue, weakness or nausea, and within 24 hours of chemotherapy.

Osteoporosis

Osteoporosis is a bone disorder in which bone density decreases to a critically low level, often resulting in fracture. Osteoporosis is primarily a disease of older Caucasian women, although elderly men may also be at risk. It has been estimated that 15% of women age 70 years and more than half of women age 80 years will suffer a bone fracture as a result of osteoporosis. The annual costs of medical care and loss of independence due to osteoporotic bone fractures are estimated in the billions of dollars in many North American and European countries.

As was noted in Chapters 3 and 5, bone is a dynamic tissue that is constantly remodelled. During ageing, the process of bone degradation exceeds the rate of synthesis of new bone, gradually reducing bone density and the loads that can be tolerated. The most commonly fractured sites are the vertebrae and the hip.

Older postmenopausal women are most at risk of osteoporosis because bone synthesis is reduced with low oestrogen (female hormone) levels; bone loss is much slower in older men than women of the same age. Medical treatment of osteoporosis may often include replacement of female hormones in postmenopausal women. Although there is a strong genetic predisposition to osteoporosis, lifestyle factors such as regular exercise and adequate nutrition, especially before adulthood, are important in the prevention of osteoporosis later in life.

Exercise and osteoporosis

Mechanical loading is necessary for maintenance of bone density. Bone atrophies (loses mass and density) in the absence of loading. For example, it is well known that bone loss occurs in astronauts spending long periods in space and in patients after prolonged bed rest. Weight-bearing activity and muscular contractions that increase mechanical loading of bone increase bone density. Bone density is highest in athletes who participate in weight-bearing sports such as running and in activities inducing muscular hypertrophy such as weight lifting.

Because bone density can be influenced by load-bearing exercise, much attention has been focused on whether physical activity can prevent or counteract the loss of bone with ageing. Achieving a high bone mass early in life (during adolescence) is of primary importance in preventing osteoporosis later in life. It is recommended that young women especially aim to achieve and maintain optimal bone mass (the 'bone bank' notion introduced in Chapter 5) through a combination of adequate intake of calcium, a well-balanced diet, and regular weight-bearing exercise as recommended for good health (see Chapter 13).

Although research is not yet conclusive, it appears that, even in older individuals, bone density may be increased by regular moderate exercise that includes load-bearing and muscle-building exercises such as walking and low-intensity resistance training (Box 14.2).

Exercise prescription to prevent osteoporosis

The most effective exercise program for increasing bone density and preventing osteoporosis is yet to be determined. However, research suggests that the general recommendations for exercise for health (see Chapter 13) also apply to prevention of osteoporosis. Exercise programs should include aerobic-type training using load-bearing exercises such as walking, jogging and aerobic dance, in combination with low-intensity resistance training such as circuit training on weight machines or with light, free weights. There is some indication that moderate exercise (less than 70 per cent maximum heart rate) is as effective as strenuous exercise in maintaining bone density. Exercises to improve motor skills such as balance and reaction time may also help to prevent bone fractures by reducing the risk of falls in the elderly.

Arthritis

Arthritis represents a variety of conditions characterised by inflammation, pain and reduced mobility in the joints of the body. Although difficult to quantify, it has been estimated that more than 50 per cent of the general population may suffer from arthritis in its various forms at some time. The two most common forms are osteoarthritis and rheumatoid arthritis.

Osteoarthritis, the most common form, is characterised by progressive degradation of cartilage in the joint, leading to local inflammation and pain. The incidence of osteoarthritis increases with age, so that nearly 85 per cent of individuals between the ages of 70 and 79 years have some form of osteoarthritis. Rheumatoid arthritis, the second most prevalent form of arthritis, is an autoimmune disease, in which the body's immune system mounts an inappropriate attack on its own cells, mainly in the joints. Migration of immune cells to the joints and the substances these cells produce cause inflammation and pain within the joints, which may also affect nearby tendons and muscles. Rheumatoid arthritis occurs more frequently in women than in men.

Exercise and arthritis

It has long been thought that sports participation, especially in load-bearing activities such as jogging, increases the incidence of osteoarthritis. However, recent well-controlled studies show that, in the absence of joint injury, the incidence of osteoarthritis is no higher in athletes such as runners or joggers than in non-athletes. Osteoarthritis may result from any injury to the joint, including sports-related injuries; athletes with a history of joint injury, such as football and basketball players, tend to have a higher incidence of osteoarthritis later in life. It is the joint injury itself, and not physical activity, that increases the risk of osteoarthritis.

Exercise does not appear to influence the incidence or development of rheumatoid arthritis. However, recent research has shown beneficial effects of moderate exercise on physical work capacity and muscular strength, without increasing joint pain, in patients with rheumatoid arthritis.

Exercise prescription in arthritis

Regular moderate exercise is recommended for individuals with either form of arthritis. There are three main types of exercise commonly prescribed for the arthritic individual: mobility exercises to maintain or increase range of motion around the joints, muscle strengthening exercises to increase the weight-bearing capacity of joints, and general fitness exercises to increase cardiovascular endurance for prevention of conditions such as heart disease and osteoporosis.

The general guidelines for exercise for good health, as outlined in Chapter 13, apply to the individual with arthritis. The exercise program may need to be modified, taking into consideration the specific joints involved and a generally low initial fitness level. Range of motion exercises should be performed daily. Walking, cycling and water-based activities such as 'aqua-aerobics' and swimming are recommended as low-impact activities that will not stress the joints. Interval exercise training is often tolerated better than continuous training, especially during the initial stages of a program.

Infectious illness in sport

Are athletes susceptible to frequent illness?

There is a general perception among athletes, coaches and team physicians that athletes are susceptible to frequent illness, especially upper respiratory tract infection (URTI; such as colds and sore throats). Athletes perceive they are especially susceptible during times of stress, such as periods of heavy training and during major competition. Newspapers frequently report on top athletes who have not performed as expected during major competition due to viral illnesses such as the common cold or influenza.

There are several recent research papers on distance runners that confirm that some athletes are more susceptible to URTI as a result of training and competition. Figure 15.2 shows data from one study on runners looking at the incidence of URTI over an entire year. The incidence of infection increased with average weekly training distance, such that runners averaging more than 27 km/week were three times more likely to exhibit URTI than those running less than 14.5 km/week. Other studies have suggested that the higher exercise intensity is during training and competition the higher the incidence of URTI among endurance athletes.

In contrast to the high incidence of illness among endurance athletes, it is believed that a moderate amount of exercise as recommended for good health may actually protect against URTI. It has been suggested that, in the average person, regular moderate exercise may stimulate the immune system, whereas very intense exercise coupled with the psychological stress of competition experienced by elite athletes may compromise immunity to infection. Fortunately, URTI is a relatively minor illness, and appears to be the only illness to which athletes are more susceptible.

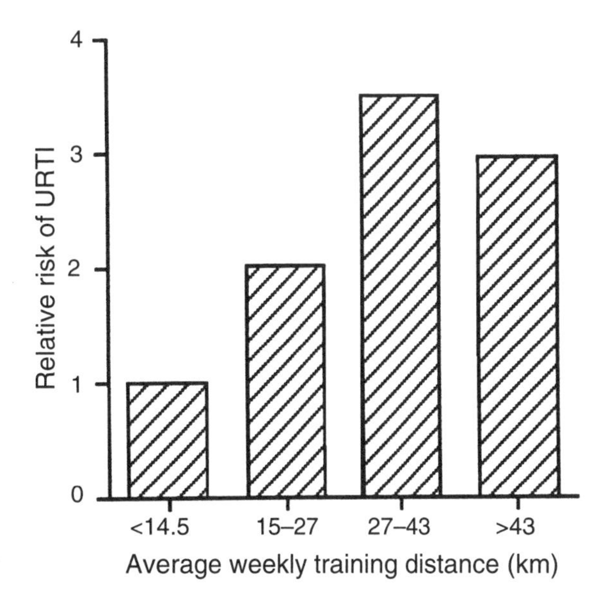

Figure 15.2 Annual incidence of upper respiratory tract infection (URTI) in male runners. Incidence of URTI increased with average weekly running distance. Runners averaging more than 27 km per week were three times as likely to develop URTI than those averaging less than 14.5 km per week
Source: Heath, G.W., Ford, E.S., Craven, T.E., Macera, C.A., Jaackson, K.L. & Pate, R.R. (1991), Exercise and the incidence of upper respiratory tract infection, *Medicine and Science in Sports and Exercise*, 23: 152–57, 155. (Copyright 1991 by American College of Sports Medicine; adapted by permission.)

Further reading

Historical perspectives

Berryman, J.W. and Park, R.J. (eds) (1992), *Sport and Exercise Science: Essays in the History of Sports Medicine*, University of Illinois Press, Urbana and Chicago, Illinois.

Chapters 12 & 13: Basic concepts of exercise metabolism and physiological adaptations to training

Costill, D.L. & Hargreaves, M. (1992), Carbohydrate nutrition and fatigue, *Sports Medicine*, 13, 86–92.
McArdle, W.D., Katch, F.I. and Katch, V.L. (1991), *Exercise Physiology: Energy, Nutrition, and Human Performance* (3rd edn), Lea & Febiger, Philadelphia
Rushall, B.S. & Pyke, F.S. (1990), *Training for Fitness and Sport*, Macmillan, Sydney.
Wilmore, J.H. and Costill, D.L. (1994), *Physiology of Sport and Exercise*, Human Kinetics, Champaign, Illinois.
Wilson, G. (1992), Strength training for sport, *State of the Art Review* Number 29, Australian Sports Commission, Canberra.

Chapter 14: Changes in physiological capacity and performance throughout the lifespan

Malina, R.M. & Bouchard, C. (1991), *Growth, Maturation and Physical Activity*, Human Kinetics, Champaign, Illinois.
Rowland, T.W. (1990), *Exercise and Children's Health*, Human Kinetics, Champaign, Illinois.
Skinner, J.S. (ed.) (1993), *Exercise Testing and Exercise Prescription for Special Cases* (2nd edn), Lea & Febiger, Philadelphia.
Wells, C.L. (1992), *Women, Sport and Performance: A Physiological Perspective* (2nd edn), Human Kinetics, Champaign, Illinois.

Chapter 15: Applications of exercise physiology to health

Pollock, M.L. & Wilmore, J.H. (1990), *Exercise in Health and Disease: Evaluation and Prescription for Prevention and Rehabilitation* (2nd edn), WB Saunders, Philadelphia.
Skinner, J.S. (ed.) (1993), *Exercise Testing and Exercise Prescription for Special Cases* (2nd edn), Lea & Febiger, Philadelphia.

Section 4 Neural bases of human movement: The subdiscipline of motor control

Introduction

In the preceding sections on functional anatomy, biomechanics and exercise physiology we have gained some understanding of the material structure, design and the energetics of the human machinery for movement. To take our analogy of the human body with a motor vehicle one step further, this section of the book, in introducing the subdiscipline of motor control, examines the control mechanisms we have for movement. Understanding the control of human movement requires an understanding of the functions of the brain and nervous system in a way roughly comparable to studying automotive electronics in order to understand the control systems within a motor vehicle. The analogy of the neural system with an electrical system is a powerful one, and one that has, in a general way, guided much theorising in the motor control field.

What is motor control?

Motor control is that subdiscipline of human movement studies concerned with understanding the processes that underlie the acquisition, performance and retention of motor skills. Motor development and motor learning are specialised areas of focus within the subdiscipline of motor control. *Motor development* (Chapter 18) deals with motor control changes throughout the lifespan, specifically those changes in the acquisition, performance and retention of motor skills that occur with growth, development, maturation and ageing. *Motor learning* (Chapter 19) deals with motor control changes that occur as a consequence of practice (or adaptation), focusing literally on how motor skills are learned and the changes in performance, retention and control mechanisms that accompany skill acquisition. Over the years a host of other terms such as motor behaviour, psychology of motor behaviour, motor learning and control, and perceptual–motor skill acquisition, have also been used to describe the subdiscipline, or parts of it, but the use of these terms will be avoided here as they only act to confuse rather than clarify the subdiscipline scope and structure.

The word 'motor' in the terms 'motor control', 'motor development', 'motor learning' and 'motor skills' literally means movement. Motor skills are those goal-directed actions that require movement of the whole body, limb or muscle in order to be successfully performed. Consequently the motor control subdiscipline has a broad focus and range of application, from the study of movements as simple as unidirectional finger or eye movements to ones as complex as those involved in fundamental actions such as walking, running, reaching, grasping and speaking; in workplace tasks such as welding, typing and driving; in artistic tasks such as dancing and playing musical instruments; and in sporting tasks such as performing a complex gymnastics routine or hitting a fast moving tennis ball. Despite their obvious diversity, all of the motor skills used in these tasks share in common their purposefulness, their voluntary nature, their dependence to some degree or other on learning and the fact that the quality of task performance is directly dependent on the quality of the movement produced.

Typical issues and problems addressed

One tangible way of gaining a feel for the breadth of the motor control subdiscipline is to list a small subset of the many questions currently under examination by motor control researchers. Discussion of some of these questions is made in the four chapters within this section.

- *How are skilled movements remembered?*
  For example:
  - Why is it we can remember how to ride a bicycle or how to swim after many years without practising the skill?
  - What elements of movement are stored in memory?
  - How can memory for movement be enhanced?
  - Do skilled performers have better memories for movement?
- *What is the most effective set of conditions for learning a new motor skill?*
  For example:
  - What type of feedback information is best for learning?
  - Can we learn when we are fatigued?
  - Should practice emphasise consistency or variability?
  - Does learning one skill aid in the learning of another related skill?
- *How do skilled performers succeed where lesser skilled performers fail?*
  For example:
  - Do experts have faster reaction times than lesser skilled?
  - Can skilled performers be identified at an early age?
  - How do expert game players manage to 'read the play'?
  - What does it mean when skills become automatic?
- *How is movement control affected by fatigue, injury and disability?*
  For example:
  - Why do stroke sufferers have difficulty with speech and gait?
  - What causes clumsiness in some children?
  - What is the best way to recover movement control in an injured joint?
  - How and why does alcohol affect movement control?

Issues addressed by motor control research scientists can clearly range from very basic questions (such as those related to how the nervous system implements the neuromuscular changes needed to make a movement purposefully faster, more forceful or with altered sequencing or timing) through to very applied questions (such as those related to optimising performance or the rate of skill learning or re-learning).

Levels of analysis within motor control

The single greatest difficulty for scientists attempting to understand how the control of movement occurs is that the processes in the brain and central nervous system that control movement (and which are modified with maturation and adaptation) are not directly observable. Knowledge about motor control is gained indirectly from inferences about control mechanisms derived from the observation, description and measurement of observable movement performed under a variety of carefully selected experimental conditions. The more different levels of analysis of the motor system are undertaken, the greater is the certainty with which inferences about motor control can be made.

Figure 1.1, which schematically represents knowledge organisation within the discipline of human movement studies, indicates that the subdiscipline of motor control draws methods, theories and paradigms from a range of cognate disciplines including physiology, mathematics, computer science, psychology and education. Of these influences the most powerful ones in both an historical and contemporary sense are the influences from physiology, especially neurophysiology, and psychology, especially experimental psychology. Indeed, within the modern field of motor control, it is still possible to identify separate, yet complementary neurophysiological and psychological approaches to the examination of movement control, development and learning.

The *neurophysiological approach* to motor control focuses on understanding the functioning of the components of the neuromuscular system, especially the functional properties of the movement receptors, the nerve pathways, the spinal cord and the brain. Its aim, in a crude sense, is to describe the basic 'wiring', interconnections and organisations of the neuromuscular system in order to understand the neural architecture or, to use a computer analogy, the 'hardware' of the motor system. Physiologists studying the motor system use a variety of methods including the tracing of nerve connections (using histological procedures such as staining or labelling where neurotransmitters are traced chemically), the measurement of metabolic activity in specific areas of the brain (by radioactively labelling a metabolic source such as glucose), the examination of the functional anatomy of the brain using modern imaging techniques (such as computerised tomography or (CT scan); magnetic resonance imaging (MRI); and positron emission tomography (PET)), the measurement of electrical activity in the brain (via electroencephalography (EEG)) and in muscle (via electromography (EMG)), the evaluation of the behavioural effects of brain damage or selective lesions in the neural system and the observation of the behavioural effects of selective electrical, magnetic or chemical stimulation of given nerve pathways in the brain. The nature of many of these techniques is that they can only be applied to anaesthetised animals or, in the case of examining the impact of brain damage, to patient populations, although improving technology, especially in terms of brain imaging techniques, allows increasing recordings and observations from the living, undamaged human brain. Basic information about motor control derived from neurophysiological studies is presented in Chapter 16.

The *psychological approach* to motor control focuses not so much on the physical structure of the components of the neuromuscular system but rather aims to develop conceptual models to describe and explain the collective behaviour of the some 100 000 000 000 000 (10^{14}) neurons and neural connections that together compose the motor system. The favoured approach of psychologists examining movement control is to use experimental tasks with altered movement demands to test their conceptual models. The validity of the conceptual models of the perceiving, deciding and acting stages involved in movement control are evaluated from measures of both movement outcome and pattern. The usual movement outcome (or product) measures are those of movement speed and/or accuracy whereas movement patterns are typically described using one or more of the various measures of kinematics, kinetics and electromyography as described earlier for the subdisciplines of biomechanics and functional anatomy. In recent times experimental psychologists have teamed up with computer scientists in an attempt to develop powerful computational models and simulations of the functioning of the neuromuscular system. Some of the basic findings concerning motor control derived from an experimental psychology perspective are presented in Chapter 17.

Historical perspectives

The neurophysiological and psychological approaches to motor control have quite independent histories. A number of important findings with respect to the neurophysiological basis of movement and movement control were made throughout the 1800s. Foundations for a neurophysiological basis of movement control can be found in Charles Bell's 1830 text *The Nervous System of the Human Body* in which the motor function of the ventral roots of the spinal cord and the sensory function of the dorsal roots were described and later, in the studies from 1856 to 1866 of Hermann von Helmholtz, in which nerve conduction velocity was estimated and studies of reaction time were begun. Subsequently the spring-like characteristics of muscle were described (by Weber in 1846), the electrical excitability of the brain was discovered (by Fritsch & Hitzig in 1870) and studies of the sensory and motor functions of the brain were started (by Beevar & Horsely in 1887).

Undoubtedly the major historical contributor to the understanding of the neurophysiology of the motor system was Sir Charles Sherrington (1857–1952) whose work on reflexes is still widely credited today. From Sherrington's work, especially his classic 1906 text entitled *The Integrative Action of the Nervous System*, arose a number of key concepts such as synaptic transmission, reciprocal inhibition, final common pathway and proprioception, which are cornerstones of modern neurophysiology. Another influential figure in modern motor control theories was the Russian physiologist Nicolai Bernstein (1897–1966), whose integrative work on phase relations, functional synergies and distributed control in natural actions such as locomotion only appeared posthumously in the English-language literature. A wealth of knowledge on the neurophysiology of the motor system appeared in the post-war era when improved electrophysiological recording techniques and neurophysiological mapping techniques facilitated new precision and insight into the structure and function of many levels of the neuro-motor system. Of particular note in this period was the work of the Canadian neurosurgeon Wilder Penfield who, together with Rasmussen, used electrical stimulation studies to map the topographical organisation of the motor and sensory cortices of the brain.

The most noteworthy and influential early studies of movement from a psychological perspective were the basic studies on the control of arm movements by Woodworth in 1899 and the applied studies on the skill acquisition of Morse code operators by Bryan and Harter published in 1897 and 1899. Little motor control research of any kind took place in the first part of the 20th century and when motor control research reappeared in the 1930s and 1940s it was very much oriented toward solving practical problems associated with specific motor tasks. Practical concerns that were highlighted in this period included personnel selection and training for war tasks such as flying, steering and weaponry skills plus optimal approaches to teaching and coaching in physical education and sport. An identifiable research field of movement control within psychology, aimed at understanding fundamental processes in movement control, did not emerge until the 1950s. This followed the major conceptual advance in experimental psychology in the late 1940s, triggered by Craik (1947, 1948), Wiener (1948) and Shannon and Weaver (1949), which viewed the brain and nervous system as processors of information in a manner akin to sophisticated, high-speed computers. The 1960s and 1970s research studies in motor control were consequently dominated by information-processing models of motor control and attempts by psychologists such as Paul Fitts to quantify the information-processing capabilities of the motor system. The theoretical contributions of the psychologists Franklin Henry (see again Box 2.2), Jack Adams, Steven Keele and Richard Schmidt were particularly prominent in this period. Schmidt founded the *Journal of Motor Behavior* in 1969, giving the subdiscipline its first specialist journal. With the passage of time the historical distinctions between the neurophysiological and psychological approaches to movement control have become less pronounced. The two approaches are complementary, one providing knowledge of the structure of the neuromuscular system, the other theories of its collective behaviour and organisation, and increasingly synthesis is sought between the two approaches.

Professional organisations and training

People interested in motor control come from a diversity of backgrounds. At any symposium or conference on motor control it is not unusual to find psychologists, neurophysiologists, computer scientists, engineers, neurologists, and therapists, in addition to researchers with backgrounds in human movement studies. It is perhaps not surprising therefore to find no single professional subgroup representing motor control internationally. The major meetings for scientists interested in motor control take place under the umbrellas of international groups such as the International Society for Event Perception and Action (ICEPA) and strong national groups such as, in North America, the North American Society for the Psychology of Sport and Physical Activity (NASPSPA), the Canadian Society for Psychomotor Learning and Sport Psychology (SCAPS), the Society for Neuroscience, and the Psychonomic Society; and, in Europe, L'Association des Chercheurs en Activités Physiques and Sportives (ACAPS). Increasingly motor control scientists are also attending conferences in biomechanics, such as those hosted by the International Society of Biomechanics. In line with its eclectic nature, motor control research is published in a diversity of journals including specialist journals such as *Human Movement Science, Journal of Motor Behavior,* and *Motor Control*; neuroscience journals such as the *Journal of Neurophysiology, Experimental Brain Research,* and *Brain and Behavioral Sciences*; and experimental psychology journals such as *Journal of Experimental Psychology: Human Perception and Performance, Quarterly Journal of Experimental Psychology,* and *Acta Psychologica.*

CHAPTER 16

Basic concepts of motor control: Neurophysiological perspectives

- The nervous system as an elaborate communications network
- Components of the nervous system
- Neurons and synapses as the building blocks of the nervous system
- Sensory receptor systems for movement
- Effector systems for movement
- Motor control functions of the spinal cord
- Motor control functions of the brain
- Integrative neural mechanisms for movement control

Humans are capable of providing movements that are truly incredible in their diversity, complexity, precision and adaptability. To support such an array of movement capability we need a neuromuscular system that is highly organised yet shares the flexible, adaptable and complex properties of movement itself. The purpose of this chapter is to:

- describe the basic components of the neural system and their functions
- examine the structure and function of nerve cells and their interconnections as the basic building blocks of the nervous system
- overview the major receptor systems that provide the sensory or perceptual information needed for the control of movement
- outline the nerve pathways within the spinal cord and the motor control functions and capabilities of the spinal cord
- outline the motor control functions of the higher centres of the central nervous system, especially the brain
- briefly consider some of the major disorders of movement and their neurophysiological causes.

The nervous system as an elaborate communications network

The nervous (or neural) system is designed in some ways like a modern electronic communication network such as the telephone system. It has receivers (receptors), which pick-up important signals; muscles (as effectors), which are able, when instructed, to bring about planned actions; and a vast array of electronic cables (neurons) and interconnections, which allow for nearly infinite linking of receptors to effectors and therefore permit information flow from one region of the network to another. The routine operations of the communication system are achieved at a local level (the spinal level) whereas the overriding authority (the brain) makes executive and policy decisions as to what tasks the communication network should attempt to achieve. The 'language' of the communication system is the coded electronic bursts (nerve impulses) that travel throughout the various links in the network.

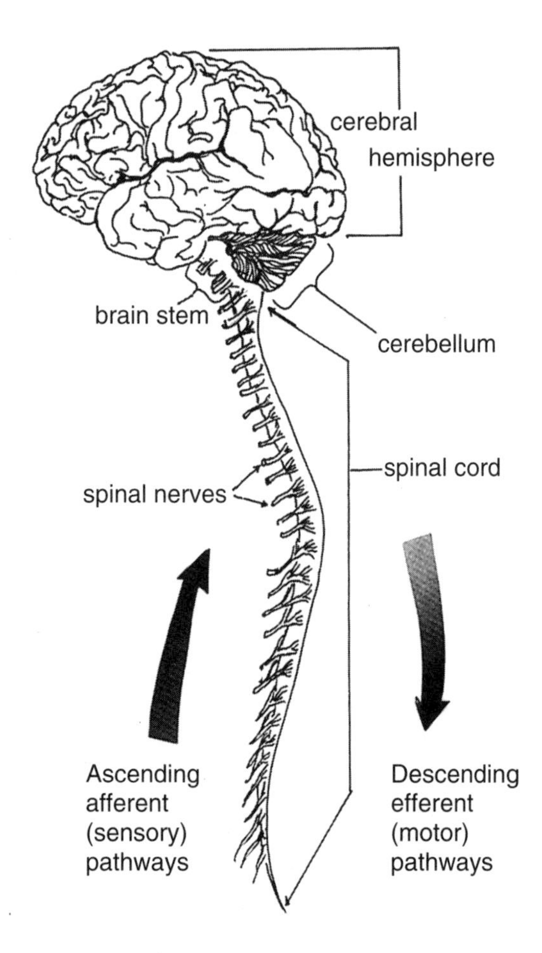

Figure 16.1 The central nervous system consisting of the brain and the spinal chord with its ascending sensory and descending motor pathways

The communication system as a whole is never static as new connections are constantly being made, damaged connections replaced and repaired and frequently used connections expanded and upgraded to improve their capacity and rate of transmission. In the nervous system, as in the communication network, this capacity for constant change (what is termed *plasticity*) is essential for accommodating growth, development and adaptation.

Just as we are able to use systems such as the telephone effectively without direct knowledge, or in many cases any understanding, of the structure and function of its component parts, as humans we are able to use our neuromuscular systems to produce all manner of skilled movements with a minimum of awareness of either the neuromuscular system's component parts or its method(s) of operation.

Components of the nervous system

As the analogy with a communications network suggests, the main physical components of the nervous system are the sensory receptors, the motor units and the nerves (neurons) and their junctions (the synapses), which permit communication between the sensory receptors, the motor units and other neurons. There are in all some 10^{12} to 10^{14} neurons within the human nervous system, *each* of which may have as many as 10^4 synaptic connections with other neurons, receptors or motor units. As a collective network the nervous system has two major subdivisions. The central nervous system (Figure 16.1) consists of the brain and the spinal cord and is responsible for overseeing and monitoring the activation of all sectors of the body, including the muscles. The peripheral nervous system carries information from the sensory receptors to the central nervous system and commands from the central nervous system to the muscles. The sensory information from the receptors reach the brain via ascending afferent (from the Latin *afferere* meaning 'to carry toward') pathways whereas the commands from the brain to the muscles are carried via descending efferent (from the Latin *efferere* meaning 'to carry away') pathways.

The signals from the brain may act to either excite or inhibit those neurons (the motor neurons) that synapse directly with muscle fibres within muscle. Complex connections between neurons within the central nervous system provide for specialised control of movement and for the storage of information essential for memory and learning. In the sections that follow, we look in greater detail at the main components of the nervous system, and their role in the control of movement.

Neurons and synapses as the building blocks of the nervous system

Structure and function of neurons

The neuron or nerve cell is the basic component of the neuromuscular system and provides the means of receiving and sending messages (or information) throughout the entire system. Although neurons vary substantially in both size and shape, depending on their specific function and location within the nervous system, most neurons share a similar structure in terms of having a cell body to which are connected a single axon and (typically) many dendrites (Figure 16.2). The cell body, containing the nucleus, regulates the homeostasis of the neuron. The dendrites, collectively formed into a dendritic tree, connect with and receive information from other neurons and, in some cases, sensory receptors. The axon is responsible for sending information away from the neuron to other neurons. Collateral branches off the main

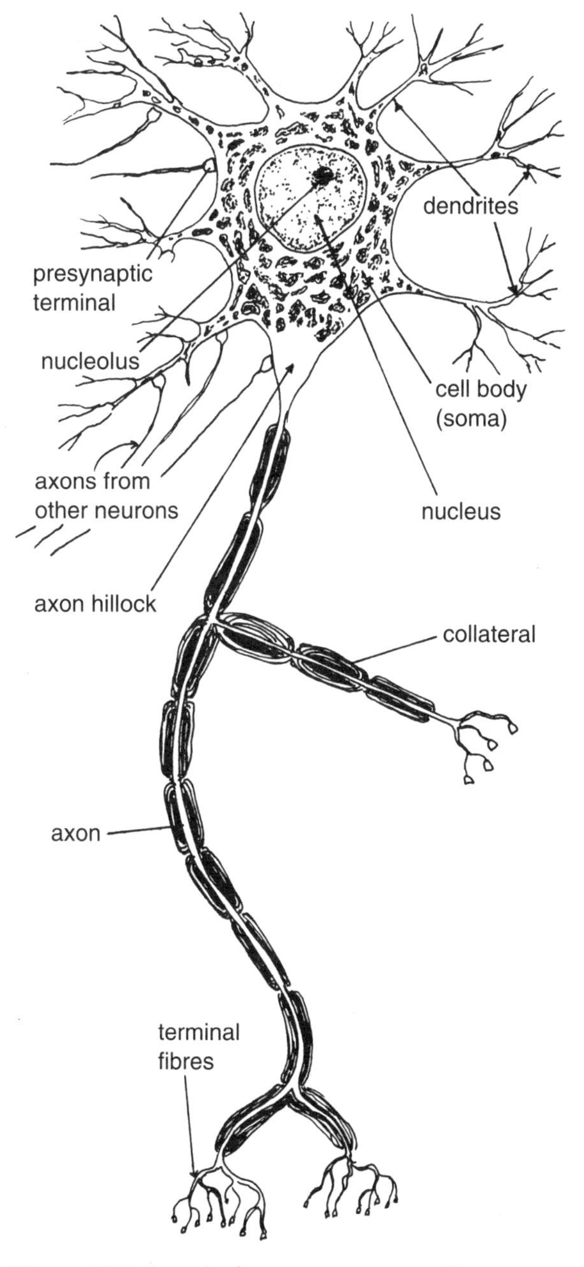

Figure 16.2 A typical neuron or nerve cell

axon permit communication of the nerve impulses from any particular neuron to more than one target neuron. Any single neuron can influence the activity of up to 1000 other neurons and is itself influenced by the excitatory and inhibitory impacts of some 1000–10 000 other neurons.

There are a number of different types of neurons with, in each case, their structure being dictated by their function (Figure 16.3). The sensory or afferent neurons are relatively linear in structure with a single axon connecting the sensory receptor ends to the cell body. The motor or efferent neurons vary in structure according to their location. The alpha motor neurons of the spinal cord possess many dendritic branches and a relatively long axon, also heavily branched to innervate multiple (100–15 000) skeletal muscle fibres. The gamma motor neurons, as we shall see in the next section, innervate contractile (intrafusal) fibres located within the muscle receptors. Consequently gamma motor neurons, which constitute about 40 per cent of the total motor neurons within the spinal cord, have smaller, considerably less branched, axons than do alpha motor neurons.

The cell bodies of both the alpha and gamma motor neurons are located within the spinal cord. The pyramidal cells, located within the motor cortex of the brain, are so-named because of their shape, which derives from their branching tree of dendrites all of which funnel down to a single slender axon. Pyramidal cells send motor commands over the long distances from the brain to the spinal cord and may have axons up to 1 m in length. The Purkinje cells

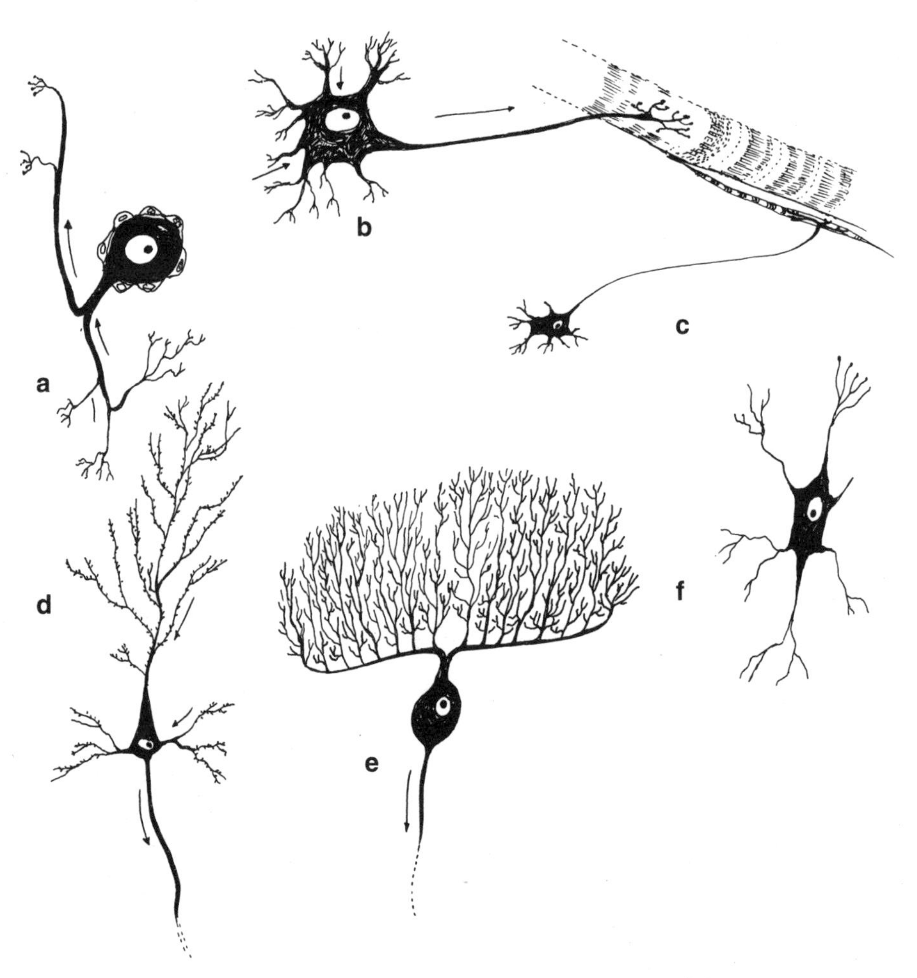

Figure 16.3 Functional types of neurons: sensory neuron (a); alpha motor neuron (b); gamma motor neuron (c); pyramidal cell neuron (d); Purkinje cell neuron (e); and interneuron (f)

within the cerebellum also have a single thin axon to which information is sent from an incredibly rich, systematically organised set of dendrites that provide these neurons with a characteristic tree-like appearance.

Interneurons are of a variety of shapes but typically have multiple dendrites and branching axons that permit the connection of multiple neurons with multiple neurons. The structure of interneurons and their connections facilitates both the convergence of multiple input messages onto a single output cell or set of cells and the divergence of a single input message to a number of different motor neurons. Interneurons originate and terminate within either the brain or spinal cord. All neurons within the central nervous system are surrounded by and outnumbered by other cells called *glia* or *glial cells*, which provide, among other things, the metabolic and immunological support for the neurons.

Neurons carry messages from their dendrites to the terminal fibres of their axons through a series of electrical pulses, produced in the axon hillock (see again Figure 16.2). The electrical pulse produced by the axon hillock is dependent on the spatial and temporal distribution of the pulses impinging on the cell body from its dendritic tree. Signals arriving early and originating from dendrites close to the axon hillock carry more weight than signals from distant neurons arriving late. If the summed weight of the impulses reaching the neuron exceeds its threshold voltage the axon hillock triggers a pulse (the cell 'fires') and this pulse is propagated along the axon to its terminus. The rate at which it is transmitted varies, being greatest in axons of large diameter, which are insulated by the fatty substance, myelin. Each neuron is therefore more than simply a conductor of electrical signals; it constantly undertakes complex processing of the input signals it receives from other sources.

Structure and function of synapses

The passage of information from one neuron to another is via the synapses (which are in many ways the equivalent in the nervous system of the joints in the musculoskeletal system). Synapse is a term coined by Sir Charles Sherrington and derives its origin from a Greek word meaning 'union'. At the synapse the axon of one neuron comes in close proximity, but not direct physical contact, with the receptor surfaces of one or more other (postsynaptic) nerve cells (Figure 16.4).

The electrical activity in the presynaptic neuron is transmitted across the gap (the synaptic cleft or junction) to the postsynaptic neuron via either the direct spread of electrical current or, more frequently, by the action of a chemical mediator, called a neurotransmitter. In the case of chemical transmission the nerve impulse in the axon of the presynaptic neuron triggers the release of a neurotransmitter from tiny storage sacs (vesicles) within the presynaptic membrane into the synaptic cleft. Specialised receptors on the membrane of the postsynaptic neuron detect the presence of the neurotransmitter triggering either a heightened excitatory or inhibitory response in the postsynaptic neuron (depending on whether the synapse is excitatory or inhibitory). The trans-

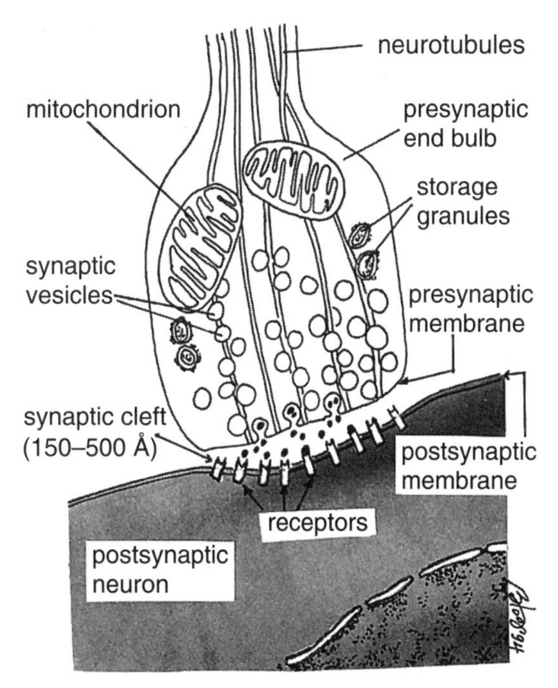

Figure 16.4 The microscopic structure of a typical synapse

mission of information from one neuron to another therefore typically requires a transduction of an electrical signal to a chemical one (at the presynaptic neuron), the diffusion of the chemical transmitter across the synaptic cleft, and then the transduction of the chemical signal back to an electrical one (at the postsynaptic neuron). There are a number of different neurotransmitters, of which acetylcholine (ACh) (an excitatory neurotransmitter) is the best known.

The option of the synaptic connector being either excitatory or inhibitory provides the foundation for more complex functional connections within the nervous system, such as reciprocal inhibition, which, as we shall see in a subsequent section, form the cornerstone of many reflex activities.

Sensory receptor systems for movement

The main sensory information to guide the selection and control of movement comes from vision and proprioception. Visual information is derived from the light-sensitive sensory receptors located within the retina of the eye. Proprioception (from the Latin *proprius* meaning own) is information about the movement of the body itself and is provided via kinesthetic receptors located in the muscles, tendons, joints and skin, and vestibular receptors for balance located in the inner ear. (The word *kinesthesis* is derived from two Greek words meaning 'to move' and 'sensation'.)

Although the sensory receptors for the many facets of vision and proprioception, as well as the receptors for other senses such as hearing, taste and smell, vary dramatically in their specific structure, all sensory receptors share the common function of transducing physical energy from either beyond the body (such as light or sound waves) or within the body (such as tendon tension) into coded nerve impulses. These nerve impulses can then be transmitted from one part of the body to another via the nervous system or integrated from one sensory system to another. In this regard the sensory receptors are very much like transducers in electronics, converting the information they receive into electrical pulses that can be transmitted along the many neural pathways that exist within the human body. Humans, like all other animals, are only sensitive to a limited range of the physical signals within the environment in which we live. Ultraviolet and infrared wavelengths of light, for example, that we know exist in our surrounding environment, are not perceived by us without the assistance of mechanical devices because these signals fall beyond the range of sensitivity of our visual system.

The visual system

Our rich visual perception of our surrounding environment is achieved through a very complex set of neural processes.

There are four major basic components or stages within human vision:

- Refraction of light rays and the focusing of images onto the sensory receptors of the eye (accomplished through the structural design of the eye).
- Transduction of light energy into nerve impulses (accomplished through photochemical processes in the retina).
- Early neural processing and transmission of impulses to the brain (accomplished through the neural structure of the retina and the optic nerve).
- Processing of the visual information in the brain, resulting in visual perception as we know and experience it (accomplished at a number of higher centres in the brain).

The anatomy of the eyeball is such (Figure 16.5) that light enters the eye through the transparent cornea, is filtered through the action of the iris, which controls the amount of light entering the eye (by adjusting pupil diameter) and is then focused, by the action of the lens, onto the retina at the back of the eyeball. The shape of the lens can be changed through the action of the ciliary muscles to ensure that the image falling on the retina remains in sharp focus. The retina consists of a very thin, but distinctly layered arrangement of

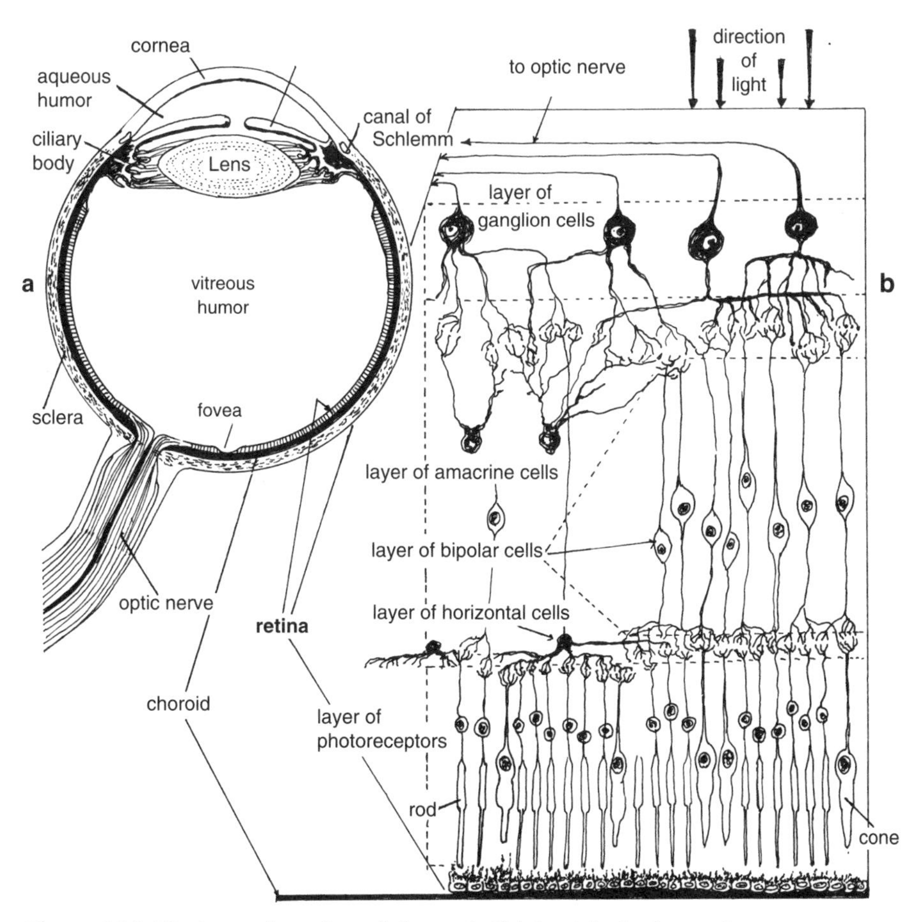

Figure 16.5 Horizontal section of the eyeball (a), with the layered microstructure of the retina (b)

cells lining about 180° of the inner surface of the eye. Light reaching the retina, after first passing through the refracting media of the cornea, the aqueous humor (a weak saline solution), the lens and the vitreous humor (a glassy gel substance), passes through a number of layers of cells to reach the photoreceptors. The photoreceptor cells (the rods and cones) contain chemicals that are sensitive to light and they transmit nerve impulses through their axons to other cells in the retina in a way that corresponds with the light falling on them. The rods are most sensitive to light, do not respond to colour, and are therefore the primary receptors for night vision. The cones, in contrast, require high levels of illumination to function but enable us to have colour vision. There is a higher density of both types of photoreceptors around the fovea, giving this area (corresponding to some 2° of the centre of our visual field) the highest level of sensitivity (acuity).

Nerve impulses arising from the photoreceptors are passed through a number of other layers of interneurons within the retina before being sent to the brain via the optic nerve. The arrangement of nerves in the horizontal, bipolar, amacrine and ganglion cell layers of the retina allows for early processing of the visual signal, especially in terms of averaging signals over a range of photoreceptors plus enhancing contrast between adjacent areas of the visual field (Figure 16.5).

Visual signals from the retina are carried via the optic nerve along two major pathways,

Table 16.1 Key structural and functional distinctions between the focal and ambient visual systems

Attribute	Focal vision	Ambient vision
Retinal location	fovea	full retina
Visual field	central vision	central and peripheral vision
Nerve pathway and terminus	lateral geniculate nucleus → visual cortex	superior colliculi
Principal function	object recognition (what is it?)	spatial orientation (where is it?)
Awareness	conscious	non-conscious
Luminance	performance deteriorates as light levels decrease	performance comparable under high and low light levels
Spatial frequency	high spatial frequencies (minimum blur) necessary for focal discrimination	low spatial frequencies (minimum resolution) sufficient for spatial orientations

Source: Based on information in Leibowitz, H.W. & Post, R.B. (1982), The two modes of processing concept and some implications, (pp. 343–63), in J. Beck (ed.), *Organization and Representation in Perception*, Erlbaum, Hillsdale, N.J.

distinct both in structure and function (See Table 16.1). Some 70 per cent of the connections from the optic nerve are to an area of the mid-brain (called the lateral geniculate nucleus) and from there to the visual cortex, which is located toward the back of the cerebrum (see Figure 16.6 for details). This pathway, contributing to focal vision is specialised for recognising

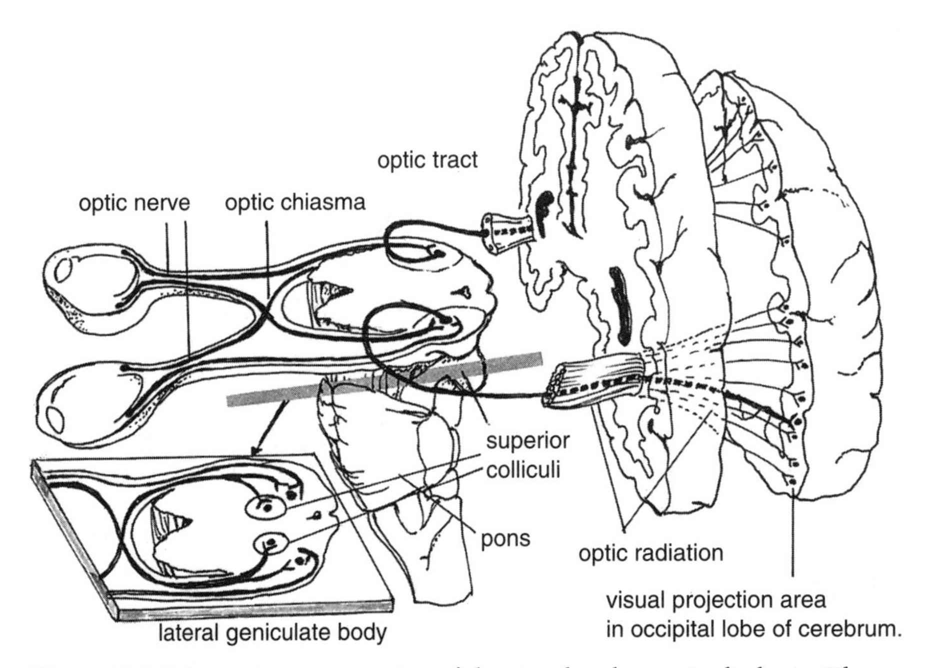

Figure 16.6 Schematic representation of the visual pathways in the brain. The insert on the lower left corner provides a horizontal cross-sectional view of the lateral geniculate body and the superior colliculi

objects, distinguishing detail and assisting in the direct visual control of fine, precise movements (such as those involved in threading a needle). Most of the remaining nerve fibres from the optic nerve terminate in another section of the mid-brain called the superior colliculi. This pathway, contributing to ambient vision, receives information from the whole of the retina, including the peripheral retina, and is concerned with the location of moving objects within the whole visual field. This pathway is especially implicated in providing information about our position in space and our rate of movement through the environment. Damage to the focal vision pathway results in an inability to identify objects but not the ability to locate them. The converse is true of damage to the ambient vision pathway.

The kinesthetic system

In addition to information provided through vision, information about the 'sense of movement' is also derived from specialised receptors located within the muscles, tendons, joints and skin.

Muscle receptors

The principal source of sensory information from skeletal muscle is provided by the muscle spindle. The muscle spindle is unique as a receptor in that it also contains muscle fibres and hence also has movement capabilities. Muscle spindles are located within all skeletal muscles, although they are particularly abundant in small muscles (such as those in the hands) used to control fine voluntary movements. Muscle spindles provide the central nervous system with information about the absolute amount of stretch plus rate of change of stretch in a particular muscle. This, as we shall see in subsequent sections, is invaluable both in the reflex control of movement and in the control and monitoring of voluntary movements.

Understanding the control capabilities of the muscle spindle requires an understanding of its unique anatomy (Figure 16.7). Under normal circumstances the contraction of any given skeletal muscle is achieved by a burst of neural activity from an alpha motor neuron, which causes uniform contraction across the whole length of the large-diameter muscle fibres, called extrafusal fibres. Lying in parallel to the extrafusal fibres, and connected to them at their endpoints, are smaller diameter muscle fibres called intrafusal fibres (which form the basis of the muscle spindles). The intrafusal fibres differ from the extrafusal fibres in a number of important ways, namely:

- they are smaller and, by themselves, are incapable of directly causing whole muscle contraction
- they are not innervated by alpha motor neurons from the local spinal level but are innervated independently by gamma motor neurons whose activity is controlled from descending pathways from the brain
- when innervated they contract only at their endpoints and not uniformly across their whole length
- they have sensory receptors located along them and afferent connections back to the spinal cord.

The sensory information from the muscle spindle comes from two sources: primary endings, located within the non-contractile central portion of the spindle and connected to the central nervous system by type Ia afferent neurons; and secondary endings, located on the contractile end portions of the spindle and connected to the central nervous system by type II afferent neurons. As the primary endings respond to stretch, the Ia afferent neurons send impulses back to the central nervous system under conditions where either: (i) the whole muscle is stretched, or (ii) contraction of the ends of the intrafusal fibres by the gamma motor system is not matched by an equal shortening of the extrafusal fibres, under the control of the alpha motor neuron system.

Tendon receptors

The sensory receptors located within tendons (the attachments of muscles to bones) are known as Golgi tendon organs. These receptors

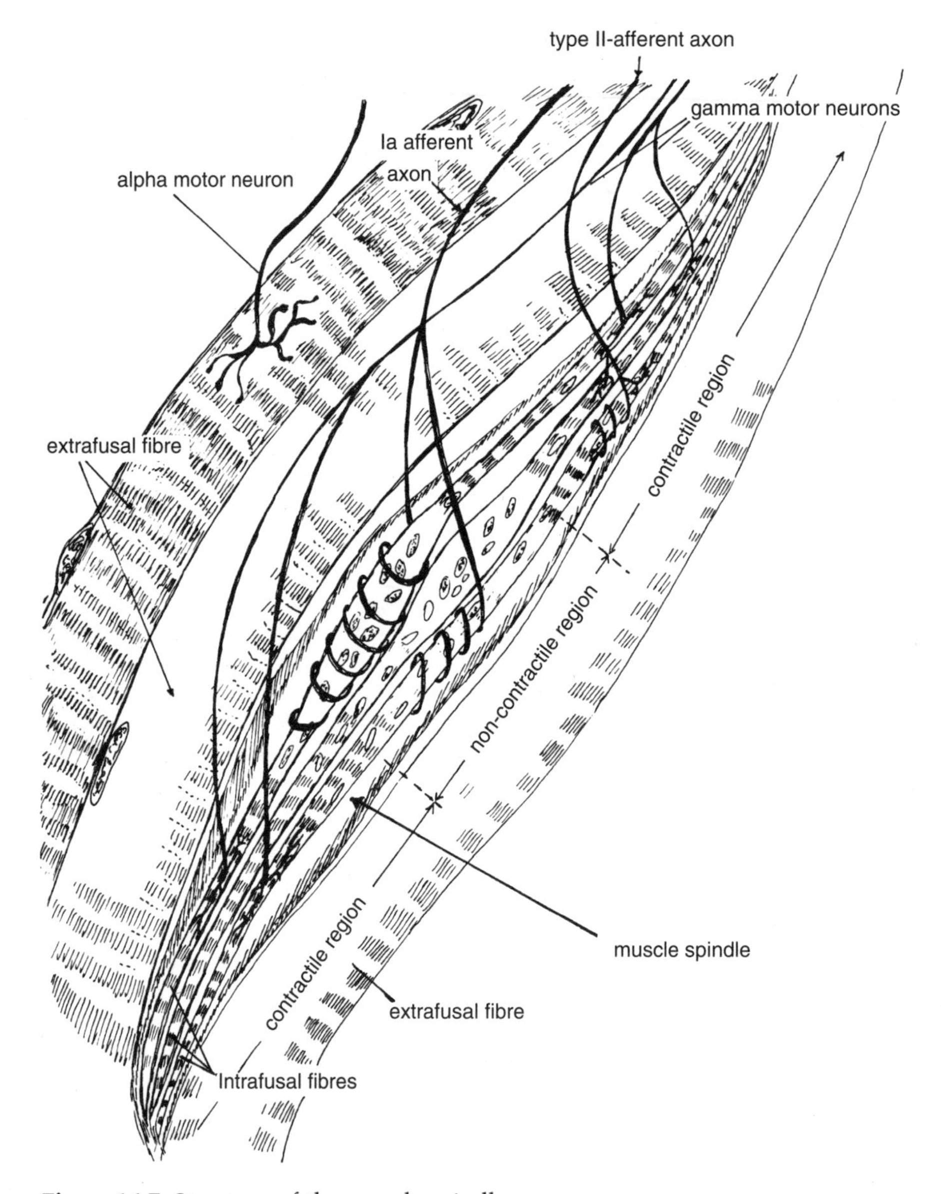

Figure 16.7 Structure of the muscle spindle

lie close to the surface of the musculotendinous tissue and send their impulses back to the spinal cord by type Ib afferent fibres (Figure 16.8). The Golgi tendon organs are sensitive to the amount of tension developed in the tendon. Tendon tension increases when a muscle contracts but decreases when a muscle is relaxed; therefore the Golgi tendon organs act in a manner that counterbalances the action of the muscle spindle. The Golgi tendon organs fire maximally when the muscle spindle is inactive and minimally when the muscle spindle is active.

The Golgi tendon organs appear to serve two major functions with respect to movement control. The first function is a protective one to signal dangerously high tensions in muscle. The Ib afferent neurons are so connected that excessive excitation of the Golgi tendon organs acts to inhibit further muscle contraction (by inhibiting the alpha motor neuron innervating the particular muscle), thus preventing damage

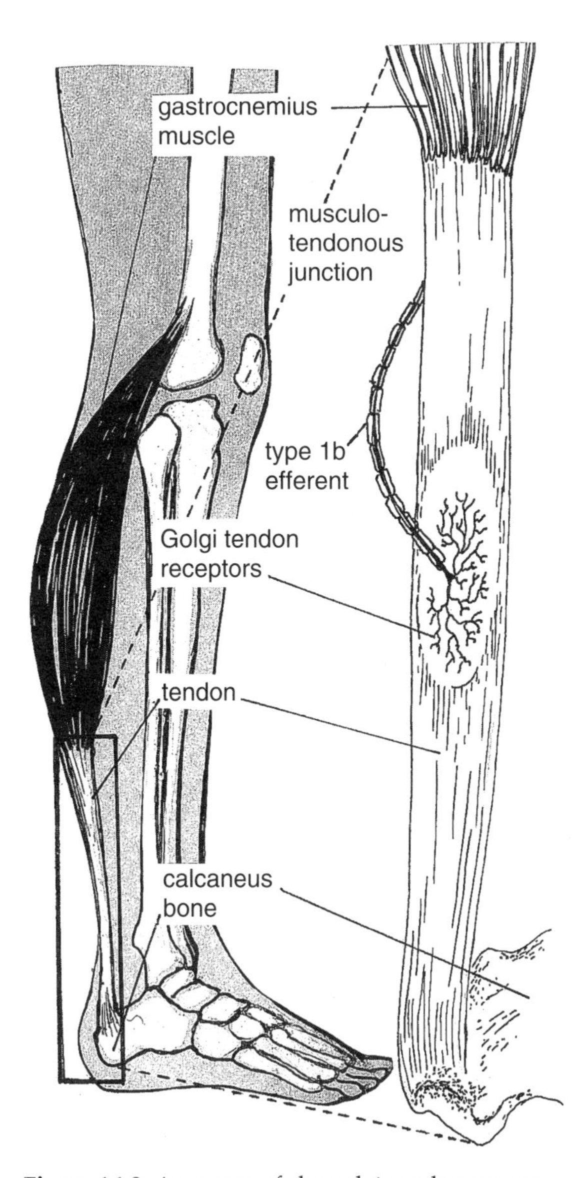

Figure 16.8 Anatomy of the golgi tendon organs

to the musculo-tendinous juncture. In this respect the Golgi tendon organ operates somewhat like a fuse within an electrical circuit. The second function, only recently discovered, is that the Golgi tendon organs also provide sensory feedback to the spinal cord even at low levels of tension, thereby providing fine-tuned feedback information that can potentially assist in continuous ongoing control throughout a movement.

Skin (cutaneous) receptors

The skin is an extremely complex and vital organ of the body and contains several different types of receptors that can provide useful sensory information for the control of movement (Figure 16.9). Detection of the deformation of the surface of the skin caused by movement or weight bearing, for example, may be a valuable source of information for monitoring and controlling voluntary movements. Meissner's corpuscles and Merkel's discs (on hairless parts of the skin such as the palms of the hands or the soles of the feet) and free nerve endings, Ruffini corpuscles and nerve endings wrapped around hair follicles or other parts of the skin all provide sensory information about light touch or low-frequency vibration. Receptors such as the Pacinian corpuscles, located deeper in the skin, respond more to deep compression and high-frequency vibration, especially the onset and offset of such events.

Like other receptors we have examined, the cutaneous receptors are not uniformly distributed throughout the body, being more densely distributed in regions such as the finger tips that are used for fine, precise movements. Cutaneous sensitivity varies throughout the body in relation to the number of receptors per unit area. Cutaneous receptors clearly have an important role in movement control because it is well known that motor performance deteriorates if these receptors are damaged. Patients with damage to the cutaneous receptors in the soles of their feet, for example, experience difficulty in maintaining balance. Likewise, engineers developing robots to perform movement tasks have also discovered that robots lacking touch receptors have great difficulty in performing any tasks requiring fine precision.

Joint receptors

There are three different types of receptors located in the tissues surrounding and composing joints and each of these bears similarity to the kinesthetic receptors located elsewhere in the body. There are modified Ruffini corpuscles and modified Pacinian corpuscles, not dissimilar to those found in the skin, located within the joint capsule itself, and Golgi

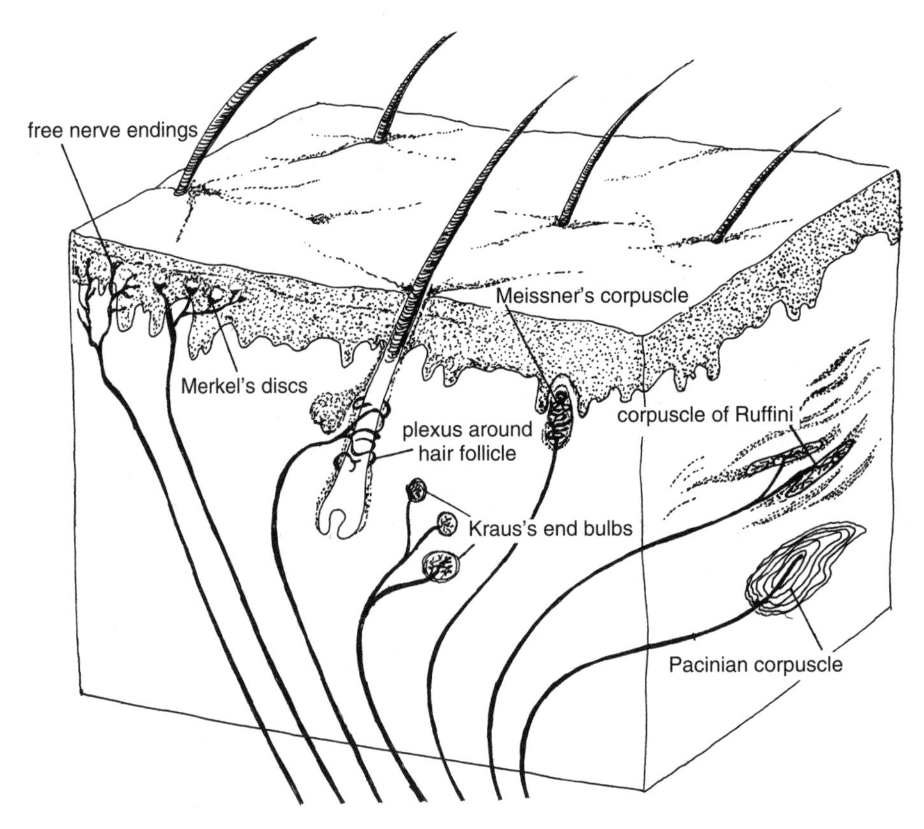

Figure 16.9 Sensory receptors in the skin

organs, not dissimilar to those found in the tendons, located within the ligaments that bind the joint together. Although there is some debate over their function, it appears that the main collective role of the joint receptors is to signal extreme ranges of motion at the joint. The joint receptors are therefore able to play a role in protecting the joint from injury by signalling to the central nervous system when the full range of motion of a joint is being approached.

The vestibular system

Whereas the various kinesthetic receptors provide valuable information about the state of individual muscles, joints and movement segments, the performance of many skilled movements also requires information about the orientation of the whole body in space. This is particularly true of movements in activities such as gymnastics, trampolining or diving. Some of this information about whole-body orientation can be provided by the visual system but much of it is provided by a uniquely designed receptor system (the vestibular apparatus) located adjacent to the inner ear (Figure 16.10).

The vestibular apparatus consists of two different types of receptors: the semicircular canals (the superior, horizontal and posterior canals), which respond to angular acceleration in three different planes, and the otolith organs (the utricle and the saccule), which respond to linear acceleration. Each of the semicircular canals is located at right angles to the other two, allowing for separate information to be sent to the brain on the horizontal, lateral and vertical angular acceleration of the head. In contrast the utricle provides sensory information on the linear horizontal acceleration of the head, and the saccule information on the linear vertical acceleration of the head. The vestibular apparatus is centrally involved in balance such that any dysfunction of the

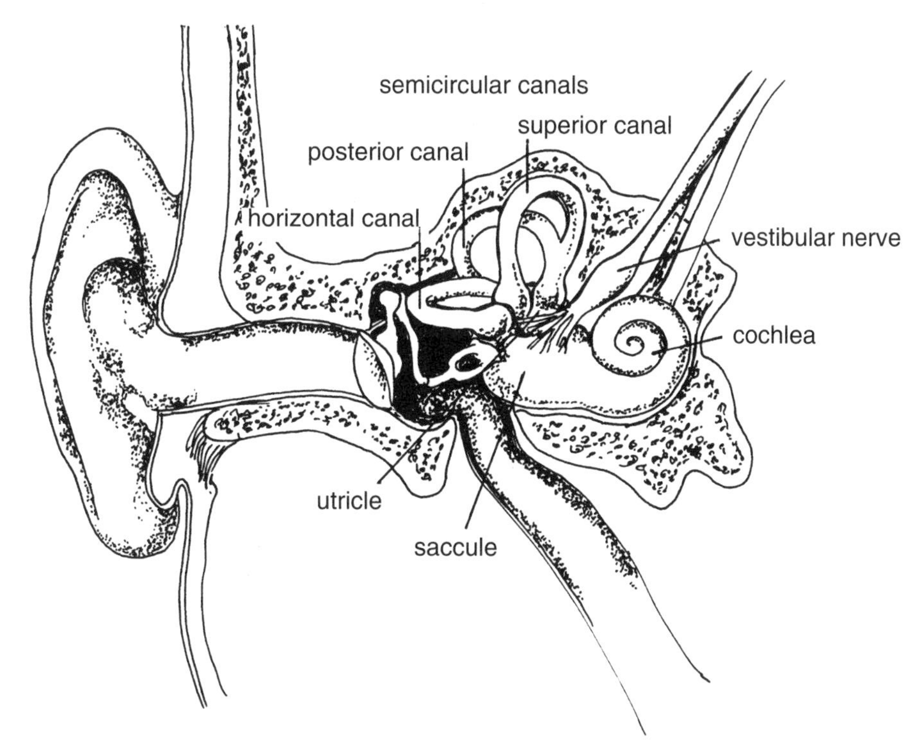

Figure 16.10 Anatomy of the vestibular apparatus

vestibular apparatus, such as occurs with some ear infections, can lead to loss of balance control. The vestibular apparatus is also in close reflex connection with the visual system. Discrepancies between the information provided by the visual and vestibular systems, such as occur in many simulated rides in amusement parks, can give rise to phenomena such as motion sickness.

Inter-sensory integration and sensory dominance

In many cases common environmental events are experienced by a number of different sensory systems in the body and the challenge for the central nervous system is to integrate these different sources of information. In the maintenance of normal upright balance, for example, sensory information needs to be integrated from the visual receptors, the many different kinesthetic receptors and the vestibular system in order to ascertain whether balance is being maintained correctly or being lost. That integration is possible at all is a consequence of the sensory receptors transducing their very different sources of physical stimulation into the common 'language' of nerve impulses and the vast array of neural interconnections and pathways within the brain and central nervous system that allows signals from diverse locations to converge. While the information coming from the different sensory systems is usually in agreement, in some circumstances the information supplied to the central nervous system may be in conflict (for example, the visual system may indicate balance is being lost while the kinesthetic system indicates balance is being retained). The brain and other sections of the central nervous system therefore require some systematic means of resolving this conflict. In humans any inter-sensory conflict is always resolved in favour of vision, which is referred to as the dominant sensory modality. This can, on occasion, lead to misperception and in turn misguided action if the sensations provided by the visual system are inaccurate or misleading (see Box 16.1).

Box 16.1 Visual dominance in balance control

An excellent demonstration of the dominance of vision over information from other sensory systems has been provided by the Edinburgh psychologist David Lee and his colleagues. Lee had his subjects stand upright within (what appeared to be) an enclosed room and gave them the apparently simple task of maintaining their balance so as to keep their head and whole body as still as possible. Under control conditions all the information coming from the visual system as well as from the proprioceptive system would indicate reliably to the subjects that

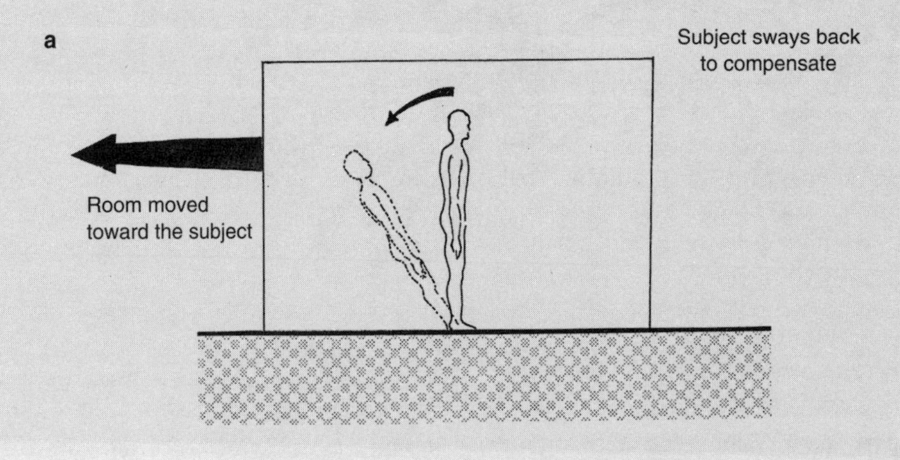

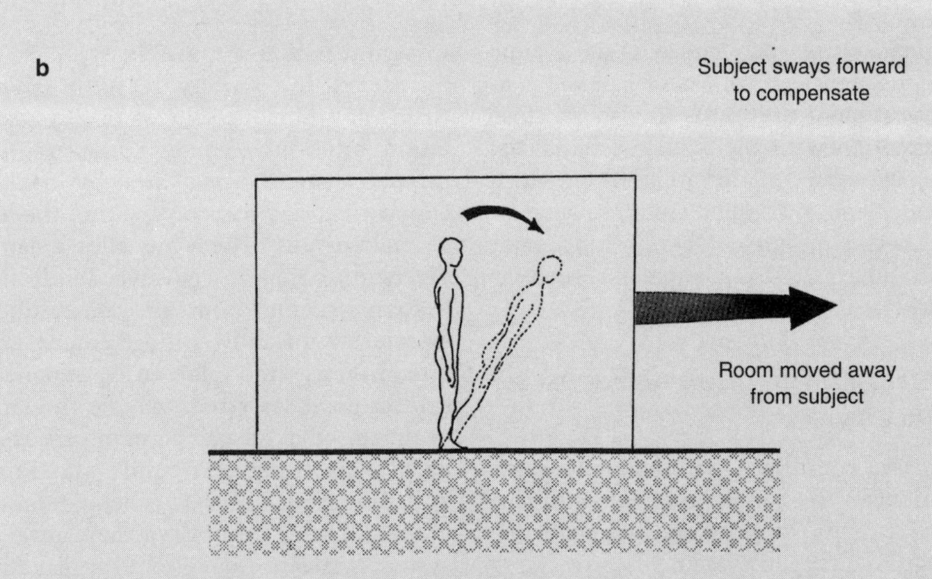

Figure 16.11 Postural adjustments induced by visual movement of room walls towards (a) and away (b) from the standing subject
Source: Lee, D.N. & Thomson, J.A. (1982), Vision in action: The control of locomotion, (pp. 411–33), in D.J. Ingle, M.A. Goodale & R.J.W. Mansfield (eds), *Analysis of Visual Behavior*, MIT Press, Cambridge, MA. (Copyright 1982 by MIT Press; adapted with permission.)

they were stationary. Unbeknown to the subjects, the 'room' surrounding the subjects, but not the surface on which the subjects were standing, was able to be moved so that the front wall could be subtly, but systematically, moved either toward or away from the subjects. In such cases the visual system would sense a loss of balance (overbalancing forwards when the room is moved toward the subject and the converse when the room is moved away) even though the independent information from the vestibular apparatus and the various kinesthetic receptors would indicate no loss of balance. In such instances of intersensory conflict the subject's response is consistently to make compensatory postural adjustments. That such responses are made in the direction opposite to the perceived overbalancing clearly indicates that the visual information is the information the central nervous system 'believes' (Figure 16.11). Even small room movements of as little as 6 mm can induce marked postural sway in adults and complete loss of balance in young children, demonstrating the dominance of vision (even under situations where the information provided by vision can be shown to be incorrect).

Sources: Lee, D.N. & Aronson, E. (1974), Visual proprioceptive control of standing in human infants, *Perception and Psychophysics*, 15, 529–32; Lee, D.N. & Lishman, J.R. (1975), Visual proprioceptive control of stance, *Journal of Human Movement Studies*, 1, 87–95.

Effector systems for movement

The motor unit, as the functional unit of interaction between the nervous system and the muscular system, has been considered elsewhere in this book in the context of being the source of neural input to the muscular system (see especially Figure 3.13). Consequently, in this section we shall only briefly reiterate some of the key features of the motor unit, this time in the context of the motor unit's role as the ultimate endpoint for the output of the neural system.

A motor unit consists of a single alpha motor neuron plus all the skeletal muscle fibres (extrafusal fibres) it innervates. This may range from as few as one or two fibres for the small muscles of the eye, which control precise movements, to up to a 1000 in some of the larger postural muscles of the lower leg. As a general rule the fewer muscle fibres there are within a motor unit the more precise is the possible control. Observable contraction of a muscle or motion of a joint crossed by a muscle requires the activation of a number of different motor units. With practice and appropriate feedback it is possible to learn to selectively recruit single motor units within a given muscle, but it is not possible to voluntarily activate only some of the muscle fibres within a single motor unit. In natural movements motor units are typically recruited in order of size, with the motor units containing the smaller, less forceful muscle fibres being recruited first.

Motor control functions of the spinal cord

So far in this chapter we have examined the structures and processes by which the human body is available to receive sensory information of relevance to movement and in turn transmit information through the motor units, to produce observable movement. We have yet to examine how the central nervous system links (appropriately) this input and output information.

The central nervous system is somewhat hierarchical in structure in that the higher levels of the system, especially the brain, are responsible for higher order creative and executive mental (cognitive) and motor control functions whereas the lower levels of the system, especially the spinal cord, are responsible for more routine, repetitive control functions. As the brain and spinal cord work together in the performance of most skilled movements, understanding the neural control of movement requires understanding of the motor control functions of each level of the central nervous system and the way in which these levels interact.

In this section we examine the basic structure of the spinal cord and its motor control capabilities. It will be revealed that the spinal cord alone is responsible for the control of reflex movements (rapid movements occurring below the level of consciousness) and for the maintenance of voluntary movements, initiated by higher centres within the brain. Much of the knowledge about the motor control capabilities of the spinal cord comes from studies of reflexes in humans and other animals and from studies of the movement capabilities of spinalised animals (animals in which the nerve pathways from the spinal cord to the brain have been severed).

Structure of the spinal cord

As with virtually all other structures in the human body, the anatomical design of the spinal cord can be readily appreciated if its basic functions are first understood. The spinal cord serves two basic functions. Its first role is as a dual transmission pathway to carry both input information from the sensory receptors to the brain and output information, in the form of motor commands, from the brain to the muscles. (See again Figure 16.1.) Its second role is to support reflexes at the local spinal level to provide rapid, essentially automatic responses to noxious (or potentially dangerous) stimuli and to ensure the successful execution of movements already underway.

The spinal cord is about as thick as an adult's little finger and runs from the base of the spine to the point at which it joins the brain at the brainstem, located at the base of the skull (Figure 16.1). Because of its importance to normal communication functions in the body, the cord, like a well laid telephone cable, is well protected throughout its length by the bony structures of the spine. The spinal cord runs throughout the length of the spine within the protection of a canal formed within the vertebral (spinal) column (Figure 16.12).

A total of 31 pairs of spinal nerves are attached to the spinal cord, with each nerve attached to its side of the cord by two roots. The anterior (or ventral) root of each spinal nerve carries the efferent or motor information away from the spinal cord to the muscles whereas the posterior (or dorsal) root carries the afferent or sensory information from the periphery back to the spinal cord. The anterior root consists almost exclusively of the axons of alpha motor neurons, the cell bodies of which are located within the ventral horn of the spinal cord. The posterior root contains both the axons of the sensory neurons and their cell bodies, the latter clustered together to form the dorsal root ganglion. The spinal cord itself, in cross-section, reveals an outer covering of white matter surrounding a central mass of grey matter, roughly approximating the shape of the letter 'H'. The white matter of the spinal cord consists primarily of nerve fibres and the grey matter the cell bodies of neurons. Nerve fibres within the white matter are frequently bundled together to form tracts that carry impulses up and down the spinal cord.

Spinal reflexes

A reflex (from a Latin word meaning 'bending back') is the simplest functional unit of integrated nervous system behaviour. The various reflexes scattered throughout the spinal cord provide the foundation on which (directly) all involuntary and (indirectly) all voluntary movement is based.

A minimum of four basic nerve units are needed to form a reflex arc, these being:

- a sensory receptor (to detect a pertinent stimulus)
- an afferent (or sensory) neuron (to transmit the sensory information to the central nervous system)
- an efferent (or motor) neuron (to transmit the output information from the central nervous system)
- an effector, typically a motor unit (to produce a movement response).

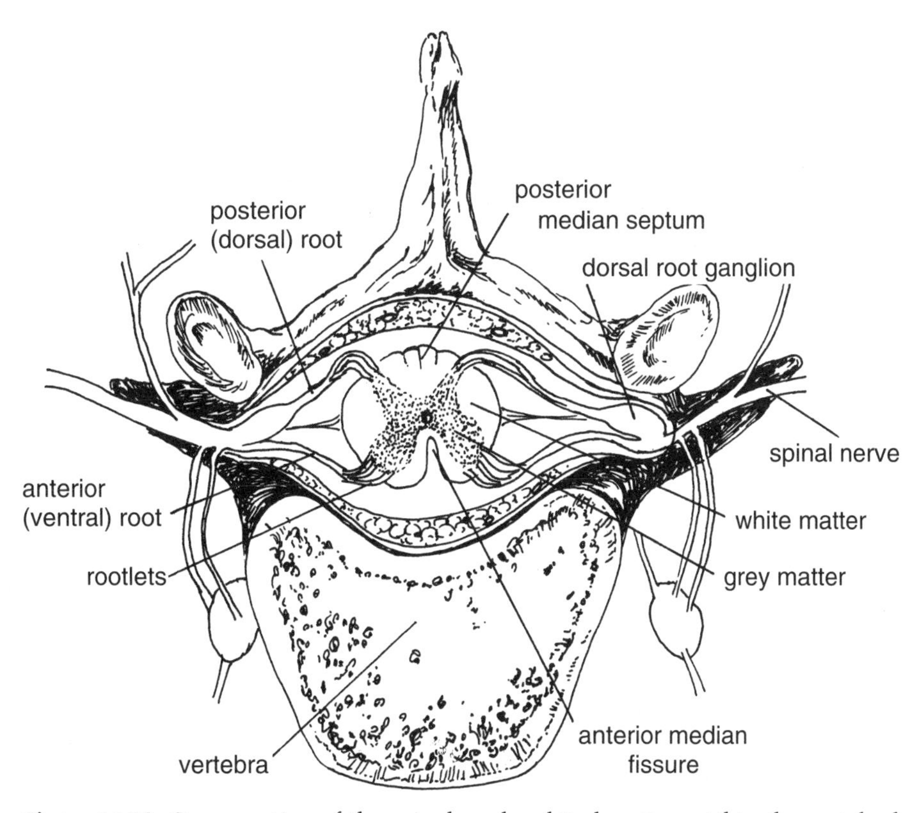

Figure 16.12 Cross-section of the spinal cord and its location within the vertebral column

The simplest of all reflex systems, which has only two neurons and hence only one synapse, is called, for obvious reasons, a monosynaptic reflex. Most reflex arcs are polysynaptic, containing multiple synapses, with the sensory and motor neurons not synapsing together directly but rather the nerve impulses being passed from one to the other through a series of interneurons. The time it takes for a reflex system to work (its so-called 'latency' or 'loop time') is measured from the time of stimulation to the time a response can be recorded in the muscle fibres. Not surprisingly, reflex loop time is longer the more interneurons (and synapses) there are within the reflex arc.

The stretch reflex

The best example of a monosynaptic reflex within the spinal cord is the simple stretch reflex, also known as the myotatic (muscle-stretching) reflex. The stimulus for the stretch reflex is excessive stretch on muscle as detected by the muscle spindles (see again Figure 16.7). This excessive stretch may arise from a number of sources, such as the unexpected addition of a weight (Figure 16.13*b*), postural sway (Figure 16.13*c*) or, as is often the case in a clinical setting, unexpected stretching of the quadriceps muscle when a doctor, using a rubber mallet, delivers a sharp tap to the patellar tendon. In all cases the excessive stretch detected by the muscle receptors results in a nerve impulse being sent to the dorsal root of the spinal cord via the afferent neuron. This neuron then synapses, within the spinal cord, directly to an alpha motor neuron, which transmits its nerve impulse back to the extrafusal fibres of the stretched muscle (Figure 16.13*a*). This, typically, is sufficient to cause the stretched muscle to contract, thus alleviating the stretch stimulus.

In the case of the patella tap test for nerve function used by general practitioners and

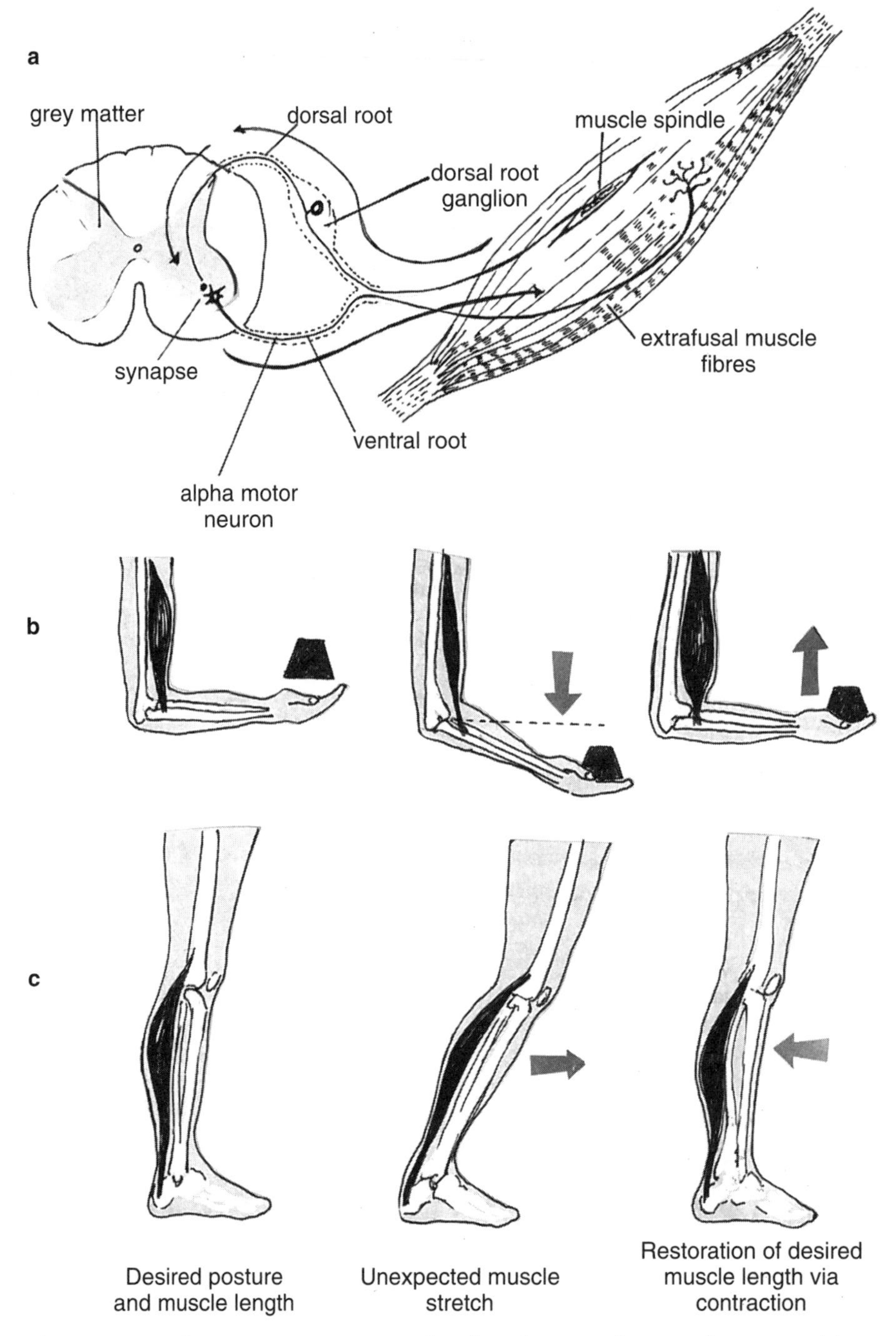

Figure 16.13 The monosynaptic stretch reflex showing the neural circuitry (a) and typical situations where the reflex would be activated (b, c)

neurologists, it is the alpha motor neuron activation of the quadriceps muscle that causes the characteristic and forceful knee extension (kicking) response. If the initial response is insufficient to alleviate the stretch two things will occur: (i) the same reflex arc will be activated a second time (as the stimulus still exists), and (ii) commands will be sent (via interneuron connections) to other segments of the spinal cord, including higher centres. Typical of most hierarchical organisations, the latter will produce a more powerful, but slower response. Whereas the simple monosynaptic stretch reflex may have a loop time as short as 30 ms, reflex responses to alleviate stretch that involve higher segments of the spinal cord, and perhaps even regions of the brain (often referred to as long-loop reflexes), may have latencies two or three times this duration.

The flexion reflex

A common purpose of many spinal reflexes is to provide rapid protection for the body against potentially injurious stimuli while preserving, at a premium, whole-body balance. The flexion reflex is a polysynaptic reflex that causes withdrawal of limbs away from potentially injurious stimuli. In this reflex, excitation of either a pain or heat receptor in the skin excites, through interneural connections, those motor neurons innervating flexor muscles crossing joints adjacent to the stimulus, and this results in the limb being flexed away from the noxious stimuli. Equally importantly this rapid limb withdrawal through flexion is facilitated by the extensor musculature being 'turned off' at the same time (Figure 16.14). This is achieved through interneuronal inhibitory connections between the afferent neuron and the extensor motor neuron. This action provides a specific example of a general spinal reflex phenomenon called reciprocal inhibition (the neural control phenomenon that ensures that agonist and antagonist muscles do not typically co-contract in opposition to each other).

The crossed extensor reflex

The crossed extensor reflex often functions in conjunction with the flexion reflex to maintain postural stability and, if necessary, help a person push away from a painful stimulus (Figure 16.14). Through interneuronal connections that again exploit the excitatory–inhibitory potential of different synaptic connections, the crossed extensor reflex ensures that not only is the limb closest to a painful stimulus flexed away from it but that the limb on the opposite side of the body extends (through extensor excitation and flexor inhibition). This reflex provides a good illustration of how nerve impulses pass not only to and from the spinal cord at a given segment level and up and down the spinal cord but also across the spinal cord from one side of the body to the other.

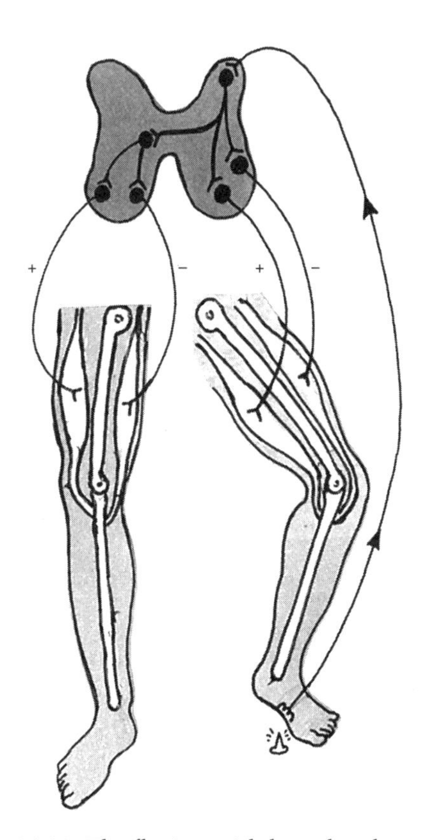

Figure 16.14 The flexion withdrawal and crossed extensor reflexes. A painful stretch causes excitation (+) of the ipsilateral flexors and contralateral extensors and inhibition (–) of the ipsilateral extensors and contralateral flexors

The extensor thrust reflex
The extensor thrust reflex is one of the more complex spinal reflexes; it aids in supporting the body's weight against gravity. Cutaneous receptors in the feet sensitive to pressure, through a vast array of inter-neural connections, cause reflex contraction of the extensor muscles of the leg. This reflex provides the foundation for standing balance without dependence on brain mechanisms.

Spinal reflexes for gait control
All forms of human gait (such as crawling, walking and running) are characterised by continuous patterns of limb flexion and extension, with each limb one half-cycle different from the other. The flexion and crossed extensor reflexes, with their reciprocal innervation of flexor–extensor pairs across matching sets of limbs, and the extensor thrust reflex, with its pathways for balance preservation, provide strong building blocks for basic gait control and maintenance.

Studies conducted on animals, in which spinal connections to the higher centres of the brain have been severed, have demonstrated that the spinal cord has an inherent rhythmicity that plays a major role in gait control. Although the spinal cord, *per se*, seems incapable of initiating gait (this appears to require either motor commands from the brain or very strong sensory information from the cutaneous receptors of the feet), the spinal cord seems well capable, through its various reflex pathways, of preserving gait once it is initiated, even to the point of controlling a transition from one gait form (such as walking) to another (such as running). Clearly spinal reflexes play a major role, not only in involuntary protective actions, but also in the control of fundamental motor activities such as gait.

The role of reflexes in voluntary movement control

It is obvious from the preceding section that reflexes clearly play a major role in involuntary movement control, that is, the control of movements below the level of our conscious awareness. Most neurophysiologists and motor control theorists believe that the spinal cord, through its reflex arcs, also plays a major role in ensuring that voluntary movements planned and initiated in the brain are executed as planned. Voluntary movements must, to some degree, use reflexes as their 'building blocks' because the final pathway for all motor commands, regardless of their origin, is through the alpha motor neurons at the spinal cord level to the muscle fibres. Voluntary movements simply involve the spinal reflex pathways being modified or used in ways specified by commands arising from higher centres of the central nervous system.

A good example of the way the reflex structure of the spinal cord can be integrated with the higher level control provided from the brain is seen through examination, yet again, of the muscle spindle (Figure 16.7) and the stretch reflex (Figure 16.13). In its simplest usage we have seen in the preceding section how the stretch reflex provides a means of protecting the muscle against damage from excessive lengthening. In a more functional manner, however, the collaborative activity of the muscle spindle, its sensory neurons, and its alpha and gamma motor neurons can be so organised as to ensure that voluntary movements are executed as planned.

The progress of any particular movement can be monitored and controlled through a process of alpha–gamma co-activation. In this process actual muscle length is determined by contraction of extrafusal muscle fibres controlled by the alpha motor neurons that originate at spinal level. Intended muscle length is set by contraction of the ends of the intrafusal muscle fibres under the control of the gamma motor neurons that originate from the level of the brain. As the name implies, for any given movement, such as holding a weight in a constant position or maintaining upright stance (Figure 16.13), alpha–gamma co-activation results in the simultaneous activation of both the extrafusal fibres (by the alpha system) and the intrafusal fibres (by the gamma system).

If the movement goes as planned, the change of muscle length of both the extrafusal and intrafusal fibres will be identical and no additional sensory impulses will be sent back from the muscle spindle to the spinal cord. If the movement does not proceed as planned (for example there is insufficient extrafusal fibre innervation to shorten the muscle), the sensory receptors on the intrafusal fibres will be placed on stretch and this will evoke, through the usual stretch reflex, additional alpha motor neuron activation to cause the muscle to contract. The alpha–gamma co-activation process therefore provides a good example of how movement plans from the higher centres of the central nervous system can be enacted, using spinal mechanisms to ensure that these movements are executed as planned.

Motor control functions of the brain

The human brain possesses a level of complexity and organisation that is beyond comprehension and perhaps unmatched by anything else in the universe. The brain serves many higher order functions, only some of which are directly related to motor control. In this section we shall examine the location and function of the main areas of the brain identified as having a significant role in motor control. These areas are the motor cortex (located immediately forward of the central sulcus in the frontal lobe of the cerebrum), the cerebellum (located off the brain stem and below the occipital lobe of the cerebrum), the basal ganglia (located within the inner layers of cerebrum) and the brainstem (located forward of the cerebellum and continuous with the spinal cord and the cerebrum) (see Figure 16.15). These areas are in constant communication through a rich, interconnecting network of nerve pathways, the major ones of which are shown schematically in Figure 16.16.

The motor cortex

The cerebral cortex is the outermost layer of the cerebrum of the brain, is some 2–5 mm deep, has an (unfolded) surface area of some 2–3 m^2 and contains over half of the total neurons in the human nervous system. The cerebral cortex is divided into two halves, which appear essentially symmetrical although they are somewhat different in function. These are the left and right cerebral hemispheres, which join at the midline through a thick sheet of interconnecting nerve fibres called the corpus callosum.

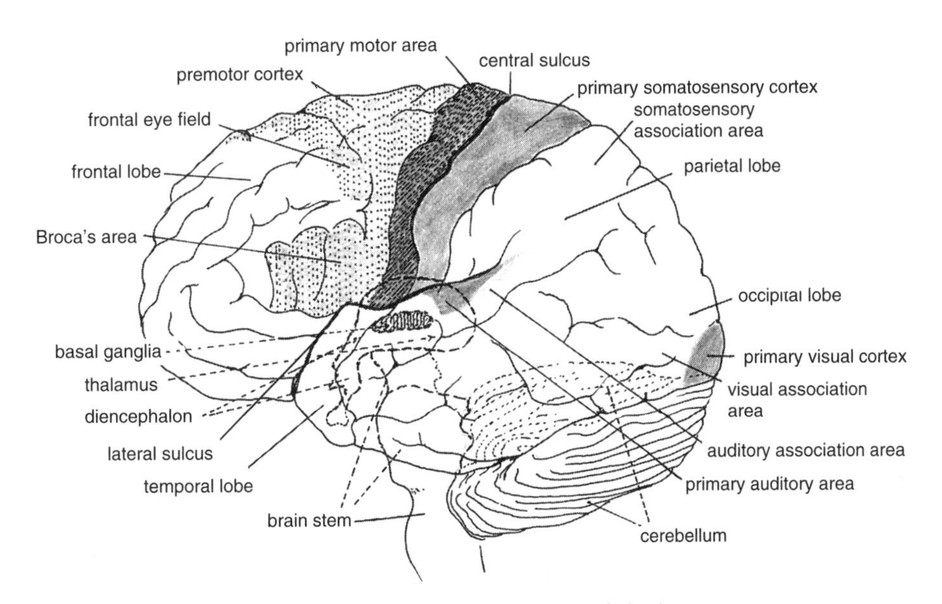

Figure 16.15 Location of the principal motor areas of the brain. Structures located within dashed lines lie underneath or within the external surface of the brain

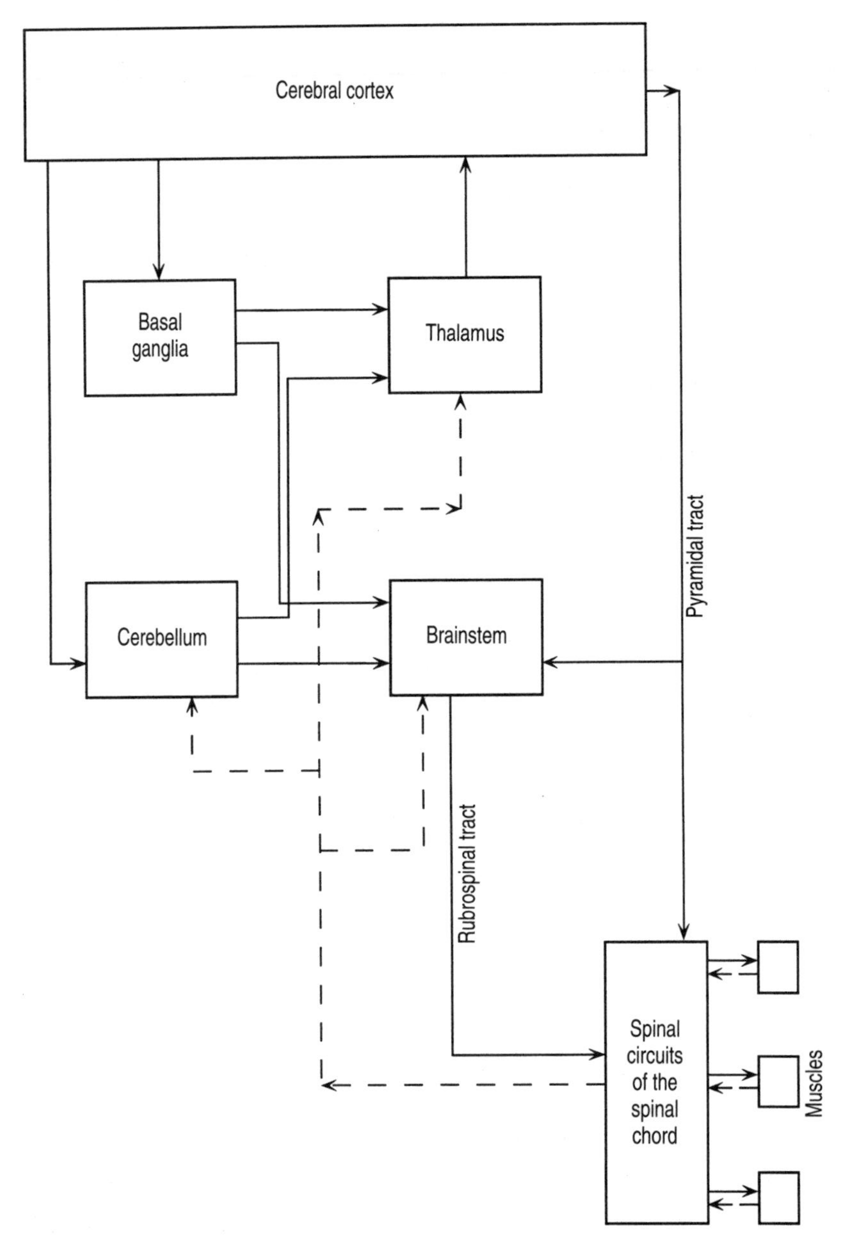

Figure 16.16 Schematic representation of the major motor pathways. Direction of information flow is shown by arrows, with afferent input denoted by dashed lines
Source: Greer, K. (1984), Physiology of motor control, p. 33 in M.M. Smyth & A.M. Wing (eds), *Psychology of Human Movement*, Academic Press, London. (Copyright 1984 by Academic Press.)

Each cerebral hemisphere contains a motor cortex (lying immediately forward of the central sulcus; Figure 16.15), a pre-motor cortex (lying just forward of the motor cortex) and a supplementary motor area (lying on the medial wall of the cerebral hemispheres and forward of the motor cortex). Each of these structures, located within the frontal lobe of the cerebrum, is intimately involved in the production and control of skilled movement.

The motor cortex and its associated areas are systematically organised such that each part of the motor cortex controls specific muscles or muscle groups within the body, to the point that all muscles are topographically represented in the brain (Figure 16.17). In pioneering studies of the human brain by two Canadian neurosurgeons, Penfield and Rasmussen, muscle maps of the motor cortex were developed by applying weak electrical pulses to distinct areas of the motor cortex and observing the muscle contractions that resulted. These electrical mapping studies revealed a number of important things, namely:

- all muscles are not represented proportionally in the motor cortex on the basis of their size; rather representation is proportional to the precision requirements of different parts of the body, with the muscles of the hand and the mouth occupying nearly two-thirds of the total area of the motor cortex
- the representation of distal musculature, such as that crossing the joints in the hands and feet, is entirely contralateral; consequently the left motor cortex controls the right hand and vice versa
- muscles located more proximally are represented in the motor cortex of both the cerebral hemisphere on the same (ipsilateral) and opposite (contralateral) side of the body
- as stimulation is moved forward into the premotor cortex, gross movements of muscle groups rather than fine movements of discrete muscle groups are observed.

Perhaps, most importantly, these electrical mapping studies suggest that the motor cortex

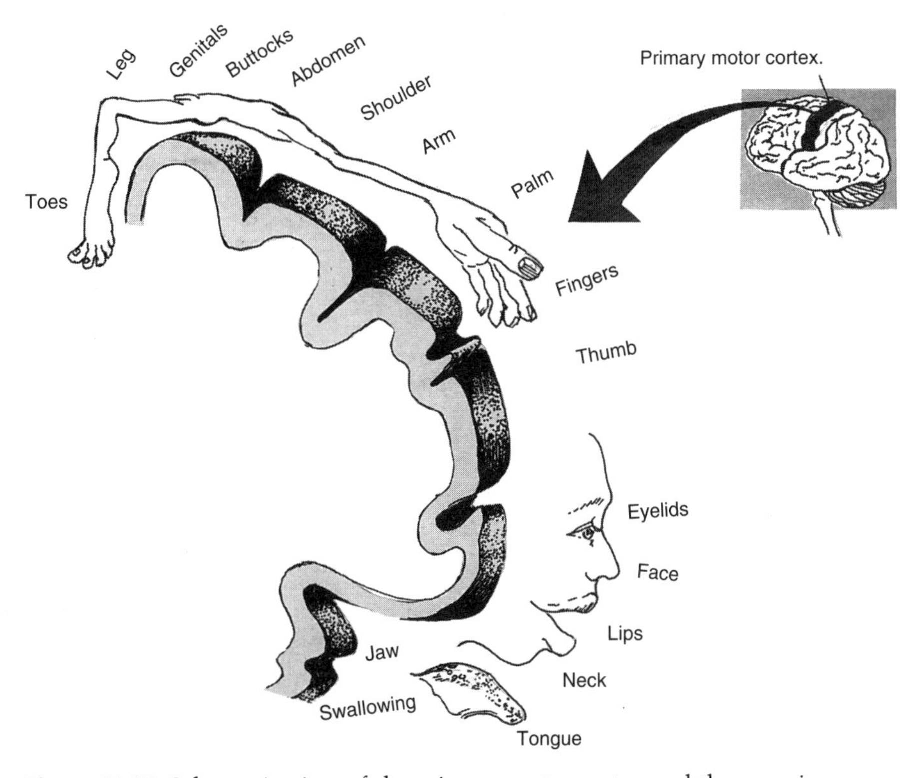

Figure 16.17 Schematic view of the primary motor cortex and the mapping between electrical stimulation of the motor cortex and regions of the body in which movement consequently occurs. The insert shows the location of the primary motor cortex within the brain

acts as something of a relay station, being the final neural station for the organisation and release of the co-ordinated motor commands to be sent out to the specific muscles or, more correctly, muscle fibres.

The motor cortex has two principal means of relaying its commands to the muscles. The most direct route is via the pyramidal tract (or cortico-spinal tract), which allows neurons from the motor cortex to synapse directly in some cases (and through a minimum of inter-neurons in most cases) with the alpha motor neurons at the spinal level. This tract carries impulses that are primarily excitatory in nature. Alternative routes, known collectively as the extrapyramidal tract, allow nerve impulses from the motor cortex to reach the spinal level through a range of pathways via the cerebellum, basal ganglia, thalamus and brainstem (Figure 16.16). Outputs from these pathways are primarily inhibitory in nature.

Damage to the motor cortex results in a loss of fine movement control, especially in the fingers and toes. Damage to the pre-motor cortex results in a disruption to gross movements involving a number of muscle groups. Damage to the supplementary motor area disrupts the performance of many tasks requiring bimanual co-ordination.

The cerebellum

As can be seen in Figure 16.15, the cerebellum attaches to the brain stem and is located behind and below the cerebral hemispheres. Like the cerebrum, the cerebellum has an outer cortex, divided into two distinct but interconnected hemispheres. Beneath the cortex are four deep-cerebellar nuclei. The cerebellum receives input information from a vast array of areas in the cerebral cortex (including the motor areas), from various areas in the brainstem, from the vestibular apparatus and, via the spinal cord, from the kinesthetic receptors located on the same (ipsilateral) side of the body.

The cerebellum itself has a very regular anatomical structure based primarily around two different types of afferent fibres (called climbing fibres and mossy fibres) and one output fibre (the Purkinje cell; see again Figure 16.3). This structure enables the cerebellum to perform a number of very complex signal-processing operations fundamental to many aspects of motor co-ordination. Indeed the cerebellum has frequently been referred to as 'the seat of motor co-ordination'. The cerebellum has two major outputs—one to the thalamus and one to the brainstem (Figure 16.16).

A number of major motor control functions have been attributed to the cerebellum, all broadly related to the translation of abstract movement plans into specific spatial and temporal patterns that can be relayed to the muscles via the motor cortex. Principal cerebellar functions appear to be the regulation of muscle tone, the co-ordinated 'smoothing' of movement, timing and learning. Patients with cerebellar damage demonstrate one or more of the symptoms of low muscle tone, inco-ordination (especially in standing, walking, speaking or performing precise aiming movements), poor temporal control of muscle recruitment, and difficulty in learning new movements or adapting old ones. Fast, ballistic types of movement appear to be particularly affected.

The basal ganglia

The basal ganglia comprise a group of interconnected nuclei located deep within each of the cerebral hemispheres and close to the thalamus. The basal ganglia receive input from two major sources, from the motor areas of the cerebral cortex and from the brainstem, and, similarly, send their output to two different locations, the thalamus and the brainstem. Therefore, like the cerebellum, the basal ganglia, while not synapsing directly with spinal neurons, are able to influence alpha motor neuron activity through both the pyramidal tract and the rubrospinal tract (Figure 16.16). The basal ganglia work together as a loosely connected unit, although each of the component nuclei are quite different and generally connected in an inhibitory fashion with each other.

Insights into the function of the basal ganglia in motor control have come primarily from studies of patients suffering from two identifiable diseases of the basal ganglia. Parkinson's disease is a degenerative disease resulting from deficiency in the natural neurotransmitter substance dopamine, which assists in carrying nerve impulses from one nuclei within the basal ganglia to another. Parkinsonian patients typically demonstrate a range of motor symptoms including shuffling, uncertain gait, limb tremor, difficulty in initiating movement and high degrees of muscle stiffness. Huntington's disease is a hereditary degenerative disease resulting from damage to the dendrites that produce one of the neurotransmitters used to communicate between selected nuclei within the basal ganglia. Patients with this disease suffer from uncontrollable, involuntary rapid flicking movements of the limbs and/or facial muscles.

Despite a knowledge of the obvious movement problems caused by basal ganglia dysfunction, the precise function of the basal ganglia in movement control remains elusive. Some favoured suggestions include the control of slow movements, the retrieval and initiation of movement plans, and the scaling of movement amplitudes, as required in daily tasks such as handwriting.

The brain stem

The brain stem contains three major areas that have significant involvement in motor control. These are the pons, the medulla and the reticular formation. The brainstem's principal function, as revealed by Figure 16.16, is to act as a relay centre, especially for the transmission of information to and from the cerebral cortex.

The pons and medulla, as the main structures within the brainstem, receive input from the cerebral cortex, cerebellum and basal ganglia as well as all the sensory systems. These structures then integrate this information for output to the spinal cord for use in the control of many involuntary movements, such as those related to posture and cardiorespiratory activity.

The brainstem functions not only in the control of muscle tone and posture but is also fundamental to the operation of a number of supraspinal reflexes. Prominent among these are the righting reflexes, which are designed to maintain the orientation of the body with respect to gravity, and the tonic reflexes, such as the tonic neck reflex, which are concerned with the maintenance of the position of one body part (such as the neck) in relation to other body parts (such as the arms and legs). (See Table 18.3 for greater detail on these reflexes.) Damage to the pons or medulla disrupts the control of involuntary movements and key orienting reflexes and endangers the control of vital physiological systems.

The reticular formation is a network of neurons that extends throughout the brainstem and, through its ascending connections to the cerebral cortex, has a major role in regulating the activity of the cortex. The ascending reticular formation controls the activation of the cortex in this way and therefore, in turn, the state of arousal experienced by the person. (The issue of arousal is examined in more detail in Chapter 20.) The descending fibres of the reticular formation input directly to the spinal reflexes and may modify reflex activity at this level as is necessary to ensure that basic postural needs are met.

Integrative neural mechanisms for movement

Given the complexity of both the human nervous system and the movement it produces, it is perhaps not surprising how little is yet known about the neural mechanisms underlying movement control in the brain and how much remains to be discovered. At this point only some speculations can be advanced on the likely flow of neural information through various brain structures and the functional consequences of such information flow. The pre-frontal cortex appears to be central to overall movement planning, the basal ganglia and cerebellum to the programming of specific motor commands and the motor cortex to the release of organised commands to the muscles, via the spinal

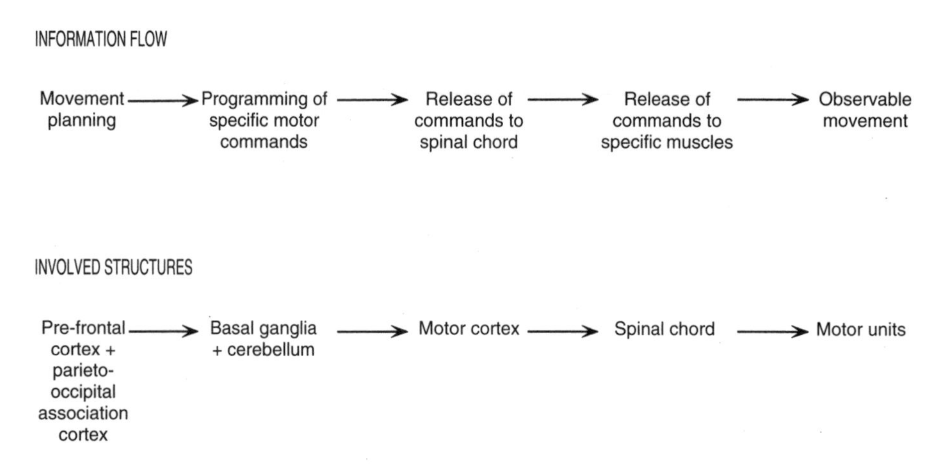

Figure 16.18 A speculation on some of the major functional roles of brain structures in movement control

pathways (Figure 16.18). Readers should be aware, however, that neurophysiologists interested in motor control are still many years away from a complete integrative model of the brain mechanisms for motor control. One approach that may hasten understanding may be to look alternatively or, better still, simultaneously at motor control from a conceptual (psychological) perspective in addition to a neurophysiological one. The next chapter does this by examining basic psychological perspectives on motor control.

CHAPTER 17

Basic concepts of motor control: Psychological perspectives

- The importance of models of motor control
- Key properties to be explained by models of motor control
- Information processing models of motor control
- Some alternative models of motor control

In the previous chapter we examined in some detail the structure and function of the main components of the neuromuscular system. Although this neurophysiological approach to understanding motor control provides us with valuable information about the receptors and effectors for movement and about the major pathways that connect the two, the sheer size and complexity of the nervous system (with its some 10^{14} neurons each with up to 10^4 synaptic interconnections) mean that it is impossible to easily or fully appreciate how movement is controlled by studying nerve pathways only. What are needed to complement the neurophysiological knowledge are conceptual theories and models that describe and explain the overall control logic used by the nervous system to collectively acquire, perform and retain motor skills. Such theories and models have typically originated from the work of experimental psychologists who have focused on the broad overall functioning of the motor system rather than the more specific detailed anatomy and physiology of individual physical components of the motor system.

In this chapter we examine the role of conceptual models from experimental psychology in understanding motor control, outline the key properties of skilled human movement that must be explained by such conceptual models and then describe, in some detail, one popular model (an information-processing model) of motor control, highlighting as we do some of its basic assumptions and practical implications.

The importance of models of motor control

In all branches of science models serve the important purpose of aiding in the understanding and advancement of theory. Models are analogies to theory that enable a theory to be visualised and understood, frequently by drawing comparison with the operation of simpler, everyday systems with which we are familiar. Systems of infinite complexity, such as the physical system of electricity or the biological system of the heart and lungs, can be more easily understood through the use of simplifying models such as those of water flow or the action of a pump. The value of models therefore is their potential to simplify a complex system to a level at which understanding can be achieved and experiments to further knowledge can be formulated. In a system as complex as the human motor system there appears to be great value in developing conceptual models as a means of aiding and advancing our understanding of how the motor system works.

It needs to be noted with caution, however, that a model is not a theory and therefore should not be taken too literally. A model is also only worthwhile to the extent to which it accurately captures the key characteristics of the system we are ultimately trying to understand. Just as good models can aid understanding, poor models can hamper understanding. As more becomes known about a particular system the shortcomings of old models are frequently realised and new models proposed in their place. As we shall see in this chapter one particular model (the information-processing model) has dominated most thinking to date about how movement is controlled but recently some of the limitations in this model have become apparent and alternatives have been suggested.

Key properties to be explained by models of motor control

A starting point for the development of a model of any system is consideration of the key features of the system that the model must be able to encapsulate. These key features, in a sense, form the constraints for the model. In human motor control there exists an impressive array of unique properties that must be adequately explained by any worthwhile model and/or theory. Some of the principal motor control properties that require explanation are:

- *Motor equivalence* is the capability of the motor system to perform a particular task, and produce the same movement outcome, in a variety of ways. Even actions as apparently simple and repetitive, and as consistent in outcome, as writing one's own signature on a piece of paper can be achieved through recruiting different motor units or even different muscle groups. Motor equivalence is a consequence of the many degrees of freedom (joints, muscles, motor units) we are able to independently control. Any plausible model of motor control must be able to account for how the nervous system rapidly and apparently effortlessly selects just one combination of joints, muscles and motor units from all the options available to it in order to perform a particular task effectively and efficiently.
- *Serial order* is the capability of the motor system to structure movement commands in such a way as to reliably produce movement elements in their desired sequence. Correct sequencing of movement components is fundamental to the performance of virtually all skilled actions and errors in sequencing inevitably result in errors in performance. Serial order errors in speech give rise to spoonerisms ('mumam hovement' instead of 'human movement') and, in typing, to transposition errors (for instance, 'cat' typed as 'cta'). In gross motor skills, such as throwing, mis-ordering of the recruitment of large, proximal muscles (such as those crossing the trunk) and smaller distal muscles (such as those crossing the wrist joint) will undermine the summation of speed principle introduced in Chapter 11 and, through this, act to limit performance. A useful model of motor control must therefore be able to account for how serial order is generated in movement sequences (and hence also how errors may sometimes arise).
- *Perceptual–motor integration* is the capability of the motor system to produce movements closely matched to the current environmental demands (perceived by the performer). Skilled movement is always subtly adjusted to meet changing environmental situations. For example, the skilled tennis player is able to

adjust racquet swing if the ball deviates unexpectedly in flight and all of us adjust our gait patterns if the surface we are walking on becomes irregular. Such adjustment can only be achieved if there is a tight coupling and integration between perception and action. Such coupling must therefore be a key element of any satisfactory model of motor control. As the role of particular muscles in either producing or opposing limb movement frequently varies according to contextual factors such as joint position or orientation of the limb with respect to gravity, the central nervous system, in issuing motor commands, must be continuously and accurately informed about the body's posture and position in space by its many perceptual systems.

- *Skill acquisition*, as we shall see in Chapter 19, is the capability of the motor system to learn and improve, given appropriate conditions of practice. Explaining skill acquisition (or motor learning) requires a motor control theory to be able, in turn, to explain how experience is stored and how, once acquired, movements are able to be modified to meet task conditions never previously encountered. A viable model of motor control must therefore be able to adequately account for the paradoxical capabilities of skilled performers to produce movements that are both *adaptable* yet *consistent*.

A number of different models of motor control have been proposed over the years, varying in the extent to which they attempt to both explain the key properties of movement as well as incorporate what is known from neurophysiology about the structure of the motor system. Those models of motor control that have been most widely developed and used can be generally described as information-processing models.

Information-processing models of motor control

Basic assumptions

Experimental psychologists have, for a long time, used the analogy of the nervous system as a computer as a means of simplifying consideration of complex neural processes of the type underpinning motor control. The information-processing model simply views the human as a sophisticated processor of information, in a way paralleling the processing of information by a computer. By first considering the (relatively) simple operations of a computer we can then perhaps try to better understand the much more complex operations of the human motor system.

A computer is basically a dedicated electronic device that, through its stored programs, is able to convert input information of one form or another into output of a specific, desired, form. The input information may come from data stored on a disk or may be input direct from a keyboard (as occurs when we type in letters or numbers) or from some other information-acquisition system (such as the EMG system depicted in Figure 3.14). The output information may also be of various forms. It may be in the form of text (letters and numbers) or graphics appearing on the computer screen or sent to a printer (the 'hard' copy) or it may be in the form of electronic commands sent on to other computers or devices controlled by the computer.

The conversion of input information to output information is not a passive process but rather an active reorganisation of information specifically controlled by the commands within the computer program(s). The type of processing the computer is capable of doing, the speed with which it can complete its operations, its capacity to store information in memory and ultimately the quality and diversity of the output it can produce is limited by two interacting factors. These are the physical construction of the computer's electrical circuits (its hardware) and the computer programs written specifically for the computer, which reside in the computer's memory (its software).

How does the central nervous system act like a computer system? The input information for movement control is the sensory information sent to the central nervous system from (primarily) the visual, kinesthetic and vestibular receptors (as detailed in the previous chapter). The output is the patterns of movement we observe, and can describe biomechanically, that arise as a consequence of the co-ordinated set of motor commands sent from the central nervous system to selected muscle groups. Input information is converted to output information through a number of information-processing (or computational) stages that take place within the brain and other regions of the central nervous system. The success of the movement that results (the output) depends primarily on the computational 'programs' within the central nervous system that are responsible for selecting and then controlling the movement (Figure 17.1). Some of the programs such as those controlling balance and gait may be 'hard-wired' into the central nervous system (especially the spinal cord) from birth and therefore may be considered to be part of the 'hardware' of the central nervous system. Other skills require programs that are not innate but rather must be developed and constantly modified and improved through repeated use and practice. These programs are therefore analogous to specialised software written for a particular computer's hardware to fulfil a specific unique task and stored within the system's memory. In the nervous system this storage may require some changes to 'hardware' as new neural connections are formed.

Just as we cannot understand how a computer works and controls its output simply by inspecting what it produces as output, we cannot expect to understand movement control by simply describing the observable movement patterns produced as output by the motor system. It is important to recognise that movement does not simply occur spontaneously as a consequence of unplanned muscular activity; rather, movement is the end product of a long series of information-processing stages (or computations) that take place beyond observation within the confines of the central nervous system. To understand movement we must therefore try to understand the processes and computations that occur in the central nervous system and form the link between sensory input and observable movement output. Research scientists, adopting an information-processing model of motor control, have directed much of their energies to attempting to uncover the computational code and programs the central nervous system has inherited and developed for movement, trying to elaborate on the processing stages used to link input and output information, and attempting to determine the capacities and limitations of the various processing stages.

Stages within a typical information-processing model

Most information-processing models of movement control assume that there are distinct and sequential stages through which information must pass (or, more correctly, be processed) from input to output. Figure 17.2 presents a typical information-processing model. It shows environmental and internal information, physically present in such forms as light and sound waves and muscle lengths and tensions, being picked up (transduced) through the various sensory receptors described in Chapter 16 and then being transmitted along the afferent pathways to the central nervous system. It is this information that then provides the input for central nervous system processes that ultimately produce, as output, motor commands that are transmitted along the efferent pathways out to the muscle fibres where they individually cause muscular contraction and collectively generate observable movement patterns. Feedback from the movement itself is monitored via the afferent pathways and can be used to either correct errors in the movement (if the movement is sufficiently slow) or make improvements to the commands for the next time that the same or a similar movement is to be produced.

The stages proposed in Figure 17.2 are themselves unremarkable and are entirely consistent with the structure and function of the receptors and effectors for movement described in the previous chapter. What remains unclear is the nature of the central processing stages and it is in the specific nature of these stages that most information-processing models differ. Most models, however, accept that there must be at least three sequential processing stages occurring in the brain and other areas of the central nervous system before the initiation of any movement. We will, for simplicity, refer to these stages as perceiving, deciding and acting. In the *perceiving stage* the focus is on deter-

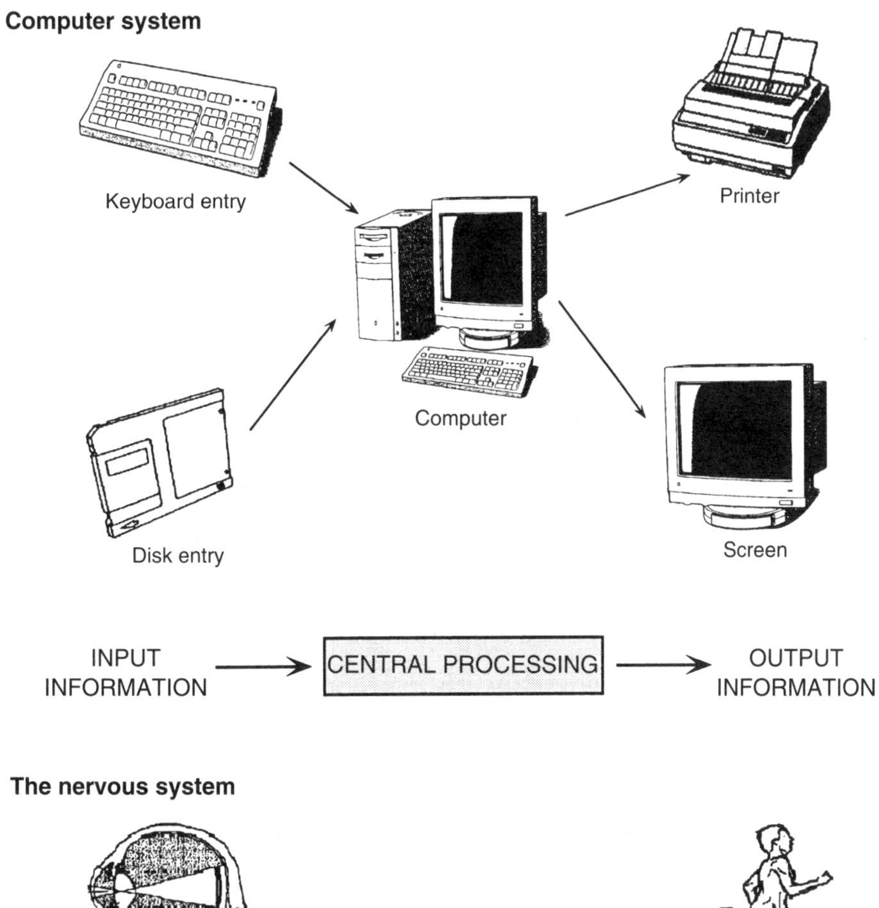

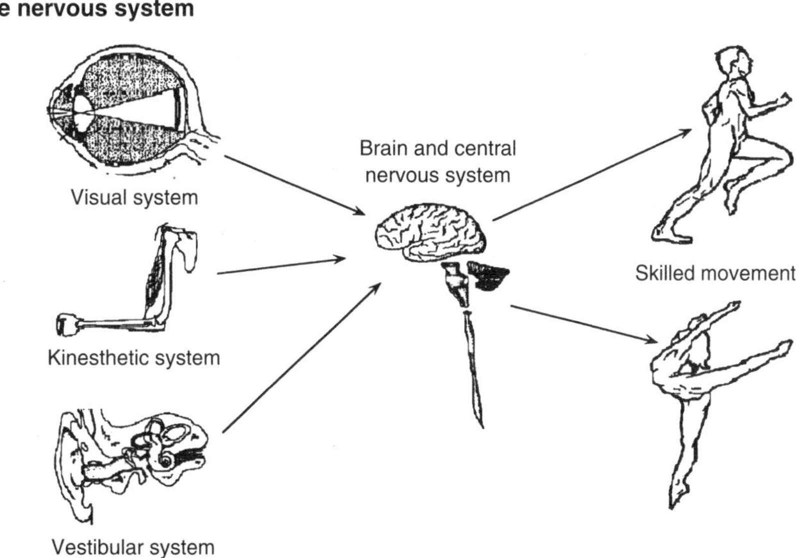

Figure 17.1 Parallels between the information-processing operations of a computer and those of the central nervous system

mining what is currently happening and what is about to happen both in the external and internal environments; in the *deciding stage* the focus is on deciding what action, if any, is needed in response to current and future events; and in the *acting stage* the focus is on organising and executing the required movement response in terms of the sequence and timing of the motor

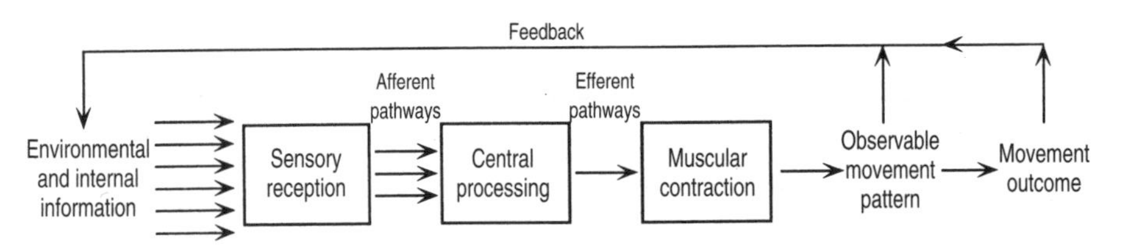

Figure 17.2 A typical information-processing model of motor control

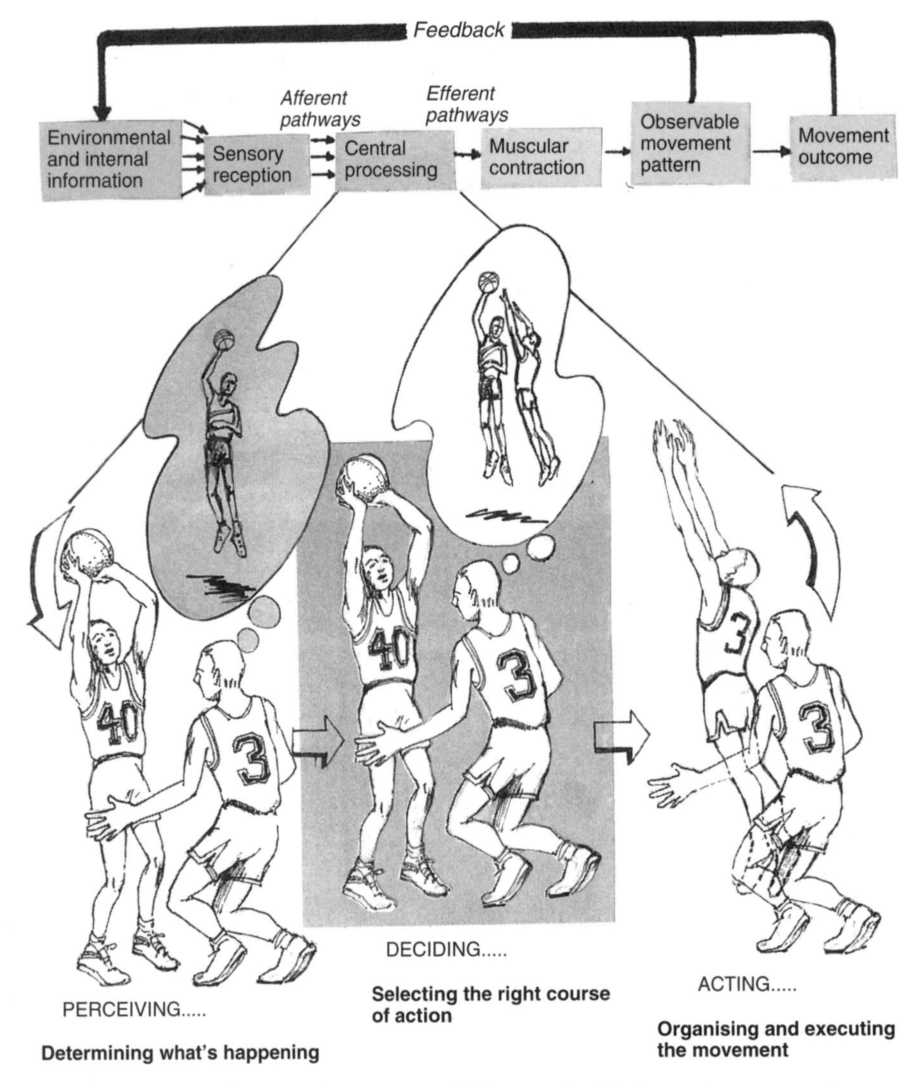

Figure 17.3 Central processing stages within a typical information–processing model of motor control

commands that have to be sent to the muscles (Figure 17.3).

The operation of each of the stages within the total composite information-processing model and the dependence of each of them on the preceding stage is perhaps best illustrated by an example. The example we consider here is one from the sport of basketball but equally the example could have been taken from any motor skill performed in any setting. The basketball guard in possession of the ball is surrounded by a nearly infinite array of physical signals (for example light and sound waves, pressure signals, vibrational signals and chemical signals), only a very limited range of which can be detected by his or her eyes, ears, vestibular, kinesthetic and other receptors. Despite this limited range of sensitivity, the brain and spinal cord of the player are nevertheless bombarded every millisecond by an enormous array of input information from the billions of sensory receptors in the body. There is, for example, visual information about the player's own location on the court relative to the basket plus information about the location, velocity and direction of motion of all four teammates and all five opponents; auditory information from the bouncing ball, the calls of team mates, opponents, coaches and spectators; kinesthetic information from stretches on muscles, strains on joints, tension on tendons and pressure on the soles of the feet; tactile information from the hand dribbling of the ball; proprioceptive information from the vestibular apparatus; plus smell (olfactory), taste (gustatory) and other information. Some of this information is highly relevant to the task at hand and much of this information is quite irrelevant.

The first important process the brain and central nervous system must perform is that of perception and this involves selecting out only the most relevant information for further processing and then using this information to determine both what is currently happening and, implicitly, what is about to happen in the near future. In the case of the point guard this will involve making important judgments about external events as well as more internal ones. Externally the player will need to determine, among other things and primarily through vision, the location and posture of defending players, the structure of the offence as well as the defence, the position and direction of movement of any unguarded team mates, their position on the court and especially their proximity to the basket, and the relative heights and agilities of team mates and opponents. Internally the player will need to be aware of his or her own body posture and balance, and monitor the tactile and auditory sensations from the ball to ensure that it is well in control.

Having determined what is currently happening, and having predicted future events, the second important stage of information processing involves the process of decision-making, that is, determining what, if any, new action or response is required. In many movement tasks, such as sports tasks, the decision-making process equates to 'picking the correct option' from a range of possible response options and clearly the quality of the decision made will be determined, in part, by the accuracy of the preceding perceptual judgment. In basketball the point guard has at least six broad response options to choose between on any occasion (continuing to dribble the ball, shooting the ball at the basket or passing to one of four possible team mates) although each option involves a number of sub-decisions. For example, if the decision is to pass or shoot what type of pass or shot is most appropriate? Determining what option to select will generally depend not only on the current perceptual information but also on other situational information such as the state of the game (the score and the time remaining), the player's confidence and knowledge about the respective capabilities of matched team mates and opponents.

Once the desired action (for example a jump shot) has been selected, the player must then organise the movement before it can be initiated. This organisation involves sending from the brain motor commands that specify the order and timing of motor unit recruitment. If these efferent commands are not appropriately structured the resulting movement pattern may lack the force, timing or co-ordination necessary to successfully realise the objective of the movement (in this case the shooting of a goal). All three central processes of perception, response selection (decision-making) and response organisation and execution (acting) are completed before any observable muscular contraction takes place or whole-body movement occurs. Feedback during the shooting movement may assist in adjusting the motor commands although the skill of shooting is, in all probability, too short in duration for feedback-based corrections to have time to

Type of evaluation	*Was the movement executed as planned?*		
	Outcome	YES	NO
Was the goal accomplished?	YES	Got the idea of the movement	Surprise!
	NO	Something's wrong	Everything's wrong

Figure 17.4 The relationship between information about movement execution and information about movement outcome as a basis for guiding future attempts at a skill
Source: Gentile, A.M. (1972), A working model of skill acquisition with application to teaching, *Quest*, 17, 3–23. (Copyright 1972 by Quest Board) p. 9

be effective. Visual feedback derived from the completed action does, however, provide a valuable source of information for the performer to help in future repetitions of the same or similar actions. If the shot is too short, for example, this information can be used to ensure that more motor units are recruited the next time the player opts to take a shot from a similar position on the floor. Comparison of visual information about the outcome of the movement with kinesthetic information from the execution of the movement provides a valuable means of 'calibrating' the force production system, enabling the player to 'find his or her range' (Figure 17.4).

In the sections that follow we shall examine the key central processes of perceiving, deciding and acting in a little more detail by considering their component processes and something of what is known of their capacities and limitations for processing information.

Perceiving—determining what is happening

Underlying processes

Perceiving involves determining what is occurring in the outside world (for example where are each of my team mates located?), what is occurring within our own bodies (for example am I in or out of balance? what position are my arms in?) and the current and ongoing relationship between these internal and external 'worlds' (for example where is my foot in relation to the side-line?). Perception, as we have noted earlier, is more than simply the passive reception of sensory information by our various receptor systems. It is rather an active process through which we interpret and apply meaning to the sensory information we receive. As our prior experiences, accumulated knowledge, expectations, biases and beliefs all contribute to perception, it is therefore not surprising that two people presented with the same pattern of stimulation (for instance looking at the same picture or experiencing the same kinesthetic sensations) will often perceive and report different things. This is true not only of simple visual images, of the type frequently used by psychologists (Figure 17.5), but also, as we shall see in Chapter 19, for the more complex images typical of natural movement tasks and, as we shall see in Chapters 20 and 21, for perception of social environments such as sport and exercise settings.

Perception involves a number of sub-processes. Central ones include:

Figure 17.5 A typical simple but ambiguous figure used by psychologists to demonstrate the subjective nature of perception. The pattern may be differentially perceived as either a vase or two faces looking at each other

- *detection* (the process of determining whether a particular signal is present or not)
- *comparison* (the process of determining whether two stimuli are the same or different)
- *recognition* (the process of identifying stimuli, objects or patterns)
- *selective attention* (the process of attending to one signal or event in preference to others)

Processing limitations

All of the sub-processes that make up perception are limited in their capacity to process information and can therefore potentially act to limit performance on any particular motor task. In detecting stimuli humans are limited not only by the range of physical stimuli to which their various sensory receptors can respond but also by the difficult task of distinguishing the firing of one or more sensory neurons from a background of general neuronal activity. Human ability to detect stimuli consequently varies from situation to situation, from one sensory system to another, and with factors such as arousal (see Chapter 20). Humans are also limited in their ability to compare and detect differences between two or more stimuli. In judging the approach velocity of objects (such as when balls are thrown toward a person to catch), for example, the object must be increased in velocity from, on average, 4.4 m/s (16 km/h) to 5 m/s (18 km/h) before the change in speed can be reliably detected. The sensitivity for detecting differences varies from one sensory system to the next, with the visual and auditory systems being able to reliably detect the smallest changes in stimulation.

In attempting to recognise stimuli, objects or events (such as recognising different positions of the arm kinesthetically), laboratory experiments have shown that humans are limited to storing some seven items before they start to make errors of identification (Figure 17.6). If more than seven items have to be recognised at any one time recognition errors occur. In natural settings the number of patterns we are able to recognise (such as friends' faces or offensive patterns in basketball) is much greater than seven but nevertheless finite. Recognition in natural settings is enhanced through the use of multiple attributes (such as hair colour, length, style, eye colour, nose size and ear type in recognising human faces; or player location, posture and size in recognising basketball offensive patterns).

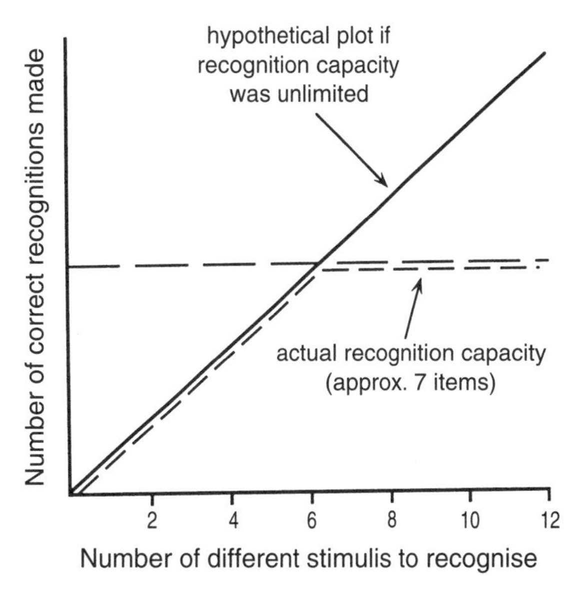

Figure 17.6 Information-processing limitations in the capability to identify different stimuli or patterns
Source: Marteniuk, R.G. (1976), *Information Processing in Motor Skills*, p. 65. (Copyright 1976 by Holt, Rinehart & Winston.)

Selective attention is both a limitation and advantage to human performance. We all know from personal experience that our processing capacity is limited in that we cannot (typically) listen to two separate conversations or attend to two separate visual signals (such as events in the left and right extremes of our field of view) simultaneously. If a movement task requires information from two or more separate locations to be processed at the same time performance of the task will generally be difficult. Being able to selectively attend to typically only one thing at a time can be an advantage, however, in that it provides a means of preventing irrelevant or potentially distracting stimuli from using up some of our valuable processing capacity. A golfer focusing on a putt, or a microsurgeon focusing on a suture, benefit from being able to apply all their attention to the specific movement and thus effectively block out surrounding noise and other potentially distracting events.

We shall see in Chapter 19 that selective attention, like recognition and other aspects of perception, can be trained. We shall also see in Chapter 19 that perceptual skill is very situation specific and that consequently to understand perception for movement it is necessary to examine the specific input information that exists for the movement. Although tests of perception range from the very simple (such as the simple visual acuity tests used in the issuing of driver's licences) to very complex, the more a test of perception precisely simulates the usual information processing required in the motor skill of interest, the more valuable is the information it can reveal about motor control.

Deciding—determining what needs to be done

Underlying processes

Decision-making is essentially the process of response selection—picking the right movement option to match the current circumstances. The quality of the decision made will obviously depend on the quality of the preceding perceptual judgments as well as the knowledge of the costs and benefits associated with each particular option. The latter is heavily dependent on the extent of the individual's experience. The speed and accuracy of decision-making about movement is also influenced by such things as the number of possible options (or response choices) that exist, the costs associated with making incorrect decisions, and the total time available to make decisions. Some activities, such as playing golf, offer essentially unlimited periods of time in which to select the correct action whereas in other motor skills, such as playing tennis, the time constraints on decision-making are severe.

Processing limitations

The measure used to determine how quickly people can make decisions is called choice reaction time (CRT). In the laboratory CRT is measured by presenting subjects with an array of stimuli (usually lights) each of which has its own associated response (frequently a button press) (Figure 17.7). The subjects' task is to view the stimulus array and respond, as soon as possible after a stimulus light is illuminated, by depressing the response button that corresponds to the illuminated light. CRT is then recorded as the time elapsed between the illumination of the stimulus light and the initiation of the button press. The limitations that exist on rapid decision-making can be examined by recording CRT while the number of possible stimulus–response pairs is systematically varied. (A task in which there are six possible stimulus–response pairs would form a reasonable analogue of the response choice facing the basketball guard in our earlier example.)

Studies in which the number of stimulus–response alternatives have been varied are consistent in their findings. When there is no uncertainty (there is only one possible stimulus and response) the reaction time to the appearance of this stimulus is about 200 ms (one-fifth of a second). This is also the delay in responding observed in sprint events between the sound of the gun and the beginning of movement, and in hand–eye co-ordination tasks when there is unexpected movement of the object that is to be intercepted (Box 17.1). As the amount of uncertainty is increased by adding more and more

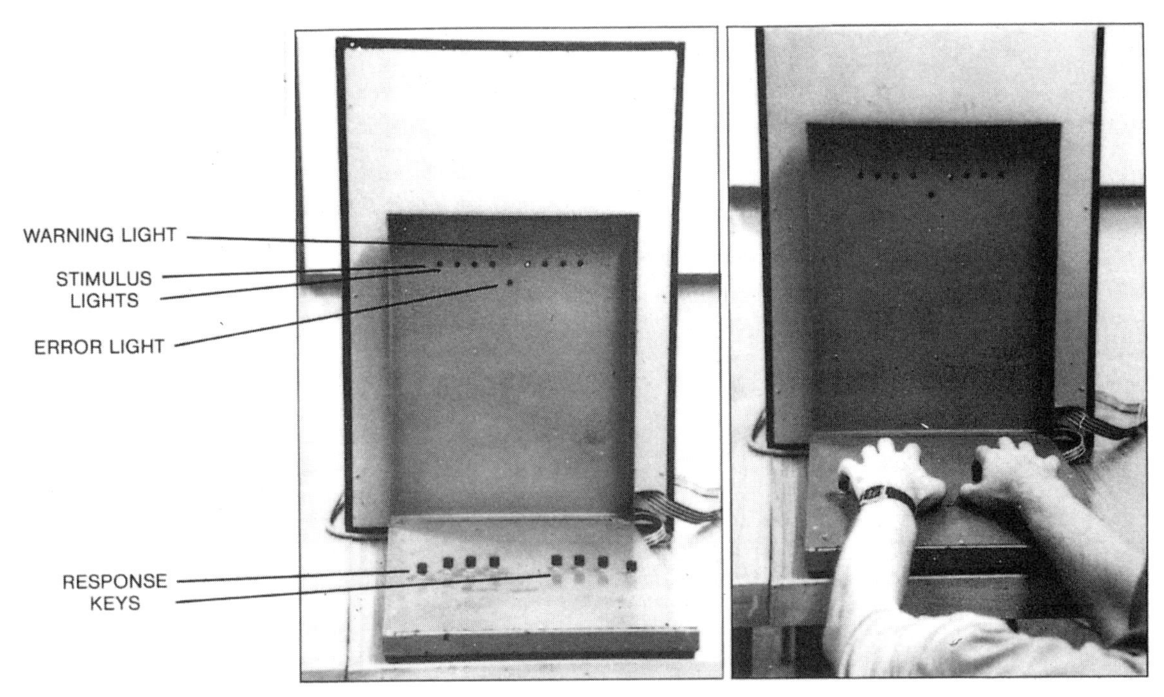

Figure 17.7 Typical laboratory apparatus for the measurement of choice reaction time
Source: Abernethy, B. (1991), 'Acquisition of motor skills,' p. 79, in F.S. Pyke (ed.), *Better Coaching*, Australian Coaching Council, Canberra. (Copyright 1991 Australian Coaching Council Incorporated; reprinted with permission.)

possible options, CRT increases substantially (Figure 17.8). Each time the number of possible stimulus–response alternatives doubles CRT increases by a constant amount, such that the increase in CRT from a two-choice situation to a four-choice situation approximates the increase in CRT from a one-choice to a two-choice situation and from a four-choice to an eight-choice situation.

This relationship between CRT and number of stimulus–response alternatives is frequently exploited in a range of movement tasks. Designers of cars and machinery try to minimise the number of options on machine controls to reduce the decision-making time of users of the equipment. Skilled sports players attempt to familiarise themselves with the preferred options and patterns of play of their opponents as a means of speeding up their own rates of responding. The basketball players who recognise that their opponent can

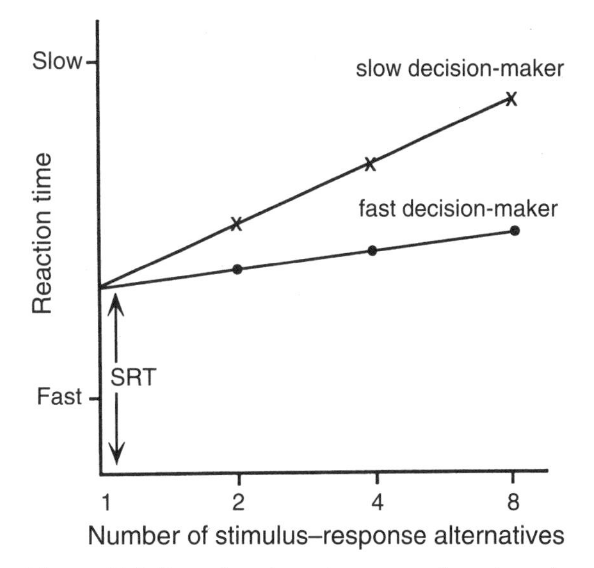

Figure 17.8 Reaction time shown as a function of the number of stimulus–response alternatives for two different subjects

Box 17.1 Measuring reaction time in a natural setting

In a controlled laboratory setting it takes the average person about 200 ms to respond to an unanticipated stimulus. Peter McLeod from the Department of Psychology at the University of Oxford, England was interested in ascertaining if processing delays of this same order exist in natural movement tasks (and therefore act as a limiting factor to skilled performance) or whether, in the natural setting, reaction times might be much shorter.

McLeod designed an innovative experiment to measure the reaction time of cricket batsmen. Cricket batsmen perform the extremely time-constrained task of striking a fast-moving ball as it bounces off a batting surface (the pitch). By placing strips of wooden dowelling under the pitch around the region where the ball would typically bounce, McLeod created a situation in which some balls (landing between the dowelling strips) bounced normally while others, landing close by but on the edge of the dowelling strips, deviated unexpectedly in a lateral direction. By comparing the path of the bat—through standard biomechanical methods—between trials in which the ball bounced normally and those in which it deviated unexpectedly McLeod was able to measure the minimum time it took the batsman to initiate a corrective movement response to the unanticipated stimulus (the deviation of the ball).

The reaction times measured were precisely in the 200 ms range as observed using typical laboratory measures (Figure 17.11). In natural tasks skilled performers must therefore develop strategies that enable them to cope with this delay in responding that is inbuilt in their nervous systems.

Source: McLeod, P.N. (1987), Visual reaction time and high-speed ball games, *Perception*, 16, 49–59.

only dribble the ball with their right hand will be able to respond more rapidly to their opponent's moves than a player who considers that both left and right hand alternatives may occur. Conversely the player who is able to execute, with equal skill, a wide range of options (through having equal shooting, passing and dribbling skills on both sides of the body) can maximise the amount of information which has to be processed by an opponent and can consequently slow the speed of decision-making of their opponent substantially.

Close inspection of Figure 17.8 reveals two independent components to decision-making. One component, given by the intercept of the CRT–alternatives line with the *y*-axis, corresponds to reaction time when there is only one option, and therefore no uncertainty. This component, known as simple reaction time, is not influenced by practice and reflects individual differences in the time it takes for the afferent nerve impulses to reach the brain, be registered there and then efferent commands sent to the muscles. The second component, given by the slope of the CRT–alternatives line, is a measure of decision-making rate; it estimates the average increase in CRT that occurs for each new additional stimulus–response option. The slope of the line is steep for individuals who are slow decision-makers and approaches zero for individuals who are very fast decision-makers. Advance knowledge about the probabilities of different events occurring can reduce the amount of information to be processed and make for faster decision-making.

Acting—organising and executing the desired movement

Underlying processes

Once a particular movement response has been selected, there still remains significant responsibility for the central nervous system in ensuring that the selected movement response is actually executed as desired. At least three further sub-processes are involved at this stage in the processing of information for movement control; these are:

- *movement organisation*—planning out carefully the sequencing and timing of the efferent commands to be sent out to selected motor units
- *movement initiation*—transmitting the required motor commands to the muscles
- *movement monitoring*—adjusting the movement commands on the basis of sensory information about the movement's progress.

Processing limitations

The speed and accuracy with which movements can be executed and controlled depend on a number of factors including the complexity of the movement (the number of joints, muscles and motor units involved plus the difficulty their co-ordination may pose for the maintenance of posture and balance), the time constraints imposed on the movement, and the acceptable margins for error in the movement. Movements of relatively long duration (greater than one-third of a second) can use feedback generated during the movement itself to assist in their control and precision. Control based on the monitoring of feedback is known as closed-loop control (Figure 17.9). In contrast, very rapid movements require all the efferent commands to be structured in advance. This type of control is known as open-loop (or programmed) control.

For movements controlled in a closed-loop manner, the time taken to complete the movements (movement time) is directly dependent on the difficulty of the movement. Movements with high-precision demands (such as movements to small targets) and/or traversing a large distance take much longer than movements made over a short distance to large targets. The relationship between movement time and movement difficulty is a lawful one governed by the amount of information that must be processed (Figure 17.10). For movements controlled in an open-loop manner, the time taken to initiate the movements (that is, reaction time) is directly proportional to the amount of pre-planning that must take place. This is greater for more complex movements.

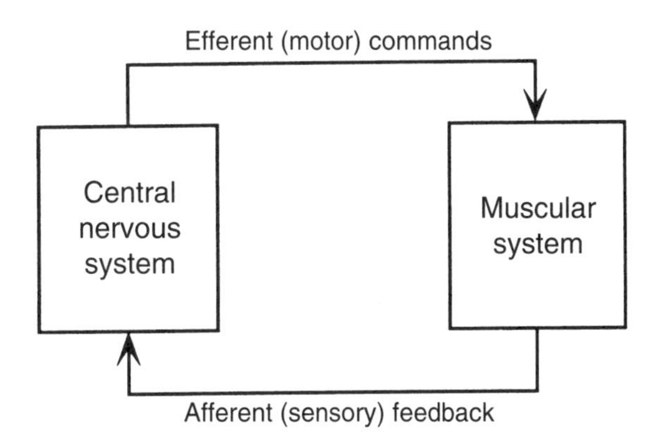

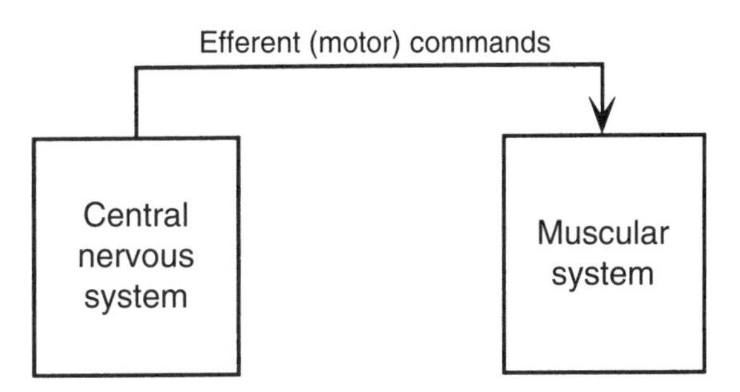

Figure 17.9 Closed–loop and open–loop systems for movement control

Some implications

Having now examined, in some detail, the central processing stages proposed by an information-processing model of motor control it is now possible to consider some of the implications and insights such a model may provide.

One insight that emerges from consideration of the central processing stages relates to the sources of error in motor skill performance. If our hypothetical basketball guard had attempted

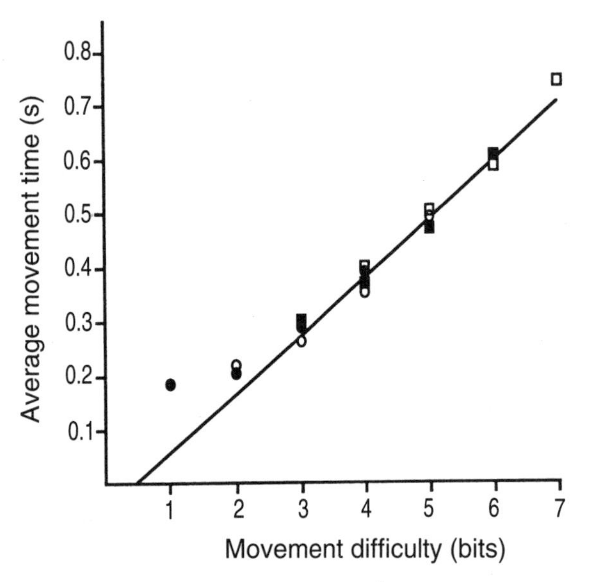

Figure 17.10 Average time to make a movement plotted as a function of the difficulty of the movement. The task involved moving as rapidly as possible for a 10 s period between two targets. Movement difficulty was manipulated by changing the distance between the targets and the size of the targets
Source: Data are plotted from Fitts, P.M. (1954), 'The information capacity of the human motor system in controlling the amplitude of movement', *Journal of Experimental Psychology*, 47, 381–91. (Reproduced with permission of the American Psychological Association.)

to shoot a jump shot but had only succeeded in losing the ball or producing an inaccurate shot it would be very tempting to attribute this error simply to poor movement execution technique and try to remedy this by having the player practise his or her shooting skills. However, it should be apparent from our preceding discussions that an error in performance could as equally result from poor perception and/or poor decision-making as from poor movement execution. Given the sequential nature of central information processing, a perceptual error (such as incorrect judgment of the distance from the basket or failure to detect the movement of one of the defensive players) or a decision-making error (such as shooting from a position where passing would have been a better option) can just as much cause an ineffective shot as poor execution of a correctly selected movement. Likewise a therapist attempting to recover the normal gait of a stroke patient or assist the skills development of a clumsy child needs to be aware that deficiencies in movement control can occur even when the ability to actually execute a selected movement may be essentially normal.

A second, related implication of the information-processing model is that although the three central stages of information processing are not directly observable they may nevertheless need to be trained just as much as the more observable components of performance such as technique, strength, speed, endurance and agility. In many motor skills the perceptual and decision-making aspects of movement control may act as the limiting factors to performance and therefore warrant systematic training. This is especially true of activities, such as the fast ball sports, in which decisions must be made in a very short time on the basis of only limited information (Figure 17.11). Conditions such as ageing, injury or disease, which slow the speed with which movements can be produced, also exacerbate the necessity for good perceptual and decision-making skills to offset delays imposed by longer movement times.

Some alternative models of motor control

Although modelling the motor control system as an information-processing (or computational) system appears to be a useful way of starting to think about the neural control of movement, it is certainly not the only way. In the past 10–15 years a number of limitations and assumptions within the information-processing model have been highlighted and some alternative models suggested. Critics of the information-processing model have been concerned by the implicit assumption that every movement is somehow represented and stored in the central nervous system. They have argued that simply assigning

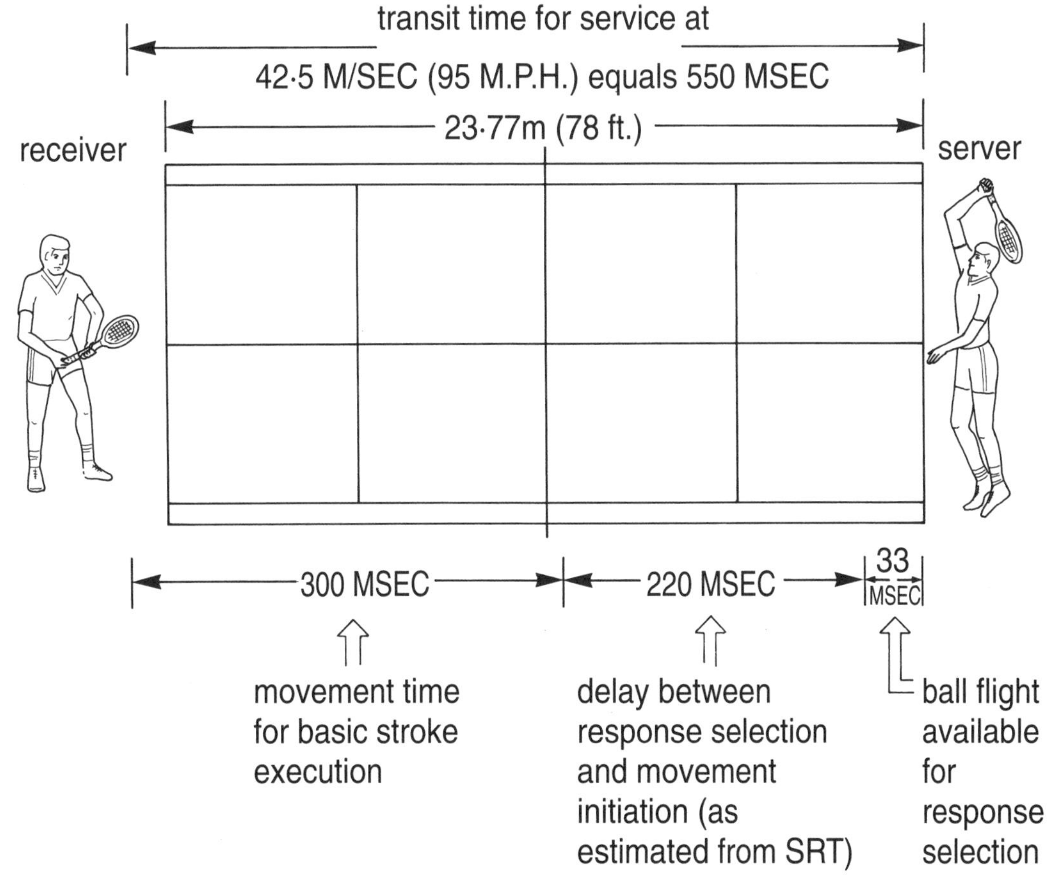

Figure 17.11 An example of the time constraints in a fast ball sport. The player receiving service in tennis may have only some 30 ms to decide what stroke to play
Source: Abernethy, B. & Russell, D.G. (1983), 'Skill in tennis: Considerations for talent identification and skill development,' *Australian Journal of Sport Sciences*, 3(1), 3–12. (Copyright ASMF & ACHPER, 1983; reprinted with permission.)

the responsibility of movement organisation to the brain does not explain movement control but simply creates the (unacceptable) need for an intelligent little person (a 'homunculus') somewhere within the brain! Critics of the information-processing model have been concerned also about the assumption that the brain and nervous system directly control the specific motor commands for all aspects of a movement, noting rather that many of the physical properties of the musculoskeletal system (such as the spring-like characteristics of muscle and the natural oscillatory frequencies of limbs) can themselves contribute significantly to movement control without the necessity for any central nervous system involvement.

Dynamical models of movement control propose that movement patterns are not represented anywhere in the nervous system by way of a plan or program but rather emerge naturally (or self-organise) out of the physical properties of the musculoskeletal and related systems. The emergence of complex patterns of organisation out of the motor system, such as the complex movement patterns that characterise gaits like walking and running, are considered to be no more in need of a pattern representation or template than are non-biological systems, such as chemicals, which, under appropriate environmental conditions, are able to organise and reorganise into complex patterns, without the need for a nervous system. For example, under appropriate environmental conditions (level of heat) water may undergo complete pattern reorganisation without the form of the new pattern being anywhere explicitly stored or represented. As some aspects of movement control share these same characteristics of pattern reorganisation at critical levels of control parameters (see Figure 17.12) there is considerable current research interest in trying to explore this model of movement control further by determining the control parameters for a range of human movements and exploring other parallels between movement pattern formation and pattern formation in physical systems (see Box 17.2). As more is understood experimentally about movement control, new models that progressively encapsulate more of the many essential and unique characteristics of movement listed earlier in this chapter can and will be developed.

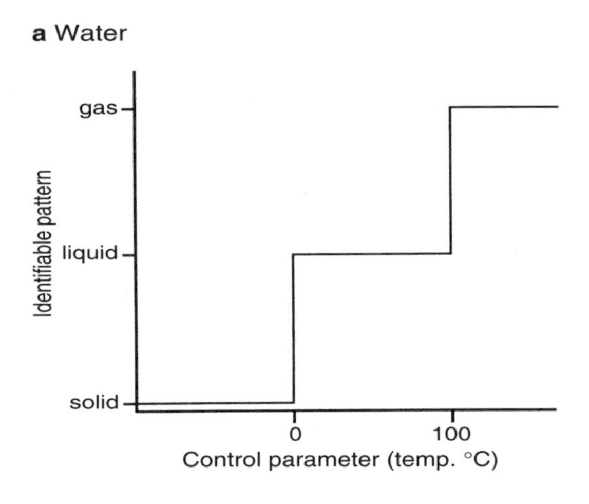

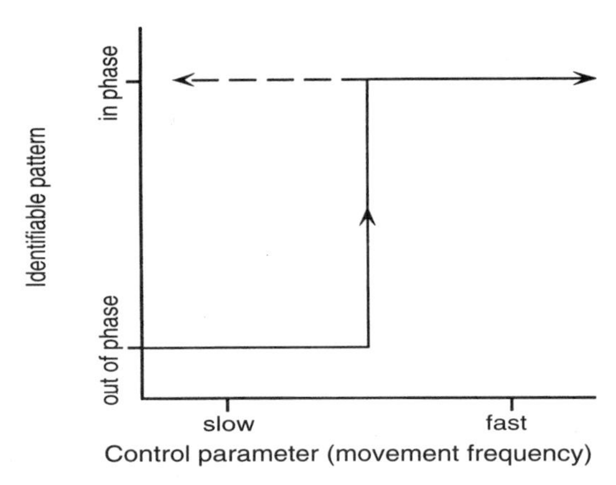

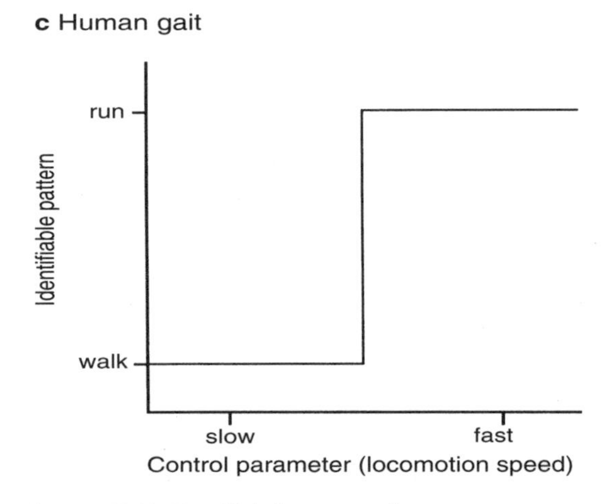

Figure 17.12 Parallels between the pattern transitions observed in non-biological systems and those occurring in some human movements

Box 17.2 Pattern transitions in human hand movements

In purely physical systems (such as chemical and laser systems) spontaneous transition from one form of organisation or pattern to another occurs when key environmental conditions (or thresholds) are reached. For example, we all know well that the water molecule spontaneously reorganises its structure as it is heated past the critical control temperatures of 0° and 100°C (Figure 17.12). The study of such transitions is called synergetics. Synergetic transitions share a number of common characteristics including sudden jumps in organisation around critical levels of the control parameter, increased variability in structure as the transition point is approached (a property called critical fluctuations) and a delay in returning to stability if the system is perturbed in some way when it is near a transition point (a property called critical slowing down).

In an extensive series of studies, Scott Kelso from Florida Atlantic University in the United States, Herman Haken from the University of Stuttgart, Germany and a number of co-workers set out to determine whether transitions in movement patterns also share these synergetic characteristics. They used as their task paired movements of the index fingers with the index fingers prepared either out-of-phase or in-phase (Figure 17.13). When the fingers started out-of-phase and the subjects were required to move the fingers progressively at a faster rate a critical frequency was reached at which the fingers moved spontaneously from out-of-phase to in-phase. The phase relation between the two fingers was found to increase in variability and be slower to respond to perturbations as the transition point (a point of pattern instability) was approached. Thus the critical fluctuations and critical slowing down properties seen in other synergetic systems also exist in some aspects of the human motor system. The in-phase co-ordination was found to be a stable one, which was able to be preserved across all finger movement frequencies.

Sources: Haken, M., Kelso, J.A.S. & Bunz, H. (1985), A theoretical model of phase transitions in human hand movements, *Biological Cybernetics*, 51, 347–57; Kelso, J.A.S. & Schöner, G. (1988), Self-organization of co-ordinative movement patterns, *Human Movement Science*, 7, 27–46.

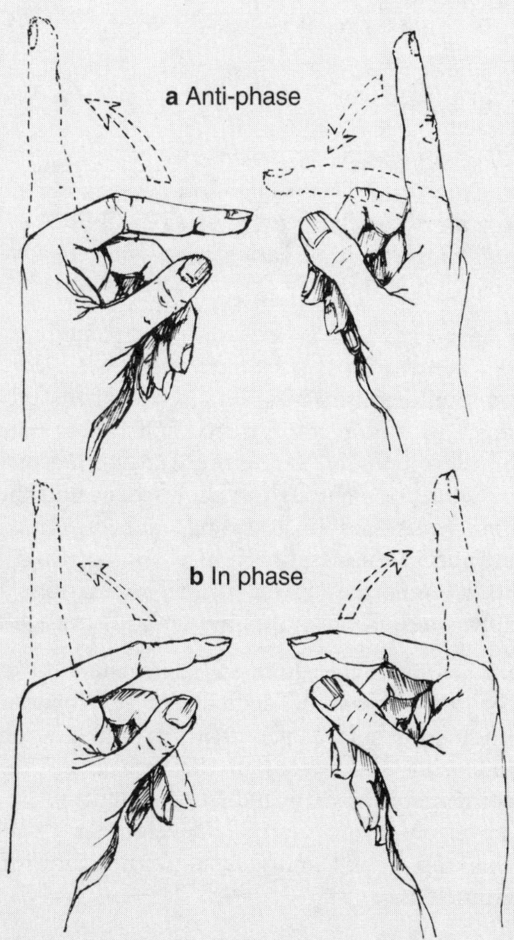

Figure 17.13 When the rate of paired figure movement is progressively increased a spontaneous shift from anti-phase (one finger extended while the other is flexed) (a) occurs to in-phase (b)

CHAPTER 18

Motor control changes throughout the lifespan

- Changes in observable motor performance
- Changes at the neurophysiological level
- Changes in information-processing capabilities

In Chapters 16 and 17 we examined some basic neurophysiological and psychological concepts of motor control. In this chapter these basic concepts are used to examine motor control changes across the lifespan. The specialised field of study concerned with the description and explanation of changes in motor performance and motor control across the lifespan is typically referred to as motor development. While the study of motor development has historically focused on the period from conception through to adolescence and the changes and stages through which the developing human progresses in attaining adult levels of motor performance, it is now also increasingly interested in the deterioration of motor skills that are apparent in the elderly. Such a broadened focus is important given ageing populations worldwide, and given that many changes in the performance and control of motor skills are age-related and occur throughout the entire lifespan.

Studies of motor development have generated knowledge about many things including:
- the normal rate and sequence of development of fundamental motor skills that arise out of the interaction of biological maturity and environmental stimulation
- individual differences in the rate of development of specific skills (and motor performance comparisons of early and late developers)
- deviations from normal development in special populations (such as the intellectually challenged, and individuals with conditions such as cerebral palsy, Down's syndrome or clumsiness).

Such knowledge is important practically in assessing the normative development of children, in screening for neurological and motor disorders as well as identifying and nurturing exceptional talent, in informing the designers of remedial and therapeutic programs, and in ascertaining the readiness of individual learners for new challenges. For both the young and the elderly, knowledge from the field of motor development is valuable in highlighting the role of regular practice and physical activity in the acquisition and retention of movement skills.

Motor development knowledge is typically derived from research studies of one of two main types: cross-sectional studies, in which motor control and performance are compared between different people of different ages, and longitudinal studies, in which the motor control and performance of the same set of subjects are traced over a number of years as the subjects mature and grow older. The changes in motor control and performance that occur across the lifespan may be described, recorded and explained at a number of different levels of observation. In this chapter we first examine observable changes in motor skills across the lifespan and then, in keeping with the previous two chapters, we examine underlying changes taking place at the neurophysiological level and at the level of information-processing capabilities.

Changes in observable motor performance

Motor development in the first 2 years of life

General development principles

Significant advances in voluntary motor control occur over the first 2 years of life, with these advances following two general principles. The *cephalo-caudal principle* highlights that development proceeds in a 'head-down' manner, with control developing first in the muscles of the neck, then followed in order by control of the muscles of the arms, trunk and legs. The *proximo-distal principle* highlights that development progresses from the axis of the body outwards with control being achieved over muscles crossing the trunk (and axial skeleton) before control is achieved over the muscles controlling the limbs (and appendicular skeleton). These cephalo-caudal and proximo-distal changes in motor control are also paralleled by changes in muscle tone. At birth, muscles controlling the movement of the axial skeleton typically have low tone whereas those controlling the movements of the limbs have high tone (or stiffness). As the infant matures this situation reverses, with the axial muscles gaining tone (to assist in maintaining posture against gravity) and the muscles of the limbs decreasing in tone (to facilitate more efficient, voluntary control of limb movements). Reflexes present at or soon after birth either disappear, are reduced in strength (attenuated) or are modified during the first 24 months of life.

Motor milestones for normative development

In order to eventually achieve movement independence it is vital in the first years of life that human infants develop control of their general body position (or posture), develop an ability to move (locomote) throughout their environment, and develop an ability to reach, grasp and manipulate objects using their hands. Because posture, locomotion and manual control are so vital to the infant's development, a number of researchers have studied, and documented in great detail, the ages and stages that infants pass through in the acquisition of these essential motor skills. Knowing the average and range of chronological ages at which key developmental stages are typically reached in these skills provides a set of motor milestones that can be used to monitor the motor development of each individual child, and to detect, at an early age, possible movement and neurological difficulties.

Development of the control of posture occurs through three main stages (head control, sitting and standing), each of which contains within it a number of identifiable progressions (Figure 18.1). Head control is achieved first with a major milestone being the ability to hold the head steady while being carried. This control is typically achieved at 2–3 months after birth. After head control is achieved, sitting without support is then typically achieved at around 5 months. By 7 months infants can find their own way to a stable sitting posture without being

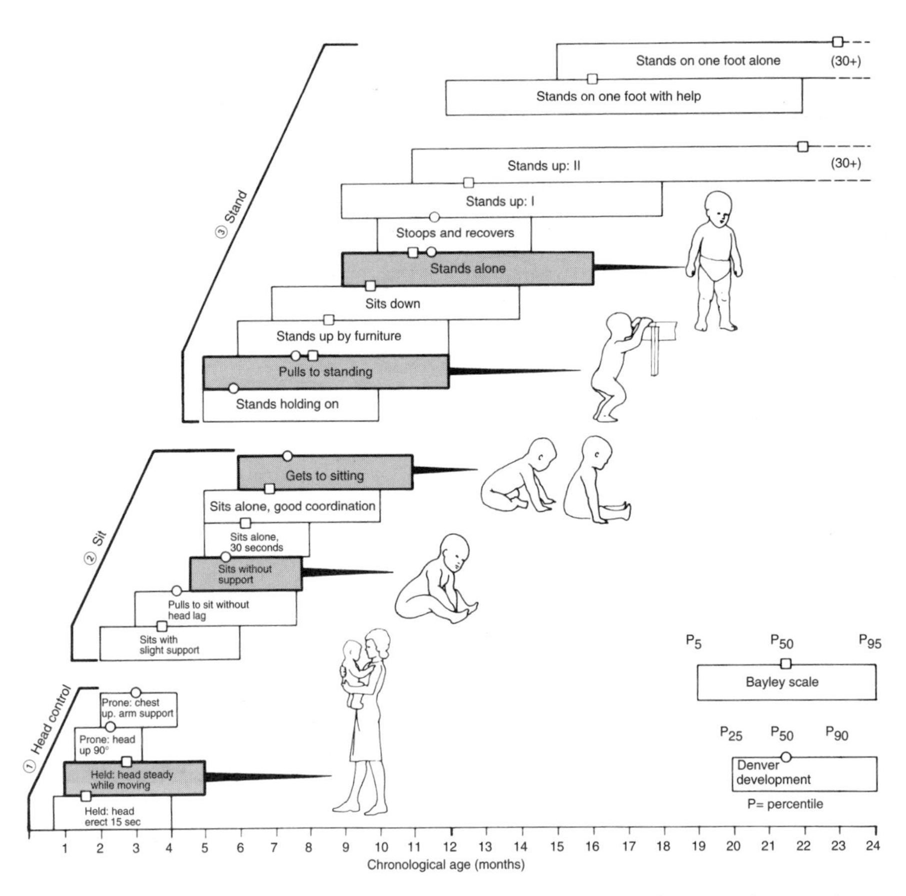

Figure 18.1 Identifiable progressions in the development of postural control
Source: Keogh J. & Sugden, D. (1985), *Movement Skill Development*, p. 32.
(Copyright by Macmillan Publishing Company, 1985.)

reliant on being placed in this position. Key milestones in the attainment of postural control for standing occur later, with babies, on average, being able to pull themselves to supported standing by 7–8 months and then stand alone, by age 11 months.

A number of the major motor milestones for the achievement of the locomotion skill of walking occur in parallel with postural control developments. Locomotion in a horizontal position (rolling, creeping and crawling) is achieved before upright walking (Figure 18.2). Major motor milestones in the pre-walking stage include attaining the ability to roll over from their backs to their stomachs and reverse (this occurs, on average, around 6 months of age) and attaining the ability to progress forward either by crawling (with arms pulling and legs pushing) or creeping on all fours. Creeping and/or crawling typically appears when babies reach 7 months of age. Upright, unassisted walking appears, on average, at the end of the first year, although walking may first appear anywhere in the range from 9 to 17 months. A general expectation is that most babies will achieve the basics of walking during the first 6 months of their second year and then refine this skill considerably by the end of their second year.

The rudiments of mature grasping and reaching control of the hands are in place typically by the end of the 1st year of a child's life, although the functional use of the hands for self-help activities, such as using a spoon, and more complex manual acts such as writing or tying shoe laces, continues to be refined for many years (Figure 18.3). Key milestones in the

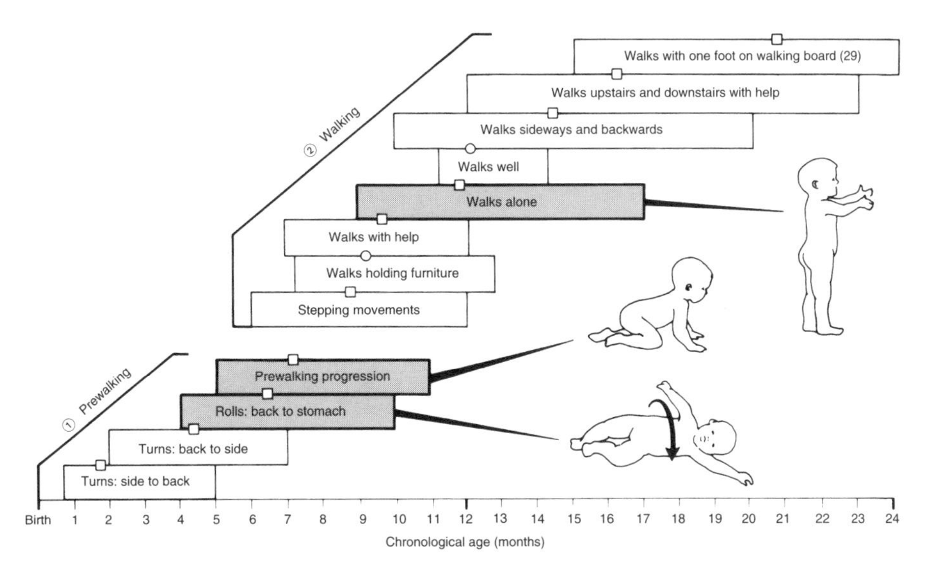

Figure 18.2 Identifiable progressions in the development of locomotion
Source: Keogh, J. & Sugden, D. (1985), *Movement Skill Development*, p. 38.
(Copyright by Macmillan Publishing Company, 1985.)

control of grasping and object manipulation by the hands occur, on average, at 3–4 months (when they first pick up large objects such as cubes), at 6–7 months (when opposition of the thumb is achieved) and at 9–10 months (when the pincer grasp for picking up small objects is first mastered). Reaching skills, controlled more by the proximal muscles of the shoulder and the arm than by the more distal muscles of the wrist and the intrinsic muscles of the hand, are achieved somewhat earlier than manipulation skills. The major motor milestone of reaching to touch a desired object is attained, on average, after some 3–4 months. Control of basic reaching and grasping lays the foundation for the acquisition of the many complex manual movements that are fundamental to many of our uniquely human actions, gestures and communications.

Motor milestones in special populations

Some special populations who suffer movement difficulties as adults are also typically slow in reaching a number of the major milestones for motor development. Children with Down's syndrome (a chromosomal abnormality that affects many aspects of development) are systematically late in achieving virtually all major motor milestones. For example, Down's syndrome children walk unsupported on average only after 2 years whereas in children without this syndrome walking, as we noted in the previous section, is generally achieved after 1 year. Mental retardation of all forms also acts to delay the attainment of motor milestones, with children with the lowest intelligence quotients (IQs) being the least likely to stand or walk within the first 12 months (Figure 18.4). Motor milestones may therefore act as early markers of potential adult movement problems as well as possible neurological problems, and therefore may serve as valuable diagnostic and detection tools. Children described by parents and teachers as clumsy but lacking any apparent neurological problems may also be identified early through assessment of motor milestones and offered appropriate additional training and intervention.

The notion of critical periods

The use and identification of motor milestones is based on the observation that the majority of children pass through the same basic stages in achieving mastery over those motor skills (such

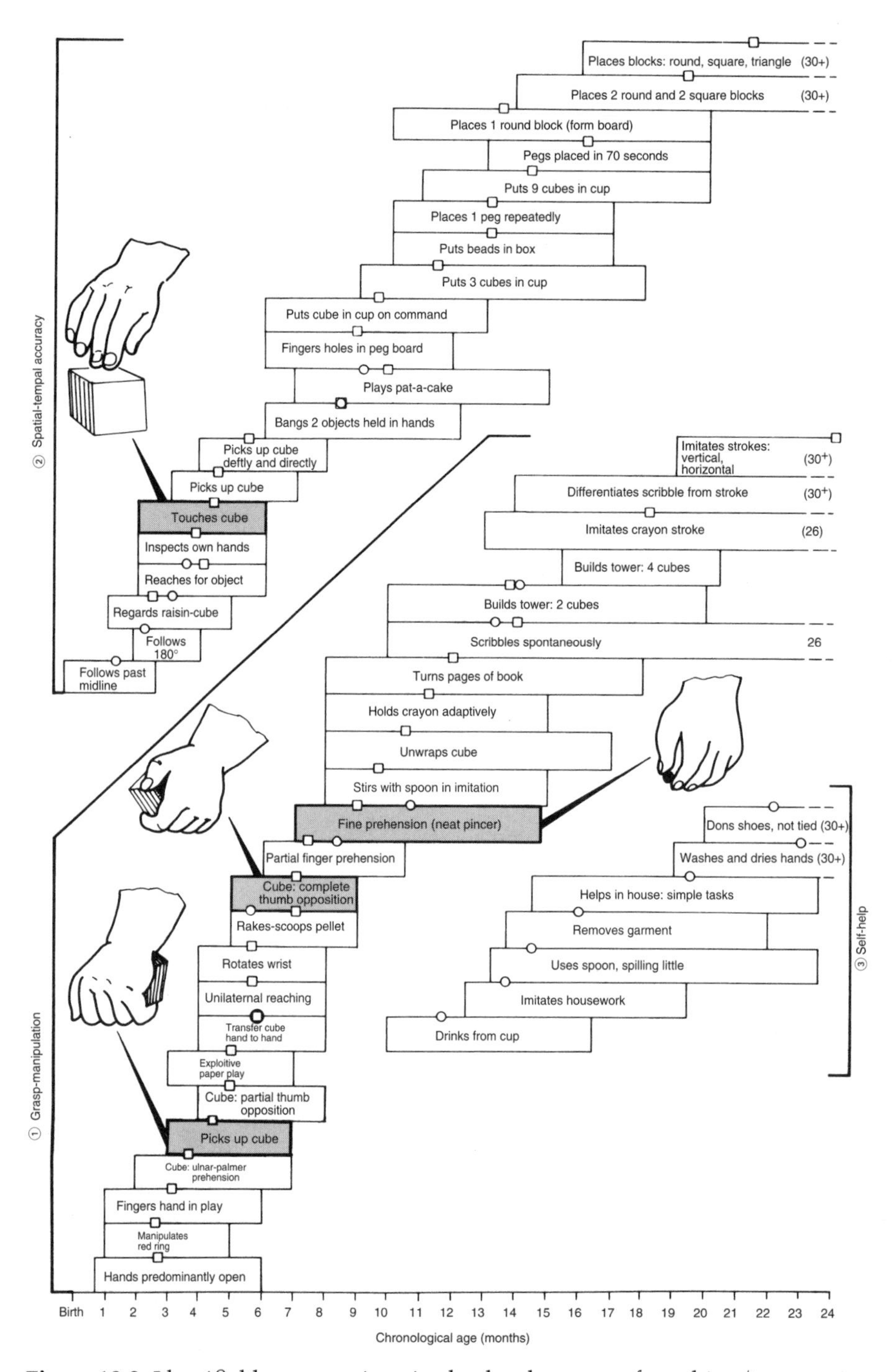

Figure 18.3 Identifiable progressions in the development of reaching (upper panel) and grasping (lower panel)
Source: Keogh, J. & Sugden, D., (1985), *Movement Skill Development*, p. 46. (Copyright by Macmillan Publishing Company, 1985.)

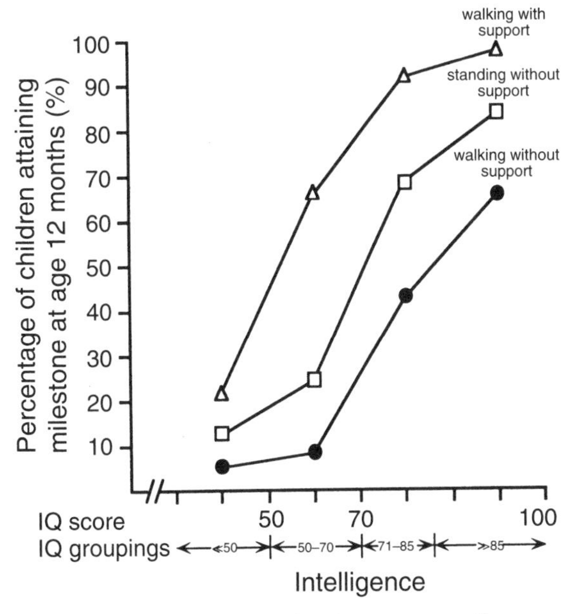

Figure 18.4 The relationship between intelligence and the attainment of specific motor milestones by the end of the first year of life. An IQ of 100 corresponds to the population mean
Source: Drawn from data reported in Makinen, H., Von Wendt, L. & Rantakallio, P. (1984), Psychomotor development in the first year and mental retardation: A prospective study, *Journal of Mental Deficiency Research*, 28, 219–25.

as posture, locomotion and reaching and grasping) that are fundamental to survival. (However, there are notable exceptions. For example, it is not uncommon for children to walk without ever first learning to crawl.) The basic similarity in stages across children suggests a significant inherited or genetic influence on the emergence of those motor skills essential for survival. An obvious question therefore is whether these normal progressions can be altered by environmental factors (for example can an enriched environment accelerate the rate of motor development or an impoverished one retard it?) and whether there are critical periods in which the child is most sensitive to learning a particular skill? There is clear evidence of critical periods in other aspects of human development. Language, for example, is most easily acquired in childhood and if it is not acquired at this time it is very difficult to pick up subsequently.

The evidence for critical periods in motor development is not clear cut. Some studies of children from cultures that restrain their infants from movement during part of the 1st year of life have indicated lower than expected levels of adult motor skill whereas other studies have demonstrated no such detrimental effects, with early delay in motor milestone achievement being quickly recovered at later years. Clear evidence is available, however, to indicate that enriched environments, in which there is extensive stimulation and many opportunities for motor exploration and play, can speed up the rate at which motor milestones are achieved. What is less apparent is whether or not the early achievement of milestones translates into improved performance at adult levels. Children who are relatively late in achieving some milestones may attain other skills rapidly provided they are exposed to them during a critical period (Box 18.1). Programs to enhance early motor development (such as Gymboree programs, infant swimming programs, and the Suzuki method for learning to play musical instruments) are all based on the assumption that early attainment of milestones is beneficial to longer term motor skill acquisition.

Practical applications

Motor milestone scales (such as those presented in Figures 18.1–18.3) provide a useful means of comparing the rate of motor development of any specific child against that of children of the same chronological age. As the ranges of ages on each milestone in Figures 18.1–18.3 indicate, there is enormous normal variability on all these measures and a child advanced on one milestone may be below the average on another. The normal range of variability also tends to increase as the child gets older and the task more difficult, probably reflecting the cumulative effect of environmental rather than genetic influences. The observation that environment can influence motor development is important as it suggests, among other things, that early intervention for children with movement difficulties offers the best option for effective development. Motor milestones may therefore, within the recognition of the scope of normal

Box 18.1 Critical periods and enriched environments — the story of Johnny and Jimmy

In 1935 the American developmental psychologist Myrtle McGraw conducted one of the classical studies on motor development, examining the role of enriched environments and experiences on the rate of acquisition of a range of movement skills. To separate the role of environment from any hereditary factors that may influence motor development, McGraw studied intensively the early development of two male twins, Johnny and Jimmy, who were raised in different environments. Johnny was given exposure to free play with a wide variety of toys as well as extensive stimulation, practice and experience on a wide variety of movement activities. Jimmy was given few toys and spent the majority of the time in his cot. The twins were given different movement experience conditions from 21 days to 22 months of age and their capacity to acquire and retain various motor skills observed and described throughout this period and at a number of times after (up to age 6 years).

McGraw found, among other things, that critical periods existed for the learning of certain skills. Johnny, for example, was given considerable practice and instruction on riding a tricycle at age 11 months but showed no sign of any acquisition of the skill of tricycle riding until around 19 months of age. Jimmy, in contrast, was first exposed to a tricycle at age 22 months but learnt the skill almost immediately. This suggests that a certain amount of maturational readiness is necessary before some skills can be learnt. Whereas in the case of learning the voluntary skill of tricycling Johnny's early movement experiences were of no apparent advantage, on many other motor tasks Johnny not only acquired the skills earlier than Jimmy but this superior performance persisted for a number of years after the enrichment period had finished. McGraw reported that overall Johnny appeared to have superior movement competence and confidence, demonstrating the important role that early movement experiences may have on the acquisition and retention of a number of motor skills. Following 22 months of minimal motor activity Jimmy was able to 'catch up' to Johnny in performance on some but certainly not all motor skills.

Given that enriched environments are now known to be important for motor development, studies that deliberately deprive a child of exposure to such an environment would, in current times, be most unlikely to be permitted to proceed on ethical grounds.

Source: McGraw, M.B. (1935), *Growth: A study of Johnny and Jimmy*, Appleton-Century, New York. (Reprinted by Arno Press, 1995.)

variability, be used validly by teachers, therapists and parents as a screening tool for early detection of possible motor and neurological problems. Further, creating an environment that provides maximal opportunities for movement in the early years would appear to be a concern that should be treated seriously by all parents.

Development of fundamental motor patterns in childhood

From age 2 onwards children develop a range of fundamental motor skills that form the basis for specialised adult motor behaviour. As with the postural, locomotor and manual control skills from the first 2 years of life, substantial information is available detailing the ages and stages at which children from 2 through to 7 years of age and older acquire basic locomotor skills (walking, running, jumping and hopping) as well as non-locomotor skills (such as throwing, catching, hitting and kicking). In most cases information is available to outline developmental trends not only in performance (product measures) but also in the movement patterns used (process measures). In this section we will provide two examples of developmental stages in fundamental motor patterns, one from the locomotor skill of running and the other from the non-locomotor skill of overarm throwing.

Ages and stages in the development of a locomotion skill

A number of children begin to run at around age 18 months and most run by the time they are 2 years old. The running pattern of a child goes through a number of developmental stages before becoming mature, on average, by age 4–6. Initial running is characterised by short, uneven steps, a wide base of support and no easily observable flight phase (the phase where neither foot is in contact with the ground). This first stage of running usually occurs before adult mastery of walking is achieved. In mature running, strides are uniformly long and there is an identifiable flight phase characterised by a full recovery of the non-support leg (Figure 18.5 and Table 18.1). The early attainment of a mature running pattern is important given the central role running plays in many childhood games and activities. Running speed typically continues to improve throughout, not only childhood, but also adolescence as strength is gained and subtle refinements in technique produce improvements in efficiency, as outlined in Chapter 14.

Ages and stages in the development of a non-locomotor skill

The overarm throwing pattern is a discrete action used to propel objects and implements, such as balls and javelins. It is a basic movement pattern fundamental to many other more specific sports skills such as throwing a cricket ball or baseball, serving a tennis ball, spiking a volleyball or passing a waterpolo ball. Overarm throwing performance (regardless of whether the throwing is for distance, accuracy or form) continues to improve throughout childhood into adolescence. Even though rudimentary throwing patterns are in place certainly by the 2nd year of life, throwing velocity continues to improve steadily right throughout the primary school years (Figure 18.6).

As with running and all other fundamental motor skills, the transition from initial throwing form to mature throwing patterns involves a number of stages. Initial attempts at throwing are characterised by an absence of backswing, weight transference and upper and especially lower body involvement. At intermediate levels there is increased weight transference, trunk rotation and the involvement of a forward step, albeit from the leg ipsilateral (on the same side) to the throwing arm. Only at the mature stage does the non-throwing hand lead toward the target, the trunk rotate fully, and the lower body contribute to the summation of speed (as discussed in Chapter 11) through a forward step of the leg contralateral (opposite) to the throwing arm (Figure 18.7 & Table 18.2). By age 12 most boys have acquired the mature throwing pattern whereas only a minority of girls have. In overarm throwing, unlike any of the other fundamental motor skills examined, there appear to be differences in skill performance between boys and girls that cannot be simply accounted for by boys having had more practice.

Practical applications

The information available on the ages and stages in the development of fundamental motor skills is of potentially great value to teachers of motor skills, especially at the primary school level, for at least three reasons. First, it may provide a reasonably objective method for monitoring the motor development of individual children and of detecting any potential movement problems (for example, most test batteries for movement clumsiness are either based on or include fundamental motor skill items). Second, it may provide the teacher with a guide as to forthcoming progression in movement sequence development and therefore may provide a basis for accelerating acquisition of specific motor skills. For example, given the information provided in Figure 18.7 and Table 18.2, it may be wise to emphasise an instruction such as 'point at the target' to facilitate the transition from the intermediate to the mature stage of throwing. Equally an instruction such as 'step toward the target with your left leg' may be inappropriate for someone at the initial stage of the throwing skill as an intermediate stage involving stepping with the other (right) leg is

Table 18.1 Characteristics of different stages in the development of running

A. Initial stage.

1. Short, limited leg swing.
2. Stiff, uneven stride.
3. No observable flight phase.
4. Incomplete extension of support leg.
5. Stiff, short swing with varying degrees of elbow flexion.
6. Arms tend to swing outward horizontally.
7. Swinging leg rotates outward from hip.
8. Swinging foot toes outward.
9. Wide base of support.

B. Elementary stage.

1. Increase in length of stride, arm swing, and speed.
2. Limited but observable flight phase.
3. More complete extension of support leg at takeoff.
4. Arm swing increases.
5. Horizontal arm swing reduced on backswing.
6. Swinging foot crosses midline at height of recovery to rear.

C. Mature stage

1. Stride length at maximum; Stride speed fast.
2. Definite flight phase.
3. Complete extension of support leg.
4. Recovery thigh parallel to ground.
5. Arms swing vertically in opposition to legs.
6. Arms bent at approximate right angles.
7. Minimal rotary action of recovery leg and foot.

Common problems:

1. Inhibited or exaggerated arm swing.
2. Arms crossing the midline of the body.
3. Improper foot placement.
4. Exaggerated forward trunk lean.
5. Arms flopping at the sides or held out for balance.
6. Twisting of the trunk.
7. Poor rhythmical action.
8. Landing flat-footed.
9. Flipping the foot or lower leg in or out.

Source: Gallahue, D.L. (1989), *Understanding Motor Development: Infants, Children, Adolescent* (2nd edn), p. 238. (Copyright 1989 by Benchmark Press Inc.)

Figure 18.5 Stages in the development of a mature running pattern. (See Table 18.1 for greater detail on the characteristics of each stage)
Source: Gallahue, D.L., (1989), Understanding Motor Development: Infants, Children, Adolescents (2nd edn), p. 239. (Copyright by Benchmark Press, Inc, 1989.)

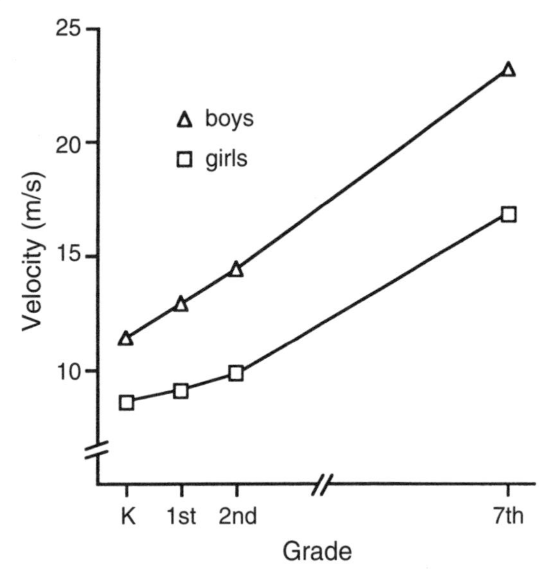

Figure 18.6 Improvements in overarm throwing velocity with age
Source: Drawn from data reported in Halverson, L., Roberton, M.A. & Langdorfer, S. (1982), Development of the overarm throw: Movement and ball velocity changes by seventh grade, *Research Quarterly for Exercise and Sport*, 53, 198–205.

typically involved as a transition from the initial to the mature pattern of throwing. Third, assessment of fundamental motor skills can provide an indication of the readiness of children for involvement in more structured activities such as sports where performance is based around proficiency in one or more fundamental motor skills. There is growing concern worldwide about declining fundamental motor skill proficiency among school-aged children (Box 18.2).

Motor performance in the elderly

Although the peak performance of motor skills, at least as evidenced by sports performance such as world records, appears to be achieved typically from the late teens through to the mid-30s, most movement skills, with regular practice, can be retained at a high level throughout most of adulthood. In older adulthood, however, there is some decline in motor performance in skills as fundamental as balance and locomotion that can seriously impair the movement capabilities of the elderly. Some motor skill deterioration with age appears inevitable although the extent of the decline can clearly be limited by regular practice.

Changes in balance and posture

Falls are a major cause of injury and loss of independent living capability in the elderly. Falls may be particularly incapacitating for older women, as we have seen in Chapters 6 and 15, because of their susceptibility to osteoporosis and hip fractures. As a major cause of falls is loss of balance, scientists who study the ageing process (gerontologists) have been particularly interested in studying balance and posture control in the elderly. Such studies have revealed an increase in postural sway as people get older (with this effect being most pronounced in women), and slower and less expert recovery of unexpected losses of balance by older people. Poorer vision, increased use of medication and changes in gait pattern that result in the foot being lifted less off the ground, also contribute significantly to increased loss of balance and falls among older people.

Changes in walking patterns

A number of discernible changes in walking patterns occur with ageing. In comparison with the gait of younger adults, older subjects typically:

- lift their feet less
- walk more slowly
- have shorter stride lengths
- have a reduced range of motion at the ankle
- have greater 'out-toeing' in their foot placement.

The net purpose of all these changes is to try to increase stability, although this is inevitably to the detriment of mobility. Interestingly, with age, the motor patterns of the elderly regress in many aspects to characteristics seen in the immature gait of children. This is especially true of the out-toeing of the foot, which is a movement pattern used by both young children and the elderly to improve lateral stability.

Figure 18.7 Stages in the development of a mature overarm throwing pattern. (See Table 18.2 for greater detail on the characteristics of each stage)
Source: Gallahue, D.L. (1989), Understanding Motor Development: Infants, Children, Adolescents (2nd edn), p. 257. (Copyright by Benchmark Press, Inc, 1989.)

Changes in more complex motor skills

Like fundamental motor skills, the performance of acquired specialised motor skills also declines with age although the extent of deterioration does not appear to be uniform across all types of skills. Motor tasks that have to be performed under time stress, require complex decision-making or involve psychological 'pressure' (or high anxiety; see Chapter 20) typically show the most marked deterioration with ageing. Importantly, motor skills can still be learnt and improved even in late adulthood, so extensive practice on new motor skills can more than offset many of the systematic effects of the ageing process.

Table 18.2 Characteristics of different stages in the development of overarm throwing

A. Initial stage.

1. Action is mainly from elbow.
2. Elbow of throwing arm remains in front of body; action resembles a push.
3. Fingers spread at release.
4. Follow-through is forward and downward.
5. Trunk remains perpendicular to target.
6. Little rotary action during throw.
7. Body weight shifts slightly rearward to maintain balance.
8. Feet remain stationary.
9. There is often purposeless shifting of feet during preparation for throw.

B. Elementary stage.

1. In preparation, arm is swung upward, sideward, and backward to a position of elbow flexion.
2. Ball is held behind head.
3. Arm is swung forward, high over shoulder.
4. Trunk rotates toward throwing side during preparatory action.
5. Shoulders rotate toward throwing side.
6. Trunk flexes forward with forward motion of arm.
7. Definite forward shift of body weight.
8. Steps forward with leg on same side as throwing arm.

C. Mature stage.

1. Arm is swung backward in preparation.
2. Opposite elbow is raised for balance as a preparatory action in the throwing arm.
3. Throwing elbow moves forward horizontally as it extends.
4. Forearm rotates and thumb points downward.
5. Trunk markedly rotates to throwing side during preparatory action.
6. Throwing shoulder drops slightly.
7. Definite rotation through hips, legs, spine, and shoulders during throw.
8. Weight during preparatory movement is on rear foot.
9. As weight is shifted, there is a step with opposite foot.

Common problems:

1. Forward movement of foot on same side as throwing arm.
2. Inhibited backswing.
3. Failure to rotate hips as throwing arm is brought forward.
4. Failure to step out on leg opposite the throwing arm.
5. Poor rhythmical coordination of arm movement with body movement.
6. Inability to release ball at desired trajectory.
7. Loss of balance while throwing.
8. Upward rotation of arm.

Source: Gallahue, D.L. (1989), *Understanding Motor Development: Infants, Children, Adolescent* (2nd edn), p. 256. (Copyright 1989 by Benchmark Press Inc.)

Box 18.2 Fundamental motor skill proficiency in Australian children

In a recent study by a group of researchers at the Royal Melbourne Institute of Technology, led by Dr Jeff Walkley, the fundamental motor skills of a sample of 1182 Victorian schoolchildren were assessed. To ensure the sample tested would be representative of the whole population of Australian children, the children tested were drawn from four grades (Grades 2,4,6 and 8) and from a range of schools varying in their location, size and structure. The children were videotaped performing five motor skills that are fundamental to performance in a wide range of physical and sports activities. These skills were overarm throwing, catching, forehand striking, two-hand side-arm striking and instep kicking. An assessment was made for each child on each skill as to whether a mature motor pattern had been achieved.

The findings were basically similar for all five skills. As is illustrated for the overarm throwing case (Figure 18.8), the percentage of children who have achieved mature movements by Grade 6, and especially by Grade 8, is alarmingly low. At the age when involvement in organised sport is likely to first take place, less than half of the boys, and less than a quarter of the girls, have mastery over the fundamental skills on which most sports are based. It is interesting in this context to note the high number of young children who drop out of organised sport because they do not perform basic skills well enough to experience regular (or, in many cases, any) success (see also Box 20.1). Of perhaps of even greater concern is the apparent regression in basic motor skill proficiency from Grade 6 to Grade 8. A likely cause is declining involvement in physical activity and practice of motor skills as other distractions compete for the child's leisure time. If this is indeed the case it is a strong argument for the need to increase the time and priority allocated to physical education within both primary and secondary schooling.

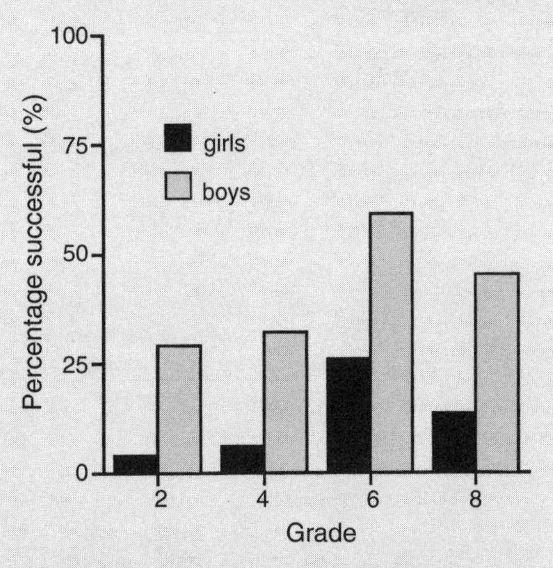

Figure 18.8 Overarm throwing performance of a sample of 1182 Australian school children of different ages. The dependent measure is the percentage of boys and girls at each age level who have attained a mature throwing pattern
Source: Holland, B., Walkley, J., Treloar, R. & Probyn-Smith, H. (1993), Fundamental motor skill proficiency of children, *ACHPER National Journal*, 40(3), 13. (Copyright ACHPER 1993. Reprinted with permission.)

Source: Walkley, J., Holland, B., Treloar, R. & Probyn-Smith, H. (1993), Fundamental motor skill proficiency of children, *ACHPER National Journal*, 40(3), 11–14.

Practical applications

Although it is clear that some decline in motor performance is inevitable with ageing it is also clear that a number of very positive things can be done to offset these effects. Regular exercise (to offset losses in strength, flexibility and endurance), regular practice (to consolidate and improve existing movement patterns) and the development and practice of 'smart' strategies (such as pacing effort to conserve energy, anticipating to avoid reaction-time delays and using encoding strategies to improve memory) can all assist in the retention of a high level of motor performance through into senescence.

Changes at the neurophysiological level

Examination of the neurophysiological changes in the motor system that accompany development and ageing can offer some useful insight into the mechanisms underpinning the observable motor performance changes across the lifespan. Knowing how the nervous system changes with age provides a basis for understanding changes in the information-processing 'hardware' available to support the control of movement at different ages.

Major physical changes in the central nervous system

The growth and development of the nervous system is controlled by a complex interaction between genetic and environmental factors. Early nervous system development is primarily genetically regulated and involves prenatally (before birth) the formation of nerve cells and postnatally (after birth) the branching and insulation (myelination) of the dendrites and axons of these nerve cells. The critical period for the development of the nervous system is from conception through to the end of the 1st year and during this period in particular the structural development of the nervous system is vulnerable to environmental influences. For example, sedative drugs (barbiturates) in the blood stream of the mother prenatally, or malnutrition postnatally, can dramatically impair the normal development of the nervous system.

At birth the brain weighs some 300–350 g or about 25 per cent of its adult weight. It reaches half its adult weight by around 6 months, 75 per cent by $2\frac{1}{2}$ years and nearly 100 per cent by 6 years. As all the neurons the central nervous system will ever possess are present at birth, the increased brain mass in the early years of life results from increases in size and branching within the neurons, myelination and growth of the supporting glial cells. Rates of growth vary dramatically in different regions of the central nervous system. Within the brain the cerebral cortex is identifiable from about 8 weeks after conception and the two cerebral hemispheres are formed, but not functional, at birth. The spinal cord, while present, is small and quite short at birth and matures, through myelination, in a top-down (cephalo-caudal) manner. The pyramidal tract is myelinated and functional by about 4–5 months, coinciding with the first appearance of voluntary movement control in the infant.

The process of myelination involves surrounding the axons of the nerve cells with a fatty sheath (myelin), which acts to insulate the nerve and substantially increase its speed of impulse conductance, its capability for repetitive firing and its resistance to fatigue. Myelination of the sensory and motor neurons begins 5–6 months before birth and is completed within the first 6 months postnatally. Higher centres of the central nervous system, especially the cerebellum and cortex, begin and complete myelination much later than subcortical regions of the system. Damage to the myelin sheath, such as occurs in the disease multiple sclerosis, causes tremor, loss of co-ordination and, on occasions, paralysis.

There is a progressive loss of nerve cells throughout life (at a rate of some 10 000 per day) such that by age 65–70 some 20 per cent of the total neurons present at birth are lost. Glial cells, on the other hand, increase with age. The net effect is a decrease in brain weight in the elderly. General slowing of sensory and motor function with ageing results from a reduced impulse (signal) strength relative to background neural activity (noise).

Changes in the sensory receptors and sensory systems

The visual system

The eye, like the brain, undergoes most of its growth before birth, even though the size of the eye at birth is only about half of its final size at maturity. The retina is fairly well developed at birth and the myelination of the optic nerve has started at this time and is complete some 1–4 months after birth. The neural pathways to the visual cortex are functional at birth. Visual acuity (sharpness of vision) is poor at birth and at the first month of life the human infant has an acuity for stationary objects about 5 per cent of that of mature adult levels. (The infant of 1 month sees the same degree of visual detail from a distance of 6 m as an adult would see from approximately 250 m!) Acuity improves rapidly during the first years of life, with adult levels being attained by about age 10 for the viewing of stationary objects (Figure 18.9) and age 12 for the viewing of moving objects (Figure 18.10). This improved acuity results, to some degree, from an improved capability to adjust the shape of the eye's lens (called accommodation) but primarily from improved neuronal differentiation in the retina and connections in the cortex. The visual system matures at a rate sufficient to provide the visual information needed to guide movement at each stage of development.

With ageing there are a number of declines in visual function that can, in turn, adversely affect movement capability. By about age 40 there are clinically significant losses in the ability to accommodate to near objects. Material that can be read at 10 cm at age 20 must be progressively moved away to distances of 18 cm at 40, 50 cm at 50 and 100 cm at 70 years of age to retain sharp focus. In addition to acuity losses with ageing, there is reduced sensitivity to glare, declining sensitivity to contrast, narrowing of

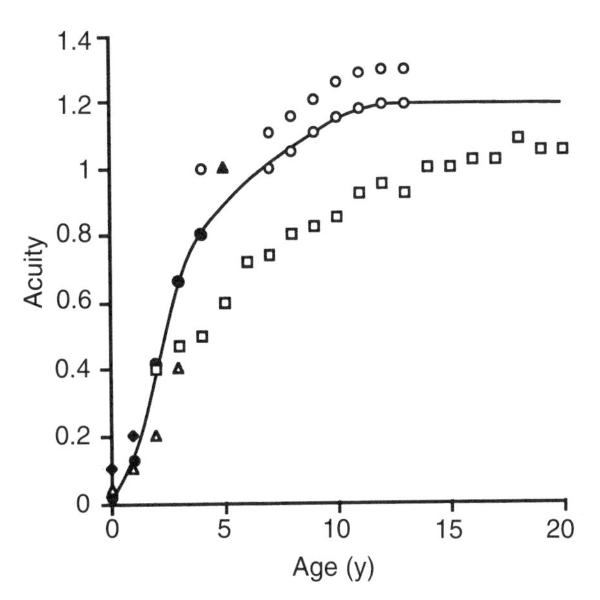

Figure 18.9 Changes in static visual acuity from birth to age 20. Data points are aggregated from eight studies and the average acuity is shown by the continuous line. A value of 1.0 represents average adult acuity
Source: Drawn from data reported in Weymouth, F.W. (1963), Visual acuity of children in M.J. Hirsch & R.E. Wick (eds), *Vision of Children*, Chilton, Philadelphia.

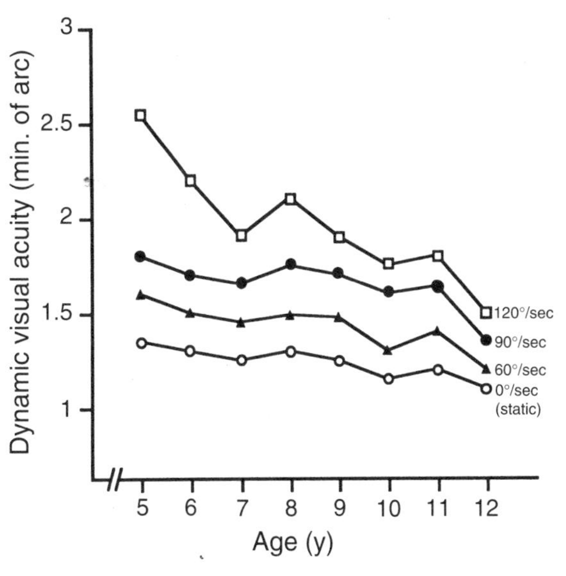

Figure 18.10 Changes in dynamic visual acuity as a function of age and the angular velocity of the target being viewed. The lower the score the better the visual resolution of the moving target which is achieved
Source: Drawn from data reported in Cratty, B.J., Apitzsch, E. & Bergel, R. (1973), Dynamic Visual Acuity: A Developmental Study.

the visual field size and increased visual difficulty at low light levels. All of these changes make it increasingly difficult to perform movement tasks (such as driving or catching) that require precise visual information and the accurate judgment of the location and speed of moving objects.

The kinesthetic and vestibular system

Because of their central role in many reflex systems essential for the survival of the newborn, the kinesthetic receptors develop early and are functional essentially from birth. Cutaneous receptors in the mouth are functional from as early as 7–8 weeks and those in the hand from 12–13 weeks after conception. Muscle spindles are evident in muscles of the upper arm from 12–13 weeks post-conception although the main development of the muscle spindles, as well as the Golgi tendon organs, joint receptors and cutaneous receptors, occurs in the period 4–6 months after conception (5–3 months before birth). The vestibular apparatus is completely formed 2–3 months after conception and may function reflexively from this point onwards. The kinesthetic and vestibular systems are thus prepared early in life to support infant activity before the visual system matures.

Relatively little is known about the effect of ageing on the kinesthetic and vestibular receptors themselves although it is apparent that there are functional losses in the elderly with respect to balance control and sensitivity to touch, vibration, temperature and pain as detected through the cutaneous receptors. It is known that the number of Meissner corpuscles in the skin (see again Figure 16.9) decreases with age, for example, and those that remain undergo changes in size and shape.

Changes in the effectors (muscles)

The growth and ageing of muscle has been considered previously in Chapters 5 and 14. The important points to remember in the context of the development of motor control is that it appears that the number of muscle cells, like nerve cells, is essentially determined at, or soon after, birth and that consequent changes in muscle strength and control arise from alterations to the muscle fibres rather than an increase in their number. As children mature their muscle fibres become wider and longer, and this comes about through an increase in both the number and length of the contractile units (the myofibrils) within each muscle fibre (Figure 3.10). The reductions of muscle size that occur in the elderly appear to be a consequence of reduced effectiveness of the nerves that activate the muscles and the selective loss of type II fibres primarily (Chapter 14).

Changes in reflex systems

Receptors and effectors, as we saw in Chapter 16, are the key elements of reflex systems. The development, modification and, in many cases, extinction of reflexes has, however, a complexity that goes beyond that of the maturational changes of the sensory and effector components of the various reflex systems. Reflexes present at birth or soon thereafter can be broadly distinguished in terms of whether their principal function is to assist survival of the newborn or to lay the foundations for the development of voluntary movement control.

Primitive reflexes

The human newborn is extremely vulnerable because of its limited mobility and capacity for voluntary movement. Consequently, in the early stages of life it must depend heavily on adult caretakers and some reflexes for survival and protection. Those reflexes present at birth that function predominantly for protection and survival are referred to collectively as primitive reflexes. Examples of primitive reflexes are the sucking reflex, which enables the newborn to instinctively gain nutrition from the mother's breast, the searching or rooting reflex, which assists the newborn in locating the nipple, and the Moro reflex, which assists with initial respiration (Table 18.3). The primitive reflexes, which are all powerful at birth, typically weaken with advancing maturity to the point of either being completely inhibited or at least highly localised by 3–4 months after birth. This coincides with the increased maturity of the cortex, suggesting a transition from involuntary

Table 18.3 Some selected primitive, postural and locomotor reflexes

Reflex	Starting position (if important)	Stimulus	Response	Time	Warning signs
Primitive reflexes:					
• Asymmetrical tonic neck reflex	supine	turn head to one side	same-side arm and leg extends	prenatal to 4 months	persistence after 6 months
• Palmar grasping		touch palm with finger or object	hand closes tightly around object	prenatal to 4 months	Persistence after 1 year; asymmetrical reflex
• Moro	supine	shake head, as by tapping pillow	arms and legs extend, fingers spread; then arms and legs flex	prenatal to 3 months	presence after 6 months; asymmetrical reflex
• Sucking	–	touch face above or below lips	sucking motion begins	birth to 3 months	–
• Babinski	–	stroke sole of foot	toes extend	birth to 4 months	persistence after 6 months
• Searching or rooting	–	touch cheek with smooth object	head turns to side stimulated	birth to 1 year	absence of reflex; persistence after 1 year
• Startle	supine	tap abdomen or startle infant	arms and legs flex	7–12 months	
Postural reactions:					
• Body righting	–	turn legs and pelvis to other side	trunk and head follow rotation	2–6 months	–
• Neck righting	supine	turn head sideward	body follows head in rotation	2–6 months	–
• Labyrinthine righting reflex	supported upright	tilt infant	head moves to stay upright	2–12 months	–
• Pull-up	sitting upright, held by one or two hands	tip infant backward or forward	arms flex	3–12 months	–
• Parachute	held upright	lower infant toward ground rapidly or tilt forward to prone position	legs and arms extend and abduct	4 months into 2nd year	–
Locomotor reflexes:					
• Crawling	prone	apply pressure to sole of one foot or both feet alternately	crawling pattern in arms and legs	birth to 4 months	–
• Walking	held upright	place infant on flat surface	walking pattern in legs	birth to 5 months	–
• Swimming	prone	place infant in or over water	swimming movement of arms and legs	11 days to 5 months	–

Source: Haywood, K.M. (1986) *Lifespan Motor Development*, pp. 78–79. (Copyright 1986 by Kathleen M. Haywood.)

control toward greater voluntary control. Persistence of reflexes for extended periods after their expected time of disappearance is used by paediatricians (doctors who specialise in the care of children) as a sign of neurological problems.

Postural and locomotor reflexes

Postural reflexes, such as the body righting, neck righting and parachute reflexes (Table 18.3), serve the function of keeping the head upright and the body correctly oriented with respect to gravity. These reflexes are not present at birth but generally appear after 2 months around the time early stages of postural control are being established (see again Figure 18.1). They disappear after the 1st year of life and are progressively replaced by more voluntary movement control. The locomotor reflexes, such as the walking and swimming reflex (Table 18.3), are present from birth or soon thereafter. These reflexes disappear after some 4–5 months and before voluntary walking or swimming are attempted (Figure 18.2). While the exact role of the postural and locomotor reflexes in the development of future voluntary movement control is not entirely clear, it appears that these reflexes collectively play a role in preparing the nervous system and its pathways for the emergence of the voluntary fundamental motor skills discussed earlier in this chapter. Localised reflexes such as the stretch reflex described in Chapter 16 persist throughout the lifespan and appear to alter relatively little in old age.

Changes in information-processing capabilities

The young and the elderly must process information through the same basic central processing stages of perception, decision-making and movement organisation and execution as used by adults and as outlined in Chapter 17. What change with development, and with ageing, are the speed, efficiency and sophistication with which information can be processed.

Developmental improvements in information-processing capability

Perception

At least three general principles can be observed in the development of perceptual skills. One principle is that the maturation of perceptual skills continues well after the sensory system and receptors have matured structurally. We noted earlier that in the visual system the optic nerve is fully myelinated and pathways from the eye to the visual cortex are functional by a few months after birth yet visual acuity does not reach adult levels until around age 10 (Figure 18.9). Similarly the kinesthetic 'hardware' for life is essentially complete at birth yet improvements in the ability to make precise kinesthetic judgments continue at least until age 8 and typically beyond (Figure 18.11*a*).

A second principle is that the more complex the perceptual judgment that has to be made, the longer it takes the developing child to reach adult levels of performance. In the visual system, acuity for moving objects matures after acuity for stationary objects (compare Figures 18.9 and 18.10). Similarly, relatively simple visual–perceptual judgments such as those involved in comparing object size or depth, differentiating an object from its background or distinguishing a whole image from its component parts, reach maturity within the first decade of life whereas more complex judgments, such as those involved in anticipating the direction of an opponent's stroke in racquet sports, may continue to improve well into the third decade of life. As with vision, kinesthetic judgments of complex movement patterns mature much later than simple acuity judgments (compare Figures 18.11*a* and 18.11*b*).

A third principle relates to the integration of information between the different sensory systems. Perhaps surprisingly, the integration of visual and kinesthetic information does not follow the individual maturation of the visual and kinesthetic systems but rather occurs simultaneously with it. For simple tasks, such as shape or pattern recognition in which the child is permitted active kinesthetic exploration of the shape, the integration of vision and

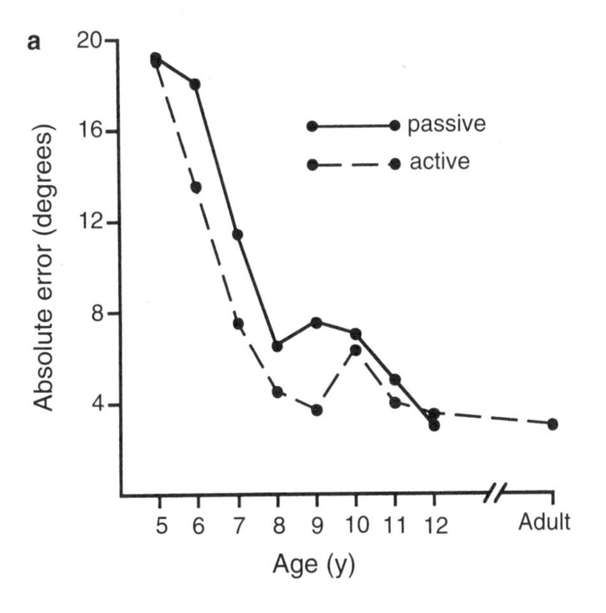

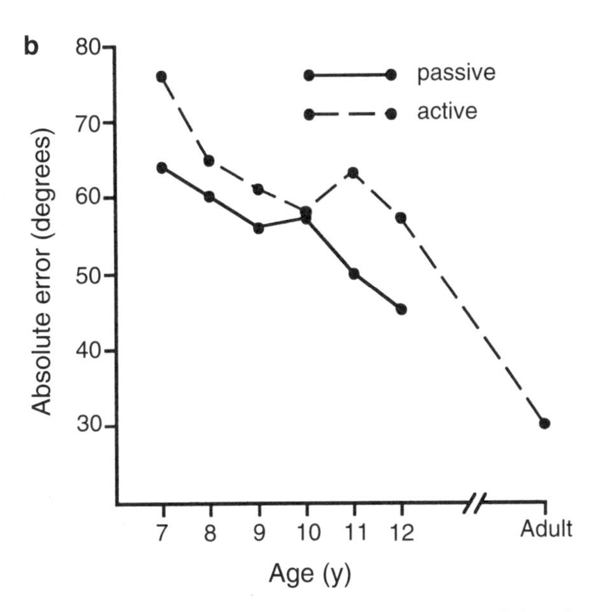

Figure 18.11 Changes in kinesthetic acuity (a) and kinesthetic memory for complex movement patterns (b) as a function of age. Active movements are those controlled by the subject whereas in passive movements the limb is moved to different positions by the experimenter
Source: Bairstow, P.J. & Laszlo, J.I. (1981), Kinaesthetic sensitivity to passive movements and its relationship to motor development and motor control, *Developmental Medicine and Child Neurology*, 23, 606–16. (Copyright 1981 by Spastics International Medical Publications; reprinted with permission.)

kinesthesis is mature by around age 8. By this age children exploring an object through touch alone are then able also to identify the object visually from among other possible shapes or objects. More complex visual–kinesthetic integrations, such as those that occur in hitting and catching tasks, where movements must be initiated and guided kinesthetically to coincide with the arrival of a ball (perceived visually), take longer to mature although the attainment of adult levels of performance do not lag behind the separate maturation of the visual and kinesthetic systems (Figure 18.12).

Decision-making

Reaction time to a single, unanticipated stimulus (simple reaction time) decreases rapidly (that is, reaction time gets faster) until the mid-teens to the late-teens whereupon adult levels are attained (Figure 18.13). Being more complex, reaction time for tasks involving decision-making (choice reaction time) takes somewhat

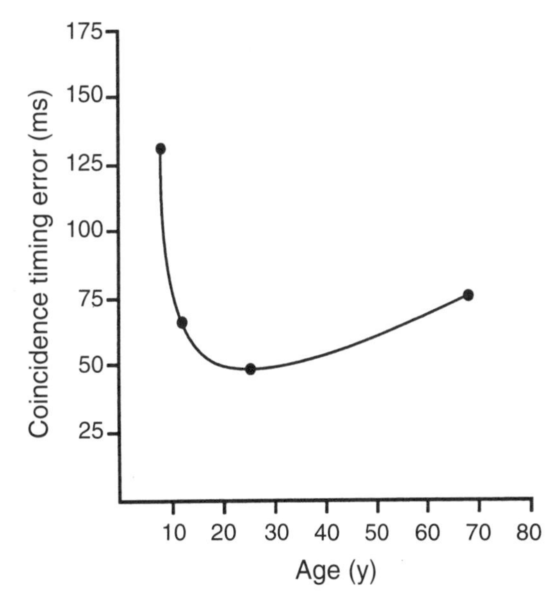

Figure 18.12 Changes in performance on a complex coincidence-timing task as a function of age
Source: Drawn from data reported in Haywood, K.M. (1980), Coincidence–anticipation accuracy across the lifespan, *Experimental Aging Research*, 6, 451–62.

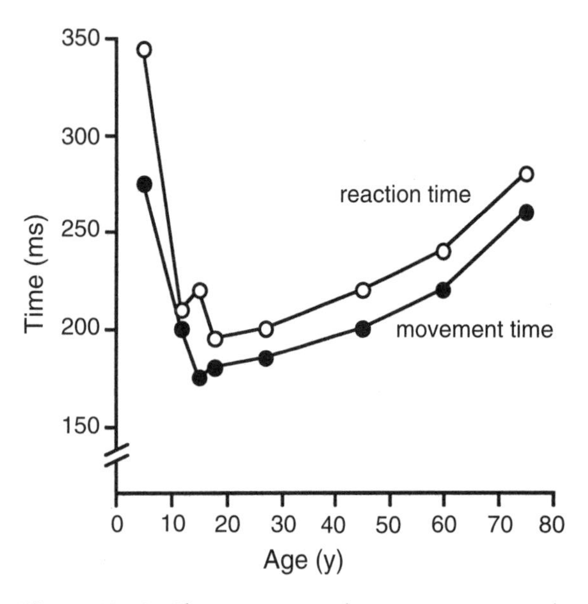

Figure 18.13 Changes in simple reaction time and movement time across the lifespan
Source: Hodgkins, J. (1962), 'Influence of age on the speed of reaction and movement in females', *Journal of Gerontology*, 17, p. 385. (Copyright © 1962 by The Gerontological Society of America; reprinted with permission.)

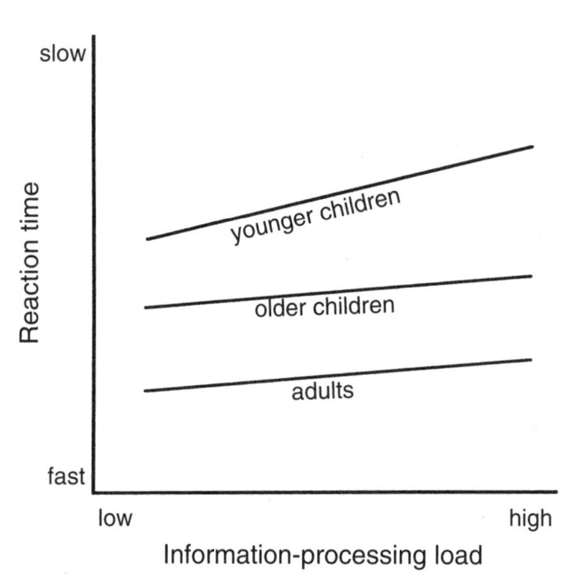

Figure 18.14 Choice reaction time as a function of information-processing load and age
Source: Keogh, J. & Sugden, D. (1985), *Movement Skill Development*, p. 337. (Copyright by Macmillan Publishing Company, 1985.)

longer to mature. Children are slower than adults in making decisions for a number of reasons in addition to their slower simple reaction times. Compared with both older children and adults, younger children have a slower information-processing rate (as shown by a steeper slope in Figure 18.14) and they also frequently process more information in selecting any particular response. This occurs because of their relatively poor ability to separate out relevant from irrelevant information. It is not surprising, therefore, that children's performance in tasks that require rapid decision-making, such as a number of the team sports, show rapid and continued improvements with practice into the mid-20s and beyond.

Organising and executing movement

The time taken to make simple movements varies across the lifespan in a manner comparable to that for simple reaction time (Figure 18.13). Movement time for simple actions reaches its minimum around the mid-teens and remains at that level, typically, until the mid-30s. More complex movements (movements with greater programming and/or control requirements) are made more slowly by young children than by adults although it is not clear whether this is due to a slower information-processing rate for movement control in children (as would be true for the dashed line shown in Figure 18.15) or not. Children show a reduction, compared with adults, in their ability to perform one or more tasks concurrently with a movement task, suggesting that movement is controlled more automatically by adults. A large part (perhaps all) of the information-processing capacity of children is needed for the control of even apparently simple movements, leaving little free capacity to allocate to the performance of other simultaneous tasks.

Declines in information processing with ageing

Much remains yet to be learnt about the changes in information-processing capability that accompany ageing. What is most apparent

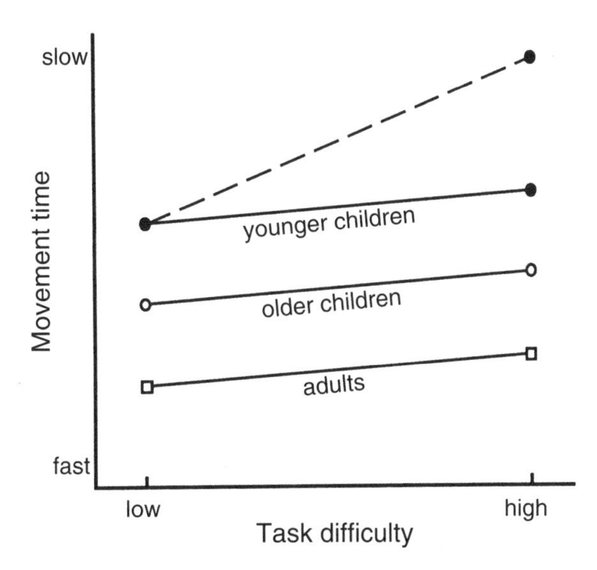

Figure 18.15 Movement time as a function of task difficulty and age
Source: Keogh, J. & Sugden, D. (1985), *Movement Skill Development*, p. 355. (Copyright by Macmillan Publishing Company, 1985.)

is the systematic decline in both the speed of reacting and the speed of moving, which appears to occur from about the 30s onwards (Figure 18.13). This reduced speed of responding becomes pronounced in the elderly and is apparent in a wide range of tasks from walking through to driving a car or typing. The loss of speed in responding is more pronounced in complex than in simple tasks. The phenomenon of slowing with age follows a *last-in-first-out rule* in that simple actions (such as reflexive movements) acquired early in life are more resistant to slowing and loss than complex, co-ordinated skills learnt voluntarily later in life. While the declines in reaction and movement time with ageing appear somewhat inevitable, the rate of decline may be slowed considerably by regular physical activity.

The slowing of simple reaction time with age appears to primarily reflect the neurophysiological changes in the central nervous system, especially nerve cell loss and changes in some of the sensory receptors. In choice reaction time tasks, slowing reflects a lowered signal-to-noise ratio in the central nervous system plus a cautious strategic change toward emphasising accuracy to the detriment of speed. The slowing of movement with ageing reflects similar central rather than peripheral factors. With ageing there is a reduction in the number of functional neuromuscular units and, from age 20–60, a 15–35 per cent increase in the amount of neural stimulation needed to excite a muscle to contraction. These changes, together with a loss of nerve cells, effectively lower the signal-to-noise ratio, slowing the rate at which movements can be not only initiated but also completed. As new movement skills are also apparently more difficult to learn for older adults, an important preparation for later adulthood is the learning of skills early in life, which can be maintained with relative ease through ongoing practice throughout the lifespan. Chapter 19 examines how skills are learned and acquired through practice.

CHAPTER 19

Motor control adaptations to training

- Changes in observable motor performance
- Changes at the neurophysiological level
- Changes in information-processing capabilities
- Factors affecting the learning of motor skills

In this chapter we use the basic neurophysiological and psychological concepts of motor control introduced in chapters 16 and 17 to examine the changes in motor control that occur as a consequence of practice or training. The specialised field of study concerned with the description and explanation of changes in motor performance and motor control with practice is typically referred to as motor learning (or occasionally perceptual–motor skill acquisition).

Learning is the change in the underlying control processes that is responsible for the relatively permanent improvements in performance that accompany practice. Like many aspects of the motor control field, learning is difficult to study because it is a process that cannot be directly observed or measured. Scientists attempting to understand motor learning are dependent on making accurate inferences about learning from observable (and measurable) changes in performance. However, as it is possible to acquire significant skill without necessarily improving performance, changes in the underlying control processes will not always be directly or faithfully reflected in observable performance. The difficulty scientists have in accurately measuring learning is shared with practitioners, such as physical educators, music educators and coaches, charged with the responsibility of trying to objectively measure skill learning.

Studies of motor learning have generated knowledge about many things including:

- the characteristics of expert performers
- the effectiveness of different types and schedules of feedback for skill learning
- the relative merits of different types of practice
- the transfer of skill from one practice setting to another
- the retention of skills over time
- the re-learning of skills following traumas, such as joint injury, or central nervous system damage, such as occurs with stroke.

This knowledge is clearly important practically for any profession involved with the assessment, measurement and improvement of motor skills.

As was the case with motor development examined in the previous chapter, knowledge about motor learning comes from both cross-sectional and longitudinal types of studies. In cross-sectional studies the motor control and performance of people with different levels or types of practice are compared, whereas in longitudinal studies the motor control and performance of the same set of people is examined on a number of occasions throughout the learning or retention of a particular skill. Comparison of skilled (expert) and lesser skilled (novice) performers is an example of the first type of study, whereas training studies are typical of the second type of study. Motor learning can be described and explained at a number of different levels and, again in keeping with the previous chapter, this chapter examines the changes that take place with practice at the observable motor performance level, at the neurophysiological level and at the level of information-processing capabilities.

Changes in observable motor performance

Characteristics of skilled performers

Skilled performance is the learned ability to achieve a desired outcome with maximum certainty and is often achieved with a minimum outlay of time, energy or both. Comparison between the observable characteristics of expert performers and those of people less skilled on the same task typically reveals a number of differences. Experts, in contrast to the lesser skilled, are frequently characterised as:

- 'having all the time in the world'
- 'picking the right options'
- 'reading the situation well'
- 'being adaptable'
- 'moving in a smooth and easy manner'
- 'doing things automatically'.

Skilled performers are apparently able to balance well the conflicting needs to be:

- fast yet accurate
- consistent yet adaptable
- maximally effective yet with a minimum of attention and effort.

How skilled performers are able to achieve this balance requires an understanding of the changes that occur with practice at the level of the underlying neurophysiological and psychological processes.

Stages in the acquisition of motor skills

It may take literally millions of trials of practice to proceed from being a novice performer through to being an expert. Although skill acquisition is a continuous process learners pass through at least three identifiable stages during the long transition from novice to expert.

Stage 1: the verbal–cognitive phase

In the verbal–cognitive phase the movement task to be learned is completely new to the person. Consequently the learner is preoccupied at this stage with trying to understand the requirements of the task, especially what needs to be done in order to perform the skill successfully. All of the learner's limited information-processing capacity is directed to such issues as where to position the whole body and limbs; where best to gain ongoing feedback about performance; or what does the correct movement feel like. Consequently the major activity at this stage of performance is thinking and planning of movement strategies (that is, cognition) and great benefits can be gained from good (verbal) instruction and especially demonstrations. The movements used to make initial attempts at the new task are typically pieces of movement patterns from existing skills, which are joined together to meet the challenges of the new task. In other words 'old habits' are reshaped into new patterns. Instruction that highlights the similarities (and also the differences) between the new skill and skills already learnt may therefore be beneficial in speeding up the rate at which the new skill is learnt. Performance fluctuates dramatically in the early stage of learning new skills as a wide range of movement strategies are tried and many discarded.

Stage 2: the associative phase

In the associative phase of learning, performance is much more consistent as the learners settle on a single strategy or approach to the task. Learners consequently spend the majority of their time and effort 'fine tuning' the selected movement pattern rather than constantly switching from one movement pattern to another. With the basic knowledge of the requirements of the task established, learners become better able in the associative phase to both produce the movement pattern they had planned and adjust to changes in the conditions in which the movement is to be performed. These developments ensure increased levels of task success. In contrast to the verbal–cognitive phase where the instruction provided by others may be the single most beneficial thing for skill acquisition, in the associative phase there is no substitute for specific practice on the task itself. Progressive increases in task complexity (for example by adding more difficult keys for a pianist, more difficult terrain or traffic conditions for a car driver, or more opponents or time constraints for a basketball player) provide a valuable means of fostering the systematic continuation of skill development.

Stage 3: the autonomous phase

With sufficient practice on a particular skill some learners will reach the third stage of learning—the autonomous phase. The autonomous phase is so named because at this stage of learning the performance of the skill appears largely automatic. The movement is apparently able to be controlled without the person having to pay attention to it. At this stage of learning movements are performed consistently with such precision and accuracy that there is no longer a need to constantly monitor feedback to ensure the movement is correctly performed. Open-loop control therefore largely replaces closed-loop control (see again Figure 17.10) and skilled performers have spare attention that can be allocated to other tasks. Consequently, one sign of the expert performer, as we shall see in a later section, is the ability to do two or more things at once. For example, typists in the autonomous phase of learning are able to conduct sensible telephone conversations with minimal interference to their concurrent typing speed or accuracy. The only major drawback with reaching the autonomous stage of learning is that performers at this stage find it difficult, if not impossible, to change their movement pattern if, for example, an error in technique becomes ingrained in their movement pattern. Experts also often cannot describe verbally how they perform the skilled movements they do; in the autonomous phase movement control operates below the level of consciousness. Even at the autonomous phase of learning there exists room for improvement in both motor performance and control. As we shall see in a later section, there is no reason that learning of movement skills cannot occur continuously even after the autonomous phase of learning is reached.

The specificity of motor skill

Skill acquisition is highly specific and there is little or no transfer of training from one motor skill to another. Level of performance on any one particular motor skill is of no use in predicting the rate of learning or the ultimate level of performance of an individual on any other motor skill. Despite its intuitive appeal, there is no evidence that motor skill is in any way generalisable. Someone expert at one activity has the same probability of being a novice on another skill as a person unskilled in the expert's activity of specialisation (Figure 19.1). The verbal–cognitive, associative and autonomous stages of learning must therefore be passed through for *each* skill for which an individual wishes to become highly proficient.

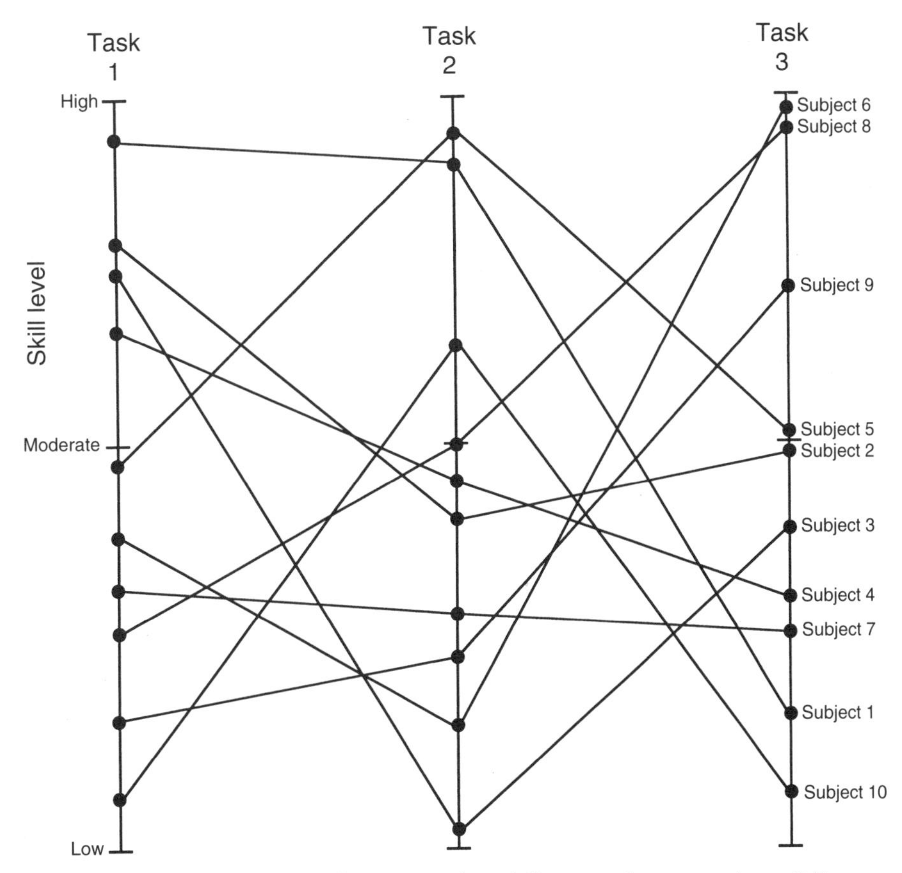

Figure 19.1 Hypothetical performance of 10 different subjects on three different motor skills. If skill were generalisable the same rank order of performance would appear for the subjects on all three tasks. As skill is highly task-specific different rank orders of performance occur on each task

Changes at the neurophysiological level

Learning, as we noted in our earlier definition, is reflected in a relatively permanent change in performance. This relatively permanent change in observable performance must be underpinned by biological changes at the level of the nervous system. Consequently, extensive efforts have been made by neuroscientists in searching for neural correlates of learning and memory.

Challenges for a neurophysiological account of learning

Understanding the physiology of learning and memory represents perhaps the ultimate challenge to research workers in the neurosciences. The task is difficult for a number of reasons. First, and foremost, both learning (and memory) and the brain (and the rest of the central nervous system) are incredibly complex, so matching the complex phenomena of learning and remembering with the complexity of structure of the brain and central nervous system is always going to be an incredibly difficult, if not impossible, task. Any attempt to locate neural changes underpinning learning is complicated by the possibility that learning a particular motor skill may be associated with structural and functional changes in only a few of the brain's many billions of neurons or, more likely, may be associated with relatively subtle changes in the relationships between a large number of neurons distributed at diverse locations throughout the brain. The neural changes that provide the biological basis for learning may take place over very short or quite lengthy time periods, and techniques are only now being developed that may be sufficiently sensitive to record some of the many subtle neural changes that accompany the acquisition of skill.

Given the complexities within both the learning phenomena and the structure of the brain and the technical difficulties in observing and recording possible neural events associated with learning, most existing neurophysiological studies of learning have, by necessity, used animals (other than humans) and very simple learning tasks (rather than the complex ones typical of human movement). Learning can be of many different types yet, to date, the majority of learning studies from a neurophysiological perspective have focused on simple stimulus–response learning or conditioning (Figure 19.2). Such studies have typically sought to find a neural mechanism for memory, as memory for events and experiences past is seen as fundamental for learning. Without the organism being changed by past experience, so that future behaviour can be influenced by experience, learning would be impossible. The inferences that can be made about the neurophysiological basis of human motor learning at this stage are therefore only very preliminary and necessarily somewhat speculative.

Plasticity as the basis for learning

Although psychologists frequently conceptualise memories as records stored in filing cabinets or other similar archives, this is not the way experience is retained within the nervous system. Experience is not stored literally but rather influences the way we perceive, decide and organise and execute movement by physically modifying the neural pathways and circuits responsible for perceiving, deciding and acting. Remembering and learning is therefore only possible because the nervous system is highly plastic; that is, it is able to dynamically alter its structure to accommodate the new functions it is required to perform. While retaining some degree of plasticity throughout life, the brain is especially plastic during the early years and during the critical periods for skill acquisition described in the previous chapter. If, for any reason, a region of the brain is injured or damaged during the first few years of life, other regions of the brain are often able to take over the functions normally performed by the injured region. There is much less plasticity in later life, making full recovery of function following

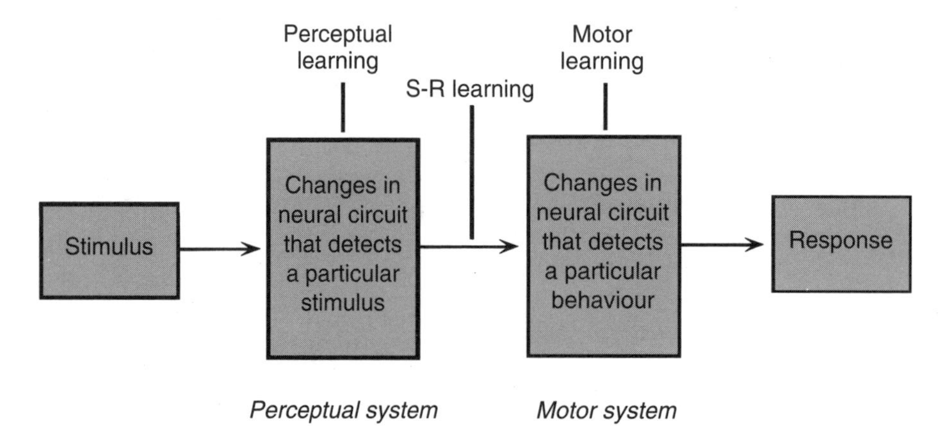

Figure 19.2 Schematic view of the different types of learning examined by neurophysiologists
Source: Carlson, Neil R. (1994), *Physiology of Behavior* (5th edn), p. 436. (Copyright 1994 by Allyn and Bacon; reprinted with permission.)

traumas, such as stroke, difficult in the elderly. (Stroke is a trauma arising from a loss of blood (and oxygen) supply to the brain. It causes damage to one or more regions of the brain and commonly results in observable deficits in motor skills such as speech and gait).

The nervous system may undergo a number of structural and functional changes either as a cause, or a consequence, of learning. Learning may result in a change in the response characteristics of neurons, the establishment of new synaptic connections (as well as the atrophy and eventual disappearance of old, unused connections) (Figure 19.3), and changes in the response characteristics of synapses. All of these changes are inter-related and contribute to synaptic plasticity. Collectively, changes in the nature of synaptic structure and function across one or more synapses modify neural circuitry, which, in turn, can modify perceiving, deciding and acting. These neural changes are, in themselves, underpinned by biochemical and molecular changes. For example, the reinforcement or repetition of successful responses is associated with increased receptivity to the chemical neurotransmitter dopamine. (We have previously seen in Chapter 16 that this chemical is important in transmitting messages within the basal ganglia (among other places) and that deficits in this chemical within specific regions of the basal ganglia appear responsible for Parkinson's disease.)

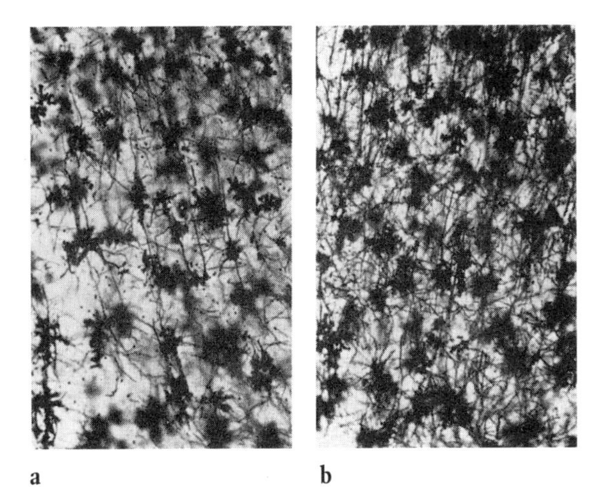

Figure 19.3 Golgi-Cox-stained microscopic sections of the somatosensory cortex of untrained (a) and trained (b) rats. A greater density of synaptic connections is apparent in the trained rats

Source: Spinelli, D.N., Jensen, F.E., & Di Prisco, G.V. (1980), *Early experience effect on dendritic branching in normally reared kittens*, *Experimental Neurology*, 62, 1–11. (Copyright 1980 by Academic Press; reprinted with permission.)

Synaptic plasticity and long-term potentiation

It has been recognised for quite a long time that synaptic changes must be fundamental to learning and memory. In the 1940s the Canadian neuropsychologist Donald Hebb proposed that the strengthening of synapses provides the neural foundation for learning and that a synapse is strengthened by simultaneous activity in both presynaptic and postsynaptic neurons (see again Figure 16.4). More specifically, Hebb hypothesised that if a synapse repeatedly becomes active at around the same time that the postsynaptic neuron fires, structural and/or chemical changes will take place in the synapse to strengthen it. Techniques were not available to test these proposals experimentally at the time Hebb proposed these effects.

More recent neurophysiological studies have demonstrated one mechanism through which the type of synaptic strengthening hypothesised by Hebb could be achieved. When intense, high-frequency electrical stimulation is applied to afferent neurons in some regions of the brain, a subsequent increase in the magnitude of the neural response in this region of the brain is frequently seen when a standard (test) stimulus of known electrical strength is subsequently given. This increased response may last for up to several months and is referred to as long-term potentiation. Long-term potentiation provides a measure of increased synaptic efficiency in that a test stimulus of the same magnitude elicits an increased response, suggesting that the effectiveness of the synapse(s) involved has been improved. The exact cellular and biochemical mechanisms underpinning long-term potentiation are still not clear but are thought to involve the combined effects of depolarisation of the postsynaptic membrane and the activation of unique receptors sensitive to a neurotransmitter called glutamate. Long-term potentiation has been observed primarily in one specific area of the brain (the hippocampus) but has also been observed at other sites. As disruption to long-term potentiation impairs learning of the type usually attributed to the hippocampus it seems reasonable to conclude that improved synaptic efficiency through long-term potentiation is an important neural foundation for skill acquisition.

Changes in information-processing capabilities

Comparative studies of expert and novice performers on different motor skills reveal a number of changes in information-processing capability with the acquisition of skill. Changes are apparent in each of the central processing stages of perception, decision-making and movement organisation and execution described in Chapter 17, as well as in the observable movement patterns and outcomes produced (see again Figure 17.3). In all cases expert–novice differences are only systematically present for the processing of information specific to the motor task for which expertise has been developed.

Sensory reception

The sensitivity of the key sensory systems for movement (the visual, kinesthetic and vestibular systems) appears to change relatively little as a consequence of learning, practising and improving specific motor skills. Although it may be possible to improve characteristics like visual or kinesthetic acuity by specifically training these attributes, as a general rule sensitivity to the sensory information needed to control movement is not the limiting factor to motor performance. Perhaps surprisingly, expert performers, from a range of motor skills, are not characterised by above average levels of visual acuity or kinesthetic sensitivity, at least when these characteristics are measured using standardised tests (Figure 19.4). What appears to be more related to skilled performance is how the basic information provided by the various sensory receptors is subsequently processed, interpreted and used by the central nervous system.

Perception

With practice and the acquisition of expertise several systematic changes take place in the way performers perceive environmental and internal events. Among other things experts are superior to novices in recognising patterns and predicting (anticipating) forthcoming events. The experts' superiority is not a general one, however, and holds only for patterns and events drawn specifically from the experts' domain of

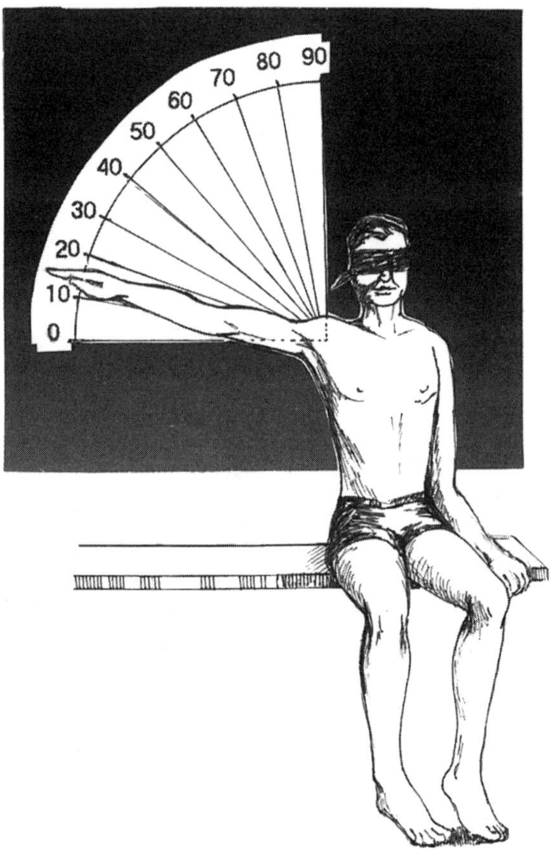

Figure 19.4 A standardised test of kinesthetic acuity. In this test acuity is determined by the subject's accuracy in reproducing the joint angle of a previously experienced movement

expertise. Expert pianists, for example, are better at recognising patterns within pieces of music than novice pianists but are typically no better than novices in recognising patterns present in skills other than music.

The improved pattern recognition that accompanies skill acquisition is especially evident in team sports, such as hockey, basketball, netball, volleyball and the football codes, where selecting the correct movement response is dependent on being able to quickly and accurately recognise the defensive and/or offensive patterns of play of the opposing team. A number of studies have been conducted in which full colour slides have been taken of an opposing team during an actual game and then these slides shown back to subjects for brief periods (usually 5 s) after which they have to recall the position of all the players (both attacking and defensive) shown within the slide. When the slides that are shown depict structured patterns of play the recall of expert team sports players is superior to that of lesser skilled players. This perceptual advantage disappears when the players shown on the slides are in random positions (or in other words there is no familiar pattern present). This finding, illustrated in Figure 19.5, is a systematic one and demonstrates that the experts' superior perception is due to skill-specific experience and not a consequence of expert performers having generically superior perceptual skills that hold across all situations. To use the computer analogy introduced in Chapter 17, the main change with the acquisition of expertise on a task is improvement in the specific programs (or 'software') used to process the input information provided by the sensory receptors.

Many motor skills, especially skills such as those involved in fast ball sports, involve substantial time constraints that pose major demands on human information-processing capabilities (see again Figure 17.12). However, as we have noted earlier in this chapter, one of the distinguishing characteristics of expert performers is their ability to give the impression of having 'all the time in the world'. One important means by which expert performers overcome the time constraints placed on them is to anticipate likely events from any advance information available to them. For example, in a range of ball sports, expert performers are better able to predict in advance forthcoming events on the basis of information available from the actions and postures of their opponents prior to ball flight. Experts are also able to pick up information from cues additional to those used by novices (Box 19.1). Expert–novice differences in the ability to anticipate occur even when the two skill groups may be looking at the same features of their opponent. This demonstrates that the limiting factor in the perceptual performance of untrained subjects is not the ability to pick up the necessary sensory information but the ability to interpret, understand and use it to guide decision-making and movement execution.

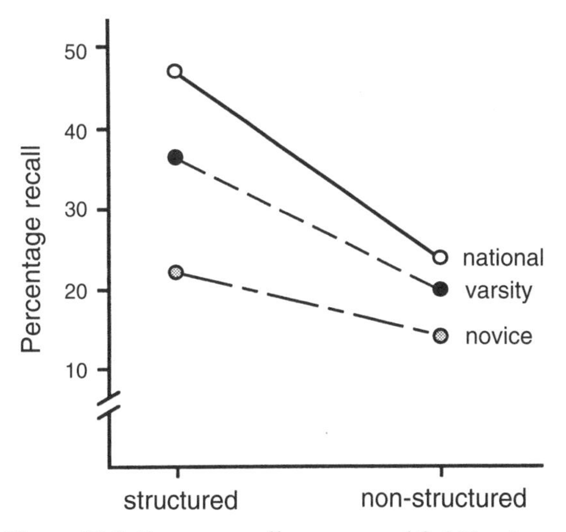

Figure 19.5 Pattern recall accuracy of field hockey players of different skill levels for displays with and without structure
Source: Skill in field hockey: The nature of the cognitive advantage by Janet L. Starkes, *Journal of Sport Psychology*, (Vol. 9, No. 2), p. 152. (Copyright 1987 by Human Kinetics Publishers; reprinted by permission.)

Decision-making

A capability to make decisions both quickly and accurately would be clearly beneficial for the performance of many motor skills. As we

Box 19.1 The anticipatory skills of expert and novice racquet sports players

To examine how well expert and novice badminton players could anticipate the direction and force of their opponents' actions from cues available before their opponents actually hit the shuttle, Bruce Abernethy, of the University of Queensland, and David Russell, from the University of Otago, developed a series of film tasks. These film tasks showed an opponent's hitting action from the perspective of a defending player. One set of film tasks was cut-off at different time periods either before or after the opponent had struck the shuttle (Figure 19.6) and the task of the expert and novice players who saw this film was to predict the stroke direction and force from the information available to them. Another set of film tasks was systematically cut-off at the time the opponent struck the shuttle but throughout the lead-up action, vision to various areas of the opponent's body and racquet was selectively masked (Figure 19.7). The first set of tasks showed that experts were superior anticipators across all the cut-off conditions and that the period between the first and second cut-off times provided advance information that experts could use but novices could not. The second set of tasks showed that experts picked up advance information from the motion of the opponent's racquet and the arm holding the racquet whereas novices could only use cues from the motion of the racquet. The experts' use of arm motion as a cue is significant as most arm motion occurs earlier in the hitting sequence than racquet motion, thereby providing a mechanism for earlier information pick-up.

Source: Abernethy, B. & Russell, D.G. (1987), Expert–novice differences in an applied selective attention task, *Journal of Sport Psychology*, 9, 326–45.

saw earlier in Chapter 17, a shorter choice reaction time on any particular occasion may be achieved with practice in one of two ways—either by facilitating the overall decision-making rate (decreasing the slope of Figure 17.8) or by decreasing the total amount of information that has to be processed before making a decision. Choice reaction time tasks, which use stimuli and responses that are skill-specific (rather than the general ones shown in Figure 17.7), provide some support for the proposition that expert performers have faster decision-making rates than novice performers. Nevertheless the more potent strategy used by skilled performers to decrease their decision-making time is to reduce the absolute amount of information that they have to process. Expert performers do this by using knowledge, acquired through experience, of options and event sequences that are either not possible or, at least, improbable. Experts are more accurate than novices in predicting the probability of particular events occurring and this knowledge is invaluable in reducing the amount of information to be processed and, in turn, allowing decisions to be made more rapidly.

Movement organisation and execution

A number of changes in the way movement is organised and executed are apparent with training. The rate of processing movement information is improved with practice and this is due, at least in part, to transition away from an exclusively feedback-dependent closed-loop type of motor control early in learning toward greater use of open-loop control (cf. Figure 17.9). Open-loop control is via motor programs, which are prestructured sets of commands that allow a movement sequence to be executed without reliance on feedback. As skills are learnt, bigger and better motor programs are believed to be formed, bringing progressively more movement elements under the control of a single program. The collective effect of a reduced need to constantly monitor feedback and a reduction in

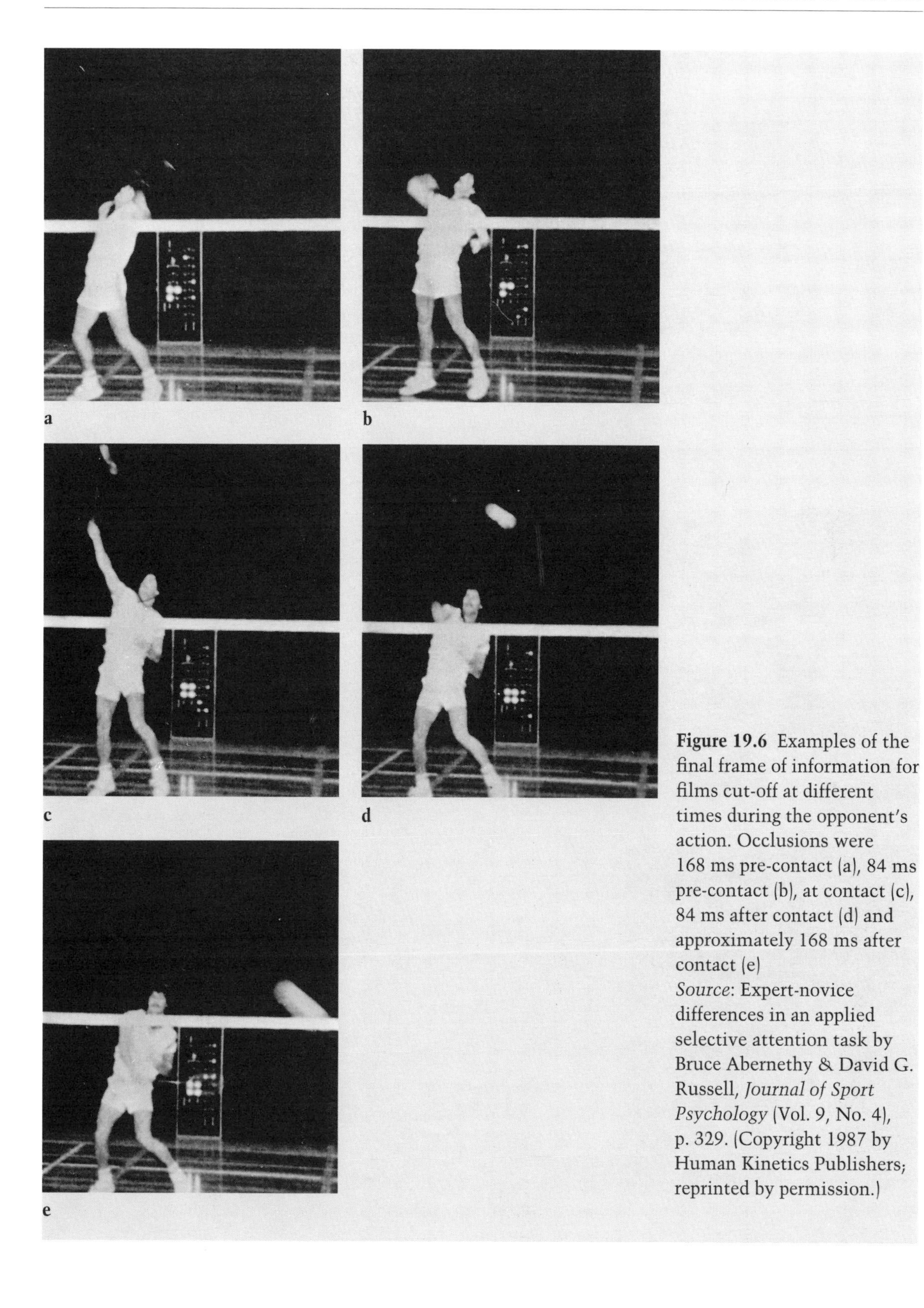

Figure 19.6 Examples of the final frame of information for films cut-off at different times during the opponent's action. Occlusions were 168 ms pre-contact (a), 84 ms pre-contact (b), at contact (c), 84 ms after contact (d) and approximately 168 ms after contact (e)
Source: Expert-novice differences in an applied selective attention task by Bruce Abernethy & David G. Russell, *Journal of Sport Psychology* (Vol. 9, No. 4), p. 329. (Copyright 1987 by Human Kinetics Publishers; reprinted by permission.)

Figure 19.7 Examples of the final frame of information for films in which visibility to different cues was masked. Occlusions were of the racquet and arm (a), the racquet (b), the player's head (c), lower body (d) and random features in the background (e)
Source: Expert–novice differences in an applied selective attention task by Bruce Abernethy & David G. Russell, *Journal of Sport Psychology* (Vol. 9, No. 4), p. 336. (Copyright 1987 by Human Kinetics Publishers; reprinted by permission.)

the number of separate programs needed to produce a particular action, is a reduction in the demands that the organisation and execution of movement places on the information-processing resources available within the central nervous system. Consequently, highly skilled performers, have significant amounts of spare information-processing capacity that can be allocated to the performance of other tasks. Expert performers are therefore markedly superior to lesser skilled performers in the performance of a second task concurrent with their usual movement control skills (Box 19.2).

Observable movement pattern and movement outcome

By definition the movements produced by skilled performers more consistently, and precisely, match the requirements of the movement task than those of lesser skilled performers. In other words expert performers develop, with practice, a greater capability to produce exactly the movement outcomes needed for successful performance. These successful movement outcomes are achieved through movement patterns that also show some reliable differences between expert and novice performers. Movement patterns, as we have seen in Chapters 7 and 8, can be best described in terms of their kinematics, kinetics and underlying neuromuscular patterns using the techniques of biomechanics.

Studies overlapping the subdisciplines of biomechanics and motor control have revealed a number of important changes in observable movement patterns with growing expertise (see also Chapter 11). In terms of kinematics the movement patterns of expert performers are characterised by greater consistency, both in terms of overall movement duration and in terms of movement trajectories and displacement-time characteristics (Figure 19.9). This greater consistency for experts in the observable movement pattern is necessarily also reproduced in the underlying movement kinetics. The force–time curves are typically not only more consistent on a trial-to-trial basis for expert performers but also show a clearer, more distinct pattern of force pulses. Expert performers make greater use of the external forces (such as gravity and reactional forces) available within movements and restrict the injection of muscular force generated by the body to only those points in the movement where it is needed and can act most effectively. Novices tend to supply muscular force more often throughout a movement, often either inefficiently in opposition to external body forces or as an unnecessary supplement to external forces. It is therefore not surprising to observe that neuromuscular recruitment patterns (as revealed from electromyography; Figure 3.14) become more discrete with practice and there is a general reduction in recruitment as muscular contraction extraneous to the movement of interest is eliminated (Figure 19.10).

Implications for training

The value of knowledge about expert–novice differences in information-processing capability is that it provides a guide as to where energy and attention should be directed in practice and training. Training based on improving those information-processing factors known to be related to the expert's advantage on a task appear to be more sensible than training focusing on factors that provide little or no discrimination between experts and novices. This logic is unfortunately not always fully appreciated or considered in the design and recommendation of practice. For example, the generalised visual and kinesthetic training programs that become popular from time to time, and are designed at improving motor skills through improving the general sensitivity of the sensory systems for movement, are most unlikely to be beneficial for skill learning because they do not, under most circumstances, train any of the limiting information-processing factors for skill performance. The available evidence on motor expertise also clearly suggests that the training of perceptual and decision-making skills will be, in many cases, equally, if not more, important than the training of movement execution skills yet this is also not frequently reflected in current training practices.

Box 19.2 Doing two things at once—the dual task performance of experienced and student pilots

One method of determining how automatic the control of movement skills has become is to have a person do an additional task at the same time that they perform the movement skill of interest. This method is known as the *dual-task* or *secondary-task* method. John Crosby and Stanley Parkinson of Arizona State University examined the skill automaticity of experienced (instructor) aircraft pilots and student pilots using this method. The primary task was a ground controlled approach (an instrument landing procedure used in inclement weather conditions) performed on a flight simulator. The measure of primary task performance was the cumulative tracking error throughout the simulated flight. The secondary task was a memory search task. In the secondary task the pilots were first presented with a set of numbers to be remembered; single numbers were then given and the pilot's task was to respond as rapidly and accurately as possible as to whether or not the single number was a member of the earlier presented set. The numbers were presented visually on a display mounted in the cockpit and response was via pushing either a 'yes' or a 'no' button. Reaction time was then taken as the measure of secondary task performance.

Predictably, Crosby and Parkinson found that the instructor pilots were more accurate than student pilots in their primary task performance and that their flight simulator performance was affected less by the imposition of the secondary task than was the simulator performance of the student pilots. The relative performance of the two groups on the secondary task was particularly revealing. While reaction time on the memory search task scarcely altered for the experienced pilots from the task being performed alone to it being performed in combination with the primary task, for the student pilots secondary task performance was markedly poorer when it had to be performed concurrently with the flying task (see Figure 19.8). The control of the primary task of simulated flying required less of the information processing capacity of the experienced pilots—leaving them more spare attention to allocate to the secondary task.

Source: Crosby, J.V., & Parkinson, S.R. (1979), A dual task investigation of pilots' skill level, *Ergonomics*, 22, 1301–1313.

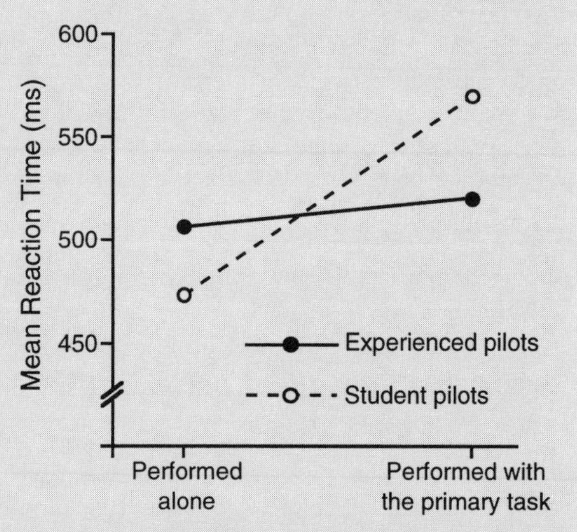

Figure 19.8 Mean reaction time on the secondary (memory) task for experienced and student aircraft pilots.
Source: Drawn from data reported in Table 1 of Crosby, J.V., & Parkinson, S.R. (1979), A dual task investigation of pilots' skill level, *Ergonomics*, 22, 1301–1313.

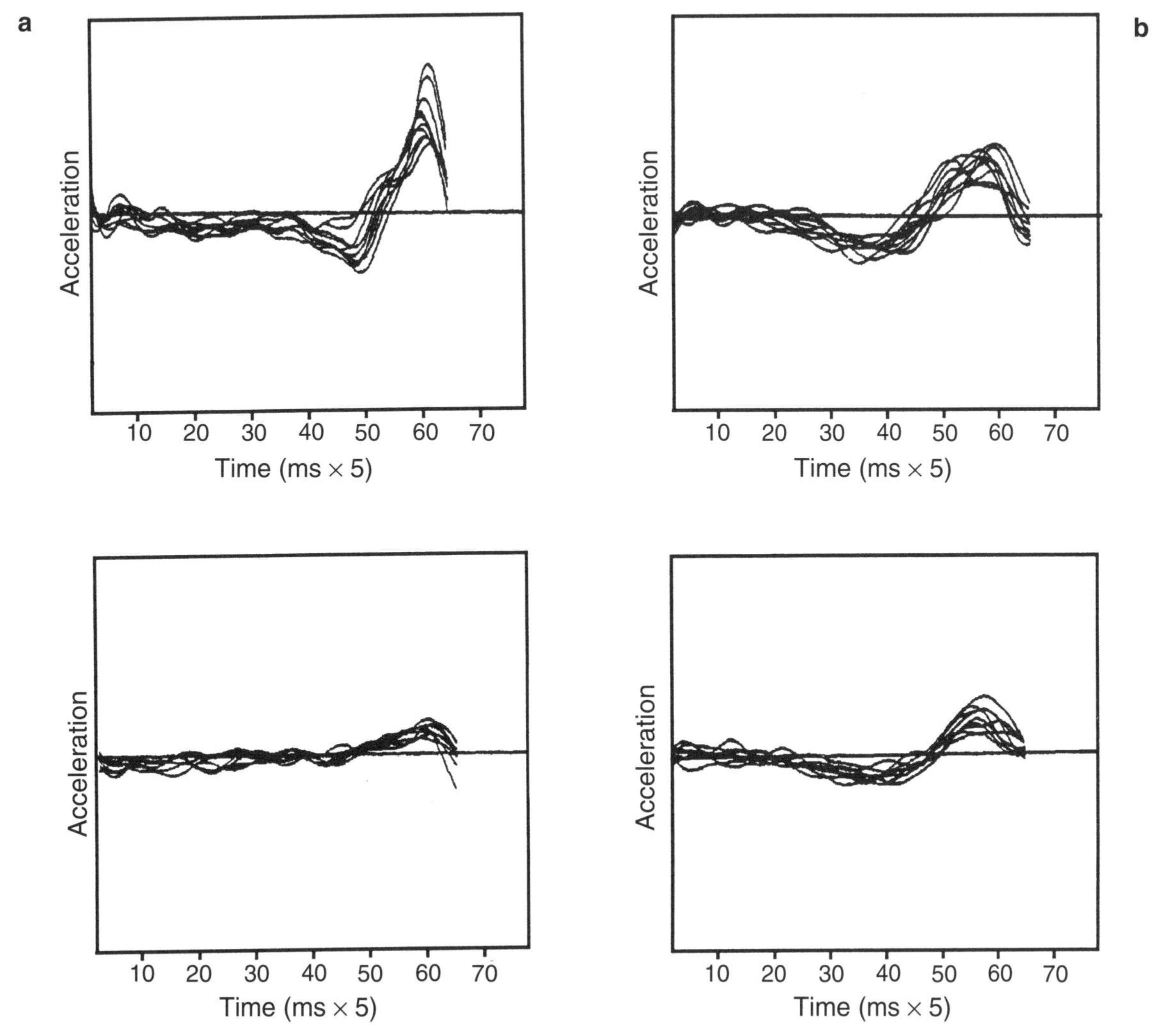

Figure 19.9 Acceleration–time curves from an expert (a), and a novice (b) squash player executing 10 fast strokes (top panels) and 10 slow strokes (bottom panels). The greater consistency of the expert is particularly evident for the slow strokes
Source: Wollstein, J.J. & Abernethy, B. (1988), Timing structure in squash strokes: Further evidence for the operational timing hypothesis, *Journal of Human Movement Studies*, 15, 70–71. (Copyright 1988 by Teviot Scientific Publications.)

Given that good practice is clearly fundamental to motor skill learning, the next section focuses on some of the major factors known to affect the learning of motor skills.

Factors affecting the learning of motor skills

An age-old adage about skill learning is that 'practice makes perfect'. Although it is certainly true that extensive amounts of practice are necessary for high levels of skill to be developed (Table 19.1), practice alone is a necessary but not itself a sufficient condition for learning. The 'practice makes perfect' adage, therefore, while essentially true, needs to be qualified in a number of ways. These qualifications reveal much about the factors that affect the acquisition of movement skills.

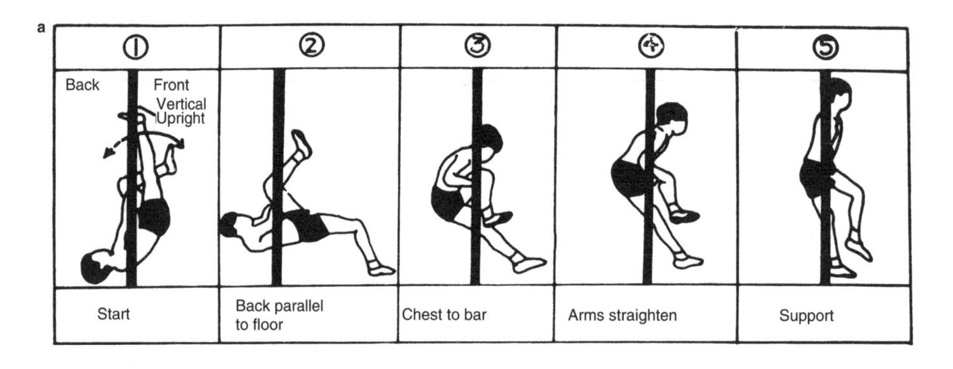

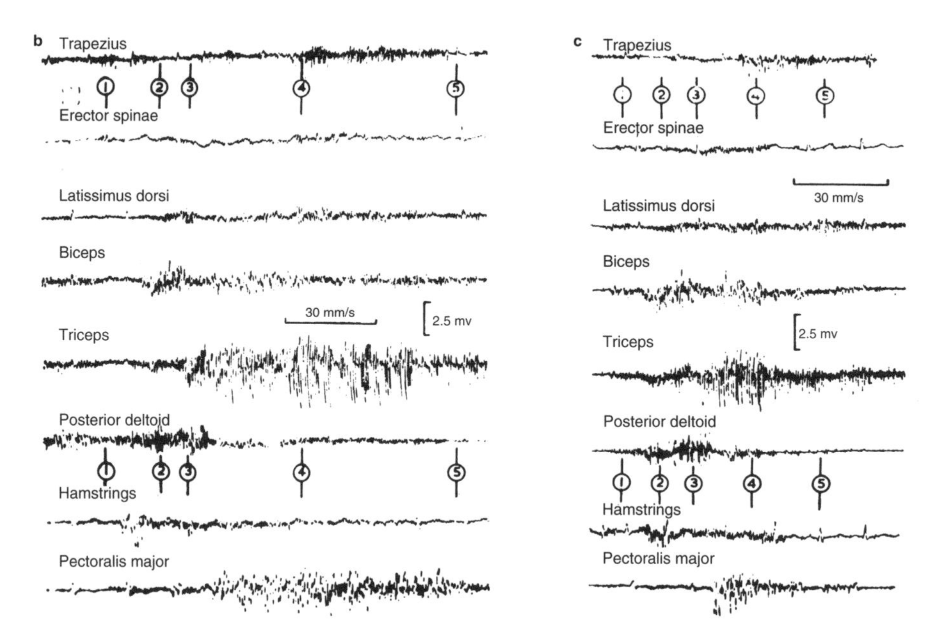

Figure 19.10 Changes in the amount and timing of the electrical activity in the muscles of a novice (b) and an expert (c) gymnast performing the five gymnastic movements shown in (a)

Source: Kamon, E. & Gormley, J. (1968), Muscular activity pattern for skilled performance and during learning of a horizontal bar exercise, *Ergonomics*, 11, 345–57. (Copyright 1968 by Taylor & Francis; reprinted with permission.)

Skills never become perfect

One important qualification that needs to be made to the 'practice makes perfect' adage is to recognise that although motor skills improve with practice there is no reason to suggest that they ever become perfect, that is, that there is no room for further improvement through learning. In even extremely simple tasks, like hand-rolling cigars in a factory, improvements in performance are still apparent after as many as 100 million trials of practice (Figure 19.11)! In more complex motor skills, where there are many more components that can be potentially improved with practice, improvements are likely to extend over an even greater time scale. There is no evidence that skill learning ever ceases, provided practice is ongoing. The levelling out

Table 19.1 Estimated number of practice repetitions undertaken by skilled performers in a variety of sports activities

Skill	Performer	Repetitions for skilful performance	Basis for estimate
Passing	quarterback	1.4 million passes	15 years × 200 days × 4 hours × 2/minutes
Punting	football player	0.8 million kicks	200/day × 5 days × 45 wk × 15 years
Shooting	basketball player	1 million baskets	estimate of practice time
Pitching	baseball pitcher	1.6 million throws	3/minute × 80 minutes × 300 days × 10 years
Gymnastics	14 year-old female gymnast	?	8 years daily practice

Source: Kottke, F.J., Halpern, D., Easton, J.K., Ozel, A.T. and Burrill, B.S. (1978), The training of coordination, *Archives of Physical Medicine and Rehabilitation*, 59, p. 571. (Copyright 1978 by the American Congress of Rehabilitation Medicine; adapted by permission.)

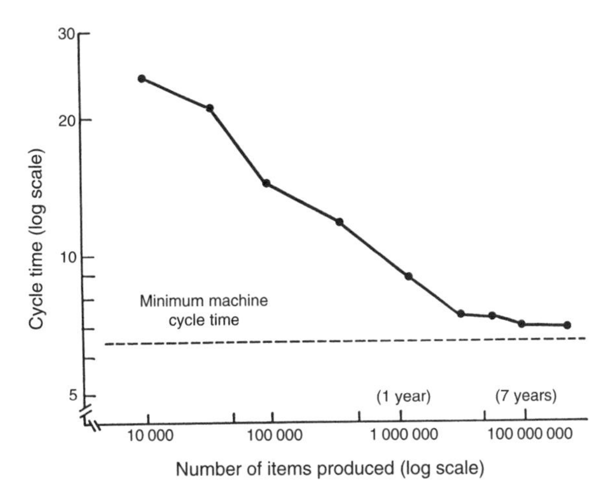

Figure 19.11 Time taken to hand roll a cigar as a function of amount of practice for factory workers *Source*: Crossman, E.R.F.W. (1959), A theory of the acquisition of speed skill, *Ergonomics*, 2, 157 (Copyright 1959 by Taylor & Francis Ltd.)

of performance observed after a number of years of practice or performance in various motor skills is more likely attributable to either psychological or physiological factors or to measurement difficulties. In light of this, a more appropriate adage may be 'practice makes better'.

Feedback is necessary for learning

Although practice is necessary for learning, practice alone does not guarantee learning. In particular, learners must be able to regularly derive feedback information about their performance for practice to be effective in improving learning. If learners are not able to gain information about the success or otherwise of each attempt they make at a new task, learning will be impaired and indeed may not occur at all.

In the study shown in Figure 19.12 subjects had to learn to perform an accurate arm positioning movement on the basis of kinesthetic information alone. The subjects were divided into four groups, who differed in the number of successive trials on which they were given feedback information. Performance was found to be directly related to the amount of feedback given; the group given feedback on 19 of the 20 trials performed the best and the group given no feedback the worst. Significantly, the group given no feedback apparently learned nothing at all over the duration of the training period. This can be ascertained from the observation that this group, when eventually given feedback, improved their performance at a rate essentially identical to that seen over the initial trials of practice for the group given feedback on every trial. All 20 trials of practice done without feedback were therefore apparently of no benefit for learning. This therefore suggests another modification of the 'practice makes perfect' adage. 'Practice, the beneficial results of which are known, makes better' may more accurately encapsulate the nature of the relationship between practice and learning.

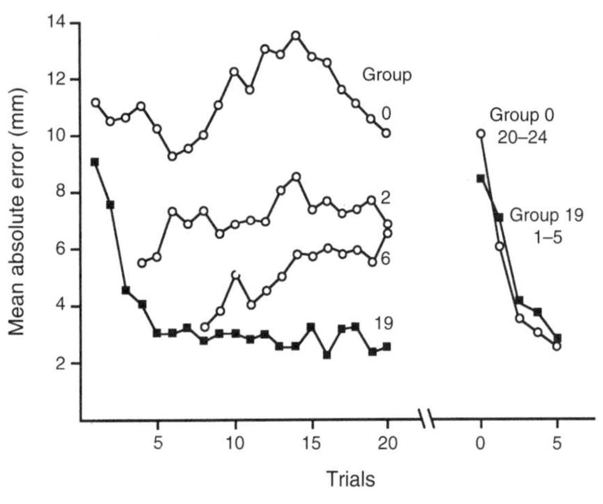

Figure 19.12 Error in reproducing a simple movement as a function of the amount of feedback received. Group O had no feedback, Group 2 feedback for the first 2 trials only, Group 6 the first 6 trials and Group 19 the first 19 trials
Source: Bilodeau, E.A., Bilodeau, I.M. & Schumsky, D.A. (1959), Some effects of introducing and withdrawing knowledge of results early and late in practice, *Journal of Experimental Psychology*, 58, 143. (Copyright 1959 by the American Psychological Association; reprinted with permission.)

Not all feedback is, of course, equally effective. As a general rule feedback information is more effective the more specific it is and the more it provides information from the learner's perspective rather than from an external, third-person perspective. This is true regardless of whether the feedback information is derived by the learners themselves, gained from media such as video, or provided by a teacher, coach or instructor. Feedback information must be limited to key features to avoid overloading the information-processing capacity of the learner. For this reason summary feedback presented at the end of a block of trials rather than after each individual trial may be advantageous. This approach also encourages learners to develop the ability to extract their own feedback information (rather than relying on feedback from external sources) and this ability is essential for learners' progression toward the autonomous phase of learning.

The type of practice is important

Just as different types of feedback vary in their effectiveness for learning so too do different types of practice. The teacher or coach of motor skills charged with the responsibility of designing training or practice regimes for maximal effectiveness must consider a range of issues such as the length of the rest intervals between practice attempts, the extent to which fatigue should be included or avoided in the practice sessions, and the degree to which the practice should be repetitive as opposed to variable. A common guiding principle, which essentially holds across all motor skills, is one of specificity. This is the notion that skills should be practised under conditions that most closely replicate the information-processing demands of the situation in which the skills must ultimately be performed. Consequently the best type of practice will differ for different motor skills. Importantly, in some cases, the best type of practice may differ from commonly and traditionally accepted methods of practice for a particular skill or set of skills. This latter point may be well illustrated by consideration of practice for the motor skills involved in the sport of golf.

Playing golf, like undertaking many other human activities, requires mastery over a number of different motor skills. To play golf successfully a player must not only be able to drive the ball a long way with wooden and long iron clubs but must also be able to accurately pitch and chip the ball, play from sand bunkers and putt with precision; and must be able to adapt these skills on a shot-by-shot basis to accommodate such things as the lie of the ball, the force and direction of the wind and the position of the hazards and obstacles. A key issue for the player and coach is how might these skills be best practised? The traditional form of structured practice for golfers is to take a bucket of balls to the practice range and hit the same club over and over again until the skill is executed effectively and ingrained. This form of practice by which each component skill is practised repetitively is referred to as blocked practice. Most practice that involves drills is of this type. This type of practice can be contrasted with random practice, in which clubs and shots are practised in essentially random order, in a manner not dissimilar to the situation that occurs in actually playing a round of golf. A random practice schedule might involve, for example, hitting in order a driver, 7 iron, sandwedge and putter. In such a practice method the same specific task is never repeated on two successive practice trials. Normally, different clubs would be used on successive trials but if the same club (for example the putter) were to be used twice in a row it would be from a different position or distance. The critical question is which type of practice is most effective for learning and performing golf skills?

Some insight into this question may be gained from examining the results of a study on blocked and random practice shown in Figure 19.13. In this study subjects had to learn three different rapid hand and arm movement tasks. One group of subjects learned the tasks through blocked practice and the other through random practice. The total amount of practice was the same for the two groups of subjects, all that varied was the type of practice. During practice, the blocked group performed best. This is not surprising given that they were constantly exposed to the same task requirements trial after trial. What is important, however, is how well the two groups were able to retain what they had practised (remembering that learning is a *relatively permanent* change in the ability to perform skills). The skills were retained best (and hence, by inference, learnt best) by the group who had experienced random practice. Blocked practice was particularly ineffective when subsequent performance of the practised skill was under random conditions. This is a crucial observation as this is precisely the situation that exists for the golfer who does repetitive practice at the driving range and then

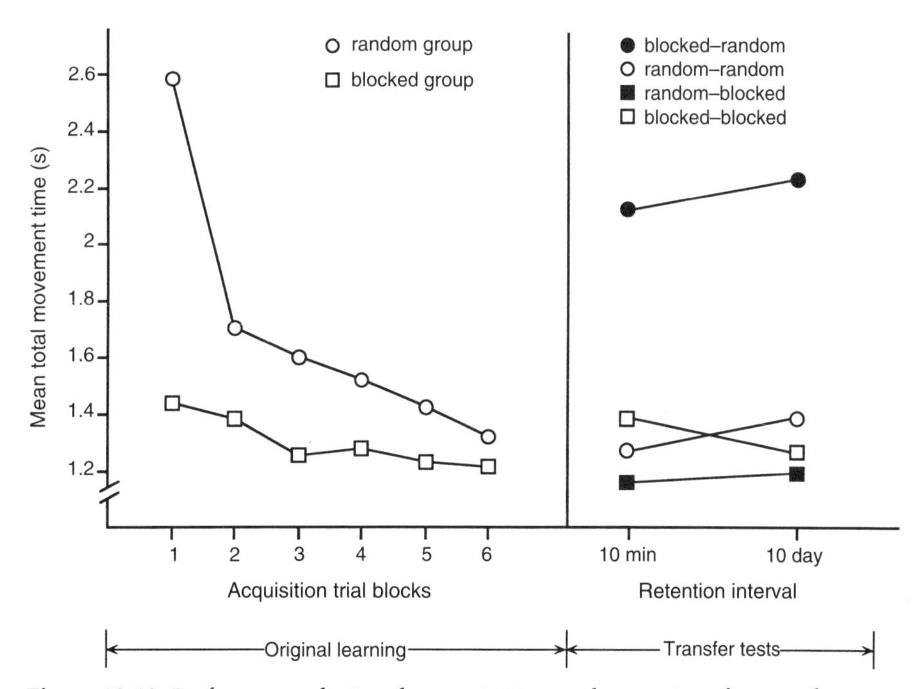

Figure 19.13 Performance during the acquisition and retention of a complex movement task for groups practising under either random or blocked conditions
Source: Shea, J.B. & Morgan, R.L. (1979), Contextual interference effects on the acquisition, retention, and transfer of a motor skill, *Journal of Experimental Psychology: Human Learning and Memory*, 5, 183. (Copyright 1979 by the American Psychological Association; reprinted with permission.)

attempts to use these skills under playing conditions. Thus, while blocked practice may be valuable in the very early stages of learning, it is important to recognise that blocked practice may give an inflated view of actual skill level and may be a less than optimal preparation for a task that ultimately requires component skills to be executed in essentially random order. This example illustrates the importance of designing practice in a way that best simulates the demands of the actual performance setting.

Learning depends on learner readiness

In Chapter 18, in discussing the notion of critical periods, we saw that if the learner is not developmentally ready, practice may be spectacularly ineffective in improving skill, yet when the learner reaches an appropriate developmental stage skills may be acquired with surprisingly few trials of practice [Box 18.1]. While developmental factors have the most pronounced effects in moderating the effectiveness of practice, other psychological factors present throughout the lifespan are also important in determining the extent to which practice translates into actual learning. Foremost among these are the motivation of the learner and the level of arousal and anxiety. These concepts are considered in the next section of the book, which introduces the subdiscipline of sport and exercise psychology.

Further reading

Chapter 16: Basic concepts of motor control: neurophysiological perspectives

Carlson, N.R. (1994), *Physiology of Behavior*, Allyn & Bacon, Boston.

Greer, K. (1984), Physiology of motor control, pp. 2–46, in M.M. Smyth & A.M. Wing (eds), *The Psychology of Human Movement*, Academic Press, London.

Rosenbaum, D.A. (1991), *Human Motor Control*, Academic Press, San Diego, Chapter 2.

Sage, G.H. (1984), *Motor Learning and Control: A Neuropsychological Approach*. Wm C. Brown, Dubuque, Iowa.

Chapter 17: Basic concepts of motor control: psychological perspectives

Abernethy, B. (1991), Acquisition of motor skills, pp. 69–98, in F.S. Pyke (ed.), *Better Coaching*, Australian Coaching Council, Canberra.

Magill, R.A. (1993), *Motor Learning: Concepts & Applications* (4th edn), Wm C. Brown, Dubuque.

Marteniuk, R.G. (1976), *Information Processing in Motor Skills*, Holt, Rinehart & Winston, New York.

Rosenbaum, D.A. (1991), *Human Motor Control*, Academic Press, San Diego.

Schmidt, R.A. (1991), *Motor Learning & Performance: From Principles to Practice*, Human Kinetics, Champaign, IL.

Chapter 18: Motor control changes across the lifespan

Haywood, K.M. (1986), *Lifespan Motor Development*, Human Kinetics, Champaign, IL.

Keogh, J.F. & Sugden, D.A. (1985), *Movement Skill Development*, Macmillan, New York.

Payne, V.G. & Isaacs, L.D. (1991), *Human Motor Development: A Lifespan Approach* (2nd edn), Mayfield, Mountain View, CA.

Sugden, D.A. & Keogh, J.F. (1990), *Problems in Movement Skill Development*, University of South Carolina Press, Columbia, SC.

Thomas, J.R. (ed.) (1984), *Motor Development During Childhood and Adolescence*, Burgess, Minneapolis, MN.

Chapter 19: Motor control adaptations to training

Abernethy, B., Wann, J. & Parks, S. (in press), Training perceptual–motor skills for sport, in B.C. Elliott (ed.), *International Handbook of Sports Science*, vol. 3, *Applied Sport Science: Training in Sport*, Wiley, Chichester.

Christina, R.W. & Corcos, D.M. (1988), *Coaches Guide to Teaching Sport Skills*, Human Kinetics, Champaign, IL.

Magill, R.A. (1993), *Motor Learning: Concepts and Applications* (4th edn), Wm C. Brown, Dubuque, IA.

Schmidt, R.A. (1991), *Motor Learning and Performance: From Principles to Practice*, Human Kinetics, Champaign, IL.

Section 5
Psychological bases of human movement: The subdiscipline of sport and exercise psychology

Introduction

What is sport and exercise psychology?

Sport and exercise psychology is the subdiscipline of human movement involving the scientific study of human behaviour and cognition (thought processes) in the context of physical activity. Within the subdiscipline, the distinction between sport psychology and exercise psychology is not always clear. Sport psychology obviously is concerned with human behaviour in the sport environment, whereas exercise psychology focuses on the exercise environment. Issues relating to competition and sporting performance traditionally fall under the jurisdiction of sport psychology. Exercise psychology typically involves the study of why people do or do not exercise and the psychological effects of exercise on the person. Some overlap between sport psychology and exercise psychology is inevitable. Many competitive athletes participate in exercise that is not sport specific to improve various aspects of fitness. Similarly, some outdoor recreation activities could be classified as either sport or exercise.

There are two major questions that sport and exercise psychologists address:

- What effect does participation in physical activity have on the psychological make-up of the participant?
- What effect do psychological factors have on physical activity participation and performance?

Areas of interest related to the first question include the effects of exercise on psychological well-being and the effect of participating in youth sport on the development of character. Examples of domains related to the second main question are the effects of anxiety on competitive performance and the effects of self-confidence on participation in physical activity.

Sport and exercise psychology focuses on both the individual as a single entity and the individual as part of a group. Because most physical activity takes place in the presence of others, it is difficult to ever entirely remove the possible influence of other people on the behaviour of a single individual. However, the psychology chapters in this book emphasise the biophysical aspects of the subdiscipline and primarily focus on sport and exercise psychology as applied to the individual. The social psychological or group focus of sport and exercise psychology is featured in the companion text on the sociocultural foundations of human movement.

Typical issues posed and problems addressed

A few examples of topics dealt with in sport and exercise psychology have already been given in the description of the two main questions addressed by the subdiscipline. Additional issues in the field include:

- the effect of personality on participation or performance
- what people should focus on when exercising or competing in sport
- the relationship between anxiety and performance
- the enhancement of sporting performance or exercise adherence through the development of psychological skills
- the development of self-confidence through participation in physical activity
- the effects of external rewards on motivation
- how to reduce aggression in sport
- the influence of an audience on performance
- determinants and consequences of team cohesion
- the impact and development of coaches and fitness instructors
- psychological predictors of athletic injuries
- strategies for controlling drug abuse in sport
- exercise addiction.

Levels of analysis

Because sport psychologists do not have the technology to directly record individuals' thoughts or feelings at a specific moment in time, they need to rely on the use of multiple levels of analysis to make inferences about mental processes. The most commonly used levels of analysis are the behavioural or observational level, the cognitive level, and the physiological level. The behavioural or observational level involves watching individuals and recording what they do. For example, in studying anxiety during competition, psychologists might use the behavioural or observational level of analysis for indications of anxiety such as irritability, the return to old habits, yawning, the trembling of muscles, the inability to make decisions or the inability to concentrate. The study of competitive anxiety using the cognitive level of analysis typically entails the use of questionnaires or inventories. The physiological level of analysis involves the direct measurement of physiological variables. For the example of studying competitive anxiety this might consist of measurement of heart rate, respiration rate or muscle tension. There is frequently a lack of coherence across different levels of analysis within sport and exercise psychology. Therefore, making generalisations on the basis of a single level of analysis may be misleading. For this reason it is important to use multiple levels of analysis within this subdiscipline.

Historical perspectives

Sport and exercise psychology is a much newer area of study than some of the other biophysical subdisciplines of human movement studies. The first recognised studies in sport psychology appeared in the late 1890s. These studies were very isolated and included the investigation of topics such as reaction time, audience effects, mental practice, and the personality of athletes. In North America the first systematic research in sport psychology is attributed to Coleman Griffith, who wrote books on the topic, worked as a practitioner with athletes, and established a sport psychology laboratory at the University of Illinois in the 1920s. At the same time organised sport psychology was beginning in Eastern Europe with the establishment of the Institutes for Physical Culture in Moscow and Leningrad. Extensive research began in the Soviet Union in the 1950s to help cosmonauts control bodily functions and emotional reactions while in space. Later these methods were applied to Soviet and East German elite athletes. Although very little of the sport psychology research and practices from within Eastern Europe were available outside the region, it is apparent that there was strong government support (and control) of applied research on the enhancement of the performance of elite athletes.

In North America very little was accomplished in the area of sport psychology after Griffith until the 1960s when numerous research scientists became actively involved in the area and professional organisations began to form. Many of these research workers had been trained in motor control but saw the need to develop sport psychology independent of the study of motor control. In 1965 the International Society of Sport Psychology was formed in Rome, with professional societies for sport psychology and motor control in the United States and Canada forming soon after (1967 and 1969 respectively). Although the emphasis in Eastern Europe remained on field research, North America emphasised laboratory research. It was not until the 1980s that greater importance was placed on applied and field-based research outside Eastern Europe. As both the research and practice of sport psychology burgeon in most countries throughout the world, financial constraints and government changes may, at least temporarily, limit the pursuit of sport psychology in areas that were previously part of the Soviet Union and were historically strong in applied sport psychology.

Although studies on exercise and psychological factors have been intermingled within sport psychology studies during the past 25 or 30 years, exercise psychology has emerged as a specialist area only recently. In 1988 the *Journal of Sport Psychology* became the *Journal of Sport & Exercise Psychology*. The first textbooks in exercise psychology were not published until the 1990s. In most professional organisations exercise psychology is seen as a subdivision of sport psychology. However, just as sport psychology and motor control have split into two distinctive fields within the USA, the future may see the professional division of sport psychology and exercise psychology.

Professional organisations

The International Society of Sport Psychology (ISSP) was organised in 1965; its stated purpose is to promote and disseminate information about sport psychology throughout the world. In addition to publishing the *International Journal of Sport Psychology*, the ISSP holds a World Congress in Sport Psychology once every 4 years. The location of the Congress varies. For example, the 7th, 8th, and 9th World Congresses were held in Singapore, Portugal, and Israel, respectively.

Numerous regional organisations of sport psychology have also been formed. For example, in Europe there is the European Federation of Sport Psychology, which is officially called the Fédération Européenne de Psychologie du Sport et des Activités Corporelles (FEPSAC). There are also several national sport psychology organisations in Europe, such as the Associazione Italiana Psicologie delle Sport (Italy), the Sociedade Portuguesa de Psicolgia Desportiva (Portugal), and the Société Francaise de Psychologie du Sport (France).

In North America, two major associations have emerged. The primary goal of the North American Society for the Psychology of Sport and Physical Activity (NASPSPA) is the advancement of the knowledge base of sport psychology through experimental research. The Association for the Advancement of Applied Sport Psychology (AAASP) was formed to promote the field of applied sport psychology. NASPSPA is traditionally academic in nature, whereas AAASP is concerned with ethical and professional issues related to the development of sport psychology and to the provision of psychological services in sport and exercise settings. AAASP promotes the development of research and theory, but also focuses on intervention strategies in sport psychology.

The specific organisations listed above are not intended to be representative of all sport psychology associations, societies, and organisations. Sport psychology is growing rapidly in many countries in addition to those in Europe and North America—such as Africa, South America, Australia, and Asia.

CHAPTER 20

Basic concepts of sport psychology

- Personality and sports participation and performance
- Motivation and sports participation and performance
- Arousal, anxiety and sports participation and performance
- Mental rehearsal, imagery and sports performance

As mentioned in the section introduction, sport psychology covers many different topics. An introductory text such as this one obviously does not allow for complete coverage of the area. Instead, four of the major domains of the field are introduced in this chapter. In terms of the analogy of the car to the human first introduced in the preface, sport and exercise psychology, with its focus on mental processes and behaviour, is about investigating the driver of the car. What assumptions might be made about an individual who owns a brand new convertible sports car versus someone who drives an old, beat-up Ford? It might be presumed that these individuals have different personalities. In addition to presenting information about personality and sport, this chapter will also consider a few of the factors that might influence the performance of the driver. Specifically, motivation, anxiety, and imagery are briefly introduced and considered in this chapter.

In keeping with the strong tradition in psychology of focusing not only on average (group) behaviour but also on individual differences, liberal use will be made in this, and the subsequent chapters on sport and exercise psychology, of applications of key concepts to (hypothetical) individuals. The approach to illustrating the key concepts will therefore be somewhat different, and more personalised, than that used in the preceding chapters.

Personality and sports participation and performance

What is personality?

Personality has been defined in various ways, but certain elements are common to all definitions. In its simplest form, personality is the composite of the characteristic individual differences that make each of us unique. Let us think about two people we know who act very differently. What is it about them that makes them different? These differences are what make each of us unique. In portraying the differences between the two people we have considered, we have depicted aspects of their personalities.

Trait framework of personality

One framework for studying personality, the trait framework, suggests that everything we do is the result of our personalities. In other words, our behaviour is determined completely by our personalities. This can be formalised by the equation:

$$B = f(P)$$

where B stands for behaviour, f stands for function of, and P stands for personality. According to the trait framework, each individual has stable and enduring predispositions to act in a certain way across a number of different situations. These predispositions, or traits, predict how we will respond. For example, if Peter had the trait of shyness he would be expected to be reserved or timid when joining a new team. If Sue, on the other hand, had the trait of being outgoing, she would be expected to be extroverted and sociable when first meeting with new teammates.

The trait framework of personality suggests that personality traits can be objectively measured. Since traits are considered to be enduring and stable, inventories are used in an attempt to measure personality characteristics.

Are athletes different from non-athletes?

Traditional sport personality research has administered personality inventories to groups of athletes and non-athletes and then examined the results for any differences. Some studies have found differences that indicate that non-athletes are more anxious than athletes, and that athletes are more independent and extroverted than non-athletes.

There have, however, been problems with some of this traditional sport personality research. Most personality inventories were designed for a specific purpose with a specific population in mind. Several of these inventories were created for use with clinical populations and therefore are inappropriate for use with non-clinical populations such as athletes. Additionally, some of the questionnaires used in the research have not been shown to be valid and reliable. In other words, it has not been demonstrated that the questionnaires actually measure what they say they will measure or that if the same questionnaire is given to a person more than once that the same results will be obtained.

Aside from possible difficulties with the inventories used in the research, other complications arise when the definitions of athletes and non-athletes are considered. Trying to generalise findings across studies is very difficult because the terms have not been defined in the same manner in the different studies. How does one define an athlete? In some American studies athletes have been defined as those competing in intercollegiate sport. Does this mean that individuals participating in club sports are not athletes? Other studies only consider individuals to be athletes if they have achieved a particular level of performance, but not if they train and compete regularly at a lower level of proficiency. Would people who run on their own many times per week be considered athletes if they did not participate in competitions? What about someone who competes in social tennis matches only once or twice per year? As you can see, even if valid, reliable, and relevant personality inventories can be obtained, actually determining who is an athlete and who is not makes comparing the personalities of athletes and non-athletes very difficult.

Interaction framework of personality

Another model for studying personality is the interaction framework. The interaction framework still recognises that personality traits influence behaviour; however, the situation or the environment is also acknowledged as influencing how we act. This framework can also be expressed as an equation:

$$B = f(P \times E)$$

where E stands for environment. Within the interaction framework, the environment refers to all aspects of the situation that are external to the individual, including the physical surroundings, as well as the social milieu (other people). It is important to note that the interaction framework is just that, an interaction. It is not only that personality traits and the environment both influence behaviour, but that the two interact and affect each other as well. For example, if Peter, the shy person, was attending the first training session of a new team, there are aspects of the environment that could influence his behaviour. If the other team members were inseparable and very suspicious of newcomers, Peter's behaviour would probably be very different than if the new team members went out of their way to welcome newcomers. Similarly, if Sue, the extroverted athlete, joined a new team, her outgoing personality would influence the behaviour of the people around her, thus influencing the team environment.

The interaction framework considers not only traits, but also states. *Traits*, as previously described, are the stable enduring personality predispositions. *States*, on the other hand, refer to how someone feels at a particular point in time. States are the mood responses we have to specific situations. Traits may influence states, but they do not directly determine states. For example, Bruce may have a low level of trait anxiety. Generally speaking he is a relaxed and calm person who doesn't get anxious very easily. In most situations he is tranquil and unruffled. Bruce could be in the final of the local basketball competition and still remain calm. Certain situations, however, may cause Bruce to react with high levels of state anxiety. An illustration of this might be if Bruce found himself hurtling down the ramp of a ski jump or shooting a free throw with his team down by one point and one second left in the game.

The main difference of the interaction framework is that it acknowledges that a situation or the environment can influence how we react. Behaviour is not solely determined by personality traits. Because the situation can influence behaviour, the use of personality inventories as a method of establishing definite measures of how individuals will perform in all situations is not viable. Although these inventories may give some indication of the personality traits of individuals, measures that take into account specific situational factors are needed to determine states.

Practical implications

If it could be determined that people with certain characteristics or traits perform better in different sports, then personality inventories could be given to individuals to determine which sport suits them best. This process should also incorporate anthropometric, biomechanical, physiological and motor control facets of the individuals. The challenge, of course, is to base any decisions of this type on factors that will continue to be predictive of performance over the passage of time.

Motivation and sports participation and performance

Many coaches complain that certain individuals would be great athletes if only they were motivated. The athletes are seen to have all of the anthropometrical, physiological, biomechanical and skill components necessary for performing at a high level except that they just don't seem to care. They might show up late to training, fail to try very hard during drills or just not show up at all. Other coaches and athletes may not be so concerned about the performance of specific individuals, but are instead troubled by the large number of people who no longer participate in sport at all. Both concerns relate to the concept of motivation.

What is motivation?

Motivation is made up of three components: direction, intensity, and persistence. Direction refers to where people choose to invest their energy. There are very few unmotivated people in the world; it may just be that they are motivated in a different direction from the one we would like them to be. For example, a roommate may be motivated to go to the movies, but is not motivated to clean the house. Similarly, an athlete with great performance potential may be motivated to go to the pub instead of to training. Neither of these people lack motivation; they have just chosen a different direction in which to invest their energy.

Intensity refers to how much energy is invested in a particular task once the direction has been chosen. Two athletes may both choose the direction of training, but where one invests very little effort, the other tries very hard and works at a high level of intensity. It is worth noting, however, that the same exercise or drill will require different amounts of intensity from different individuals depending on their levels of fitness and skill.

The third component of motivation is persistence. Persistence alludes to the long-term component of motivation. It is not enough to have an athlete choose the direction of training and train at a high level of intensity during one or two training sessions. Athletes need to continue to train and participate over time. It may be preferable to have athletes train at a moderate level of intensity over an entire season, rather than have them train at an extreme level of intensity at the beginning of the season and then drop out (see Chapter 13).

Individuals interested in enhancing the motivation of sports participants should keep in mind all three components. Direction can be influenced by making the path of participation more enjoyable. Intensity can be increased by stipulating a reason for doing any particular activity. For example, how much effort are workers likely to put into a job if they think there is no reason for doing it? The same principle holds true for sport. If a sporting drill requires a lot of effort, athletes will be more likely to put in the effort if they perceive there is a reason for doing so. Persistence can be improved by providing positive feedback to participants. Athletes are much more likely to continue participating at something if they are convinced of the long-term benefits. Would students continue to study if they honestly felt that it did not accomplish anything? If, however, they felt that studying gave them greater understanding of the subject, better prepared them for more advanced subjects, led to higher marks or caused them to get positive recognition from someone they care about, they would be more likely to persist. Believing that continued participation leads to greater success increases persistence.

What is success?

Traditionally success in sport has been considered in terms of winning and losing; winning equals success and losing equals failure. If outcome is the only criterion for success, anyone who does not win fails. Using this definition, only one team in a given league can be considered successful at the end of the season, only one swimmer in a particular event is successful, and only one runner in a marathon is successful. Everyone else is a failure. Given that experiencing success enhances motivation, the outcome definition of success leaves the majority of participants lacking in motivation as they only experience failure.

Success, however, can mean many different things to different people. Some people do define success in terms of outcome. A performance is only considered to be successful if the athletes were able to demonstrate that they were better than everyone else. Nevertheless, many athletes feel a performance was successful if they improved their own performance, regardless of outcome. In other words, if technique was improved, times were decreased or some other aspect of performance enhanced, then success was experienced. In this case, comparisons are made with one's own previous performances rather than with the performances of others.

Some people do not use performance as the basis of their definition of success. Success for these people is achieved when they get social approval or recognition from others. This form of success could be experienced in different ways. For example, a hockey player may consider the season to be successful because she made a lot of friends on the team or was recognised by the media. Similarly, a basketballer may consider a game to be a success because he was complemented by the coach for the amount of effort he put into the game.

These different definitions of success are called achievement goal orientations. Very few individuals have only one achievement goal orientation. Athletes often have multiple motives for participating in sport. They may want to improve their skills as well as win. Additionally, a single individual may have different achievement goal orientations for different sports. For example, winning may be the primary definition of success in rugby for this individual, but improving specific skills may be the principal basis for determining success in soccer. See Box 20.1.

Achievement goal orientations and motivation

These different definitions of success influence motivation. For example, if a swimmer is in a race and defines success as winning, effort may be decreased if the swimmer is well ahead of the others or if the swimmer is well behind and it is perceived that there is no chance of winning. If, however, the swimmer defines success on the basis of previous personal performances, effort will be maintained no matter where the swimmer is placed in the race as success is related to improving one's own time or technique.

Coaches should also be aware that athletes often have different achievement goal orientations. Unfortunately, many coaches assume that all their athletes define success the same way they do and that their reasons for participating are also identical. This situation can be problematic if the goals of the coach and the athletes are in fact different. There is a greater likelihood of athletes dropping out of a situation if they perceive

Box 20.1 Achievement goal orientations and participation in youth sport

An individual may have different achievement goal orientations for different sports and for different contexts. For example, an individual may have different goal orientations for recreational sport than competitive sport. Achievement orientations may also change with age.

Jean Whitehead, from the University of Brighton and the Institute for the Study of Children in Sport in England, investigated the achievement orientations of competitors, dropouts, and non-participants in school and nonschool sport. Questionnaires were completed by 357 boys and 473 girls during their physical education classes. The following four achievement orientations were identified: 1. an ability orientation (doing something few others can do); 2. a task orientation (reaching a goal or meeting a challenge); 3. a social approval orientation (making other people happy); and 4. an intrinsic orientation (doing something adventurous, new, or different).

The ability orientation of competitors was higher in nonschool sport than school sport and fewer students had participated in nonschool sport than in school sport. This finding suggests that ability-oriented students choose nonschool sport as an additional or alternative context in which to demonstrate their ability. As a smaller percentage of students dropped out of nonschool sport than school sport, it appears as though the nonschool context allowed ability-oriented students to fulfil their achievement goals.

For upper school students (mean age = 15 years), competitors in school sport scored lower than dropouts in the task and intrinsic orientations and higher than non-participants in the ability and social approval orientations. In nonschool sport, competitors scored higher than dropouts in the ability orientation. In contrast, for middle school students (mean age = 11 years) competitors and dropouts did not differ in their levels of ability orientation for either school or nonschool sport. Instead, the task orientation discriminated the groups.

These results support the concept of individuals having multiple achievement goal orientations. The study also indicates that individuals' achievement orientations vary according to their ages as well as the context (school or nonschool sport).

Source: Whitehead, J. (1995), Multiple achievement orientations and participation in youth sport: A cultural and developmental perspective, *International Journal of Sport Psychology*, 26, 431–452.

that their needs are not being met. Therefore, if a coach is only interested in winning, but the athletes are interested in making friends or learning new skills, the athletes may feel that they are not getting what they want and abandon the sport altogether. Similarly, if a coach stresses individual growth and improvement by ensuring that every athlete gets similar playing time, and some of the athletes are solely interested in winning, friction and unhappiness may result.

Arousal, anxiety and sports participation and performance

What is arousal?

Arousal is traditionally considered to be physiological activation. Common methods of determining different levels of arousal include measurement of heart rate, respiration rate, muscle tension and blood pressure. This physiological consideration of arousal makes sense because when individuals are highly aroused or activated they often have tense muscles and higher blood pressures, heart rates and respiration rates than they do when they are calm. More recently, arousal has been regarded as involving mental activation in addition to physiological activation. Therefore, arousal can be considered to be the degree of mental and physical activation or intensity. An example of a low level of arousal would be the grogginess experienced when first awakening in the morning. An example of a high level of arousal would be the heightened mental alertness and physiological activation experienced before a major competition.

What is anxiety?

Confusion is often caused when people use the terms arousal and anxiety interchangeably. Anxiety is the subjective feeling of apprehension and is usually accompanied by increased arousal levels. High levels of arousal, however, are not always accompanied by anxiety. If a team just beat the defending league premiers, they would probably be fairly aroused, yet they would probably interpret this arousal as excitement rather than anxiety.

Anxiety is usually experienced when a situation is perceived to be threatening. This commonly occurs when there is a perceived imbalance between the demands of the situation and individual ability to meet those demands. For example, if Diane believes that she must make the final two freethrows in a basketball game, yet does not believe that she is very good at making freethrows, she is likely to experience anxiety. This anxiety will increase the more Diane believes that there may be negative repercussions if she misses. If, for example, Diane did not care about the outcome of the game or her own performance, and was only playing to have fun with her friends, she would probably only experience mild anxiety, if any. On the other hand, if she felt the outcome of the freethrows to be extremely important, then she would likely experience a relatively high level of anxiety. Anxiety, therefore, is determined by the perceptions of the individual. Two individuals may be in the same situation, with one perceiving it to be slightly challenging and the other perceiving it to be extremely threatening.

Trait and state anxiety

This sensation of anxiety is a good example of the interaction framework mentioned in the personality section of this chapter. The level of anxiety experienced is a result of the interaction of personal factors (for example personality, needs, capabilities) and situational factors (for example opponent(s), task difficulty, and the presence of other people). An athlete with a predisposition to perceive situations as stressful or threatening would be considered as having a high level of trait anxiety. State anxiety, on the other hand, is the experience of apprehension at a particular point in time.

Cognitive anxiety and somatic anxiety

State anxiety is a multidimensional concept. Cognitive anxiety is the mental facet of state anxiety. Worry, perceived threat and self-defeating thoughts are all examples of cognitive state anxiety. Somatic anxiety refers to physiological anxiety responses such as butterflies in the stomach and sweaty palms. Although somatic and cognitive state anxiety are related, it is possible to experience high degrees of one without experiencing high degrees of the other. For example, individual students experience the anxiety of examinations differently. Some physically get very tense, whereas others drive themselves (and everyone else) crazy with their worrying. The same is true for the experience of anxiety in sport. Some athletes have more signs of somatic anxiety, whereas others experience greater cognitive anxiety.

The arousal–performance relationship

If Don had just awakened and was feeling sluggish and tired, he probably would not perform very well if asked to engage in competitive sport. His performance would probably improve as he became more alert and awake. In other words his performance would improve with increases in arousal. This proposed linear relationship between arousal and performance is called the drive theory and is illustrated in Figure 20.1. According to the drive theory, performance will continue to improve with further increases in arousal or activation. Therefore, if the drive theory is to be believed, the way to ensure optimum performance is to arouse athletes as much as possible.

Some coaches have done some very strange things because of their belief that the more aroused their athletes are, the better they will perform. Coaches have had their players bite the heads off live chickens, castrate a bull, stage a mock gun battle in a school cafeteria (complete with fake blood), and watch films of prisoners being murdered in concentration camps. Fear, anger, and horror were seen as emotions that could raise arousal levels. All of these activities were done to arouse the athletes as part of precompetition preparation in the belief that the drive theory is correct.

Unfortunately there are still some coaches who continue to base precompetition preparation on the drive theory. A more useful method of considering the relationship between arousal and performance is to follow the inverted U hypothesis. As can be seen in Figure 20.2, when the relationship between arousal and performance is plotted according to the inverted U hypothesis, the graph forms an upsidedown U (hence the name). As with the drive theory, there are increases in performance with increases in arousal. However, these increases in performance only occur up to a certain point of arousal. If arousal continues to increase past that point, performance begins to deteriorate instead of improve. The point at the top of the inverted U is called the point of optimal arousal. It is at this point that performance is best. If arousal is below that point (under-arousal), performance will be less than optimal. If, however, arousal is past the point of

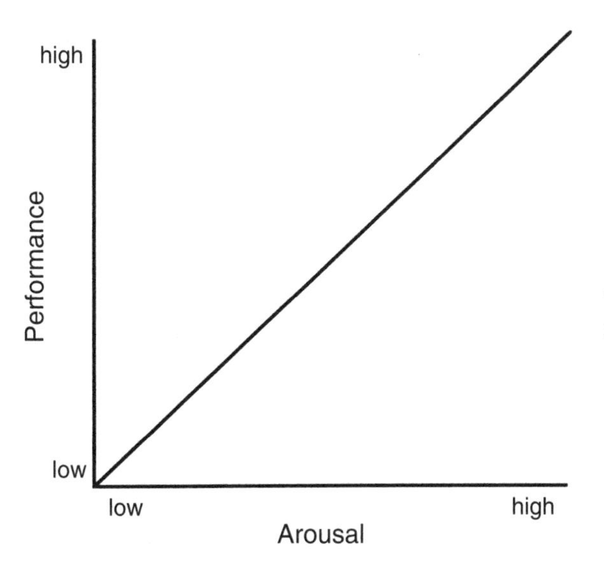

Figure 20.1 Drive theory

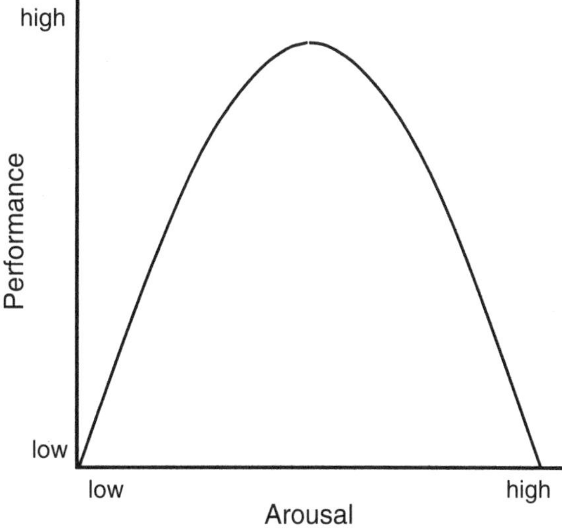

Figure 20.2 Inverted U hypothesis

optimal arousal (over-arousal), performance will also be less than optimal. Therefore, continually increasing arousal could be problematic as this could cause athletes to become over-aroused and therefore impair performance.

The challenging aspect of this relationship for coaches is that the optimal level of arousal varies across individuals (see Figure 20.3). Judy may perform her best when feeling very relaxed. However, Shirley may perform her best when feeling pumped up and activated. Obvious differences in optimal arousal can be seen when considering different sports. If Judy is putting in golf, and Shirley is attempting to lift a personal best in bench press, it is fairly obvious that Judy's optimal level of arousal would be lower than Shirley's. It should be noted, however, that optimal levels of arousal also differ across athletes within any one sport. Even if Judy and Shirley both competed in netball, they could still have distinctive levels of optimal arousal.

In practical terms, these individual differences in optimal arousal mean that the identical pre-competition build-up will not be equally effective for everyone. Athletes who arrive at the competition feeling very sluggish and under-aroused may benefit from a big motivational speech from the coach. However, if other athletes on the team are already at their optimal levels of arousal, or already over-aroused, a 'motivational' pep-talk would probably have a negative impact on their performances by causing them to be over-aroused. Some athletes actually may need to focus on relaxing and lowering their levels of arousal before competing.

The impact of anxiety on performance

As mentioned previously, state anxiety is considered to be multidimensional. Somatic and cognitive state anxiety have different impacts on performance. The relationship between somatic state anxiety and performance is almost identical to that between arousal and performance. That is, the relationship is curvilinear. As state somatic anxiety increases, there are increases in athletic performance up to a point. If somatic anxiety continues to increase, then performance gradually decreases.

Cognitive state anxiety, however, is believed to have a negative linear relationship with performance. That is, as cognitive anxiety increases, performance decreases (see Figure 20.4). In simple terms, as people begin to worry, their athletic performance begins to suffer. The more worry and distress, the worse the performance.

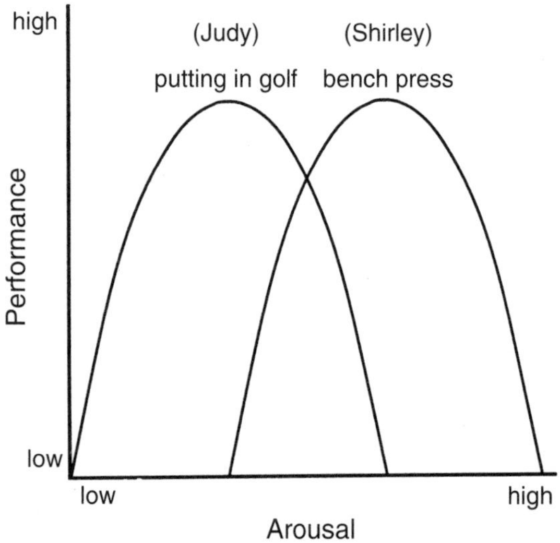

high
(Judy)
(Shirley)
putting in golf
bench press
Performance
low
low
high
Arousal

Figure 20.3 Variance in optimal levels of arousal

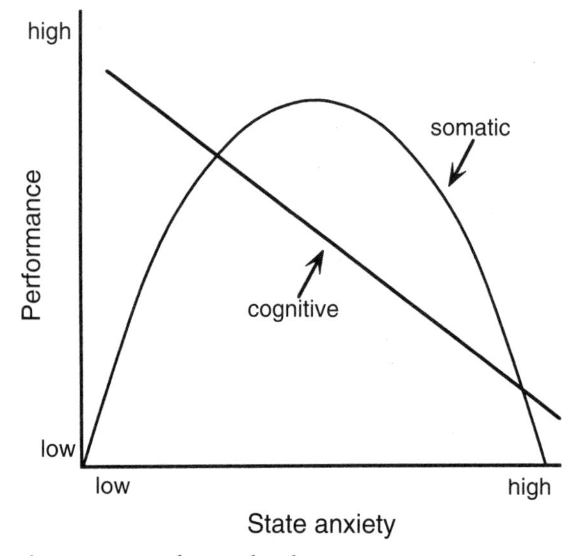

Figure 20.4 Relationship between multidimensional state anxiety and performance

Things become more complicated when we try to consider the simultaneous impact and interaction of cognitive anxiety and somatic anxiety (or arousal) on performance. When cognitive anxiety is low, the relationship between arousal and performance takes the form of the inverted U. However, when cognitive anxiety is high, the relationship between arousal and performance takes the form of the catastrophe model. According to the catastrophe model, if arousal continues to increase after the point of optimal arousal, there is a sharp and rapid decline in performance rather than a gradual decrease in performance. Basically, a catastrophe occurs.

If cognitive anxiety is high, an increase in arousal can cause a terrible performance. For example, if Rob is playing basketball and has a high level of cognitive state anxiety, an increase in arousal past the optimal point could cause him to not only miss easy shots, but also to make multiple errors in shot selection, defence, and passing. According to the catastrophe model, for Rob to return to his optimal level of performance while experiencing high cognitive anxiety, he would need to allow his physiological arousal to return to a relatively low level before gradually building his arousal and performance back to optimal levels. That is, although a relatively small increase in arousal can cause a catastrophe in performance, a minor decrease in arousal will not allow performance to return to its previous level. Instead, a significant decrease in arousal is needed before performance will return to its previous standard.

This catastrophe model, which explains the relationship between arousal and performance under high levels of cognitive anxiety, could be thought of as being like a wave crashing on the beach. Once the wave gets to its highest point, it does not just gradually decrease in height, but rather suddenly and dramatically decreases in height. The water is unable to just wash down the beach to where it broke to build up its height again. Instead, it washes back into the ocean before once again gradually building up to its optimal height.

In practical terms large and sudden decreases in performance can be avoided by minimising cognitive state anxiety or by closely monitoring physiological arousal. One way of monitoring arousal is through biofeedback. Biofeedback instruments are designed to give individuals information about their biological processes. By increasing awareness of biological processes such as heart rate, sweating or muscle tension, individuals can eventually learn to modify or control their physiological arousal. Biofeedback allows us to learn to control bodily functions that traditionally are thought to be beyond our control (see Box 20.2).

Mental rehearsal, imagery and sports performance

What is imagery?

Imagery is a skill that involves using all the senses to create or recreate an experience in the mind. Many terms have been used interchangeably with the term imagery. These include mental practice, mental rehearsal, and visualisation. Mental practice and mental rehearsal could be considered as specific forms of imagery—using imagery to mentally practice or rehearse particular skills or techniques. Visualisation and imagery are often considered to be synonymous. However, the term imagery is preferable as visualisation implies a restriction to the sense of vision, whereas imagery more easily incorporates sound, smell, taste and feel, as well as sight.

How does imagery work?

Although sport psychologists, athletes, and coaches generally agree that imagery helps individuals to learn new skills and improve performance, there is less agreement about how it works. Three of the existing theories proposed as possible explanations of how imagery works will be briefly described. No single theory has been proven to be the definitive answer to the question of how imagery works. Nevertheless, each of the theories presented here should help to create an understanding of the impact of imagery.

Box 20.2 State anxiety and sport performance—case study

In Western Australia there was a 20-year-old male state-level small-bore rifle shooter whose performance suffered due to excessive anxiety. He had trouble controlling nervousness and tension and reported high levels of both somatic state anxiety and cognitive state anxiety. His somatic anxiety was primarily manifested through excessive muscle tension and accelerated heart rate.

A group of researchers from the Department of Human Movement at the University of Western Australia obtained baseline measures of state anxiety and performance during competition. The athlete then went through a 12 session intervention program involving education about anxiety, yoga-like exercises, muscular relaxation, stopping negative thoughts and biofeedback to learn to control heart rate. State anxiety and performance were then measured again during another competition after the intervention.

Post-intervention, there were changes in many measures. Questionnaire measures of cognitive and somatic state anxiety were both lower. There was less gun vibration in the two seconds before firing, indicating that the athlete was calmer. Biochemical measures of urine demonstrated lower levels of the stress hormones noradrenaline and adrenaline, physiological measures indicating lower levels of anxiety. Self-confidence and performance both improved. There were no significant differences in heart rate or muscular tension.

Overall, the study indicates that athletes can learn to control state anxiety and enhance performance. However, a few things are worth noting. The subject in this study originally demonstrated symptoms of anxiety that were detrimental to performance. Not all signs of anxiety are necessarily detrimental to performance. Similarly, putting athletes through an intervention designed to improve performance through controlling anxiety would probably not be beneficial to athletes with no initial difficulties with anxiety. From a research angle, it is important to note the multiple measures of anxiety that were used. If only heart rate and muscular tension had been used as physiological indications of anxiety, it would have been surmised that the intervention was ineffective in helping with anxiety.

Source: Prapavessis, H., Grove, J.R., McNair, P.J. & Cable. N.T. (1992), Self-regulation training, state anxiety, and sport performance: a psychophysiological case study, *The Sport Psychologist*, 6, 213–29.

Psychoneuromuscular theory

The psychoneuromuscular theory states that when we image ourselves moving, our brain sends subliminal electrical signals to our muscles in the same order as when we actually physically move. In other words, all of the co-ordination and organisation of movement takes place in our brains. When we image movement, all of that organisation and co-ordination takes place just as if we were actually moving. Our brain sends electrical signals to our muscles concerning which muscles should contract, when and with what intensity. Therefore, whether an athlete is actually performing a movement or vividly imaging the performance of the same movement, similar neural pathways to the muscles are used. The psychoneuromuscular theory claims that it is through this mechanism that imagery may aid skill learning.

Symbolic learning theory

The symbolic learning theory is similar to the psychoneuromuscular theory except that electrical activity in the muscles is not required. According to the symbolic learning theory, imagery helps the brain to make a blueprint (or plan) of the movement sequence without actually sending any messages to the muscles involved. Imagery helps to develop a blueprint so that when action is required, the brain can follow this blueprint.

Attention–arousal set theory

According to the attention–arousal set theory, imagery works because it helps the athlete to reach the optimal level of arousal (as mentioned in the previous section) and focus attention on what is relevant. According to this theory, imagery does not send messages to the muscles or help develop a blueprint, but instead primes the athlete for performance. Physiologically, imagery helps the athlete raise or lower arousal levels to the appropriate point. Cognitively, imagery helps the athlete selectively attend to what is important for good performance, decreasing the chance of distraction.

Why use imagery?

Imagery can be used in many different ways. Athletes can use imagery to help to learn to control emotions. For example, Jay is a reasonable rugby player who contributes much to the team except for when she feels that an official has made a poor decision. She has a tendency to lose her cool, which often results in penalties. Through imagery, she can imagine herself shrugging off a poor call and continuing to focus on what she needs to do to perform well. If she practises an appropriate reaction through imagery, the desired reaction will become natural when it happens in real life.

Imagery can also be used to improve confidence. If athletes can see and feel themselves performing the way they want to immediately before they go out to perform, the better they will probably feel about themselves. Another way of thinking about this notion is to consider how previous performances relate to confidence. If Chris was about to shoot a freethrow in the final seconds of a basketball game, he would probably feel more confident if he had just made his last 20 attempts than if he had just missed his last 20 attempts. Basically, the more times something is done, the easier it is to do. This is where imagery can come in handy. Athletes can always create that feeling of confidence by imaging themselves performing the way they want to immediately before the actual physical performance.

Imagery can be used in combination with physical practice to enhance the learning of new skills and the performance of known skills. Although there is some disagreement in the literature, most of the research supports the concept that combining physical practice and mental practice is more effective than either practice alone. For example, if Bob is learning how to serve in tennis, he will probably learn the skill more quickly if he both images himself serving correctly and also physically practises serving. The combination of the two techniques together works better than either in isolation.

There are, however, examples of individuals achieving a lot with the sole use of imagery. One of these stories has been reported in many countries with different names, places and details, but with identical end results. The reports concern members of the armed forces who were taken as prisoners during war. In various circumstances these prisoners used golf as a way of keeping their minds off their current situation. They imaged themselves playing golf although they were physically confined. Some played different courses at different times of day, others continually played the same course. The end results varied only in the location or the specific situation. All of the individuals in these reports eventually were released, returned to their home countries, and then went out to a golf course and played extremely well, even though it had been years since they had actually physically played.

Although there are many other uses of imagery, only one more will be discussed here. Imagery has been shown to help with the control of pain. Although many of the examples of using imagery for pain control come from anecdotes relating to illness, sports often involve injuries and pain. The most common form of pain control imagery is probably removing oneself mentally from the situation. By conjuring up scenes that are unrelated or incompatible with pain, people can learn to distract themselves from pain. For example, instead of thinking how much her reconstructed knee hurts, Terry can instead image herself relaxing at the beach with the waves gently breaking at her feet, the sun warming her shoulders, and the breeze wafting across her face.

Advantages of imagery

Imagery has numerous advantages. It is not physically fatiguing, so practising with imagery just before competition does not cause a decrease in the energy available to the athlete. Imagery can be practised anywhere and anytime. When sitting on a bus, waiting in the marshalling area, or having a shower, imagery is always available. Imagery can also be a welcome change of pace during a training session. When an athlete is physically working very hard, it can be a useful break to take a timeout and image specific aspects of technique. Imagery also uses a language that is understood by the body. Sometimes technical corrections of a skill can be understood by the individual athlete, yet still fail to translate to the actual movement. Using imagery can enhance the translation. Additionally, in these times of expensive equipment, shoes, and gadgets, imagery is free!

Vividness and control

If imagery is to be effective, two factors need to be developed—vividness and control. The more vivid an image is the more likely the brain is going to be convinced that the image is the real thing. When actually physically participating in sport, most athletes are aware of the feel of the movement as well as sounds and sights. Some sports, such as swimming in a chlorinated pool, are also associated with smells and tastes. The more senses that can be included in an image, the more vivid, and therefore the more effective the image will be. Athletes will get more out of imagery if they image in technicolour with surround sound than if they image in fuzzy, silent, black and white. Table 20.1 lists examples of images using different senses.

Vividness alone, however, is not enough. The athlete also needs to be able to control imagery for it to be effective. If we try to image a good performance, and instead can only image mistakes and errors, imagery will most likely increase anxiety and destroy self-confidence. Following imagery scripts that are pre-written or pre-recorded can often help the athlete image the desired performance. Following a guided imagery script is often easier than trying to develop an image anew. If vividness and control can both be developed, then imagery can be an extremely productive technique for athletes to learn.

Table 20.1 Imaging with all the senses
Try to image the following sensory experiences.
Are some images more vivid than others?

See:	• a sunrise over the ocean
	• the face of a friend
Hear :	• the roar of a crowd after the home team has scored
	• a door slamming shut
Feel:	• diving into an unheated pool in winter
	• a bear hug from a close friend or relative
Smell:	• freshly mown grass
	• cigarette smoke
Taste:	• honey
	• a cool refreshing drink

CHAPTER 21

Basic concepts of exercise psychology

- Effects of psychological factors on exercise
- Effects of exercise on psychological factors

The purpose of this chapter is to examine the reciprocal links between psychology and exercise, namely the effects of psychological functions, such as motivation, on exercise and the effects of exercise on psychological factors, such as feelings of well-being, mood states and mental performance.

Effects of psychological factors on exercise

The three components of motivation, mentioned in the previous chapter in relation to sport, apply to exercise motivation too. The direction facet clearly relates to whether an individual chooses the direction of the gym or the couch, the stairs or the lift, the pool or the bath. Once the direction of exercise has been chosen, then the intensity and persistence aspects become important. For example, John and Ian both decided as a New Year's Resolution to join a gym. They have both chosen the direction of exercise. During their first day at the gym, John rides the bicycle for 30 min, tries the stepper and the rowing machine, completes multiple sets on the different weights available, and finishes off with an aerobics class. Ian, on the other hand, begins with just a few minutes on the cycle and then completes just one set of each of the different weights exercises using very light weights. John exhibited high intensity and Ian demonstrated low intensity. If we stop at this point in the example, you may conclude that John is more motivated than Ian. Your opinion may change when the third component of motivation, persistence, is considered. Two days later Ian is once again at the gym adding a couple of minutes to his time on the bicycle and sticking to his light weight workout. John, on the other hand, doesn't make it to the gym. He is home in bed so sore he has trouble sitting up. Six months down the road Ian is still regularly attending the gym. John, on the other hand believes that exercise is painful and avoids it whenever possible.

Exercise participation motivation

Exercise participation motivation refers primarily to the direction component of motivation.

Exercise participation motivation is the initiation of exercise. A variety of factors influence whether or not individuals initiate an exercise program.

Knowledge, attitudes and beliefs about exercise influence motivation towards exercise participation. Individuals who understand the importance and value of regular exercise are going to be more likely to initiate an exercise program. Similarly, if people have a positive attitude about the value and importance of regular exercise they will have greater motivation to participate in exercise than do people with a negative attitude.

Valuing the importance of exercise, however, is not the only determinant of exercise participation. Beliefs about ourselves influence motivation as well. Even if Jenny understands that exercise is important, she will be unlikely to begin an exercise program if she believes that she cannot succeed at an exercise program. If she believes that the exercise program is too difficult, requires more fitness, strength, co-ordination, or time than she has, then it is doubtful that she will join the program. This confidence in one's ability to succeed at an exercise program is called exercise self-efficacy. It is very logical that self-efficacy would influence behaviour. Let us think about it for a minute. How likely would people be to do something they were convinced they could not do, particularly if they had to pay to do it? How likely would they be to invest any energy into pursuing that activity? Most people in this situation would not even attempt the activity. People who have such feelings are described as having very low self-efficacy. As self-efficacy, or one's belief in one's own ability to succeed at a particular task, increases, so does the likelihood of undertaking that task.

What does this mean in terms of enhancing exercise participation motivation? Educating people about the importance and value of exercise can be a valuable first step, as individuals who understand the merit of exercise are more likely to adopt an exercise program. Unfortunately, imparting knowledge is not enough. Enhancing the exercise self-efficacy of individuals will increase their motivation. Demonstrating how individuals can control their own activity is useful. Some people have low self-efficacy about exercise because they believe that they are too unfit to begin exercising. They may equate exercise with young, thin, lycra-clad gym enthusiasts whom they see in the media. Programs that emphasise choice of activities, illustrate exercisers similar in age and fitness level to the potential exercisers, and reveal that exercising can be enjoyable may increase exercise participation motivation. Exercise programs that begin with activities such as walking and climbing stairs, which the individuals already know they are capable of doing, may also increase their self-efficacy, hence, increasing motivation.

Exercise adherence motivation

Although many people get motivated to begin an exercise program, many of the people who begin fail to continue. As we saw in Chapter 15, approximately 50 per cent of the individuals who begin a regular physical activity program drop out within the first 6 months. These people had exercise participation motivation, but they lacked exercise adherence motivation, the persistence angle of motivation.

Biological, psychological, sensory and situational factors all interact to influence exercise adherence. Biologically, body composition, aerobic fitness, and the presence of disease influence adherence. Unfortunately, it is usually the people who could gain the most from exercise who are the least likely to adhere. People who are overweight or obese, low in fitness, or suffer from chronic illness are less likely than thinner, fitter, and healthier people to adhere to an exercise program, and therefore may require extra effort when it comes to changing their activity patterns.

Just as with exercise participation motivation, attitudes and beliefs influence exercise adherence motivation. Attitudes and beliefs about the importance of exercise play a role. However, expectations about the impact of exercise and beliefs about the effect that exercise is having on them personally also influence adherence. For example, if David believes that major changes in fitness and body composition should occur after 6 weeks of regular exercise, and he does not perceive major improvement in his own body after 6 weeks, he may believe that exercise does not do what it should, and therefore give up. Even though it is unrealistic to believe that 6 weeks of exercise can make up for 6 years of inactivity, it is David's beliefs that influence his behaviour. Therefore, when introducing newcomers to exercise it is important that they have realistic expectations about the time and effort required and the anticipated impact of the proposed program.

In addition to attitudes and beliefs, other psychological factors also influence exercise adherence motivation. Extroverts tend to adhere to exercise programs better than do introverts. Extroverts are people who are social and outgoing. Exercise programs that are executed in the presence of other people are probably more comfortable for extroverts than introverts. Extroverts tend to enjoy the interaction with class members and exercise partners, possibly encouraging their adherence. Introverts, on the other hand, may adhere better to individual, home-based exercise programs. The bulk of the research on exercise adherence has involved programs that take place on site at a fitness facility with other people. This setting may have led to the conclusion that extroverts are better adherers than introverts. If the research had been done on independent home-based exercise programs, it might have been found that introverts were better adherers than extroverts. Practically, this information means that efforts should be made to match the social environment of the exercise program to the personality of the exerciser.

Individuals with high levels of self-motivation are more likely to adhere to exercise programs than individuals with low levels of self-motivation. It is very logical that highly self-motivated people have better adherence rates. The challenge is to help those individuals with low levels of self-motivation. One of the most effective methods of helping these individuals is to encourage their involvement in the goal-setting process.

Goal setting

Setting goals can help to enhance motivation for a number of reasons. Goal setting addresses all three components of motivation. Goals give direction by providing a target. Intensity can also be enhanced through goals. Effort levels can be strengthened through goal setting because goals provide reasons for the activity. We are all more likely to put in effort when we feel there is a reason for doing so. If individuals are given two jobs to do at work, one of which has a particular target and objective, and one that seems very vague and purposeless, into which job are they more likely to put their effort? Having goals helps to focus attention and effort. In addition, goals can augment persistence by fostering new strategies. If individuals have a goal to which they are committed and initial tactics appear unsuccessful, they will search for alternative strategies to achieve their aim. If the goal had not been set in the first place, instead of persisting with different plans of action, giving up would have been the more likely result.

Goals are also beneficial because they reflect improvement. Too often people compare their strength, fitness, flexibility or weight over the short term. Because the positive effects of exercise take time to emerge, improvements being made are often not noticed as people use a very short time-frame for comparison. If a goal is achieved, evidence of improvement exists.

Goal setting involves a number of steps: setting the goal, setting a target date by when the goal is to be achieved, determining strategies to achieve the goal, and evaluating the goal on a regular basis. If a target date is not set, then the goal is really just a dream: 'One of these days I'll ride the exercise bike continuously for one hour'. For this idea to be a goal, a specific date needs to be set by when the behaviour is to be achieved. Goal setting usually involves both long-term and short-term goals. The long-term goal provides direction. The short-term goals provide the increase in intensity and effort. People often err by only setting long-term goals. They begin to work towards achieving the goal, but success seems so far away that they give up before they get there. So if the person had decided to ride the cycle for 1 hour and was currently having trouble lasting 10 minutes, lasting 1 hour would seem virtually impossible. Achieving short-term goals along the way to the long-term goal boosts confidence and motivation because it is obvious that the effort is worthwhile as improvement is being made. Target dates are set for each short-term goal in a progressive order until the long-term goal is achieved. This pattern of goal setting can be considered as a staircase, where each short-term goal is a step on the way to the long-term goal (See Figure 21.1).

However, for goals to be effective, they need to be properly set. Goals can be considered to be good goals if they meet certain criteria (see Table 21.1). Goals need to be challenging but realistic. If goals are not challenging, they probably are not requiring any real change in behaviour and therefore will have little impact. Nevertheless, there needs to be a balance between challenge and realism. If goals are so challenging that they are unrealistic, then people are setting themselves up for failure. Continued failure leads to lowered confidence and less motivation.

Goals also need to be specific and measurable. Saying that 'I want to be fitter' or 'My goal is to be stronger' does not provide any way of knowing when success has been achieved. What is 'fitter'? How strong is 'stronger'? There needs to be some way of knowing whether or not the goal has been achieved when the target date arrives. The easiest way of doing this task is to make goals numerical. Numbers can easily be used with time spent exercising, distance travelled, repetitions accomplished, weight lifted, or exercise sessions attended.

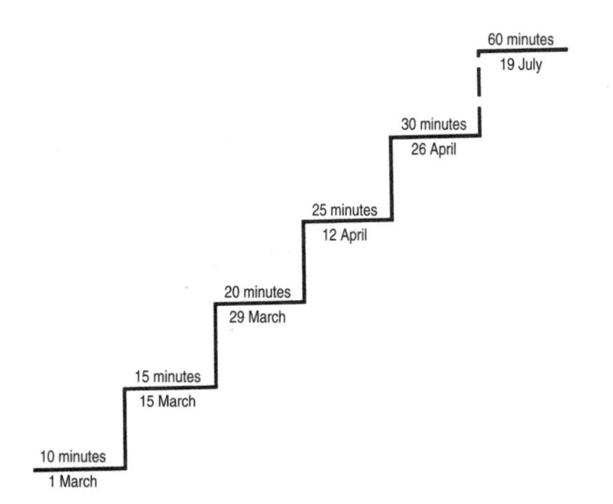

Figure 21.1 The goal-setting staircase

Table 21.1 Goals need to be . . .

Principle	Question to determine if it has been met
Challenging	Will it require effort?
Realistic	Is it reasonable?
Specific	Is it obvious what the precise objective is?
Measurable	Will there be an exact method of determining if it has been achieved?
Positive	Does it stipulate the desired behaviour?
Controllable	Does it relate to the performance, technique, or behaviour of the goal setter?

In addition to being specific and measurable, goals need to be positive. If Diane makes it her goal not to take advantage of the abdominals exercise section of her aerobics class by just reclining and resting, she will be thinking about reclining and resting. If instead she made it her goal to complete first 1 minute and then 2 minutes of the abdominal exercises she would be thinking about doing the exercises, increasing her chances of doing them. If I tell you not to think about pink elephants, what is the first thing you think about? Similarly, if you set a goal of not letting your back arch off the bench when you do bench press, you will be thinking about your back arching. When you think about your back arching, your brain may be sending messages to the muscles that make your back arch (just as in the pyschoneuromuscular theory of imagery discussed in the previous chapter). So by having 'not arching' as your goal, you may actually be increasing the likelihood of arching your back. The goal would be much more effective if it was to keep your back pushed flat against the bench (or to think about blue elephants). Goals should stipulate the desired behaviour. This positive phrasing helps you think about, plan for, and prepare to do what it is you want to do.

For goals to be effective, it is also important that the person setting the goal has control over the activity. Goals should be related to the performance, technique or behaviour of the goal setter. Goals are ineffectual when they rely on the behaviour of others. Ultimately, people only have control over what they do. Therefore, goals such as being the strongest person in the gym or getting everyone in one's family involved in regular exercise are not completely under personal control. Best efforts might result in failure because individuals cannot control others. If someone really does want to be the strongest person in the gym, goals that reflect what he or she thinks he or she needs to do to achieve that should be set. Individuals need to focus on their own behaviour. No one can help it if the defending world power-lifting champion suddenly moves to the neighbourhood and joins the local gym. We can, however, set goals about how much weight we will lift, how many sets or repetitions of what specific exercises we will do or how many times each week we will train. Similarly, if Pat really wants everyone in the family involved in regular exercise, goals related to what she will do to try to achieve that should be set. For example, she might set a goal of organising weekly family fun days that involve physical activity, creating a package of exercise options that fit their schedules or babysitting the young kids for her brother and his wife so they have the opportunity to go to the gym together. The focus is on what individuals can do themselves, rather than on what other people may or may not do.

In summary, goal setting can be a very effective method of improving exercise adherence motivation. The goals need to be challenging but realistic, specific and measurable, positive and controllable. If the goals require a lot of work or a major change in behaviour, a series of short-term goals that lead to the long-term goal should be set. For each goal there needs to be a target date and a list of strategies that can be used to achieve the goal. Finally, the goals need to be evaluated on a regular basis. If a goal has been achieved, great! People can reward themselves: give themselves a pat on the back, buy themselves that CD they've been wanting, or just feel good because of their achievement. If, however, the goal has not been achieved, they need to think about what may have happened. Possibly the goal was too big a step on the staircase, maybe a smaller step would be more appropriate. Maybe they became sick or injured and could not work towards achieving their goal. In that case, a new target date should be set. Perhaps the goal is realistic, but the strategies selected to achieve it were not suitable. In that case a new target date should be set and different strategies implemented. Goals can strengthen self-motivation, which, in turn, can enhance exercise adherence motivation.

Sensory factors

The same actual workload is going to be perceived differently by various individuals. If someone perceives an exercise program as being excessively stressful, that person is apt to drop out. Therefore, when exercise is prescribed, it is important to take into account the individual's perceptions of the difficulty of the exercise and how much distress is experienced when participating.

Problems sometimes arise when fit individuals who are used to training with their peers invite less fit (or unfit) friends to join them while exercising. Although they genuinely may have intended to help their friends by encouraging them to exercise, the opposite actually occurs. The usually sedentary friends perceive the exercise sessions as anything but enjoyable. Because of the negative experiences, they are even less likely to exercise in the future.

Exercise professionals who are qualified to prescribe exercise take into account the initial fitness level of clients. This process decreases the chance of novice exercisers experiencing excessive stress when exercising. However, when working with a large number of clients, exercise management professionals sometimes tend to generalise. People with certain fitness levels tend to be given particular exercise programs. Although individual variations in fitness are accounted for, individual differences in perceived exertion and perceived effort are sometimes ignored. Even people exercising at the same relative work intensity have diverse experiences of that exercise. It is not only the fitness levels that need to be considered, but the individual perceptions of the exercise. One of the challenges for the exercise management profession is to cater for a large number of people while taking into account not only individual differences in physiological fitness measures but also more subjective individual differences in perceptions of the exercise experience.

One determinant of how exercise is perceived is whether we dissociate or associate while exercising. Associating is attending to the body while exercising, being aware of what our muscles are feeling, how we are breathing, and even our heart rate. Dissociating is having an external focus of attention. Dissociation can involve listening to music, daydreaming, planning our day, checking out the bodies of the other exercisers or focusing on anything else that keeps the mind off the actual exertion of our own bodies. People who dissociate while exercising tend to have better rates of adherence. This phenomenon may be one reason why aerobics have become so popular. It is doubtful that many of the people currently participating in aerobics classes would continue to do the same exercises if they were done on their own in silence. Focusing on moving the right way at the right time and attending to the music, the instructor, or the other people in the class keeps the mind off how the heart is beating, the muscles are straining or the breath is gasping.

Situational factors

A number of situational factors also influence exercise adherence motivation. The size of the exercise group has been shown to influence adherence. Although the ideal group size has not been determined precisely, if the group is too large adherence may decrease—the individual may feel lost and unimportant. The larger the group the less individual feedback from the exercise leader is received by each participant. Additionally, many people find it very difficult to get to know people in a large group. With a great number of people, there is less chance that anyone will notice the absence of a single individual. Some exercisers are motivated to adhere to a program because they do not want others to think they could not cope. Smaller groups allow for the development of more social relationships and more attention and recognition from the instructor. It should be noted, however, that some exercisers, particularly those with low self-confidence, prefer the anonymity of a large class.

Ease of access to the exercise venue also affects adherence. Convenience of the program influences how often people are likely to attend. Convenience can involve a number of components. If the program is in a location close to home or to work, there is greater chance of attendance than if a long journey to the exercise site is required. Similarly, even if we drive straight past the venue when returning home from work, difficulty in finding a parking place could make us think that the hassle is not worth the bother. Exercise programs are more accessible if child-care facilities are available to parents of young children. Additionally, how pleasant the exercise environment is may sway individuals to attend more or less frequently. Although Tom may think that a smelly, confined space in need of a coat of paint is perfectly fine for lifting weights, others may be immediately dissuaded by these conditions.

Having the social support of others can also influence adherence. If a spouse, friend, or significant other is supportive of our exercise behaviour, many advantages may ensue. First of all, it is likely that we will receive positive reinforcement from the person, increasing our self-worth and feelings of competence. In addition to encouraging our exercise behaviour, others may provide informational support by giving advice or suggestions that may decrease the chance of injury or increase the benefits gained through exercise. Tangible or instrumental support can also be provided by others. For example, exercise equipment may be lent, transportation provided or child care shared. Social support can be a definite advantage.

A lack of social support would be better, however, than the presence of social disapproval. Ridicule of exercise attempts can indisputably have a negative impact on adherence. Outright resistance to exercise involvement can create additional hurdles to adherence. For example, an individual may try to make a positive health change by stopping by the local gym for a 1-hour exercise class after work three times per week. A spouse who condemns this activity can become an insurmountable barrier to adherence. This disapproval may be exhibited by the silent treatment, sarcasm, or outright rage. When one person makes a behaviour change, it may have an impact on the partner. Ride sharing arrangements may have to be changed, dinner may be an hour later, or child-care responsibilities may vary. If the partner is opposed to the lifestyle alterations, the exerciser may experience the exact opposite of social support—social disapproval.

In summary, many factors influence exercise adherence motivation. Understanding the benefits of exercise is only a preliminary to exercise behaviour. Biological factors, attitudes, beliefs, personality characteristics, goal setting, sensory perceptions, group size, program convenience, and social support all influence exercise adherence motivation. If people want to enhance the exercise adherence of themselves or others, they should consider all these factors.

Exercise addiction

Some people have no trouble adhering to exercise. In fact, some exercisers are actually addicted to exercise. Individuals can be defined as being addicted to exercise when physical or psychological withdrawal symptoms are experienced after 24–36 hours without exercise. An addiction occurs when there is dependence on or commitment to a habit, practice, or habit-forming substance to the extent that its cessation causes trauma. Exercise can be a positive addiction as there are many physical and psychological benefits of exercise. If discomfort or other negative effects are experienced when an individual lets more than 1–2 days pass without exercising, then the individual may be more likely to exercise on a regular basis. When the individual exercises on a regular basis because withdrawal is experienced without exercise, the person is considered to be positively addicted to exercise. Positive addiction also assumes that the individual has control over the exercise.

Some individuals, however, are negatively addicted to exercise. Although exercise is usually a positive and healthy habit, there are those who think that more is always better. These individuals can end up with overuse injuries and psychosocial problems. Although minimum exercise guidelines have been suggested for fitness benefits, there is little information about how much is too much. Negative addiction to exercise occurs when the individual no longer has control over the exercise. Exercise becomes a detriment when there is no decrease in exercise despite pain and injury or when exercise becomes more important than family or work.

In some cases, exercise becomes more important than anything else. Life is determined by when and where exercise is available. Minor injuries can develop into serious conditions. But in addition to the probable physical damage that can be incurred by excessive exercise, psychosocial problems turn into major predicaments. Negative exercise addicts have lost jobs and families because exercise became more important than work or a relationship. In negative addiction, exercise controls the individual.

Encouraging participation in a wide range of exercise and recreational activities can be the first step in helping individuals who are negatively addicted to exercise. If one part of the body is injured, a person can engage in some other form of exercise or recreational activity that can allow the affected body part to recover. In addition, alternative forms of recreational activity can be used to replace compulsive exercise behaviour. The wider the choice of activities, the greater the chance that the individual will interact with others, possibly becoming less self-absorbed. Also, making decisions about the activity in which to participate begins to give the individual power and control over exercise.

Effects of exercise on psychological factors

Up to this point this chapter has concentrated on how psychological factors influence exercise participation and adherence. The remainder of the chapter will focus on how exercise can influence psychological factors. Traditionally when the positive effects of exercise are discussed, the majority of the benefits are seen as being physiological. For example, as discussed in Chapter 15, regular physical exercise is associated with lower cholesterol, lower blood pressure, reduced weight and a decreased percentage of body fat. There are, however, many psychological benefits of exercise as well. Exercise is associated with reduced state anxiety, decreased depression, enhanced psychological well-being, and improved cognitive performance.

Exercise and psychological well-being

Exercise has been shown to influence how people feel about themselves. When people exercise they tend to feel more positive and self-confident. This relationship between exercise and psychological well-being has been demonstrated both in terms of long-term exercise and single bouts of exercise. Although there is general agreement that there are positive psychological effects of exercise, there is disagreement as to why this relationship exists. Some of the possible explanations appear below.

Individuals may experience a sense of mastery or achievement through exercise. They feel they are successful when they are able to walk for a longer period of time, run faster, stretch further, or lift more weight than they could before. It may be that it is this feeling of success or achievement that makes people feel better about themselves. Through mastery in exercise, people may realise they have the capacity for change. They realise they were able to change their exercise habits, their fitness levels or even their body shapes. This realisation that change is possible may transfer to other areas of life, giving individuals a greater feeling of control.

Some argue that exercise is psychologically beneficial because it provides a distraction from problems and frustrations. Instead of exercise actually having a positive effect in and of itself, it really just allows people to take 'time out' from the problems in their lives. This temporary respite from worry may be the only positive influence of exercise.

Yet another proposed explanation for the positive effect of exercise on psychological well-being is that exercise causes biochemical changes that in turn influence psychological factors. The most commonly suggested biochemical change is an increase in endorphin levels. Endorphins are naturally occurring substances with opiate-like qualities. Endorphins are important in regulating emotion and perception of pain. If exercise increases the release of endorphins, then it may be the endorphins that cause the enhanced psychological well-being. Although theoretically this argument makes sense, the relationship has not been proven. Endorphin levels vary across individuals exercising at the same relative intensity, making it difficult to prove that any changes in psychological well-being are due to exercise-induced changes to endorphin levels.

Although a definitive explanation of the positive effects of exercise on psychological well-being does not currently exist, the psychological benefits of exercise should not be overlooked. Just because we do not understand why something happens, does not mean we should fail to take advantage of the phenomenon. In addition to enhancing self-confidence, feelings of control, and general well-being, exercise also appears to reduce negative moods.

Exercise and negative mood states

Both regular exercise and single bouts of exercise effectively reduce depression. Although exercise is effective in reducing depression in mentally healthy subjects, greater decreases in depression through exercise occur in subjects requiring psychological care. Both long-duration lower intensity as well as short-duration higher intensity exercise training has been found to be effective in reducing depression. Whatever the type of exercise, the more sessions per week (within reason) or the more weeks in the exercise program, the greater the decrease in depression.

Exercise has also been shown to decrease effectively both state and trait anxiety. Individuals who are initially low in fitness or high in anxiety tend to achieve the greatest reductions in anxiety from exercise. However, unlike depression, anxiety appears to decrease only with endurance exercise. Short-duration high-intensity exercise such as weight training may in fact increase anxiety.

Exercise has been so effective in decreasing anxiety and depression that endurance exercise training has been used by clinical psychologists as a form of therapy. In some cases it is the sole psychotherapeutic tool, but more often it is used in conjunction with other modes of therapy. Running is the most common form of exercise chosen for therapy. Obviously, however, exercise is not a panacea for anxiety and depression. In fact, if it is difficult to get the average person to exercise regularly, we can imagine how difficult it can be to get a depressed person to exercise regularly.

Exercise and cognitive performance

Exercise not only influences how individuals feel, it also influences how well they think! Regular exercise is associated with improved cognitive performance. Studies have shown that regular exercisers perform better on reasoning tests, mathematics tests, memory tests, and IQ tests than individuals who do not regularly exercise. There is also some evidence that exercisers have better creativity and verbal ability than non-exercisers. Although the reasons for this relationship between exercise and cognitive functioning are not completely understood, it has been suggested that exercise in older individuals may slow neurological deterioration. For younger individuals, exercise may increase the vascular development of the brain as well as increase the number of synapses in the cerebellar cortex in the brain (of the type discussed in Chapter 19). More research is needed to ascertain if there is a cause–effect relationship. However, when we decide to go outside for that game of touch football or go down to the gym for an aerobics class, we may be helping ourselves academically as well as physically!

CHAPTER 22

Physical activity and psychological factors across the lifespan

- Changes in personality
- Exercise and life satisfaction in the aged
- Retirement from sport

Like the other biophysical subdisciplines of human movement studies, sport and exercise psychology provides basic concepts that are of value in explaining changes in human movement across the lifespan and in response to training and practice. In this chapter some applications of basic concepts from sport and exercise psychology to human movement at different points in the lifespan are explored.

Changes in personality

In Chapter 20 it was reported that some personality differences have been found between athletes and non-athletes. Although there is some question as to whether these differences are real or merely artifacts arising from problems with operational definitions and the validity of questionnaires, there is also considerable debate about what these differences might mean (assuming they are real). Some conclude that participation in sport causes personality changes. That is, by participating in sport people become more independent and extroverted. If this is the case, then if you want to help people become more independent and extroverted, having them participate in sport may be beneficial. Others, however, argue that participation in sport does not cause personalities to change. Instead, they suggest that individuals with certain types of personalities are more likely to participate in sport. Sport participants are more likely to have those characteristics of extroversion and independence before taking part in sport. In other words, certain personality traits gravitate towards sport.

So which side is right? Does participation in sport lead to certain personality traits, or do certain personality traits lead to participation in sport? Although there is no definitive answer to this question, most of the evidence suggests that the latter is more accurate. Individuals with certain personality traits seem to gravitate towards sport. Having said that, exceptions always exist. Although extroverted, independent individuals may be more likely to participate in sport, there are numerous individuals participating in sport who are either introverted or dependent on others. Additionally, there is some evidence indicating that participation does influence the personality development of young participants. So although involvement in sport as an adult probably has little if any effect on personality, involvement in sport when young may influence personality development as the individual is maturing.

Psychosocial development through sport participation

Experiences in youth sport may impact on a variety of factors that may influence personality development. Values, attitudes and beliefs can be influenced by childhood experiences in sport. Through sport children may learn about co-operation, respect, leadership, assertiveness, discipline and fair play. They may also learn that hard work results in positive achievement and develop self-esteem and self-confidence as a result.

Not all effects of participation in sport are positive, however. For example, if a child encounters numerous negative events in sporting situations, he or she is likely to avoid the sporting environment when an adult. If fitness activities such as running and push-ups are used as punishment in youth sport, children may learn to associate fitness-related activities with penalties. If running is something we must do when we have made an error or misbehaved, why should we ever do it voluntarily? Also, if youth coaches believe that winning is the single most important result of participation in sport (see again Box 20.1), children may learn about aggression and cheating instead of assertiveness and fair play. Similarly, just as children can develop positive self-esteem and self-confidence through participation in sport, they can also develop negative self-esteem and lose self-confidence if their experiences in sport are demeaning or humiliating. Sport, itself, is neither good nor bad. The psychosocial development of children through sport is largely dependent on the quality of the sport experience and this, in turn, is often determined by the quality of the adult leadership provided.

Design of youth sport

Adults need to be aware that sport programs not only have the potential to benefit the growth and development of children, but also have the potential to be detrimental. Many adults mistakenly believe that any sport experience is better than none. Although all children should have the opportunity to participate in sports, these opportunities need to allow children to experiment, make mistakes and succeed without fear or pressure. Many of the problems that can occur in youth sport, occur because of inappropriate adult behaviour. Adults sometimes design youth sport programs using the same structure as in professional adult sport. It is not uncommon, unfortunately, for children as young as 7 years to be cut from teams, dropped to reserves or permanently benched. Winning is frequently overemphasised, youngsters are occasionally harangued because they do not perform or think like adults, and parent behaviour on the sidelines sometimes keeps children from enjoying the sporting experience.

For sport to have a positive effect on the psychosocial development of children, adults need to remember that children are children, and not miniature adults. (This is a theme that has been emphasised within a number of the subdisciplines considered in this book.) When fun and development are emphasised instead of winning, many potential problems are avoided and the chances of sport being a beneficial experience are augmented. It needs to be recognised that normal growth and development includes psychological and social considerations in addition to the traditional physical emphases. When fun and development are stressed instead of winning, the negative experiences of fear, inadequacy, anxiety and inferiority are more easily averted.

Research has demonstrated that the quality or type of adult leadership in sport influences the attitudes of the children involved. Youngsters who have coaches who use more encouragement and praise and provide more technical instruction like their coaches more and think their coaches are better teachers than children who have coaches who are either negative or fail to provide information. These results should not be surprising. What is astonishing to some, however, is the fact that the children with positive coaches also like their teammates, enjoy being involved in the sport, and desire to continue participating to a much greater extent than children who play under less positive coaches. Participating in sport can be a very positive experience for youngsters. However, parents, teachers and coaches need to realise that they have to actively structure sporting circumstances to ensure that sport is perceived as a pleasure rather than as on ordeal.

Exercise and life satisfaction in the aged

Although participation in sport has a greater impact on the psychosocial development of children than adults, the impact of exercise on psychological health continues throughout life. Unfortunately, prejudicial and discriminatory views about ageing exist. This discrimination is known as age stratification. Age stratification often results in lower self-expectancies of older individuals. It is believed that they should 'act their age', which most people think means being less competitive and having poorer physical performance. Two myths contribute to age stratification: 'people need to exercise less as they age', and 'exercise is hazardous to the health of the elderly'. In fact, as we have seen already in Chapter 15, exercise becomes increasingly important for maintaining and improving the quality of life as individuals get older.

Research has revealed that exercise is related to life satisfaction in the aged. Older adults who exercise regularly report significantly enhanced health, increased stamina, less stress, increased work performance and more positive attitude towards work than those who do not exercise regularly. Exercise may also increase self-efficacy in the aged. For example, a study examining a 5 week swimming program for adults over 60 years of age found that the program had multiple positive effects. Not surprisingly, participants had greater belief in their swimming ability, and were more confident of their swimming competence. More importantly, participants also had increased generalised feelings of self-efficacy and competence. For example, after the program they felt they could do chores more easily and were more able to use public transportation, which greatly strengthened their independence.

As we noted in Chapter 21, exercise can improve self-concept and self-esteem. Exercise can also decrease stress, reduce muscle tension and anxiety and lessen depression. These psychological benefits of exercise are advantageous to people of all ages, but in some respects may be particularly useful for the elderly. Depression is the major mental health problem of the elderly. Studies have shown that participation in a mild exercise program can significantly decrease depression in relatively independent adults (aged 60–80) as well as clinically depressed residents of a nursing home (aged 50–98).

Encouraging participation of older adults in exercise and sport

Exercise in the aged is associated with enhanced self-efficacy, life-satisfaction, and happiness as well as decreased tension, anxiety and depression. However, with age stratification, exercise participation motivation and exercise adherence motivation are even poorer for older adults than for the general population. If people are concerned that exercising is undignified or inappropriate for them because they are older, or that exercising will cause a heart attack or worsen existing medical conditions, they have additional barriers to overcome. Exercise programs can help to combat age stratification by including pictures and videos of older exercisers in their advertising. Some fitness centres are targeting the upper end of the age spectrum by having classes and social events specifically designed for older members. In addition to scheduling senior weights circuits or over-fifties aerobics classes, a few programs now offer exercise classes for those with arthritis and exercise classes based around ballroom dancing. Additionally, hiring individuals who are expressly trained to assess older people with a variety of possible medical conditions and then prescribe suitable exercise for them can help to overcome some of the fears and concerns of potential exercisers. (Chapter 15 provided some guidelines for exercise prescription for the elderly.)

Obviously, not all older adults are hesitant about exercising. In fact, the growing popularity of masters sports should help to overcome some of the prejudicial and discriminatory views about the aged. The Third International World Masters Games were held in Australia in 1994. Over 20 000 competitors made these Games the largest participatory sporting competition in the world that year. Some older people definitely are active! Masters competitions are also held annually at the national level, and many countries hold ongoing local and regional competitions as well.

One of the more popular masters sports is swimming. Masters swim meets differ from traditional swimming competitions in that there are no heats. People are assigned to races within an event on the basis of previous best times. Each race therefore has people with similar performances, although a variety of ages may be represented. Results, however, are not on the basis of any single race. Instead results are posted on the basis of times within 5 year age groups (e.g. ages 45–49 or ages 50–54). Consequently a fast 65 year old may be the fourth person to finish in a particular race yet still win the event for that age group, as the first three finishers were in other age groups. One of the advantages of this system is that the emphasis is automatically placed on individual performances and participation.

Different masters sports tend to have different attitudes towards competition and training. A large percentage of masters swimming competitors, for example, train and compete on a regular basis, yet focus on participating and improving (or maintaining) their own performance. Other sports, however, only organise masters competitions at the national level. Although some of these competitors train regularly and compete locally in non-masters events, a number of them only begin exercising immediately before the competition in the hope that this will get them through the competition.

Greater availability of local masters events may encourage more regular involvement by more people. The growth of local and regional leagues and clubs for masters athletes in the United States suggests this notion is correct. The largest masters sports organisations in the United States are athletics, baseball, basketball, cycling, soccer, swimming and tennis. Once again, however, participants in different sports are inclined to have different attitudes toward competition as opposed to participation. Masters basketball players are known for being extremely aggressive and fiercely competitive. Masters baseball players, though, have the reputation for being relatively unfit. Although these players take pride in the fact that they are playing hardball instead of softball, they have referred to their Mens' Senior Baseball League (MSBL) as the Men's Big Stomach and Butts League. Obviously, not all masters sports participants take themselves seriously. Nevertheless, organised masters sports may encourage some older people to exercise who would not ordinarily do so.

Retirement from sport

For some athletes, one of the major changes experienced in their lives is when they retire from elite sport. Although sporting opportunities continue to exist for many of these athletes through participation in masters sports, the structure of their lives is often altered. No longer are they representing their country at the open level or being paid to compete by a professional sporting organisation. Although they still may be able to compete in other forms of competition, they very rarely continue to have their competitive performance as the main focus of their lives.

The effect of retirement from elite sport on the athlete will depend, in part, on the circumstances surrounding the retirement. If athletes voluntarily retire because they no longer look forward to training or competing and they no longer enjoy participating, they may adapt more quickly to retirement because they retain a sense of control. Athletes who are forced to retire because of injury or non-selection may find the process more difficult.

Retirement from elite sport can lead to depression. Some athletes have equated the experience to breaking off a close personal relationship. For some competitors self-esteem drops after retirement because they suddenly feel a loss of identity. (They often perceive a precipitous drop from 'hero' to 'zero'.) The problem is not just a lack of recognition from others. They are used to identifying themselves as athletes and have previously gained considerable self-satisfaction and self-expression through their participation. When this avenue is no longer available to them, they sometimes question their own identity.

To experience depression (or a feeling of flatness) after being removed from the environment in which one has spent a lot of time is normal. Schedules change. Goals that have been a source of energy are suddenly no longer there. If athletes finish their elite careers with a poor performance, they may feel that their previous devotion to their sport was for nothing. However, finishing with a major success does not necessarily circumvent depression. Very rarely does a brilliant sporting performance have any long-lasting bearing on life. So after months and sometimes years of building up for a major competition, there is often a let-down even after success.

Assisting the retirement process

When athletes retire from elite sport, either voluntarily or involuntarily, they can employ a number of strategies to ease the transition. Just as social support can help to enhance exercise adherence (as mentioned in Chapter 21), it can also help in a major transition period such as retirement. Athletes can go to others for emotional support (reassurance, being loved in a time of stress), esteem support (agreement, positive comparison with others), instrumental support (direct assistance), informational support (advice, suggestions), and network support (feeling of belonging to a group of people who share a common interest). Table 22.1 provides some examples of these different types of social support. Not all athletes will require all types of social support, but it is worth knowing that social support can be more than just a hug and an 'everything is going to be OK'. Making others aware of the decision to retire from elite sport is the first step towards setting up social support. Other retired athletes can also be a valuable source of social support. Sharing experiences and discussing what has and has not been effective in dealing with the retirement process can be very beneficial.

Table 22.1 Types of social support that may help athletes retiring from sport

Type of support	Examples
Emotional	being loved for who they are off the field/court/track; supporting their decision to retire from sport
Esteem	praising their non-sporting abilities (such as communication skills, organisational talents or artistic expertise)
Instrumental	introducing them to potential future employers; teaching them basic computing skills
Informational	suggesting that the skills they used to enhance their performances in sport can also be used to enhance their performances in another career; informing them of the various courses they qualify to attend that are offered locally
Network	forming a group of individuals who are either currently retiring from sport or have already successfully retired from sport; joining a social club, church group or volunteer organisation

Retiring athletes also need to have alternative activities in which to redirect their energy. Ideally this process is initiated before retirement. Some athletes may choose to become involved in other aspects of sport such as coaching, administration, or broadcasting. Others may prefer to invest their energy into their families, study, or an entirely new career. If participating in a new activity, it is important to avoid comparison with prior sports competence. This aspect is particularly relevant to athletes who were national or world champions as it is most improbable that they will be the best in the country or the world at their new careers. As was illustrated in Chapter 19, competence in any new activity is only developed over a long period of time.

Athletes may find the transition to be easier if they apply skills that they learned through their participation in sport. For example, goal setting can be used not only for fitness enhancement or technique improvement, but also for any aspect of life that requires a change in behaviour or sustained effort over time. Many elite athletes have also become very adept at controlling stress and anxiety. The skills that they employed to maintain calm and focus attention at a grand final or an international competition can be used to control stress in a new career. Similarly, if athletes have discovered that certain types of thinking are self-defeating in sporting performance, chances are good that comparable thinking would be harmful in work performance as well.

A percentage of retiring athletes may need to learn to take greater control of their lives. Depending on the sport, the coaches, managers and others may have made all the major decisions for the athletes in the past. Some coaches are very dictatorial and some sports traditionally regiment the lives of the athletes. If individuals were used to being told when to train, when to sleep, what to wear, what to eat, where to go and how to get there, they might find it difficult to be suddenly on their own outside this decision-making process. Athletes may need to learn how to plan their own lives, not just in terms of a new vocation, but also in terms of day-to-day responsibilities.

Certain athletes may need to be encouraged to 'be themselves' when they retire. They may be used to playing a role for the media and the public as an athlete. Consequently, they may feel that they have to be the 'perfect athlete' when they retire, believing that others expect them to be perfect in whatever they do. Realistically, a proportion of the public presumably appreciates it when former athletes are less than perfect, as it demonstrates the 'human' aspect of their nature. The majority of society probably cares little about how successful, or otherwise, retired athletes are in their new professions.

Whether it is gaining social support from significant others, finding new activities to pursue, learning to take greater control of one's life, applying skills previously learned in sport to new situations or changing attitudes and priorities, adapting to retirement from elite sport takes time. The adaptation will be quicker for some than others, but the process is always easier when preparations are made before retirement. Athletes who during their sporting careers enjoy participating in other activities, change their routines in the off-season, develop ideas for new careers and create positive attitudes towards retirement experience less depression when the transition is finally made. Adaptation is quicker when retirement is perceived by athletes as presenting new opportunities rather than the end of life as they know it.

CHAPTER 23

Psychological adaptations to training

- Aerobic fitness and the response to psychological stress
- Changes in personality
- Changes in motivation: staleness, overtraining and burnout
- Changes in mental skills

As mentioned in Chapter 21, both single bouts of exercise and long-term involvement in exercise have been shown to have a positive impact on mood. The present chapter focuses on the psychological effects of prolonged participation in sports and exercise. The topics to be covered include the role of fitness in the response to psychological stress, personality changes as a result of participation in sports, negative adaptation to training, and the development of mental skills over time.

Aerobic fitness and the response to psychological stress

It has been mentioned previously that exercise is effective in reducing depression (Chapter 21). Depression can be related to stress. For example, failing in an achievement situation that is perceived to be important may be stressful. This failure may lead to lowered self-esteem, which in turn may lead to depression. Involvement in regular exercise has been shown both to reduce existing depression as well as to prevent the onset of depression. Fitter people report lower levels of depression after prolonged life stress than do less fit individuals.

Depression is not the only area in which fitness level plays a role. Aerobically fit individuals are able to deal with many forms of psychological stress more effectively than less fit individuals. Probably the greatest indication of this difference is in the varied levels of cardiovascular arousal after exposure to psychological stress. When comparing highly fit individuals with individuals with low levels of fitness, the fitter individuals have smaller increases in heart rate and blood pressure in response to any particular psychological stressor. Moderate endurance training, such as participation in aerobic dance classes, has been shown to be more effective than relaxation training in reducing heart rates before, during and after the experience of psychological stress.

Changes in personality

Researchers have investigated changes in psychological factors in response to participation in sport as well as exercise. A number of studies have investigated the personality profiles of athletes with varying levels of skill. Studies that have compared athletes within a team have found no significant results. There are no meaningful personality differences between successful and unsuccessful athletes or between starters and bench players within a team. However, elite athletes (such as international-level competitors) can be distinguished from novice athletes on some personality characteristics. For example, there is some evidence that highly successful athletes are greater risk-takers than are novices; that is, they are more likely to take chances that might result in injury or failure. Nevertheless, generalisations about expert–novice differences in personality are difficult to draw because there have been many inconsistencies in the research.

First, a variety of personality inventories and questionnaires have been used, making comparisons across studies difficult. Also, some studies have focused on traits, some on states, and some have inappropriately merged the two together. The available studies are also difficult to interpret because of the different sports investigated. Generally, athletes in team sports are more extroverted, dependent and anxious than athletes from individual sports. Therefore comparing elite gymnasts with novice netball players can be puzzling. Just as there are difficulties in operationally defining athletes and non-athletes (as mentioned in Chapter 20), there are similar problems when trying to define elite and novice athletes. Is someone who competes nationally in swimming and socially in tennis a novice or an elite athlete? Finally, even if consistent differences were found between elite athletes and novice athletes, there is no indication that training, competing, and developing physical skills over a long period of time cause personalities to change. It may be that individuals with specific personality types are more likely to develop into elite athletes than are individuals with other personality types. Not enough evidence exists, however, to support the use of personality tests for screening or selecting athletes.

Changes in motivation: staleness, over-training and burnout

The physiological aspects of overtraining were mentioned in Chapter 13. Physiologically, training stress needs to be imposed to achieve training gains. Too much training stress, however, can lead to a negative psychophysiological reaction. Both psychological and physiological factors are involved in the negative adaptation to training. Although this section will focus on the psychological factors, it must be emphasised that psychological and physiological factors interact with each other. For example, physically being unable to maintain training loads can decrease our enthusiasm for training. Similarly, having low levels of enthusiasm can decrease our ability to maintain training loads.

Studying the area of training stress can be very confusing, as different terms are used to mean different things for different people. There is general agreement that there are a series of stages through which athletes progress when negatively adapting to training stress. Exercise physiologists often suggest that athletes first experience overtraining, then over-reaching, and then staleness. In the exercise physiology literature staleness is usually perceived as being equivalent to burnout.

On the other hand, many sport psychologists refer to the negative adaptation to training as the training stress syndrome. This syndrome is made up of three phases: staleness, overtraining, and burnout. Staleness is the initial failure of the person to cope with the psychological and physiological stress created by training. The body and mind try to adapt to the training demands, but the demands exceed the person's capabilities. Staleness is characterised by an increased susceptibility to illness, flat or poor performance, physical fatigue, and a loss of enthusiasm. The second phase of training stress syndrome, over-training, is then the repeated failure of the person to cope with chronic training stress. Symptoms of over-training include chronic illness and mental and emotional exhaustion characterised by grumpiness and anger; minor stresses become bothersome. Burnout, the final phase of the syndrome, occurs when the athlete is exhausted both physiologically and psychologically from frequent but usually ineffective efforts to meet excessive training and competition demands. When experiencing burnout athletes lose interest in the sport, are extremely exhausted, and generally do not care about their training or performance. In some cases there is resentment towards the activity.

Staleness can be caused by monotony, repetition and boredom, too much stress, or too much training. Since one of the characteristics of staleness is a flat or poor performance, many coaches and athletes react to staleness by increasing the training load. This further increase in stress and training demands often leads to over-training. In addition to too much stress or pressure, too much repetition or too much training, over-training can also be caused by lack of proper sleep, a loss of confidence and the feeling that one is never successful. If the training load is maintained or increased at this point, burnout may result. Burnout is the result of an ongoing regressive process. Athletes do not just awaken one morning and suddenly experience burnout. Burnout is usually caused by excessive devotion with little positive feedback over a long time, severe training conditions, unrealistic expectations, and a lack of recovery time from competitive stress. Burnout typically results in dropout. It should be noted, however, that although burnout often leads to dropout, many people drop out for reasons other than burnout.

What might help?

A number of strategies can be implemented if signs of staleness, over-training, or burnout emerge. The simplest course of action is to introduce a time-out. Taking a break from training (and in some cases competition) can refresh and invigorate athletes. For athletes to be able to recharge their batteries, it is important that the time-out is a break from everything, not just physical training. An emotional vacation is needed. Some coaches grudgingly allow their athletes to take a break physically, believing that allowing a respite from physical work is all that is needed. Athletes who during a physical pause in training continue to observe training sessions, analyse game film, or otherwise involve themselves mentally in their sport gain very little (if anything) from the physical respite. Training stress syndrome is a psychophysiological process. Although reducing the intensity or volume of training may avoid the progression of the syndrome from staleness to over-training or from over-training to burnout, rarely is it enough to eradicate the syndrome entirely. The more significant the changes in day-to-day activities, the more effective the break. If Alf takes a week off weight training, swim training, time-trials, and team meetings, yet spends the week living in his unit with his three roommates who are also team-mates, he remains in a swimming environment. With three roommates draping wet bathing suits around the place, leaving for and arriving from training, complaining about the session's format, and generally talking about swimming, Alf fails to experience a real time-out. Alf's stress levels may even increase in that situation if he feels guilty or ashamed of his inactivity.

Avoiding training stress syndrome

Coaches can help athletes avoid or overcome training stress syndrome by providing more variety in training. Because monotony, repetition and boredom can contribute to staleness and over-training, activities that break the tedium and routine of training can be beneficial. Variety can be achieved in many ways. Some examples include the following: introducing new drills; changing the training schedule; altering the training venue; encouraging fitness training through cross-training (i.e. participating in activities that are not specific to the sport); involving new faces (e.g. guest coaches, admired athletes); incorporating fun and games; and inviting and using athletes' input in the design of training.

A club swim team that was having problems with dropouts, staleness, and even behaviour problems dealt with the difficulties by introducing boardgames to Friday training sessions. Fridays were selected because attendance was worse at the end of the week. The process began with the coach bringing in an old game of Monopoly. Chance and Community Chest cards were replaced with cards that stipulated specific drills. Buying houses and hotels earned swimmers the right to specify the next activity for his or her lane. The game's popularity encouraged the coach and certain swimmers to adapt other pre-existing boardgames and create entirely new boardgames for use at the pool. Attendance records, punctuality during the week, and personal improvement in fitness and time trials were used to determine which swimmers got to select the games for each Friday session. Attendance, training productivity, and enthusiasm all greatly increased. There were fewer signs of staleness and fewer disciplinary problems. This specific example of providing variety in training illustrates how effective it can be.

The majority of the additional strategies that help to deal with staleness, over-training, and burnout involve the development of a variety of mental skills. Goal setting can help athletes pursue more realistic expectations. Training stress syndrome is exacerbated by athletes trying to achieve performances that are unrealistic and expending a lot of energy for little, if any, return. Short-term goals give meaning to training sessions, help to take minds off how long the season may be, and increase positive self-concept as feelings of success are generated every time a short-term goal is achieved.

Managing stress

Because too much stress is one of the causes of training stress syndrome, skills that help athletes to manage stress are extremely useful. The most obvious skill in this regard is the ability to relax. The two most common physical relaxation strategies are progressive muscular relaxation and breathing exercises. Progressive muscular relaxation (PMR) exercises involve contracting and then relaxing specific muscle groups systematically throughout the body. Many times if athletes are tense, telling them to relax is ineffective because they do not know how to relax. By increasing the tension in the muscle and then relaxing, athletes become more aware and sensitive to what the presence and absence of muscle tension feels like and they learn that they can control the level of tension. Many individuals find it very difficult to decrease the tension in a muscle voluntarily. However, by first increasing the tension and then relaxing, the level of tension in the muscle automatically drops below the initial level of tension. After regular practice of this exercise, athletes learn to relax the muscles without first having to tense them.

Breathing exercises are also useful for cultivating relaxation. When stressed, many people either hold their breath or breathe in a rapid and shallow manner. By learning to control breathing, stress reduction can be achieved. Controlled, deep, slow breathing is the aim of most breathing relaxation exercises. This type of breathing physically relaxes the body. The advantage of using breathing as a form of relaxation is that we are always breathing. With the possible exception of underwater hockey, athletes can always increase their awareness of their breathing even while being physically active.

Focusing on breathing not only enables athletes to control their breathing, and therefore aid physical relaxation, it also may help them to mentally relax. Stress has both physical and mental repercussions. Mentally athletes may begin to worry, create self-doubt, say negative things to themselves, and focus on factors that are irrelevant or counterproductive to performance. Concentrating on breathing is a useful way of refocusing attention and controlling self-talk. Practising functional cue words such as 'relax' or 'focus' in conjunction with exhaling helps athletes to stop negative self-talk. Breathing then becomes a technique for relaxing the body both physically and mentally.

Another skill very useful for individuals experiencing staleness, over-training, or burnout is learning to say 'no'. Although physical training is often a very common source of stress, it is rarely the only source. Most athletes who experience training stress syndrome are hard-working, idealistic individuals who strive for high achievement. Because these people are highly motivated and often have a track record of getting things done, other people frequently ask them for favours and help. These athletes want to be successful and often agree to do too much. Any single act does not require too much effort or too much time, but when added together the demand is greater than the individual resources. Every time the athlete agrees to appear at a public promotion for the sport or team, stay late at training to help another athlete, develop a text book or training video, attend a charity fundraiser or complete an extra assignment or project at work or at home, the total amount of stress is increased. When the person has the time, energy, and capability to attend to all the demands there is no problem. But for people who are in a situation in which the demands already exceed the capabilities, the inability to say 'no' exacerbates the problem. Saying 'no' sounds easy, but often if individuals know they are capable of doing what is being asked, they believe they may get positive recognition for doing the task (remember, a lack of positive feedback is a cause of burnout), and they don't want to take the chance of disappointing the individual requesting the favour. Often it is easier in the short-term to say 'yes' than 'no'.

Time-outs, greater variety in training, reduction in training, goal setting, relaxation, positive self-talk, and learning to say 'no' are all strategies that can help athletes to avoid and to deal with negative training syndrome. Severe cases of burnout may require professional counselling. However, there is no single intervention for treating athletes suffering from staleness, overtraining, or burnout. Research is currently being undertaken by sport and exercise psychologists to determine which intervention or combination of interventions is most effective under specific circumstances.

Changes in mental skills

The previous section on the use of mental skills to help to combat training stress syndrome is just one example of how the development of mental skills can help athletes adapt to training and competition. Participation in sport can lead to the evolution and improvement of an assortment of mental skills. These mental skills are largely believed to serve one function—the enhancement of performance. Although the acquisition of mental skills can help athletes to enhance their performance, they can also enhance the enjoyment of participation and provide individuals with a battery of skills that can enhance the quality of life in areas outside sport. Table 23.1 provides several examples of

Table 23.1 Mental skills that may be acquired through participation in sports and applied to other situations in life

Skill	Possible non-sport applications
Arousal increase	feeling tired and needing to make a presentation
Communication	organising a major event
Concentration	studying at home with loud roommates
Emotional control	maintaining calm when the boss is rude
Goal setting	changing health/work habits
Imagery	rehearsing difficult confrontations
Injury rehabilitation	dealing with work-related injuries
Preparation	being mentally ready for a job interview
Relaxation	getting to sleep more quickly
Self-confidence	speaking in front of a large group
Self-talk	remaining positive when things look bad
Team harmony	working on a group project
Time management	studying adequately for all exams
Travel	avoiding jet lag

mental skills that athletes may acquire through participation in sport and situations in which those skills may be applicable away from sport.

Participation in sport does not automatically lead to the acquisition of multiple mental skills. Some successful athletes have developed mental skills through trial and error. Others have copied the skills of previously successful athletes, and some have obtained them through the guidance of good coaches. Unfortunately, there are many individuals who never adequately develop these skills. A fairly large percentage of these people drop out from sport because they do not enjoy their participation in sports and/or they perceive their performances to be less than satisfactory. Providing a mental skills training program for participants not only enhances the formation of mental skills that individuals can use throughout life, it can increase the levels of both performance and enjoyment of participation. These two factors often go hand-in-hand. For example, if Laurel always gets extremely nervous before competition to the point of being physically ill, chances are she will not perform her best immediately after being sick. If, however, Laurel can learn to control her anxiety and therefore avoid being sick before competition, she will not only improve her performance, but also probably enjoy her participation much more.

For mental training programs to be effective it is important that they are designed around the needs of the participants. No single technique will work for everyone. For example, although controlled, deep breathing can be a very effective relaxation technique, a proportion of individuals will find that focusing attention on breathing actually increases anxiety as they no longer feel that breathing is natural and they develop fears of suffocation. Though this reaction is rare, it does exist in some people. Mental training programs therefore need to be flexible and individualised. It is also imperative that individuals systematically practise their mental skills in the same way they practise their physical skills. Watching and understanding the principles in an instructional video about how to pole vault does not mean that we would be able to physically pole vault with correct technique. Practice would be needed. Similarly, listening to a lecture and understanding the principles of relaxation does not mean that we will be able to relax on command in an extremely anxiety provoking situation. Extensive, systematic practice is needed to bring about this positive adaptation.

Further reading

Chapter 20: Basic concepts in sport psychology

Martens, R. (1987), *Coaches Guide to Sport Psychology*, Human Kinetics, Champaign, IL, (Chapter 2).

Martens, R., Vealey, R.S. & Burton, D. (1990), *Competitive Anxiety in Sport*, Human Kinetics, Champaign, IL.

Perry, C. & Morris, T. (1995), Mental imagery in sport, pp. 339–85, in T. Morris & J.J. Summers (eds), *Sport Psychology: Theory, Applications and Issues*, Wiley, Brisbane.

Chapter 21: Basic concepts in exercise psychology

Grove, J.R. (1995), An introduction to exercise psychology, pp. 437–55, in T. Morris & J.J. Summers (eds), *Sport Psychology: Theory, Applications and Issues*, Wiley, Brisbane.

Rejeski, W.J. & Kenney, E.A. (1988), *Fitness Motivation*, Life Enhancement, Champaign, IL.

Willis, J.D. & Campbell, L.F. (1992), *Exercise Psychology*, Human Kinetics, Champaign, IL. (chapters 1 and 2).

Chapter 22: Physical activity and psychological factors across the lifespan

Martens, R., Christina, R.W., Harvey, J.S. & Sharkey, B.J. (1981), *Coaching Young Athletes*, Human Kinetics, Champaign, IL.

Ogilvie, B. & Taylor, J. (1993), Career termination in sports: when the dream dies, pp. 356–365, in J.M. Williams (ed.), *Applied Sport Psychology: Personal Growth to Peak Performance*, Mayfield, Mountain View, CA.

Tremayne, P. (1995), Children and sport psychology, pp. 516–37, in T. Morris & J.J. Summers (eds), *Sport Psychology: Theory, Applications and Issues*, Wiley, Brisbane.

Chapter 23: Psychological adaptations to training

Henschen, K.P. (1993), Athletic staleness and burnout: diagnosis, prevention, and treatment, pp. 328–37, in J.M. Williams (ed.), *Applied Sport Psychology: Personal Growth to Peak Performance*, Mayfield, Mountain View, CA.

Morris, T. & Thomas, P. (1995), Approaches to applied sport psychology, pp. 215–58, in T. Morris & J.J. Summers (eds), *Sport Psychology: Theory, Applications and Issues* Wiley, Brisbane.

Part 3

Multidisciplinary and crossdisciplinary approaches to human movement

CHAPTER 24

Integrative perspectives

- Specialisation versus generalisation
- Examples of multidisciplinary and crossdisciplinary research

Specialisation versus generalisation

In describing the history of the discipline of human movement studies in Chapter 2, we noted that until as late as the 1960s researchers interested in human movement typically had broad interests and backgrounds and frequently examined questions that would now be considered to belong to a number of different subdisciplinary fields. For example, the pioneer researcher Franklin Henry (Box 2.2) actively researched issues in both exercise physiology and motor control as well as educating a number of graduate students who played key roles in the establishment of the subdiscipline of sport and exercise psychology. As the discipline of human movement studies has matured and there has been a proliferation of specialised scientific research on human movement, it has become increasingly difficult, if not impossible, for individual research scientists to remain expert in more than one of the subdisciplines of the field. Certainly the expansion of knowledge is such that it is now impossible for any one person to stay abreast of all the latest developments in each subdiscipline of human movement studies. Today, knowledge is becoming so specialised that expertise is generally confined to selected areas within a subdiscipline. Although it may have been realistic even 30 years ago for an exercise physiologist, for example, to be knowledgeable in all the areas of human exercise physiology, the situation is now very different, with most researchers and academics possessing only limited knowledge of each of the major subdisciplines of the field and in-depth knowledge of a few specialist areas (such as strength training, endurance training, muscle soreness, temperature regulation, immune function or overtraining).

Although the increased specialisation of the discipline has allowed greatly increased understanding of many aspects of human movement, the danger of such specialisation is the potential for fragmentation of the field, a concern discussed previously in chapters 1 and 2. With fragmentation of a discipline, researchers become so interested in their own specialist field that they fail to fully appreciate the importance and significance of knowledge from other subdisciplines and the importance of integrating information from across the different subdisciplines in order to ultimately advance understanding. Even with increasing specialisation it is important to retain a sound general knowledge and understanding of the entire field of human movement studies and an interest in, and an appreciation of, how the various systems within our bodies interact. If we are truly interested in the key disciplinary question of how and why humans move, we need to consider not just functional anatomy, biomechanics, exercise physiology, motor control and/or sport and exercise psychology in isolation; rather we must integrate the various subdisciplines to produce a coherent, global view about human movement. Increasingly we should expect to see specialists from the different subdisciplines of human movement studies working together in research teams to try to understand complex problems (of the type to be examined in this chapter) from a number of different perspectives.

Research involving more than one of the specialist subdisciplines can be classified as either multidisciplinary or crossdisciplinary, dependent on the nature and extent of the integration that occurs (Figure 1.2). In multidisciplinary research specialists from the various subdisciplinary fields all investigate a common problem but do so only from their own subdisciplinary perspective. What results is a number of different perspectives on a particular research issue but with no particular attempt to integrate these different perspectives into a consolidated viewpoint. An example may help to illustrate this approach.

Suppose a group of research workers, including a biomechanist, a sport psychologist and an exercise physiologist, are interested in determining those features that characterise elite long-distance runners. The top five male and female 10 000 m runners at a national championship are selected as subjects and a battery of tests are administered to them. The exercise physiologist chooses to measure VO_{2max} and anaerobic threshold, the biomechanist measures leg strength, anthropometry and muscle activity patterns, and the psychologist conducts tests to ascertain mental imagery ability and pain tolerance. Each then, on the basis of their data, ascertains the physiological, biomechanical and psychological predictors, respectively, of 10 000 m running performance. No particular effort is made to ascertain the relative importance of the different measures nor to determine if there is a certain combination of physiological, biomechanical and psychological factors that best distinguishes elite runners. Such an approach is termed multidisciplinary because it involves a number of the subdisciplines of human movement although each specialist area essentially works independently.

Crossdisciplinary research, on the other hand, relies on two or more specialists working together on a problem, taking their own perspective to the situation but integrating their views with the theories of others from different backgrounds. This approach is in many ways preferable to unidisciplinary and multidisciplinary approaches because it has the potential to create a greater understanding of human movement than would be possible if the topic were investigated in a fragmented way. A human being moves in a certain way because of a host of interacting factors and looking from only one perspective or level of analysis may lead to an incomplete, and perhaps flawed, understanding of the problem. In crossdisciplinary research a genuine attempt is made to cross the traditional subdisciplinary boundaries and for each subdisciplinary perspective to attempt to understand the perspective of the other(s), so that a new consolidated perspective may be developed. The biomechanist and motor control specialist who work together to decide the best type of kinematic feedback to assist learning or the exercise physiologist and psychologist who work together to promote adherence to exercise with health benefits are engaged in crossdisciplinary work.

In the remainder of this section we briefly present some examples of how the integration of the methods and knowledge of two or more of the subdisciplines can contribute to understanding of some central issues in the field of human movement studies. The issues we examine, through presentation of some existing examples of multidisciplinary and crossdisciplinary research, are: (i) injury prevention and rehabilitation, (ii) talent identification, (iii) performance optimisation, and (iv) participation maximisation.

Examples of multidisciplinary and crossdisciplinary research

Injury prevention and rehabilitation

The prevention of, and rehabilitation from, injury is a central concern for many professions related to the discipline of human movement studies. Here we examine two examples of integrative research work on this topic. The first example describes a workplace setting in which manual lifting tasks have been examined from a number of different perspectives including physiological, biomechanical and psychological. The second example is chosen from a sport and details the crossdisciplinary approach that researchers have taken in an attempt to solve the problems posed by elbow injuries in Little League baseball players.

An example from the workplace—manual lifting

Ergonomics is the field of study concerned with investigating the interface between humans and their working environment. As such, it involves the disciplines of human movement studies and engineering, as well as other areas of biological, medical, and technological sciences. A major area of concern in ergonomics is the prevention of work-related injuries. A disproportionately large number of injuries in the workplace are caused by 'over exertion' while manually lifting and carrying objects. The back is the principal site for such injuries, with over one-third of all worker's compensation claims in most Western industrialised countries being related to back injuries.

Many of these countries have produced guidelines related to the prevention of injuries during manual materials handling, especially those involving the back. The guidelines for the United States have been developed by a committee of the National Institute of Occupational Safety and Health (NIOSH). NIOSH considered information from a number of different subdisciplines of human movement studies during the formulation of their guidelines. NIOSH defines limits for manual materials handling based on three criteria, which were derived from the results of biomechanical, physiological, and psychophysical studies of lifting tasks.

The compressive force on the base of the lumbar spine is thought to be directly and causally related to some types of back injuries, and so a *biomechanical* criterion (based on a maximal acceptable lumbar compressive force) provides the limit to safety when lifts are performed infrequently. If solely the biomechanical criterion existed, an apparently logical recommendation to follow might be that small loads be lifted frequently. This, however, is likely to lead to physiological fatigue, which, in turn, is a risk factor for back injuries. Therefore for repetitive lifting tasks, it is more appropriate to use *physiological* criteria based on cardiovascular responses and metabolic fatigue to set limits for safe lifting. Had only the physiological approach been used a recommendation may have emerged that large loads be lifted infrequently, but this would clearly have been inappropriate given the knowledge from functional anatomy and biomechanics that injury may result from the excessive forces generated during heavy lifts.

The *psychophysical* criterion is based on the lifter's perception of physical work. Presumably each lifter monitors the responses of muscles, joints, and the cardiovascular and respiratory systems to estimate an acceptable work load. Each person seems to be able to effectively analyse all this information because the psychophysical approach usually defines a limit that is a good compromise between the biomechanical and physiological results. The psychophysical approach is applicable to most lifting tasks except for high-frequency lifting when the object is lifted more than six times per minute for an extended period. In this situation, the physiological limit is recommended by NIOSH because it is lower than the psychophysical limit.

Many studies have compared the psychophysical results with those from either the biomechanical or physiological approaches. These studies have led to the recommendations provided above. The only true long-term measure of the validity of the NIOSH recommendations will be their effectiveness in lowering the injury statistics following their implementation. The important point in the current context is that this approach, aimed at reducing the incidence of work-related back injuries, relies on knowledge and criteria determined from a number of the biophysical subdisciplines of human movement studies.

Source: Waters, T.R., Putz-Anderson, V., Garg, A. & Fine, L.J. (1993), Revised NIOSH equation for the design and evaluation of manual lifting tasks, *Ergonomics*, 36, 749–76.

A sporting example: Little Leaguer's elbow

In the 1960s, epidemiological surveys highlighted a growing problem among child athletes involved in Little League baseball. These surveys revealed an alarmingly high incidence of conditions such as avulsion of the ossification centre of the medial epicondylar epiphysis (see Chapter 5), widening of the epicondylar epiphysis, enlargement of the medial epicondyle, and compression of the radial head and the capitellum. These injuries were hypothesised to be related to the throwing action. In fact, in 1960, the term *Little League elbow* was coined to describe this general condition of the elbow because of the believed relationship between the condition and the pitching action.

The first series of studies by Dr. Joel Adams involved radiological examination of 162 boys, some of whom were involved in Little league baseball (80 pitchers, 47 non-pitchers) and a control group who were not. This study revealed that the Little League players showed higher incidences of abnormal bone growth and development at the elbow joint than the control group. The other major observation of this study was that there was a direct relationship between the magnitude of the changes at the joint and the amount of pitching/throwing. That is, the pitchers showed higher incidences and increased severity of injury to the medial aspect of the elbow than the other players and controls. Furthermore, symptoms are insidious in onset and typically reach their peak, requiring medical attention for the first time, in the 13- to 14-year age range, as the players move from Little League to Pony League.

Numerous follow-up studies to the seminal work of Adams have been conducted, including comprehensive radiological and clinical surveys of Little League baseballers in Eugene, Oregon, and Houston, Texas. These research programs identified early chondrosseous changes to the elbows of many Little League players but, because the subjects were in the 11- to 12-year age group, very few had been playing long enough to develop pain in the elbow (i.e., they were asymptomatic) and only a small percentage (5%) showed changes to the lateral compartment of the elbow (capitellum and radial head). Injury to the lateral aspect is much more serious than injury to the medial side, and it often leads to osteochondritis dissecans (bone death). If osteochondrosis does develop there is usually permanent damage and loss of full range of elbow extension.

Little League elbow has its highest rate of incidence at the onset of the adolescent growth spurt, with boys in the 13- to 14-year age bracket. During these years of rapid growth the bones remodel in response to the applied loads. The stresses applied during the pitching action (and during throwing in general) often cause the medial epiphysis of the throwing arm to widen and develop abnormally.

Follow-up research to the initial clinical and epidemiological studies included biomechanical investigations of the pitching technique. High speed film and more recently video analysis identified various phases of the movement including the wind-up, cocking, acceleration, and follow-through phases. It is generally believed that the end of the cocking phase and the start of the acceleration phase is the time at which the elbow is placed under greatest stress.

During these phases, the elbow is flexed and the humerus is externally rotated with the muscles on the medial side of the joint undergoing eccentric activity prior to positively accelerating the forearm toward the target. During the follow-through there are large loads placed on the lateral side of the joint as the forearm pronates after ball release. These loads are believed to be related to damage to the capitellum and head of the radius.

Another important finding of the biomechanical research has been the relationship between style of throwing and injury rate. In fact, children who pitch with sidearm motions are three times more likely to develop problems than those who use a more overhand technique. Specific pitches, notably the curve ball, require the application of different loads at the elbow than do fast balls. In fact, the altered action required for curve balls places even higher loads on the medial aspect of the elbow, which in turn heightens the prospect of injury to the medial humeral ossification centre.

Little League administrators were extremely concerned about the high injury rates among their athletes. In response to the medical and biomechanical research findings, various controls have been put in place to try to reduce the number and severity of elbow injuries to pitchers. These measures have included general advice to coaches and parents regarding the number of pitches to be thrown during training sessions, typical warm-up and cooldown activities, advice on suitable progressions and build-up in throwing during pre-season, limits on the number of innings that pitchers can throw in a game, and limits to the number of games that can be pitched within a week. Many of these rule changes were instigated in 1972 and research findings have been presented to support the effect of these changes. While it is believed in general that the incidence rates of Little League elbow declined following the rule changes, the problem is still of substantial proportions.

The most recent advice being proposed by clinicians and researchers is to limit the number of pitches thrown by young participants rather than limiting the number of innings pitched. The logic behind this advice is based on findings that, depending on the age of the players, Little League and Pony League pitchers can average anywhere from approximately 18 to 30 pitches per inning. Reasonable limits according to Joseph Congeni, a medical director at a Sport Medicine center in Ohio, are 90 to 100 pitches per game or practice, providing there is no pain. If pain is present, pitching should cease immediately.

Clearly the problem of Little League elbow has not been eliminated by the rule changes that were implemented in 1972 and research has continued to identify higher than expected rates of elbow problems in young boys and adolescents involved in baseball. Surveys conducted as recently as the 1990s have indicated that there are still many unanswered questions relating injury of the elbow to training practices and throwing technique. While the problem has been reduced by the implementation of rule changes and altered coaching practices, the problem has not been eliminated and will continue to be a fertile area of future research.

Sources: Adams, J.E. (1968), Bone injuries in very young athletes, *Clinical Orthopaedics*, 58, 129–140.

Wells, M.J. & Bell, G.W. (1995), Concerns on Little League elbow, *Journal of Athletic Training*, 30(3), 249–253.

Talent Identification

For many years in industry and in the public service, management have been concerned with selecting personnel to ensure the best possible match between the person and the task to be completed. While person-task matching remains an important concern in ergonomics, a parallel problem in sport is talent identification—the identification of talent in youngsters who have potential to become future champions in specific sports. The complex nature of elite sport performance necessitates that talent identification be broadly based on the knowledge from a number of the sub-disciplines of human movement studies. We have chosen to highlight one example of a very successful talent identification program that was practised in the German Democratic Republic (GDR). The success of this program was evidenced by the very high numbers, relative to total population, of world champions and Olympic medallists during a 20-year period through the 1970s and 1980s. The example summarised below is for rowing but many other sports and games (e.g., track and field, gymnastics, swimming) have similar programs in place.

An example from rowing

Initially, sport scientists in the GDR spent considerable time identifying those physical (anthropometric), biomechanical, and physiological characteristics typical of Olympic gold medallists and world champions in rowing. They formulated a notion that a model rower was above average height (males = 195 cm; females = 182 cm) and weight (males = 90–100 kg; females = 75–85 kg); had the capacity to perform long arcs of motion (determined by height, arm, and leg lengths); extraordinarily high endurance strength; and a very high maximum oxygen uptake (e.g., 6.0–7.0 L). While these characteristics typified adult performers who had been successful at the very highest level of the sport, they did not immediately know how to determine which junior athletes would grow into the *model rower*.

The next task was to begin to understand growth and how measurements made on prepubescent children could be used to predict adult size, strength, and endurance. First, it was recognised that only 3% of the population of the GDR are 190 cm or taller and rowing competes with other sports (e.g., basketball, handball, volleyball, track and field) for these potentially large athletes. Thus, based on height norms of the population, children between the ages of 10 and 14 were selected. These norms account for early and normally developed children. Advertising, scouting various communities, and enticing children from other sports are methods also used to ensure a sufficiently large base of children from which the eventual champions will come.

From these groups, approximately 120 to 140 talented children are selected to attend special sport schools after completing grade 7. These children make up approximately 60–70% of the rowing intake and a further 30–40% are scouted from children in grades 7 and 8 who had not previously been recruited. These children typically need to learn how to row but have come from other sport disciplines so they have been training, albeit not specifically for rowing. Approximately 40–50 males and 30 females of this junior group proceed, each year, to the senior level, providing a consistent supply of talented, well-credentialed juniors who compete with senior athletes and attempt to make the final step toward becoming a world champion.

During the junior years (ages 10–14 years), the coaches attempt to secure a continuous squad of suitable talent, develop a love of rowing, establish rowing skills and abilities, and develop basic skills that underlie all sport performance at the elite level (e.g., fitness, strength, motor abilities, skills). One of the key pillars of the GDR's talent identification program is not to overemphasise specialisation in rowing since it is recognised that children may migrate to other sports for which they are better suited, and that development of fundamental sports skills is the prerequisite to success in any sport, including rowing.

Once the children reach 14–16 years of age, the development of general fitness and coordination takes up approximately 50% of the time spent training. Strength-endurance resistance training is included and athletes do high numbers of repetitions at 50–70% of their maximum force. Jogging, gymnastics, swimming, and games also make up a large part of the training time. It is during these years that the boys and girls learn how to scull and sweep oar row. Numerous on-water tests (distance covered per stroke, control of rowing technique, stroke frequency) and on-land tests (e.g., 30 m sprint, 800 m, and 1500 m running, medicine ball leaping, and throwing) are used to monitor development and assist in isolating those individuals who show the most promise.

The final stage of the talent identification

stage occurs when those most proficient and talented at the youth level are selected for further nurturing: They either take part in a professional apprenticeship or complete high school and prepare for university. Their entire training is now specialised for rowing and they are selected for either sculling or sweep oar boats. The amount of training increases with age, in accordance with a well-defined plan.

Parallel with the various training programs, sport scientists and medicine specialists constantly monitor and test various physiological functions four times per year. Tests include incremental ergometer performance while various respiratory, cardiovascular, and haematological variables are measured. Medical and paramedical tests are also conducted regularly along with anthropometric assessment. The data are stored in a central database.

The success of this talent identification program is evident by the overrepresentation of GDR athletes in World Championship and Olympic finals. One of the interesting aspects of the research that underpins the talent identification program is that it is very collaborative in nature. That is, the work is very much crossdisciplinary and athletes are selected for rowing based on a large range of parameters, including physiological qualities,biomechanical aspects, motor skill competence, psychological skills, and medical factors.

Source: Körner, T. (1989), Talent identification and guidance scheme of the Rowing Association of the German Democratic Republic (DSRV). *Talent identification: selection, guidance, developments, feedback: Papers presented at the Oceania Region Olympic Solidarity Rowing Seminar* (pp. 1–17), Canberra: Australian Institute of Sport.

Performance optimisation

In contrast to the previous examples on talent identification, which rely largely on identifying individuals with appropriate genetic predisposition for success, this section on performance optimisation illustrates how integrated knowledge from different subdisciplines of human movement studies may be used to optimise performance within the limits of each individual's physical structure. The example we use here is that related to the exploitation of stretch–shortening cycles in muscle—cycles we have referred to in a number of chapters in Part 2 of this book.

A series of experiments conducted in Finland by Dr Paavo Komi and his colleagues in the 1970s and 1980s has led to an enhanced understanding of muscle function and techniques used in many human actions. Komi focused his work on what has now become known as stretch–shortening cycles. This term is used to describe those activities in which the muscles are stretched prior to concentrically contracting. It necessarily follows that the muscle needs to be acting eccentrically (that is, contracting, while increasing in length) for stretching to actually occur.

In the 1960s, Cavagna among others had noted two very important aspects about eccentric muscle contraction in isolated muscle, namely that greater force was able to be developed and higher efficiency resulted, compared with purely concentric contraction. Komi established that the same situation was true for intact muscles and further showed that increased or augmented work and power were available during a concentric muscle contraction if it was preceded by an eccentric phase. His work over approximately 10 years culminated in findings relating to efficiency of movement and how this increased efficiency came about.

The work is typified by the experiment described below in which a specially designed sled was used to investigate concentric and eccentric lower limb muscular work. The sled was mounted on two parallel rails inclined approximately 40° above the horizontal. A force plate was positioned at right angles to the rails at the bottom of the sled. The force plate was used to determine the forces applied to the ground while an oxygen measuring system was placed nearby allowing collection of expired gases from the subjects and hence calculation of the physiological cost of work. A person could sit on this sled, be positioned at a fixed point up the rails and released or, alternatively, be required to push themselves up the sled to a certain height. The work done in either the eccentric condition (being released from a fixed point) and in the concentric condition (pushing up to a certain height) was constant but the power and physiological costs were found to be different.

Two reasons have been proposed to account for the augmented work (or increased efficiency) of the stretch–shortening cycle. The first of these is a mechanical one whereas the second is neurally based. Simplistically, energy was believed to be stored in elastic structures of the muscles and tendons during the eccentric phase and then used to do work during the concentric phase. Concurrently, during the stretching phase, muscle spindles and Golgi tendon organs were stimulated, which resulted in reflexive innervation and therefore increased potentiation of the muscles. (See chapters 9, 11 and 16 for background information.) Komi and his fellow workers were able to establish, through a series of cleverly designed experiments, that approximately two-thirds of the increased work was due to elastic energy return and that one-third was due to increased potentiation through reflex activity.

The implications of Komi's work are far reaching since virtually all motor activities use stretch–shortening cycles. At first glance, it may appear that augmented work available through the stretch–shortening cycles would only be of benefit to power athletes who require maximal, one-off performances. In such cases, techniques reflect the fact that muscles must be pre-stretched prior to contracting concentrically. However, closer examination reveals that the use of stretch–shortening cycles is of equal importance to endurance athletes, who are clearly interested in improving efficiency. For example, in running stretch–shortening takes place during the stance phase when the calf muscles are stretched as the body rotates over the foot and the knee flexes. This energy can then be used to do work on the body during the concentric, push-off phase. Use of this elastic energy increases the economy of gait, possibly enhancing performance.

Komi's work is an excellent example of crossdisciplinary research in which physiology, biomechanics, biochemistry and neurophysiology have led to an improved understanding of muscle contraction dynamics. By using an activity that required the large muscles of the lower limbs to perform sizeable amounts of work, it was possible to induce sufficiently large differences between the eccentric and concentric conditions to be able to use an expired gas analysis system to estimate oxygen metabolism. If small limb movements only had been chosen, the precision of the gas analysis system would probably not have been high enough to obtain significant differences. By understanding the concept of mechanical work, Komi and his colleagues recognised that the same amount of work could be done on the sled by requiring the subjects to either accelerate the sled to a predetermined height or stop the sled after it had been released from that height. The only difference was that the muscles had to perform concentrically in one situation and eccentrically in the other.

Knowledge of the anatomy of muscle, tendon and neural pathways allowed these researchers to theorise on how these structures would influence the stretch–shortening cycle. That is, if they had not known about muscle spindles, Golgi tendon organs and the role of the Ia afferent neurons, they would not have been able to recognise that the increased work output following muscle stretch could be due to increased potentiation of the muscles through a reflex arc. Thus to obtain a complete picture of the ways in which animals, including humans, move, it is important to be eclectic and to investigate these multifaceted problems from numerous perspectives using a variety of techniques.

Source: Komi, P.V. (1984), The stretch–shortening cycle and human power output, pp. 27–39, in N.L. Jones, N. McCartney & A.J. McComas (eds), *Human Muscle Power*, Human Kinetics, Champaign, IL.

Maximising exercise participation in the community

As we saw in a number of the chapters in Part

2, human movement studies should have an equally important emphasis on elements of participation as on aspects of performance. To conclude our consideration of some multidisciplinary and crossdisciplinary aspects of human movement we examine in this section examples of programs aimed at maximising exercise participation in community settings. By focusing on an entire community, there is a higher probability of influencing the health behaviour of a large number of people—as well as those individuals most at risk who might not otherwise have access to such intervention. Such community intervention can also be very cost-effective compared with other types of intervention.

Before implementing community strategies to enhance participation in physical activity, it is important to first understand those factors which influence an individual's decision to initiate and maintain activity (discussed in Chapters 15, 21, 22, and 23). One of the first studies to address these issues in a community setting, by Sallis and associates, queried more than 1400 men and women in four California communities about their activity patterns over one year. Factors predicting adoption of activity differed from those predicting maintenance, indicating that different strategies must be used to attract individuals to first initiate activity and then to continue a physically active lifestyle. Few people over age 35 years adopted vigorous activity, and at all ages, maintenance was higher for moderate compared with vigorous activity, suggesting that, to be effective, community programs should focus on attracting people to moderate physical activity (this strategy has been incorporated into the 1996 U.S. Surgeon General's Report on Physical Activity, see page 247). Education level and knowledge about health and exercise strongly predicted maintenance of activity, suggesting that efforts to educate the community about the health benefits of exercise may enhance continued participation in physical activity.

Several countries have implemented national programs to increase participation in physical activity. PARTICIPaction was developed in Canada in the early 1970s to be a catalyst to motivate people to be more physically active and to become conscious of the personal and societal benefits of a physically active lifestyle; in 1972 only 5% of adults were considered physically active. PARTICIPaction worked through three main activities including "public service" (i.e., generally free) advertising via thousands of radio, television, and print media outlets; a cost-effective network through thousands of professionals in the fitness, health, nutrition, education, and human resource professions; and privately (corporate)-sponsored promotion of health and fitness events or programs. Some examples of sponsored promotion were fitness messages which appeared on milk cartons; private-sector funding for a school-based physical activity program including materials to be used by the entire family; and fitness and health promotion resources for the workplace.

In Europe, a newly formed European Network for Promotion of Health-Enhancing Physical Activity, coordinated by organisations in the Netherlands and Finland, encourages participation in physical activity through local and national community intervention programs. One example is "Netherlands on the Move!" launched in 1995. One of the first steps was to convene a national conference aimed at convincing health professionals of the health benefits of regular physical activity. Projects and resource materials include a handbook providing guidelines for organising local recreational activities for older adults, public information about the health benefits of exercise, resources detailing examples of successful initiatives, workshops, and an interactive compact disk providing information about suitable activities. Intervention projects include a large walking promotion program for an entire city; development of sports groups for people with chronic disease; and a program to promote youth participation.

Two initiatives in Finland are also associated with the Europe on the Move! program. "Steps to Positive Health" provides positive affective experiences for adults in different physical activities. Community representatives are trained to develop programs using a variety of physical activities, and these representatives then organise programs in their local communities. Another Finnish project, started in 1994, is the "Walking School" program. Based in local communities, a six-session program aims to provide both physical and intellectual stimulation while encouraging people to adopt a physically active lifestyle. Future projects include training lay people and health care professionals to promote walking as a form of physical activity, establishing different types of walking groups throughout the country, media campaigns to promote walking, and international walking events.

As discussed in Chapter 15, socioeconomic factors are strong predictors of participation in physical activity; at present participation is generally low in lower income and education groups. A recent study focused on St.-Henri, a low-income inner-city neighborhood of Montreal, Canada, with a high incidence of heart disease mortality. A five-year community-based, multi-factorial heart health promotion program was developed. Due to cost constraints, the program targeted primarily women, although it was anticipated that since women are the family "gatekeepers" for health information there would be a carry-over effect to the wider community. A walking club was developed in response to concerns expressed by community groups about the lack of low-cost options for physical activity among women. A walking kit included maps of safe routes developed with the police, a log-book to record activity, and information about warm-up and the health benefits of physical activity. Walking groups met two to three times per week for one-hour brisk walks; coordinated by a community volunteer, these walking groups continued to meet after the end of the study. This study shows that positive changes in physical activity participation are possible in relatively low-cost programs which work within existing community groups, address the issues of social support and safety, and combine educational efforts with environmental changes (such as providing walking paths).

Since patterns of physical activity are set early in life, encouraging participation among children is an effective way to ensure a life-long habit of regular physical activity. One example of a successful program in South Carolina, United States, showed how physical activity could be creatively incorporated into the school curriculum. Students, teachers, and parents participated in weekly sessions of at least 20 minutes of walking. In addition, ideas were developed for teachers to incorporate physical activity into learning experiences.

Successful efforts to encourage participation in regular physical activity depend upon a crossdisciplinary approach that encompasses not only the physiological but also the social, psychological, and economic factors influencing adoption and maintenance of exercise. Working with other health professions, the field of human movement studies can play a key role in advancing community health through effective programming to enhance participation in regular physical activity among all members of the community.

Sources: European Network for Promotion of Health-enhancing Physical Activity. (1996), *Europe on the Move! Newsletter*, 1(1), 1–8. Arnhem, The Netherlands; King, A.C. (1991), Community intervention for promotion of physical activity and fitness, *Exercise and Sport Sciences Reviews*, 19, 211–260; King, A.C. (1994), Community and public health approaches to the promotion of physical activity, *Medicine and Science in Sports and Exercise*, 26(11), 1405–1412; Netherlands Olympic Committee/Netherlands Sports Confederation. (1995–96), Netherlands on the Move! information file, Arnhem, Netherlands; Paradis, G., O'Loughlin, J., Elliott, M., Masson, P., Renaud, R., Sacks-Silver, G., & Lampron, G. (1995), Coeur en sant—St-Henri—a heart health promotion programme in a low income, low education neighborhood in Montreal, Canada: Theoretical model and early field experience, *Journal of Epidemiology and Community Health*, 49, 503–512; Sallis, J.F., Haskell, W.L., Fortmann, S.P., Vranizan, Taylor, C.B., & Solomon, D.S. (1986), Predictors of adoption and maintenance of physical activity in a community setting, *Preventive Medicine*, 15, 331–341.

Glossary

Acceleration: The rate of change of velocity.
Accommodation: The process by which the curvature of the lens of the eye is adjusted for viewing objects at different distances; achieved through control of the ciliary muscles.
Acetylcholine (ACh): A common neurotransmitter in the peripheral and central nervous systems.
Achievement goal orientations: Individual definitions of success.
Actin: Thin protein filaments of muscle.
Adaptation: The structural or functional adjustment of an organism to its environment that improves its chances of survival.
Adenosine triphosphate (ATP): 'High energy' phosphate molecule produced in all cells; cleavage of the terminal phosphate bond yields energy to fuel cellular work such as development of force in skeletal muscle.
Aerobic (oxidative) energy system: Energy system of oxidative metabolism requiring oxygen to produce ATP; provides the major source of ATP for endurance exercise lasting longer than 3 minutes.
Aerobic power: See maximal oxygen consumption ($\dot{V}O_{2max}$).
Afference: Sensory (or input) information transmitted from the sensory receptors to the central nervous system and brain.
Allometric scaling: The non-linear relationships used to describe how objects and particularly animals of differing sizes vary according to biologically sound and systematic criteria.
Alpha–gamma co-activation: The process involving simultaneous activation of both alpha motor neurons (to extrafusal muscle fibres) and gamma motor neurons (to intrafusal muscle fibres), which permits comparison of actual muscle length with intended muscle length.
Alpha motor neurons: Large nerve cells that innervate skeletal (extrafusal) muscle fibres.
Ambient vision: That aspect of vision, deriving from the whole visual field and the nerve pathway connecting the optic nerve to the superior colliculi in the midbrain, responsible for the location of objects and the perception of self-motion.
Anaerobic capacity: The total amount of work (in kilojoules) accomplished during brief high-intensity exercise such as the 30 second bicycle ergometer test; considered an indication of the capacity of the anaerobic glycolytic energy system.
Anaerobic glycolytic energy system: Metabolic energy system that uses glycogen or glucose to produce ATP and produces lactic acid as a byproduct; the major source of ATP for maximal exercise lasting between 20 seconds and 3 minutes.
Anaerobic power: Peak or maximal power (in watts) achieved during brief high-intensity exercise such as the 10 second bicycle ergometer

test; considered an indication of the capacity of the immediate energy system.
Androgyny index: The use of pelvic and shoulder widths to distinguish between adult males and females.
Angular acceleration: The rate of change of angular velocity.
Angular displacement: The smaller of the two angles between a body's initial and final positions.
Angular distance: The angular positional change of a body.
Angular speed: The rate of change of angular distance.
Angular velocity: The rate of change of angular displacement.
Anterior: Front.
Anthropometer: An instrument used to measure lengths of body segments.
Anthropometry: Study of the size, proportions and composition of the human body.
Anxiety: The subjective feeling of apprehension or worry often experienced when a situation is perceived to be threatening.
Appositional growth: Addition and/or erosion of bone at the outer and inner surfaces of the shaft, to cause changes in shaft diameter and thickness of the compact bone.
Apraxia: The inability to carry out purposeful movements in the absence of paralysis or other sensory or motor impairments; generally following damage to the cerebral cortex.
Arousal: General state of physiological alertness as controlled by activation of the reticular formation.
Arthritis: Inflammation of synovial joints.
Arthrology: The study of the joints of the body.
Articular capsule: See joint capsule.
Articular cartilage: cartilage forming the smooth bearing surface of a synovial joint.
Articulation: See joint.
Asceticism: The religious doctrine of the Middle Ages that demanded extreme physical self-denial in order to focus exclusively on spiritual matters
Ataxia: The breakdown or irregularity of muscular co-ordination, generally following damage to the cerebellum.
Atrophy: The process whereby cells decrease in size (waste away) with disuse or disease.
Auxology: The study of growth.
Axon: A single nerve fibre extending away from the cell body of a neuron and responsible for sending nerve impulses away from the cell body.
Basal ganglia: Collection of nuclei, located within the cerebral hemispheres, that are intimately involved in movement control and co-ordination.
Bending: A combination of tensile, compressive and shear forces within a structure.
Biarticular muscle: A skeletal muscle crossing two synovial joints.
Bicondylar calipers: An instrument used to measure width of bones near their ends.
Biofeedback: Immediate feedback of a biological phenomenon provided by the use of electronic recording instruments; or the use of instruments to obtain information about biological processes of which one is not normally aware; through biofeedback training individuals can learn to modify or control these biological processes.
Biomechanics: The field of study that uses the methods of mechanics to describe the motion and the effects of forces on biological specimens.
Blocked practice: A type of practice in which each component skill is practised repetitively as an independent block; practice is fully completed on one component skill before practice is started on the next component skill.
Body mass index: The relationship between body weight and the square of the height.
Brainstem: That section of the brain consisting of the medulla, pons and midbrain lying between the cerebrum and the spinal cord.
Cadaver: A human body that has been embalmed after death for the purposes of anatomical dissection and instruction.
Calcium: A mineral stored in bone and essential for muscle contraction.
Callus: Unorganised meshwork of fibres or bone laid down after a fracture; eventually replaced by compact bone.
Cancellous bone: See spongy bone.
Cardiac muscle: The specialised striated muscle forming the walls of the heart.
Cardiac output (Q): The volume of blood pumped by the heart per minute.
Cartilage: A type of connective tissue with a high water content.
Cartilaginous joint: A type of joint in which the

material between the bones is mainly cartilaginous.
Centre of mass (gravity): The theoretical point in an object at which its entire mass appears to be concentrated.
Central nervous system: The nervous system consisting of the brain and the spinal cord.
Central sulcus: A deep trench or vertical groove in the middle of each cerebral hemisphere, which separates the frontal lobe from the parietal lobe; the motor cortex lies immediately forward of the central sulcus and the sensory cortex immediately behind it.
Cephalic index: Relationship between the breadth and length of the skull.
Cephalo-caudal development: The trend for development to occur in a 'head-down' or 'head-to-tail' manner.
Cerebellum: A subdivision of the brain lying below the cerebral cortex and behind the brain stem that plays a major role in movement co-ordination.
Cerebrum: The largest and most important region of the brain, consisting of the two cerebral hemispheres.
Circuit training: Training in which the athlete moves around a circuit of different exercise stations (exercise machines or activities).
Clavicle: Collar bone.
Climbing fibre: A type of afferent nerve fibre within the cerebellum.
Closed-loop control: A type of movement control where the movement is controlled continuously on the basis of sensory feedback arising during the movement itself.
Collagen: A fibrous protein that is an important constituent of connective tissues such as bone, ligament, tendon, and cartilage.
Compact bone: Dense bone, which does not appear porous.
Complex joint: Synovial joint containing an intra-articular structure.
Compliance: The ability of a material to store energy (as strain energy) and then return it to the object that initially possessed the energy.
Compound joint: Synovial joint involving more than one pair of articulating surfaces.
Compression: A type of force in which the two ends of a structure are squeezed together.
Computerised tomography (CT): Production of images of sections through the body using an X-ray source.
Concentric: Action of a muscle in which it shortens.
Concentric muscle contraction: Dynamic muscle contraction in which the muscle shortens while developing tension.
Concurrent forces: Forces that are applied at the same point.
Contralateral: On the opposite side of the body.
Coronal plane: A plane dividing the body into front and back.
Corpus callosum: The thick band of neural tissue connecting the two cerebral hemispheres.
Cortical bone: See compact bone.
Cortico-spinal tract: See pyramidal tract.
Critical period: A period during development when the organism (or a specific system or skill) is most readily influenced by both favourable and adverse environmental factors.
'Cross-bridge' hypothesis: Explanation of mode of striated muscle contraction.
Crossed extensor reflex: A reflex that increases activation of the extensor muscles of the limb on the side opposite to a limb undergoing flexion.
Cross-sectional study: A study in which individuals of different ages are measured at about the same time.
Cutaneous receptors: Receptors located in the skin.
Deformable body: A body that changes shape when a load is applied to it; the positions of the particles of the body do not move as a single body.
Delayed onset muscle soreness (DOMS): Muscle soreness characterised by tender and painful muscles, which usually appears in the few days after unaccustomed exercise; DOMS most commonly occurs after exercise with a large eccentric component, such as downhill running.
Dendrites: The branches of a neuron that synapse with, and receive nerve impulses from, other neurons.
Dense bone: See compact bone.
Density: Mass per unit volume of a material.
Diaphysis: Shaft of a long bone.

Displacement: The straight-line distance between the initial and final positions of a body.
Distance: The distance that an object moves, along the path followed, between its initial and final positions.
Dopamine: A neurotransmitter, the absence of which in parts of the basal ganglia gives rise to Parkinson's disease.
Dorsal: Pertaining to the posterior side or back of the body.
Down's syndrome: A chromosomal abnormality associated with delays in both mental and motor development.
Dual photon absorptiometry (DPA): Determination of body composition using a radionuclide source.
Dual X-ray absorptiometry (DEXA): Determination of body composition using X-ray analysis.
Dynamics: The study of the forces and torques that produce motion in a body.
Dynamometer: An instrument or machine used to measure muscular strength.
Eccentric: Action of a muscle in which it lengthens while remaining active.
Eccentric muscle contraction: Dynamic muscle contraction in which the muscle lengthens while developing tension.
Ectoderm: Primary germ layer that gives rise to the outer skin and nervous system.
Ectomorphy: One component of a somatotype, based on a person's height and mass.
Efference: Motor (or output) information transmitted from the brain and central nervous system out to the muscles.
Efficiency (economy): The measure used to define the amount of work done per unit of cost.
Electroencephalography (EEG): The recording of electrical activity from the brain.
Electromyography (EMG): The recording of the electrical signal produced by skeletal muscle as it contracts.
Elevated post-exercise oxygen consumption (EPOC) (formerly called oxygen debt): Oxygen consumption in excess of resting level during recovery after exercise.
Embryo: A developing human during the first quarter of intra-uterine life.
Endochondral ossification: Development of bone from a cartilaginous model.
Endoderm: Primary germ layer that gives rise to the organs of the body.
Endomorphy: One component of a somatotype, based on the thickness of the skinfolds.
Endurance exercise capacity (cardiovascular or cardiorespiratory endurance): Performance measure such as the maximum time an individual can exercise at a given speed or total amount of work that can be accomplished in a given time.
Epiphyseal plate: Cartilaginous growth plate in developing long bone.
Epiphysis: Expanded end of a long bone.
Equilibrium: That state at which the sum of all forces and torques acting on a body equals zero.
Exercise adherence motivation: The drive to maintain regular physical activity.
Exercise-induced asthma (EIA): bronchoconstriction (narrowing of breathing tubes) causing reduced ventilation during exercise in asthmatic individuals.
Exercise participation motivation: The drive to initiate regular physical activity.
Exercise physiology: The subdiscipline of human movement studies concerned with understanding physiological responses to exercise.
Extension: Movement at a synovial joint in which the angle between the limb segments increases.
Extensor thrust reflex: A reflex extension of the legs in response to stimulation of the soles of the feet; assists in supporting the body's weight against gravity.
Extrafusal muscle fibres: The characteristic skeletal muscle fibres, activated by alpha motor neurons, the contraction of which causes voluntary movement.
Extrapyramidal tract: All the descending motor pathways from the brain to alpha motor neurons other than those direct connections contained within the pyramidal tract.
Extroversion: The tendency to be social and out-going.
Fibrous joint: A type of joint in which the material between the bones is fibrous.
Flexion: Movement at a synovial joint in which the angle between the limb segments decreases.
Flexion reflex: A reflex that produces flexion of the joints that withdraw an injured limb away from a painful stimulus.
Focal vision: That aspect of vision, deriving

from the central retina (fovea) and the nerve pathway connecting the optic nerve to the visual cortex, responsible for object recognition and the resolution of fine detail.

Fetus: Developing human during the last three-quarters of intra-uterine life.

Force: An interaction between two bodies, often described as a push or a pull.

Force components: The (usually) orthogonal parts of a force vector.

Force couple: That situation when two forces of equal magnitude but opposite direction are applied to an object and pure rotation results.

Fovea: The area near the centre of the retina in which cone cells are most concentrated and which therefore provides the most acute vision.

Frontal plane: See coronal plane.

Functional anatomy: The subdiscipline concerned with understanding the anatomical basis of human movement and the effects of physical activity on the musculoskeletal system.

Gamma motor neurons: Small nerve cells that innervate spindle (intrafusal) muscle fibres.

Gene: Any of the portions of the DNA molecule that contain the elementary units of heredity.

Genotype: A genetically determined somatotype.

Geometric scaling: The linear scaling typically found in Euclidean geometry.

Gerontology: The study of ageing.

Glia (or glial cells): Non-neuronal cells within the brain and spinal cord that help to regulate the extracellular environment of the central nervous system through the provision of metabolic and immunological support for the nerve cells.

Glycogen: Cellular storage form of glucose, comprised of polymers of glucose molecules.

Golgi tendon organs: Specialised receptors located within tendons that respond to tendon tension.

Goniometer: An instrument used to measure range of movement at a joint.

Gustatory information: Information derived through the sense of taste.

Hamstrings: The muscle group at the back of the thigh.

Heath–Carter anthropometric somatotype: An example of a somatotyping technique based on anthropometric measurements.

Hippocampus: A portion of the brain believed to play a central role in memory and learning.

Huntington's disease: A degenerative, inherited disease affecting neurotransmission within the basal ganglia and characterised by rapid, involuntary limb movements.

Hyperplasia: Increase in the number of cells forming a tissue.

Hypertrophic phase: In weight training this is the second phase of increase in muscle size.

Hypertrophy: An increase in the size of each cell forming a tissue.

Imagery: A mental skill involving the use of all senses to create or recreate an experience in the mind.

Immediate energy system: Metabolic energy system in which phosphocreatine is split to provide energy for ATP resynthesis; provides an immediate source of energy at the onset of exercise.

Impulsive load: A force that is applied to and then removed from an object in a very short time; impacts are impulsive types of loads.

Inertia: The resistance of a body to a change in state of its motion.

Interneurons: Nerve cells that connect one nerve cell to another.

Interval training: Form of exercise training of alternating work and rest intervals.

Intrafusal muscle fibres: The modified skeletal muscle fibres found within the muscle spindle receptors and activated by gamma motor neurons.

Introversion: The tendency to be concerned primarily with one's own thoughts and feelings rather than with the external environment.

Ipsilateral: On the same side of the body.

Isokinetic: Dynamic muscle contraction at a constant speed, which is controlled by a machine such as the KinCom or Cybex isokinetic dynamometer.

Isometric: Static muscle contraction in which the muscle length remains constant while developing tension.

Isotonic: Dynamic muscle contraction in which the muscle length changes during tension development; may be concentric (muscle shortens) or eccentric (muscle lengthens).

Joint: Union of two or more bones.

Joint capsule: Thick connective tissue membrane forming the boundary of a synovial joint and enclosing the joint cavity.

Kinanthropometry: The scientific specialisation dealing with the measurement of humans in a variety of morphological perspectives, its application to movement, and those factors that influence movement.
Kinematics: That branch of mechanics concerned with the description of motion.
Kinesiology: The scientific study of movement; often used more narrowly to refer to those subdisciplines concerned with understanding the anatomical and mechanical bases of human movement.
Kinesthesis: Sensory information provided by the receptors located in the muscles, tendons, joints and skin.
Kinetic energy: The energy of motion; further subdivided into that energy associated with linear (translational KE) and angular (rotational KE) motion.
Kinetics: The branch of mechanics that is concerned with the forces and torques that produce motion in a body.
Lactate threshold: Point of inflection of the curve of blood lactate concentration versus exercise intensity; above this point blood lactate concentration increases disproportionately with increasing exercise intensity; considered to identify an exercise intensity that can be maintained at the upper limit of aerobic metabolic capacity without major input from the anaerobic glycolytic system.
Lactic acid: Byproduct of anaerobic glycolytic breakdown of glucose or glycogen; hydrogen ions dissociated from lactic acid increase acidity (decrease pH) causing fatigue in skeletal muscle.
Lamina: Plate-like structure forming part of the arch of a vertebra.
Ligament: A dense regular connective tissue joining bone to bone.
Load: A force, a push or pull applied by one body to another.
Locomotor reflexes: Those reflexes present at birth or soon thereafter that are the primitive precursors to locomotion.
Longitudinal study: A study in which the same individuals are measured over a period of years.
Long-loop reflexes: A collective term for reflexes involving multiple synapses in which nerve impulses are passed from the local level of the spinal cord to higher levels of the cord and perhaps even to regions of the brain.
Long-term potentiation: The prolonged increase in the efficiency of a synapse that occurs as a result of repeated stimulation; thought to be a fundamental neurophysiological mechanism for memory and learning.
Magnetic resonance imaging (MRI): Production of images of sections through the body using a very strong magnetic field.
Magnetic resonance spectroscopy (MRS): Determination of chemical composition of the body using a very strong magnetic field.
Marrow: Tissue within the shafts of long bones and in the spaces within spongy bone.
Maturation: Sequence of changes occurring between conception and maturity.
Maximal heart rate: Highest possible heart rate, usually achieved during maximal exercise, estimated by the equation 220 – age in years.
Mechanics: The branch of physics focused on describing motion of objects.
Medulla: Major anatomical component of the brainstem.
Meissner's corpuscles: Specialised receptors found primarily on the hairless surfaces of skin that respond to light pressure or touch.
Menarche: Onset of menstruation.
Meniscus: An intra-articular cartilaginous structure that is simultaneously shaped like a crescent and a wedge.
Merkel's discs (or corpuscles): Specialised capsular receptors of the skin that respond to light pressure or touch.
Mesenchyme: Primitive connective tissue that gives rise to other forms of more specialised connective tissue.
Mesoderm: Primary germ layer that gives rise to the structures of the musculoskeletal system.
Mesomorphy: One component of a somatotype, based on the development of the musculoskeletal system.
Minute ventilation: Volume of air inspired by the lungs per minute.
Mixed longitudinal study: A study in which one

group is measured over a period of years, and another group of a different age is measured during the same period.
Moment of force: A torque that tends to angularly accelerate an object.
Moment of inertia: A body's resistance to rotation.
Momentum: The quantity of motion possessed by a body; is dependent on velocity and inertia.
Monoarticular muscle: A skeletal muscle crossing a single synovial joint.
Monocyte: One type of white blood cell.
Mossy fibre: A type of afferent nerve fibre within the cerebellum.
Motor control: That subdiscipline of human movement studies concerned with understanding the processes that underlie the acquisition, performance and retention of motor skills.
Motor cortex: A strip of cerebral cortex immediately anterior of the central sulcus that is responsible for the relay of many of the motor commands from the brain to the muscles.
Motor development: The field of study concerned with understanding the changes in motor control that occur throughout the lifespan.
Motor equivalence: The capability of the motor system to produce the same movement outcome in a variety of ways.
Motor learning: The field of study concerned with understanding the changes in motor control that occur in response to practice (also known as skill acquisition).
Motor milestones: Identifiable stages in the development of specific fundamental motor skills.
Motor program: A set of motor (efferent) commands, set up in advance of a movement starting, that provide the potential for the movement to be completed without relying on sensory feedback from the movement itself (see also open-loop control).
Motor unit: Consists of a single motor nerve fibre and all the muscle fibres it innervates.
Movement time: The time elapsed between the initiation and completion of a movement.
Multiple sclerosis: A disease caused by damage to the myelin sheath surrounding nerves, resulting in hardening of the nerve tissue and consequent tremor, loss of co-ordination and occasionally paralysis.
Muscle: See cardiac muscle, skeletal muscle, smooth muscle.
Muscle fibre: Muscle cell; in humans there are three types of skeletal muscle fibres that differ in their metabolic and physiological characteristics: type I (slow oxidative), type IIA (fast oxidative glycolytic) and type IIB (fast glycolytic).
Muscle spindle: Specialised receptor located within the intrafusal fibres of skeletal muscle that responds to stretch.
Muscle tone: The tension produced in skeletal muscles under resting conditions as a result of low-frequency discharge of the alpha motor neurons.
Muscular endurance: Ability of a muscle to repeatedly develop and maintain submaximal force over time.
Muscular power: Rate at which force can be applied or force times speed.
Muscular strength: Maximal force that can be produced in a single movement or contraction.
Myelin: The fatty insulating material covering the axons of many neurons and responsible for increasing conduction velocity within those neurons.
Myoblast: Muscle-forming cell.
Myofibrils: Longitudinal bundles of thick and thin contractile filaments located within muscle fibres.
Myology: Study of the muscular system.
Myosin: Thick protein filaments of muscle.
Myotatic reflex: The simple mono-synaptic 'muscle-stretching' reflex in which the excitation of a muscle spindle receptor by the imposition of stretch to a muscle causes that muscle to contract.
Neck index: Relationship between neck circumference and neck length.
Neuron: A nerve cell; the fundamental building block of the nervous system.
Neurotransmitter: A chemical agent released by one nerve cell that acts on another nerve or muscle cell by altering its electrical activity or state.
Neurotrophic phase: In weight training this is the initial phase of motor learning.

Olfactory information: Information derived through the sense of smell.
Open-loop control: A type of movement control in which the efferent (motor) commands are pre-planned before the movement is initiated and control is not dependent on any sensory feedback arising during the movement itself (see also motor program).
Optimisation: A technique used to determine the best way of doing a task.
Oxygen debt: See elevated post-exercise oxygen consumption.
Oxygen deficit: Difference between the amount of energy required during exercise and that which can be supplied by oxidative metabolism; excess ATP above which can be resynthesised oxidatively is supplied by the anaerobic energy systems (immediate and glycolytic systems).
Osteoarthrosis: Degeneration of articular cartilage.
Osteoblast: Bone-forming cell.
Osteoclast: Bone-eroding cell.
Osteocyte: Bone cell.
Osteology: Study of the skeletal system.
Osteometry: Study of the size and proportions of bones.
Osteopenia: Reduced bone density.
Osteoporosis: Reduced bone density below a certain level, which is likely to be associated with fractures.
Otolith organs: Set of two specialised receptors (the utricle and the saccule) located in the inner ear in which the movement of calcium carbonate 'stones' (the otoliths) against nerve endings signals horizontal and vertical linear acceleration of the head.
Ovum: Female sex cell containing half the normal number of chromosomes.
Pacinian corpuscles: Specialised plate-like nerve endings located deep in the skin, which respond to pressure, deep compression or high-frequency vibration.
Palpation: Using the senses of touch and pressure to identify anatomical features on a living human.
Palsy: A persisting movement disorder due to brain damage acquired around the time of birth.
Parkinson's disease: A disease characterised by tremor, rigidity and a delay in the initiation of movement caused by a deficiency in the production of the neurotransmitter dopamine within regions of the basal ganglia.
Particle: An object of infinitesimal size but whose mass appears to be located at one point.
Peak height velocity (PHV): Point of most rapid growth in height; is a major landmark in pubertal growth.
Percentile ranking: The percentage of the population of the same age who fall below an individual on a particular measurement.
Peripheral nervous system: The collection of all the nerve fibres that connect the receptors and the effectors to the brain and spinal cord.
Personality: The composite of the characteristic individual differences that make each person unique.
Phenotype: A somatotype determined by both genetic and environmental factors.
Phosphocreatine (PC): 'High energy' phosphate molecule stored in muscle cells; cleavage of the terminal phosphate bond yields energy used to rephosphorylate ADP to ATP at the onset of exercise.
Phosphorus: A mineral stored in bone.
Plasticity: Flexibility or adaptability of structure or function; often used specifically in reference to the ability of early embryonic cells to alter in structure or function to suit the surrounding environment.
Polyarticular muscle: A skeletal muscle crossing more than two synovial joints.
Ponderal index: Relationship between height and cube root of body mass.
Pons: Major anatomical component of the brainstem.
Positron emission tomography (PET): Production of images of sections through the body using positively charged electrons as the energy source.
Posterior: Back.
Postnatal: After birth.
Postural reflexes: Those reflexes present in the first year of life responsible for keeping the head upright and the body correctly oriented with respect to gravity.
Potential energy: The energy associated with the gravitational attractive force; defined by the distance between the two interacting objects.
Power: The rate of doing work.

Prenatal: Before birth.
Primitive reflexes: Those reflexes present at birth that function predominantly for protection and survival.
Progressive muscular relaxation (PMR): A relaxation technique involving the systematic contraction and relaxation of specific muscle groups throughout the body.
Proprioception: The sense of the position of the body and the movement of the body and its limbs.
Proximo-distal development: The tendency for development of control over muscles close to the body (attached to the axial skeleton) to occur before development of control over muscles of the limbs (attached to the appendicular skeleton).
Puberty: The period of rapid physical growth between childhood and adulthood.
Purkinje cells: Large branching neurons within the cerebellum.
Pyramidal cells: Nerve cells within the pyramidal tract (see below) so named because of their pyramid-like shape; they are broad at the top with a large branching tree of dendrites funneling down to a single, slender axon.
Pyramidal tract: The major descending motor pathway consisting of nerve cells having their origin in the cerebral cortex and synapsing directly with motor neurons at the spinal cord level.
Quetelet index (QI) or body mass index (BMI): Ratio of body mass in kilograms to height in metres, used for recommending appropriate body mass.
Radiology: The use of X-rays to visualise structures within the body.
Random practice: A type of practice in which, in contrast to blocked practice, component skills are practised together in an unstructured order; one trial of practice on any particular component skill can be followed, in random order, by a trial of practice on any other (or the same) skill component.
'Ratchet' hypothesis: See 'cross-bridge' hypothesis.
Reaction time: The time elapsed between the presentation of a stimulus and the initiation of a response to that stimulus.
Reciprocal inhibition: The inhibition of one motor neuron or nerve pathway by the simultaneous excitation of another having an opposing action, for example the inhibition of extensor motor neurons during the activation of flexor motor neurons.
Reflex: An involuntary movement elicited by a specific stimulus and typically completed rapidly and without conscious thought.
Refraction: The bending or change in direction of a wave, such as is experienced by light waves travelling through different media.
Reliability: How consistently measurements can be repeated.
Repetition maximum (RM): Maximum weight that can be lifted a specified number of times; for example, 3 RM = maximal weight that can be lifted only three times.
Response time: The time elapsed between the presentation of a stimulus and the completion of a response to that stimulus (that is, the sum of reaction time plus movement time).
Resultant force: The vector sum of all acting forces.
Reticular formation: A loosely defined network of cells extending from the upper part of the spinal cord through the medulla and pons to the brainstem responsible for the control and regulation of attention, alertness and arousal.
Rheumatoid arthritis: Inflammation of synovial membranes.
Righting reflexes: Reflexes designed to return the body to an upright position after any event causing a loss of balance.
Rigid body: A body whose particles remain in the same position when a load is applied; an object is considered rigid when the movement of the constituent particles is negligible in comparison with the movement of the object itself.
Ruffini corpuscles: Specialised plate-like nerve endings located in the skin that respond to pressure or touch.
Saccule: The otolith organ (see above) responsible for the detection of vertical linear acceleration of the head.
Sagittal plane: A plane dividing the body into left and right sides.
Sarcomere: The structural and functional unit of skeletal muscle, containing thick and thin filaments.

Scholasticism: The philosophical doctrine of the Middle Ages that placed emphasis on the development of the mind and scholarly pursuits to the exclusion of any concentration on the body or physical activity.
Secular trend: Changes in physical dimensions between people of one generation and following generations.
Selective attention: The perceptual process of attending to one signal or event in preference to all others.
Self-efficacy: The confidence one has in one's ability to successfully perform a particular behaviour.
Semicircular canals: Set of three specialised receptors located within the inner ear and positioned at 90° to each other that respond to angular acceleration of the body in three different planes.
Sexual dimorphism: Structural differences between males and females.
Shear: A type of force in which adjacent parts of a structure move parallel to each other in opposite directions.
Simple joint: Synovial joint involving a single pair of articulating surfaces.
Skeletal muscle: Striated muscle forming the major muscles of the trunk and limbs.
Skinfold calipers: An instrument used to measure the thickness of a skinfold.
'Sliding filament' hypothesis: Description of the mode of contraction of striated muscle.
Smooth muscle: Muscle tissue forming the walls of organs such as blood vessels.
Somatochart: A special diagram indicating the three components of a person's somatotype.
Somatotype: A short-hand way of representing the shape and composition of a person.
Spasticity: An increase in the tone (resting tension) of certain muscle groups involved in the maintenance of posture; thought to arise from damage to extrapyramidal motor fibres.
Specific gravity: The density of a material compared with the density of water.
Speed: The rate change of distance.
Spermatozoan: Male sex cell containing half the normal number of chromosomes.
Spinal nerve: One of the 31 pairs of nerves arising from the spinal cord.
Spongy bone: Less dense bone, which appears to be porous.
Sport and exercise psychology: The scientific study of human behaviour and cognition in sport and physical activity.
Stadiometer: Instrument used to measure height.
States: How individuals feel at a particular point in time; transitory feelings.
Static load: A force that remains constant.
Statics: The study of systems in which no acceleration takes place.
Stature: Total body height.
Sternum: Breastbone.
Stiffness: In the mechanical testing of biological structures it indicates the relationship between force applied to the structure and the deformation it produces.
Strain energy: The energy stored by a body as its shape is altered under the action of external forces.
Stroke volume (SV): Volume of blood pumped by the heart with each beat.
Subchondral bone: The thin layer of compact bone under the articular cartilage.
Subcutaneous: Beneath the skin.
Supplementary motor area: Area of the cerebrum located forward of the motor cortex involved in the production and control of skilled movement.
Synapse: The junction between two neurons or nerve cells.
Synergetics: The study of complex pattern formation and transitions between patterns.
Synovial joint: A type of joint containing fluid within a cavity surrounded by a capsule.
Synovial membrane: Thin membrane lining the inner surface of the joint capsule.
Tendon: A dense regular connective tissue joining muscle to bone.
Tension: A type of force in which the two ends of a structure are pulled apart.
Thalamus: A mass of grey matter near the base of the cerebrum.
Thoracic index: Relationship between breadth and length of the chest.
Tonic reflexes: Those reflexes that are the basis

for posture, concerned with the maintenance of the position of one body part with respect to others.

Torque: An eccentric force that causes or tends to cause rotation.

Torsion: A combination of compressive, tensile, and shear forces; results in twisting of a structure.

Trabeculae: Small bony rods that form the framework of spongy bone.

Trabecular bone: See spongy bone.

Tract: A large bundle of nerve fibres within the central nervous system.

Training stress syndrome: The psychological and physiological negative adaptation to training characterised by the three phases of staleness, overtraining, and burnout.

Traits: The stable and enduring predisposition each individual has to act in a certain way across a number of different situations.

Utricle: The otolith organ (see above) responsible for the detection of horizontal linear acceleration of the head.

Validity: How well a measurement instrument actually measures the property it sets out to measure; generally established by correlating the measure with an outside criterion or independent measure.

Velocity: The rate of change of displacement.

Ventral: Pertaining to the anterior or abdominal side of the body.

Vertebra: One of the bones forming the spine or vertebral column.

Vertebrate: Animal with a spine or 'backbone'.

Vestibular apparatus: Sensory system in the inner ear consisting of the semicircular canals and otilith organs and responding to angular and linear accelerations on the head.

Viscosity: The resistance to flow of a fluid.

Visual cortex: Region at the back of the cerebrum involved in the processing of visual information.

$\dot{V}O_{2max}$: Maximal oxygen consumption or aerobic power usually expressed as a volume of oxygen consumed per minute as measured in a progressive exercise test to volitional fatigue; high $\dot{V}O_{2max}$ values are generally indicative of high endurance exercise capacity.

Work of a force: The integral of force with respect to displacement.

Z disc: Boundary of sarcomere; site of attachment of thin filaments of skeletal muscle.

Zygote: Single fertilised cell resulting from the union of a male and a female sex cell.

Index

Textbooks for Related Courses

1995 • Cloth • 544 pp • Item BWEI0812
ISBN 0-87322-812-X • $49.00 ($73.50 Canadian)

Instructor's Guide
1995 • 3-1/2" disk • Available in Windows and Macintosh • **FREE** upon request to course adopters. Call for special instructions.

Using visuals, anecdotes, case studies, and stories, the authors explain key concepts in the field and how they apply to counseling, teaching, coaching, and fitness instruction. The text also features chapter overviews, summaries, review questions, and key points to guide students through the content.

The ***Instructor's Guide*** features options to search for questions from the test bank, create original questions, and edit and print in test form. Call for system requirements.

1994 • Cloth • 560 pp • Item BWIL0693
ISBN 0-87322-693-3 • $52.00 ($77.95 Canadian)

Color Transparency Package
1995 • 104-transparency set • Item MWIL0417
ISBN 0-87322-800-6 • $199.00 ($298.50 Canadian)
FREE upon request to course adopters.

Test Bank Package

Windows Item MWIL0491 • *Macintosh* Item MWIL0490
1995 • 3-1/2" disk • **FREE** upon request to course adopters. Call for special instructions.

Physiology of Sport and Exercise is superbly written, with careful attention given to explaining concepts clearly in language appropriate for introductory students. And it is full of color photos and illustrations to enhance understanding.

The ***Color Transparency Package*** includes 104 high-quality acetates on durable, heavy stock and a printed instructor's guide to the illustrations. Each transparency is numbered and in full-color with boxed and highlighted captions.

The computerized ***test bank*** contains more than 400 questions prepared by the authors themselves. Call for system requirements.

1991 • Cloth • 320 pp • Item BSCH0308
ISBN 0-87322-308-X • $38.00 ($56.95 Canadian)

Motor Learning and Performance shows the processes underlying skilled performance, how skilled performances are learned, and how to apply the principles of skilled performance and learning in teaching, coaching, and rehabilitative settings.

Practical applications, highlight sections, and hundreds of real-world examples bring the theories of motor learning and performance to life.

Instructor's Guide
1992 • Paper • 72 pp • Item BSCH0381 • **FREE** upon request to course adopters. Call for special instructions.

The ***Instructor's Guide*** provides valuable suggestions, hints, and ideas for teaching such as discussion topics, demonstrations, term paper ideas, test questions—both short-answer and correctable true/false statements—and transparency models.

30-day money-back guarantee!

To request more information or to place your order, U.S. customers call **TOLL FREE 1-800-747-4457**. Customers outside the U.S. place your order using the appropriate telephone/address shown in the front of this book.

Prices are subject to change.

BROWSE'S
INTRODUCTION TO
THE SYMPTOMS & SIGNS OF SURGICAL DISEASE